Lecture Notes in Computer Science 1004

Edited by G. Goos, J. Hartmanis and J. van Leeuwen

Advisory Board: W. Brauer D. Gries J. Stoer

Springer
Berlin
Heidelberg
New York
Barcelona
Budapest
Hong Kong
London
Milan
Paris
Santa Clara
Singapore
Tokyo

John Staples Peter Eades
Naoki Katoh Alistair Moffat (Eds.)

Algorithms
and Computations

6th International Symposium, ISAAC '95
Cairns, Australia, December 4-6, 1995
Proceedings

Springer

Series Editors

Gerhard Goos, Karlsruhe University, Germany

Juris Hartmanis, Cornell University, NY, USA

Jan van Leeuwen, Utrecht University, The Netherlands

Volume Editors

John Staples
Department of Computer Science, University of Queensland
Queensland 4072, Australia

Peter Eades
Department of Computer Science, University of Newcastle
Callaghan, New South Wales 2308, Australia

Naoki Katoh
Department of Management Science, Kobe University of Commerce
8-2-1 Gakuen-Nishimachi, Nishi-ku, Kobe 651-21, Japan

Alistair Moffat
Department of Computer Science, University of Melbourne
Parkville, Victoria 3052, Australia

Cataloging-in-Publication data applied for

Die Deutsche Bibliothek - CIP-Einheitsaufnahme

Algorithms and computation : 6th international symposium,
Cairns, Australia, December 4 - 6, 1995 ; proceedings / ISAAC
'95. John Staples ... (ed.). - Berlin ; Heidelberg ; New York ;
Barcelona ; Budapest ; Hong Kong ; London ; Milan ; Paris ;
Tokyo : Springer, 1995
  (Lecture notes in computer science ; Vol. 1004)
  ISBN 3-540-60573-8
NE: Staples, John [Hrsg.]; ISAAC <6, 1995, Cairns>; GT

CR Subject Classification (1991): F.1-2, G.2-3, I.3.5

ISBN 3-540-60573-8 Springer-Verlag Berlin Heidelberg New York

© Springer-Verlag Berlin Heidelberg 1995
Printed in Germany

Typesetting: Camera-ready by author
SPIN 10487131   06/3142 – 5 4 3 2 1 0   Printed on acid-free paper

Preface

The papers in this volume were selected for presentation at the Sixth Annual International Symposium on Algorithms and Computation (ISAAC'95), held on December 4-6, 1995 in Cairns, Australia. Previous meetings were held in Tokyo (1990), Taipei (1991), Nagoya (1992), Hong Kong (1993), and Beijing (1994).

The Symposium was sponsored by The Australian Computer Society; The University of Newcastle; James Cook University; Software Verification Research Centre; Computer Science Association (Australia); Special Interest Group on Algorithms, Information Processing Society of Japan; Technical Group on Computation, Institute of Electronic, Information and Communication Engineers of Japan; and the Combinational Mathematics Society of Australasia. We thank all of the sponsors for their support.

In response to the call for papers, more than 130 extended abstracts were submitted. There were many more acceptable papers than places available in the conference schedule, and the program committee's task was extremely difficult. Each submitted paper was reported on by at least three program committee members, with the assistance of referees, as indicated in the list at the back of this volume. The 45 papers selected for presentation had a total of 103 authors, resident as follows: Japan 29, Canada 13, Germany 11, United States of America 9, Australia 8, Sweden and Taiwan 6 each, Hong Kong, India, Korea and Switzerland 3 each, Denmark and Italy 2 each, and Hungary, Iceland, Mexico, Moldova, and Russia 1 each.

We thank all program committee members, their support staff, and their referees for excellent work within demanding time constraints; they have given the conference its distinctive character. We thank all who submitted papers for consideration; they have given the conference its high quality. Finally, we thank all our colleagues who worked hard to put in place the logistical arrangements of the conference; they have made the conference enjoyable. It is the cooperative efforts of all of these people—with their widely differing talents and great geographical spread—that have made the conference both possible and pleasurable.

August 1995

John Staples
Peter Eades
Naoki Katoh
Alistair Moffat

Conference Organisation

Symposium Chair Peter Eades, Australia

Program Committee Chair John Staples, Australia

Program Committee Co-Chairs Peter Eades, Australia
Naoki Katoh, Japan

Program Committee Mikhail Atallah, USA
Giuseppe DiBattista, Italy
Peter Eades, Australia
Norbert Eisinger, Germany
Naoki Katoh, Japan
Sing-Ling Lee, Taiwan
Bruce Litow, Australia
John Staples, Australia
Seinosuke Toda, Japan
Takeshi Tokuyama, Japan
Jeff Vitter, USA
Dorothea Wagner, Germany
Sue Whitesides, Canada
Nick Wormald, Australia
Xiang-Sun Zhang, China

Local Arrangements Gopal Gupta, Australia
Bruce Litow, Australia

Publicity Bob Cohen, Australia

Proceedings Alistair Moffat, Australia

Sponsors

Australian Computer Society
The University of Newcastle
James Cook University
The Software Verification Research Centre
The Computer Science Association (Australia)
Special Interest Group on Algorithms,
    Information Processing Society of Japan
Technical Group on Computation,
    Institute of Electronic, Information and Communication Engineers, Japan
Combinational Mathematics Society of Australasia

Referees

Abello, James
Agarwal, Pankaj
Akinari, Nakagawa
Alimonti, Paola
Alon, N.
Anderson, R.
Aurenhammer, Franz
Avenhaus, Juergen

Baginski, Boris
Bajaj, C.
Basin, David
Bertossi, A.
Bjoerner, Anders
Bradford, Philip
Brandes, Ulrik
Brauer, Wilfried
Briaud, Daniel
Burkart, Olaf
Burnikel, Christoph

Cadoli, Marco
Calmet, Jacques
Chen, Danny
Cheriyan, Joseph
Chiang, Yi-Jen
Chou, Shang-Ching
Cohen, Arjeh
Colbourn, Charles J.
Corneil, Derek
Crescenzi, Pierluigi

d'Amore, Fabrizio
Díaz, Josep
Dal Cin, Mario
Das, Gautam
Day, Khaled
Devillers, Olivier
Devroye, Luc
Dey, Tamal
de Berg, Mark
De Prisco, Roberto

de Vel, Olivier
Dietzfelbinger, Martin
Dor, Dorit
Du, Ding Zhu

Eiter, Thomas
Emiris, Ioannis
Etiemble, Daniel
Everett, Hazel

Fekete, Sandor
Felsner, Stefan
Franciosa, Paolo G.

Gao, Shuhong
Garg, Ashim
Gautschi, Walter
Gerberding, Stefan
Goemans, Michel X.
Goettler, Herbert
Goldberg, Andrew
Goldberg, Mark
Gonnet, Gaston
Gramlich, Bernhard
Grove, Eddie
Gupta, Ajay

Handke, Dagmar
Hassin, Refael
He, Xin
Hedetniemi, Sandee
Hedetniemi, Steve
Herrmann, Juergen
Hershberger, John
Hirschberg, Daniel
Hoffmann, Frank
Hu, Xiao Dong
Huttel, Hans

Jiang, Tao
Jouannaud, Jean-Pierre
Jutla, Charanjit C.

Kalkbrener, Michael
Kannan, Sampath
Kao, Ming-Yang
Kapoor, Sanjiv
Kapur, Deepak
Kapur, Shyam
Keil, Mark
Kerber, Manfred
Klein, Rolf
Kohlhase, Michael
Kozen, Dexter
Kraetzl, Miro
Kraus, Guenther
Kriegel, K.
Kriegel, Klaus
Kuechlin, Wolfgang

Lakshmivarahan, S.
Lange, Steffen
Larmore, Lawrence L.
La Poutre, Johannes A.
Leonardi, Stefano
Lesske, Frank
Letz, Reinhold
Liang, Weifa
Liebers, Annegret
Lingas, Andrzej
Liotta, Giuseppe
Loos, Ruediger
Lubiw, Anna
Lucks, Stefan
Lynch, Robert

Macarie, Ioan
Marchetti-Spaccamela, Alberto
Mayr, Ernst
Megiddo, Nimrod
Merz, Stefan
Milenkovic, Victor
Mooney, Raymond
Morain, F.
Mueller, Fritz
Muggleton, Stephen
Muller, Jean-Michel
Müller-Hannemann, Matthias
Murali, T.M.

Myers, Gene

Naor, M.
Narasimhan, Giri
Natsev, Paul
Nutt, Werner

O'Connor, L.
O'Rourke, Joseph
Olariu, S.
Ostrovsky, Rafail

Pagli, Linda
Panario, Daniel
Papadimitriou, Christos
Pischel, Markus
Plaisted, David
Pottier, Loic
Proskurowski, Andrzej

Rajasekaran, S.
Ramaiyer, Kumar
Reif, John
Ressler, Gene
Richards, Dana
Richter, Michael M.
Rinaldi, G.
Riviere, Stephane
Robert, Jean-Marc
Rossmanith, Peter
Rote, Günter

Sarnath, R.
Sarrafzadeh, M.
Sausse, Alain
Savage, John E.
Schaefer-Lorinser, Frank
Schaerf, Marco
Scheffler, Petra
Schieber, Baruch
Schirra, Stefan
Schiwietz, Michael
Schnitger, Georg
Schott, René
Schwiegelshohn, Uwe
Seberry, Jennifer
Sherbrooke, Evan

Sipala, Paolo
Smid, Michiel
Smith, Carl
Snoeyink, Jack
Socher-Ambrosius, Rolf
Sokolsky, Oleg
Souvaine, D.
Srimani, P.K.
Stanley, Richard P.
Steele, Mike
Steiger, W.
Stein, Cliff
Stephan, Frank
Stewart, Lorna

Takaoka, T.
Tamaki, Hisao
Tamassia, Roberto
Tel, Gerard
Tezuka, Shu
Thiel, Christian
Thomas, Federico
Tollis, Ioannis G.
Tripathi, Anand
Tromp, John

Vegter, Gert
Vismara, Luca
von Puttkamer, Ewald
Voronkov, Andrei

Wagner, A.S.
Wagner, F.
Wagstaff, Sam
Walsh, Toby
Watanabe, Toshimasa
Weihe, Karsten
Weispfenning, Volker
Wernisch, Lorenz
Winkler, Gerhard
Wuertz, Joerg

Yellin, Dan
Yvinec, Mariette

Zhang, Kaizhong
Zimmermann, Gerhard
Zwick, Uri

Contents

Session 3B

Session 4

Invited Presentation

Session 5

Session 11B

Algorithmic Problems Arising from Genome Informatics
(Invited Presentation)

Satoru Miyano

Research Institute of Fundamental Information Science,
Kyushu University 33, Fukuoka 812, Japan.
miyano@rifis.kyushu-u.ac.jp

Abstract

Important information about genes and proteins is encoded as nucleic sequences and amino acid sequences. The genome databases *PIR*, *GenBank*, *PDB*, *EMBL*, *PROSITE*, etc. compile these sequences together with their function and structure, from information published in the open literature or directly submitted to the database owners by researchers. For proteins, the number of amino acid sequences registered in these databases is currently about 50,000. This number is expected to exceed 100,000 within a few years.

The Human Genome Project started around 1990 in the United States, Japan, and a number of European countries, and is marked by international cooperation of researchers in a wide range of fields. The Human Genome Project is now entering a second stage, in which the goal is to determine all information about human genes by about 2005 or 2010. Genome Informatics or Molecular Bioinformatics is a new and rapidly evolving field which considers the various problems involved in processing these enormous amounts of genome data. Although great strides have been made in the last 30 years in terms of gathering and analyzing genome data, it also seems that in some ways Computer Science and Molecular Biology have developed on different evolutionary branches of the Science family tree. However, it is now understood that Genome Informatics is concerned with and embraces almost all fields in Computer Science.

This talk discusses some algorithmic problems arising from our research into developing a knowledge discovery system *BONSAI Garden* for amino acid sequences and nucleic sequences in relation to algorithm design and complexity. Algorithmic challenges for knowledge discovery may provide new directions in experimental investigations in Molecular Biology.

An Approximation Algorithm for Alphabet Indexing Problem

(Extended Abstract)

Shinichi Shimozono

Department of Control Engineering and Science,
Kyushu Institute of Technology
Kawazu 680-4, Iizuka, 820 JAPAN
E-mail : sin@ces.kyutech.ac.jp

Abstract. A K-indexing is a mapping from an alphabet Σ to a set Γ of K symbols forming the homomorphism to transform strings. Given Σ, two disjoint sets of strings P, Q over Σ and $\Gamma = \{1, \ldots, K\}$, Alphabet Indexing is the problem to find a K-indexing that transforms no two different strings taken from each P and Q into the same one. Although this problem is NP-complete, applying K-indexing to input data brings remarkable advantages in actual applications. In this paper, we introduce Max K-Indexing, a maximization version of Alphabet Indexing, that intends to maximize the number of pairs in $P \times Q$ whose strings are not transformed into the same ones. We show that the problem is MAX SNP-hard, that is, the problem seems to have no polynomial-time algorithm achieving an arbitrary small error ratio. Then we propose a simple polynomial-time greedy algorithm and show that the algorithm attains the constant error ratio $1/K$ for K-indexing. Also we define M-Tuple K-Indexing problem by extending pairs of strings in Max K-Indexing to tuples of strings over more than two sets, and show that a natural extension of the algorithm also achieves a constant error bound.

1 Introduction

In actual applications of combinatorial problems of strings, the size of the alphabet of input data has strong influence on the efficiency of algorithms even it is a constant factor in a theoretical analysis. In bioinformatical knowledge acquisition, Arikawa et al. showed in [1] that the use of a mapping reducing the size of the alphabet can improve efficiency and accuracy of machine learning algorithms. According to biochemical knowledge, they defined a mapping from 20 kinds of amino acid symbols to three categories, and transformed amino acid sequences used as training examples of the machine learning system. The mapping distinguished the transformed positive examples from negative ones, and they obtained successful experimental results for a certain protein classification problem.

From the view point of computer science, there are researches of "mechanical method" for finding such mappings, namely, alphabet indexing [10, 11]. Given an alphabet Σ, disjoint sets of strings $P, Q \subseteq \Sigma^*$ and an indexing alphabet $\Gamma = \{1, \ldots, K\}$, an alphabet indexing is a mapping f from Σ to Γ such that the transformation of strings derived from f converts no two strings $p \in P$ and $q \in Q$ into identical ones. Unfortunately, the problem for finding an alphabet indexing is NP-complete [10]. Therefore we introduced the optimization problem to find an indexing that allows some overlaps between the transformed sets of strings, and defined the optimization problem that employs the machine learning algorithm in [1]. We designed a local search algorithm for the problem, showed the PLS-completeness [10], and obtained the successful empirical results [11]. However, this optimization problem uses a hypothesis produced by a machine learning algorithm to measure the goodness of indexings, and the local search algorithm has no proved error estimations. To take advantages of an alphabet indexing, we need a simple optimization problem of alphabet indexing and a polynomial-time approximation algorithm whose error can be estimated.

In this paper, we introduce an optimization problem MAX K-INDEXING. This problem requires to find a K-indexing $f : \Sigma \to \{1, \ldots, K\}$ that maximizes the number of the pairs in $P \times Q$ whose strings are transformed into different ones by f. We show the MAX SNP-hardness of the problem in the context of class MAX SNP [2, 3, 8, 9]. This asserts that the problem seems to have no polynomial-time approximation scheme, that is, there is a certain constant bound for error ratios of approximation algorithms that run in polynomial time. Then we present a simple greedy algorithm inspired by the well-known greedy algorithm for MAXIMUM SATISFIABILITY by Johnson [7], and show that our algorithm achieves a constant relative error $1/K$ for K-indexing, with providing the worst case instance. By this new approach to the alphabet indexing problem, we can find an appropriate indexing in reasonable time not only for knowledge acquisition problems but also for the lossy text compression processed by the character-based compression [4].

Additionally, we define the problem M-TUPLE K-INDEXING by extending pairs in MAX K-INDEXING to tuples of strings. Then we show that the polynomial time algorithm extended from the greedy algorithm for MAX K-INDEXING also achieves a constant error ratio.

2 Notations

Let Σ and Γ be finite alphabets, and let s and t be strings in Σ^*. The ith symbol of s is denoted by $s[i]$. We say that s and t are the same if $s[i] = t[i]$ and $|s| = |t|$ for all $1 \leq i \leq |s|$, where $|s|$ denotes the length of s. We call a mapping f from Σ to Γ an *alphabet indexing* of Σ by Γ. Especially, an alphabet indexing that maps to $\Gamma = \{1, \ldots, K\}$ is called a K-*indexing*. For $a \in \Sigma$, we say that a symbol $f(a) \in \Gamma$ is an *index* of a. The homomorphism $f^* : \Sigma^* \to \Gamma^*$ derived from f is the transformation of strings from s in Σ^* to $f^*(s) = f(s[1]) \cdots f(s[|s|])$ in Γ^*. We say that a K-indexing f *distinguishes* $s, t \in \Sigma^*$ if $f^*(s) \neq f^*(t)$. For convenience, we regard the range of K-indexings as $\{0, 1, \ldots, K\}$ when we talk about algorithms, and by $f(a) = 0$ we represent that an index for a is undefined. We say that a K-indexing f is *partially defined* if $f(a) = 0$ for some $a \in \Sigma$.

An NP *optimization problem* Π is either a maximization or a minimization that consists of (1) a set of instances L_Π, (2) a finite set of feasible solutions $S_\Pi(\pi)$ for an instance $\pi \in L_\Pi$, (3) a measure function $\mu_\Pi : S_\Pi(L_\Pi) \rightarrow Z^+$, and (4) an optimization directive $opt \in \{\max, \min\}$. The measure of the optimal solution $OPT(\pi)$ for $\pi \in L_\Pi$ is defined by $OPT(\pi) = opt\{\mu_\Pi(s) \mid s \in S_\Pi(\pi)\}$. An algorithm A is an *approximation algorithm* if, for any instance $\pi \in L_\Pi$, it outputs a feasible solution in $S_\Pi(\pi)$. For an approximation algorithm A that solves Π, we define the *relative error*

$$\epsilon = \max_{\pi \in L_\Pi} \frac{|OPT(\pi) - \mu_\Pi(A(\pi))|}{\max\{OPT(\pi), \mu_\Pi(A(\pi))\}},$$

where $A(\pi)$ is the output of A for π. Also, we say that A is the ϵ-*approximation* for Π.

Let Π, Φ be optimization problems. We say that Π is *L-reducible* to Φ [9] if there are two polynomial time computable functions τ, ρ and constants $\alpha, \beta > 0$ such that:

1. Given an instance π of Π, function τ produces an instance $\tau(\pi)$ of Φ such that the measure of the optimum solution of $\tau(\pi)$ is at most $\alpha \cdot OPT(\pi)$, and
2. Given any solution s of $\tau(\pi)$, function ρ produces a solution $\rho(s)$ of π such that $|\mu_\Pi(\rho(s)) - OPT(\pi)| \leq \beta|\mu_\Phi(s) - OPT(\tau(\pi))|$.

L-reduction is transitive, and if problem Π is L-reducible to problem Φ and Φ can be approximated in polynomial time within relative error ϵ, then Π can be approximated within relative error $\frac{\alpha\beta\epsilon}{1-\epsilon}$. Moreover, if there is a polynomial-time approximation scheme (PTAS) [8] for Φ then there is also a PTAS for Π. A problem is MAX SNP-hard if every MAX SNP problem is L-reducible to it [9].

3   Definition and Intractability of The Problem

The problem ALPHABET INDEXING [10, 11] is formulated as follows.

Definition 1. ALPHABET INDEXING
Instance: Alphabet Σ, disjoint sets of strings $P, Q \subseteq \Sigma^*$ and indexing alphabet Γ.
Question: Is there an alphabet indexing $f : \Sigma \rightarrow \Gamma$ for P, Q such that f distinguishes p from q whenever $p \in P$ and $q \in Q$?

We showed in [10] that ALPHABET INDEXING is NP-complete for $|\Gamma| \geq 2$, and in order to deal with this intractability, we also introduced an optimization problem of an alphabet indexing that allows the transformed sets $f^*(P)$ and

$f^*(Q)$ having some overlaps. However, that problem requires a machine learning algorithm to produce hypotheses used by the measure function. Therefore we give a simpler optimization measure for indexing mappings.

Let $[P \approx Q]$ be the set of the pairs that are not distinguished by K-indexing f, $\{(p,q) \in P \times Q \mid f^*(p) = f^*(q)\}$. Let us consider the following optimization:

> To minimize the number of pairs in $[P \approx Q]$, maximize the number of the pairs that are distinguishable by f.

According to this intention, we introduce the following NP optimization problem.

Definition 2. MAX K-INDEXING

Instance: Alphabet Σ, sets of strings $P, Q \subseteq \Sigma^*$.
Object: Find a K-indexing $f : \Sigma \to \{1, \ldots, K\}$ that maximizes the number of the pairs in $P \times Q$ and distinguished by f.

Clearly no K-indexing distinguishes the pairs consisting of the same strings, and no K-indexing transforms the strings with the different length into the same ones. Therefore, from now on, we concentrate on the pairs consisting of different strings of the same length. For convenience, we refer to the set $\{(p, q) \in P \times Q \mid |p| = |q| \wedge p \neq q\}$ by $[P \times Q]$.

Compared with the ALPHABET INDEXING, the definition of MAX K-INDEXING allows that the original sets P and Q have overlaps. However, the following lemma holds.

Lemma 3. *Even for two same sets, say $Q, Q \subseteq \Sigma^*$, and $K = 2$, the problem deciding whether (Σ, Q, Q) has a K-indexing that distinguishes all pairs in $[Q \times Q]$ is NP-complete.*

Proof. We show a sketch of a reduction from 3-SAT [5]. Given an instance of 3-SAT, a set of 3-literal clauses C over variables X, we construct an instance (Σ, Q, Q) of MAX K-INDEXING as follows. Let Σ be $X \cup \{\mathtt{t}, \mathtt{f}\}$. For each clause $(l_1, l_2, l_3) \in C$, we put strings $s_1 \bar{s}_1 s_2 \bar{s}_2 s_3 \bar{s}_3 \langle i \rangle$ and $\mathtt{tftftf} \langle i \rangle$ into Q, where $s_j \bar{s}_j$ is $\mathtt{t}x_k$ (resp. $x_k \mathtt{f}$) if l_j is a positive literal x_k (resp. a negative literal $\bar{x}_k$), and $\langle i \rangle$ is a binary string coding i by viewing $\mathtt{t}, \mathtt{f}$ as $1, 0$. Then f distinguishes all originally different pairs in $Q \times Q$ if and only if $f(\mathtt{t}) \neq f(\mathtt{f})$ and $f^*(s_1 \bar{s}_1 s_2 \bar{s}_2 s_3 \bar{s}_3) \neq f^*(\mathtt{tftftf})$, and this implies that there is an assignment $f : X \to \{\mathtt{t}, \mathtt{f}\}$ satisfying all clauses in C. $\square$

Furthermore, the next lemma holds.

Lemma 4. MAX K-INDEXING *is MAX SNP-hard.*

Proof. We show an L-reduction from a MAX SNP-complete problem MAX CUT [9] to MAX 2-INDEXING. Given a graph $G = (V, E)$, MAX CUT is the problem to find a partition of the vertices S and $\bar{S} = V - S$ that maximizes the number of edges going from S to $\bar{S}$.

The algorithm τ builds an instance (Σ, P, Q) of Max 2-Indexing from an instance G of MAX CUT as follows. For every ith vertex v_i of V, the corresponding symbol v_i is in Σ. For each edge (v_i, v_j) of E (assume $i < j$), a string $(v_i)^i$ is in P and $(v_j)^i$ is in Q. The algorithm ρ is defined as follows. Given a K-indexing f, the partition $\rho(f)$ is $(S = \{a \in V \mid f(a) = 1\}, V - S)$. Then the pair $((v_i)^i, (v_j)^i)$ is in $P \times Q$ if and only if the corresponding edge (v_i, v_j) is in E, and f distinguishes a pair in $P \times Q$ if and only if the partition separates the corresponding vertices. Thus the lemma holds for $\alpha = \beta = 1$. $\square$

4   A Polynomial-time Approximation Algorithm

Lemma 4 concludes that MAX K-INDEXING has no polynomial-time approximation scheme (PTAS) unless P = NP [2, 8]. Therefore our next goal is inventing a polynomial-time approximation algorithm that achieves a fixed error ratio.

Before looking into our polynomial time greedy algorithm, note the following observation. Let p and q be strings in Σ^* and let a and b be symbols in Σ. For a pair of strings (p, q), we say that a and b are *facing* in (p, q) if, for some i with $1 \le i \le |p| = |q|$, either $(p[i] = a) \wedge (q[i] = b)$ or $(p[i] = b) \wedge (q[i] = a)$. Notice that (p, q) is distinguished by f if (p, q) has at least one facing pair of symbols a and b with different indices, $f(a) \neq f(b)$.

Algorithm $Greedy_K$ (*input*: Σ, P, Q)
 1. Let $Diff := \emptyset$, $Res := [P \times Q]$ and $f(a) := 0$ for all $a \in \Sigma$.
 2. For each $a \in \Sigma$ do
     a. For each $i \in \{1, \cdots, K\}$ compute $Saved_i :=$
         $\{(p, q) \in Res \mid$ there is a symbol b facing a in (p, q) and $0 < f(b) \neq i\}$.
     b. Find k that maximizes $|Saved_k|$.
     c. $Diff := Diff \cup Saved_k$, $Res := Res - Saved_k$ and $f(a) := k$.
 3. Output f.

This algorithm runs in polynomial time, $\mathcal{O}(K \cdot |\Sigma| \cdot |P| \cdot |Q|)$, and thus runs in $\mathcal{O}(n^3)$ with respect to the coding length of P and Q. It is still polynomial if K becomes a part of input. For this algorithm, we have the following theorem.

Theorem 5. *The relative error of* $Greedy_K$ *for* MAX K-INDEXING *is* $\epsilon = 1/K$.

Proof. Suppose that the algorithm has been applied to $\pi = (\Sigma, P, Q)$, and is now going into Step 2-(a) of lth iteration for choosing the best index to $a_l \in \Sigma$. Let $d_i = \{(p, q) \in Res \mid b$ is facing a_l in (p, q) and $f(b) = i\}$ for $1 \le i \le K$, where Res is the remained pairs after the $(l - 1)$th iteration. In another words, d_i is a subset of $Saved_k$ if $i \neq k$ and is removed from Res in step 2-(b) if $f(a_l) \neq i$. We divide $\bigcup_{i=1}^{K} d_i$ into $K + 1$ disjoint parts $D_0, D_1, \ldots, D_K$ as follows.

(1) $D_0 = \bigcup_{i \neq j}(d_i \cap d_j)$: the set of pairs distinguished by assigning any index $f(a_l) \neq 0$ to a_l.

(2) $D_i = d_i - D_0$: the set of pairs distinguished if and only if the index for a_l is not i, $f(a_l) \neq i$.

Additionally, $Ex_i \subseteq D_i$ is the set of the pairs that are not distinguished by an assignment $f(a_l) := i$ and have no symbols indexed by zero. (We ignore the symbols that are still not indexed but facing only the same symbols.) Notice that $D_i \cap D_j = \emptyset$ for all $1 \leq i \neq j \leq K$, and if i is indexed to a_l, then all of Ex_i will remain in Res and never be saved. For choosing $k \in \{1, \ldots, K\}$ to $f(a_l)$, algorithm $Greedy_K$ maximizes the size of $Saved_k = D_0 \cup (\cup_{i \neq k} D_i)$ and thus k must be a number that minimizes $|D_k|$. For an index $f(a_l) = k$ we lose the pairs in Ex_k, so we have

$$|Saved_k| = |D_0| + |\cup_{i \neq k} D_i| \geq |D_0| + (K-1) \cdot |D_k| \geq (K-1) \cdot |Ex_k| .$$

This inequality holds at any iteration. When the algorithm stopped, $Diff$ is the union of all $Saved$'s at each iteration, and Res is also the union of all Ex's. As a result, we have $|Diff| \geq (K-1) \cdot |Res|$, and by $OPT(\pi) \leq |P \times Q|$,

$$OPT(\pi) \leq |P \times Q| = |Diff| + |Res| \leq |Diff| + \frac{1}{K-1}|Diff| .$$

Therefore we have $\epsilon \leq \frac{1}{K}$ for the algorithm $Greedy_K$.

Now we show that the worst case error of our algorithm becomes $1/K$. Let m be a sufficiently large positive integer. Let us consider the following instance π' with strings of the length two:

$$\Sigma = \{a_1, \cdots, a_K, b_1, b_2, c_1, \cdots, c_m\}.$$
$$P = \{a_ib_1 \mid 1 \leq i \leq K\}, Q = \{a_ib_2 \mid 1 \leq i \leq K\} \cup \{c_ib_1 \mid 1 \leq i \leq m\}.$$

In $P \times Q$, we have (a_ib_1, a_jb_2) for $1 \leq i,j \leq K$ and (a_ib_1, c_jb_1) for $1 \leq i \leq K, 1 \leq j \leq m$. An optimal indexing is, for example,

$$f(\sigma) = \begin{cases} 1 \text{ if } \sigma = a_i \text{ with } 1 \leq i \leq K, \\ i \text{ if } \sigma = b_i, \\ 2 \text{ if } \sigma = c_i \text{ with } 1 \leq i \leq m. \end{cases}$$

This distinguishes all K^2 pairs of the form (a_ib_1, a_jb_2) by $f(b_1) \neq f(b_2)$ and all mK of the form (a_ib_1, c_jb_1) by $f(a_i) \neq f(c_j)$. Although the algorithm may assign K different symbols to $a_1, \cdots, a_K$ firstly with saving $2(i-1)$ pairs at each ith decision of index. Then the algorithm assigns different two symbols to b_1, b_2. As a result, all pairs of the form (a_ib_1, a_jb_2) are distinguished, but at least one pair of the form (a_ib_1, c_jb_1) can not be distinguished for each c_j. For $m \to \infty$, the ratio $Greedy_K(\pi')/OPT(\pi')$ goes to

$$\frac{K(K-1) + K + m(K-1)}{K^2 + mK} \to \frac{K-1}{K},$$

and thus $\epsilon = 1/K$. $\square$

5 Extending The Problem and The Algorithm

In this section, we consider a problem M-TUPLE K-INDEXING, an extension of MAX K-INDEXING, that intends to distinguish not only a pair of strings but also more than two, say M, strings in a tuple.

Let $t = (q_1, \ldots, q_M)$ be a tuple of strings $q_1, \ldots, q_M \in \Sigma^*$, and let f be a K-indexing of Σ. We say that f *distinguishes* tuple t if $f^*(q_i) \neq f^*(q_j)$ for all $q_i \neq q_j$ with $1 \leq i \neq j \leq M$. Let f and f' be a partially defined K-indexing. We say f' *derives* f if $f(a) \neq 0$ then $f(a) = f'(a)$. We say that a tuple t is *distinguishable* by f' if there is a derived indexing f that distinguishes t.

Definition 6. M-TUPLE K-INDEXING
Instance: An alphabet Σ and sets of strings $Q_1, \ldots, Q_M$ over Σ.
Object: Find a K-indexing f that maximizes the number of tuples that are in $Q_1 \times \cdots \times Q_M$ and distinguished by f.

We apply the following greedy algorithm $Greedy_K^M$, which is an extension of $Greedy_K$.

Algorithm $Greedy_K^M$ (*input*: $\Sigma, Q_1, \ldots, Q_M$)
1. Let $T := Q_1 \times \cdots \times Q_M$, $f(a) := 0$ for all $a \in \Sigma$ and $w(t) := r^{-diff(t)}$ for all $t \in T$, where $r \geq 1$ is the "preserving constant" and $diff(t)$ is the number of originally different pairs in t.
2. For each $a \in \Sigma$ do
    a. Compute $D[i]$, $p(i, t)$ for $i \in \{1, \ldots, K\}$ and $t \in T$ defined as follows:
    $p(i, t) = $ the number of the pairs in t distinguished if $f(a) := i$.
    $D[i] = \{t \in T : \text{assignment } f(a) := i \text{ retains } t \text{ being distinguishable}\}$.
    b. Choose index k that maximizes $\displaystyle\sum_{t \in D[k]} w(t) \cdot r^{p(k,t)}$.
    c. $T := D[k]$, $w(t) := w(t) \cdot r^{p(k,t)}$ for all $t \in T$ and $f(a) := k$.
3. Output f.

Notice that for $M = 2$ the algorithm $Greedy_K^M$ is equivalent to $Greedy_K$ since $diff(t) = 1$ and $p(i, t) \leq 1$.

Claim. The algorithm runs in polynomial-time.

The time needed for computing $D[i]$'s in Step 2–(a) dominates the running time of the algorithm. Let n be the coding length of $Q_1, \ldots, Q_M$, and let l be the length of the longest string in these sets. In a computation of $D[i]$, we decide whether each t is distinguishable by a partially defined K-indexing f. This can be done by checking whether a tuple has a set of facing pairs of symbols proving the distinguishability, namely a "witness." A witness must have (i) at least one facing pair in each pair of strings in t and (ii) at least one derived indexing that distinguishes all pairs of symbols. There are at most $l^{M(M-1)/2}$ sets that are candidates for a witness. In each set, we have at most $M(M - 1)$ different symbols, and we have to check at most $K^{M(M-1)}$ derived K-indexings by fixing

$M(M-1)$ indices. Therefore the computation of $D[i]$ for each $1 \leq i \leq K$ can be done in $\mathcal{O}(n \cdot l^{M(M-1)/2} \cdot K^{M(M-1)}) = \mathcal{O}(n^{M(M-1)/2+1})$, and the algorithm runs in $\mathcal{O}(n^{M(M-1)/2+2})$ time.

Notice that, if M is unbounded, then deciding whether one tuple t has a K-indexing that distinguishes t is NP-complete: It is equivalent to the problem of Lemma 3, which requests a K-indexing that distinguishes all the pairs of two same sets of strings.

For the above algorithm, we have the following result.

Theorem 7. *Generally for $r = K$, algorithm Greedy$_K^M$ computes an indexing that totally distinguishes at least $\frac{1}{r^{M(M-1)/2}}$ of all the tuples. If K is sufficiently large for satisfying $K > M(M-1)/2$, r can be reduced to $r = \frac{K}{K-M(M-1)/2}$.*

This states the absolute error and thus also states the bound of the relative error. Notice that for $M = 2$ the relative error is also the same with that of MAX K-INDEXING, $\frac{1}{K}$.

Proof. Let us note the following two observations.

- At initialization of the algorithm, Step 1, the number of the originally different pairs in tuple t, $\textit{diff}(t)$ is at most $M(M-1)/2$. Thus the initial weight sum $\sum_{t \in T} w(t)$ is at least $|T| \cdot r^{-M(M-1)/2}$.
- While the algorithm is running, the weight $w(t)$ of a tuple t does not exceed one since any tuple t with $w(t) = 1$ has no more pairs to be distinguished.

Suppose that the algorithm is deciding an index for a symbol in Step 2. At each iteration of Step 2–(b), an index k is chosen to maximize the sum of the updated weights of tuples in $D[k]$. This sum is at least $1/r$ of the sum of the original weights in $T = \cup_{j=1}^K D[j]$ for $r = K$,

$$\sum_{t \in D[k]} w(t) \cdot r^{p(k,t)} \geq r \sum_{t \in D[k]} w(t) \geq \sum_{t \in T} w(t).$$

Thus the choice of an index k does not decrease the weight sum in new $T = D[k]$ in Step 2–(c) compared with that of T at the end of the previous iteration. This fact holds for any iteration of Step 2, and the algorithm finally produces a K-indexing that totally distinguishes at least $r^{-M(M-1)/2}$ of all the tuples.

Now we show the case in which the preserving constant r can be reduced. Let $r_0 \leq M$ be a positive integer. If all distinguishable tuples belong at least r_0 of all D's, the weight sum of all D's satisfies $\sum_{j=1}^K \sum_{t \in D[j]} w(t) r^{p(j,t)} \geq r_0 \cdot \sum_{t \in T} w(t)$.

Since an index k decided at each iteration maximizes the sum of the weights of survived tuples, the weight sum in $D[k]$ must be larger than the average of all D's. Thus $\sum_{t \in D[k]} w(t)$ is at least r_0/K of $\sum_{t \in T} w(t)$. In general, any survived tuple belongs at least one of D's and thus we took $r_0 = 1$ and $r = K$ in the above discussion. Here we attempt an easy extension of r_0 for the case M is relatively small. To distinguish tuples, we have to distinguish pairs of facing symbols at most $M(M-1)/2$. Obviously a tuple shall remain to be distinguishable if no facing pair matches in its indexing. Therefore, if there are indices more than pairs in tuples, $K > M(M-1)/2$, then $r_0 = K - M(M-1)/2$. $\square$

Note that a worst case instance for this algorithm can be constructed in the similar way with MAX K-INDEXING.

The above result states that the error raises with the size of the indexing alphabet. This may seem to be curious; It is clear that using more indices makes distinguishing tuples easier. However, the statement of the theorem does not contradict with this fact: The result states "if more and more indices are needed, then distinguishing tuples is more and more difficult." If specific subproblems permit a larger factor r_0, we can fill gaps between these observations.

6 Conclusion

We introduced an optimization measure for Alphabet Indexing, and according to it we defined a maximization problem MAX K-INDEXING. For this optimization problem, we proposed a polynomial-time approximation algorithm $Greedy_K$ and showed that the algorithm achieves a constant relative error $1/K$. Also we defined M-TUPLE K-INDEXING, which requires all transformed strings in tuples mutually distinguished, and showed the algorithm $Greedy_K^M$ also achieves a constant relative error. Note that the problem defined by another variation of the condition "at least one string is distinguished from others in each tuple" is equivalently transformed to MAX K-INDEXING: It can be done by concatenating strings in each tuple to be a pair of strings.

Although we discussed only the problems with no weights on strings, weighted versions of MAX K-INDEXING and M-TUPLE K-INDEXING can be defined with additional weight functions on the sets of strings, and $Greedy_K$ and $Greedy_K^M$ can be applied to the weighted problems with slight modifications. We should notice that those algorithms achieve the same relative error for the weighted problems.

The error ratio $1/K$ of $Greedy_K$ seems not so good, especially when a few indices are available. In the recent results, there are remarkable improvements of approximation algorithms for MAX SAT and MAX CUT [6, 12]. There may be relations between MAX K-INDEXING and these problems, although the techniques in [12] can not be applied directly to our problem. Thus polynomial time algorithms that achieve improved error ratios and the approximability of MAX K-INDEXING will be considered in our forthcoming issue.

Acknowledgments: The author thanks to Magnus M. Halldorsson for his helpful suggestions and comments, and also thanks to Hiroki Arimura for his valuable advice to the latter algorithm's polynomial-time solvability.

References

1. S. Arikawa, S. Miyano, A. Shinohara, S. Kuhara, Y. Mukouchi, and T. Shinohara, *A machine discovery from amino acid sequences by decision trees over regular patterns*, New Gener. Comput., 11 (1993), pp. 361–375.
2. S. Arora, C. Lund, R. Motwani, M. Sudan, and M. Szegedy, *Proof verification and hardness of approximation problems*, in Proc. 33rd Annual Symposium on Foundations of Computer Science, 1992, pp. 14–23.
3. A. Blum, T. Jiang, M. Li, J. Tromp, and M. Yannakakis, *Linear approximation of shortest superstrings*, in Proc. 23rd Annual ACM Symposium on Theory of Computing, 1991, pp. 328–336.
4. S. Fukamachi, S. Shimozono, H. Arimura, and T. Shinohara, *Lossy text compression for string pattern matching*, Technical Report of IEICE NLC95-6, 1995.
5. M. R. Garey and D. S. Johnson, *Computers and Intractability: A Guide to the Theory of NP-Completeness*, W. H. Freeman and Company, New York, 1978.
6. M. X. Goemans and D. P. Williamson, *.878-approximation algorithms for MAX CUT and MAX 2SAT*, in Proc. 26th Annual ACM Symposium on Theory of Computing, 1994, pp. 422–431.
7. D. S. Johnson, *Approximation algorithms for combinatorial problems*, J. Comput. Sys. Sci., 9 (1974), pp. 256–278.
8. C. H. Papadimitriou, *Computational Complexity*, Addison-Wesley Publishing Company, Reading, Massachusetts, 1994.
9. C. H. Papadimitriou and M. Yannakakis, *Optimization, approximation, and complexity classes*, J. Comput. Sys. Sci., 43 (1991), pp. 425–440.
10. S. Shimozono and S. Miyano, *Complexity of finding alphabet indexing*, IEICE Trans. Inf. Sys., E78-D (1995), pp. 13–18.
11. S. Shimozono, A. Shinohara, T. Shinohara, S. Miyano, S. Kuhara, and S. Arikawa, *Knowledge acquisition from amino acid sequences by machine learning system BONSAI*, Trans. Inf. Proc. Soc. Japan, 35 (1994), pp. 2009–2018.
12. M. Yannakakis, *On the approximation of maximum satisfiability*, J. Algorithms, 17 (1994), pp. 475–502.

A Fast and Space-Economical Algorithm for Length-Limited Coding

Jyrki Katajainen[1]   Alistair Moffat[2]   Andrew Turpin[2]

[1] Department of Computer Science, University of Copenhagen,
Universitetsparken 1, DK-2100 Copenhagen East, Denmark
jyrki@diku.dk
[2] Department of Computer Science, The University of Melbourne,
Parkville 3052, Australia
{alistair,aht}@cs.mu.oz.au

Abstract. The *minimum-redundancy prefix code problem* is to determine a list of integer codeword lengths $l = [l_i \mid i \in \{1 \ldots n\}]$, given a list of n symbol weights $p = [p_i \mid i \in \{1 \ldots n\}]$, such that $\sum_{i=1}^{n} 2^{-l_i} \leq 1$, and $\sum_{i=1}^{n} l_i p_i$ is minimised. An extension is the *minimum-redundancy length-limited prefix code problem*, in which the further constraint $l_i \leq L$ is imposed, for all $i \in \{1 \ldots n\}$ and some integer $L \geq \lceil \log_2 n \rceil$. The package-merge algorithm of Larmore and Hirschberg generates length-limited codes in $O(nL)$ time using $O(n)$ words of auxiliary space. Here we show how the size of the work space can be reduced to $O(L^2)$. This represents a useful improvement, since for practical purposes L is $\Theta(\log n)$.

1   Introduction

Use of Huffman's algorithm [2] for the generation of minimum-redundancy prefix codes for a weighted set of symbols is well known. For practical use an important restriction is to limit the codewords to be at most L bits long, since implementations of data compression methods are usually designed around fixed-width registers. For example, most current computers use words of 32 bits.

Larmore and Hirschberg [3] described the first efficient algorithm for the generation of minimum-redundancy length-limited prefix codes. Their *package-merge* method requires $O(nL)$ time and $O(n)$ space, where n is the number of symbols in the alphabet and L is the length limit. This improves the method of Van Voorhis [9] (see also Hu and Tan [1]), which consumes $O(Ln^2)$ time and space. Asymptotically faster algorithms for solving the problem have been developed recently, see Schieber [7] and the references therein.

In practical applications the speed of the package-merge algorithm is not a problem and the space requirements turn out to be of greater importance. In this paper we describe an improved implementation of the package-merge paradigm—the *boundary package-merge* algorithm—that constructs a minimum-redundancy length-limited prefix code in $O(L^2)$ auxiliary space, while retaining the $O(nL)$ time bound. For most purposes the space required is negligible, since L is typically $\Theta(\log n)$. That is, we have developed an "almost in-place" algorithm for calculating length-limited codes.

2   Prefix Codes

Consider a list $p = [p_i \,|\, p \in \{1 \ldots n\}]$ of n positive symbol *weights*. For example, p might be the observed frequencies of an alphabet of symbols, as defined by some data compression method. A *code* is an integer list $l = [l_i \,|\, i \in \{1 \ldots n\}]$, where l_i is the length of the codeword to be assigned to the ith symbol of the alphabet described by p. A *prefix code* (more precisely, *prefix-free code*, but for brevity we use the contraction) is a code for which $K = \sum_{i=1}^{n} 2^{-l_i} \leq 1$. Given a prefix code l, it is straightforward to create a set of n binary codewords such that no codeword is a proper prefix of any other and such that the ith codeword is l_i bits long. In terms of codeword calculation, we can thus consider the task to be essentially done when a code l has been devised. An *L-limited prefix code*, or L-bit *length-limited* prefix code, for some integer $L \geq \lceil \log_2 n \rceil$, is a prefix code in which $l_i \leq L$, for all $i \in \{1 \ldots n\}$.

A prefix code is *minimum-redundancy* if $B = \sum_{i=1}^{n} l_i p_i$ is minimal over all prefix codes. An L-limited prefix code is minimum-redundancy if $B = \sum_{i=1}^{n} l_i p_i$ is minimal amongst all L-limited prefix codes. If the symbol weights are frequencies quantity B is the number of output bits required by the code l.

Here we examine the construction of minimum-redundancy L-limited prefix codes. The model of computation we suppose is a unit-cost random access machine, in which integers as large as $U = \sum_{i=1}^{n} p_i$ can be stored in a single word of storage and manipulated (addition, comparison) in $O(1)$ time. It is also assumed that p is presented in non-decreasing order, $p_1 \leq p_2 \leq p_3 \cdots \leq p_n$. We measure the space requirements of various algorithms by counting the number of extra words of storage required by that algorithm, above and beyond the space required by input list p, which is assumed to be free. In this framework the recursive package-merge algorithm described by Larmore and Hirschberg [3], requires $O(n)$ space, and the improved boundary package-merge described in Section 4 requires $O(L^2)$ space. This difference represents a significant improvement, since for practical use L is a small constant such as 32.

The package-merge algorithm makes use of a *tree* data structure. An *item* s is a tree, and if s is an item and t_1 and t_2 are two trees then $t = s(t_1, t_2)$ is also a tree, with item s the *root* of tree t. If $t = s(t_1, t_2)$ and items s_1 and s_2 are the roots of trees t_1 and t_2 then s is the *parent* of items s_1 and s_2 in tree t. Similarly, items s_1 and s_2 are the *children* of item s. If item s is a singleton and has no children, then it is a *leaf* item of the tree, otherwise it is an *internal* item. To be consistent with the description of Larmore and Hirschberg, we will also refer to internal items as *packages*. The *depth* of any item is one greater than the depth of its parent; the depth of a root is zero.

We also require an analogous one-dimensional structure called a *chain*. A singleton *node* x is a chain; and if y is a chain and x is a node then $x(y)$ is a chain. In this latter case x is the *head* of the chain and y is its *tail*. Note that chains may coalesce; if y is a chain and x_1 and x_2 are two nodes then $x_1(y)$ and $x_2(y)$ are both chains.

3 Package-Merge

In the package-merge algorithm of Larmore and Hirschberg [3] L lists of trees are developed, with each tree in each list having an associated *weight*. The first list is simply a list of n leaves, with the ith tree in the list having weight p_i. The second list is then developed by merging, in increasing weight order, a copy of the first list with a list of packages produced from the first list. Packages are produced from a list of trees by forming new trees $s_i(t_{2i-1}, t_{2i})$, for each $i \in \{1, 2 \ldots \lfloor m/2 \rfloor\}$, where t_h is the hth tree in the list, and m is the cardinality of the list. The weight of a package, stored in item s_i, is the sum of the weights of its children. In general, list j is developed by forming a list of packages from list $j-1$ and merging this list with a copy of the first list developed. For example, the lists developed by the package-merge method when applied to the input weights $p = [1, 1, 5, 7, 10, 14]$ with $L = 4$ are shown in Figure 1.

Define the *active leaves* to be the leaves of the first $2n - 2$ trees of the Lth list. Extracting the code from the final set of L lists is achieved by processing these active leaves—each active leaf corresponds to exactly one of the original symbols, and l_i should be set to the number of active leaves corresponding to p_i. In the above example $n = 6$, so the first 10 trees in the fourth list contain the active leaves that yield the code, as shown by the shaded region in Figure 1. The first two trees in the bottom list of Figure 1 are active leaves corresponding to p_1 and p_2, thus $l = [1, 1, 0, \ldots, 0]$ after these two trees are processed. The third tree has two leaves, again corresponding to p_1 and p_2, so $l = [2, 2, 0, \ldots, 0]$ after this tree has been expanded. After all 10 trees are processed $l = [4, 4, 3, 2, 2, 2]$, which is a minimum-redundancy 4-limited code. Note that a 3-limited code can also be calculated by processing the active leaves reachable from the first ten items in the third list, yielding $l = [3, 3, 3, 3, 2, 2]$.

It is instructive to examine how $K = \sum_{i=1}^{n} 2^{-l_i}$ changes as the code is extracted from the active leaves. Initially, $l = [0, 0, \ldots, 0]$ and $K = n$. If a tree chosen in list L is a leaf with weight p_i, then l_i increases from 0 to 1, and K reduces by 2^{-1}. If the chosen tree is a package then it will have leaves at various levels. A straightforward inductive argument shows that the reduction in K

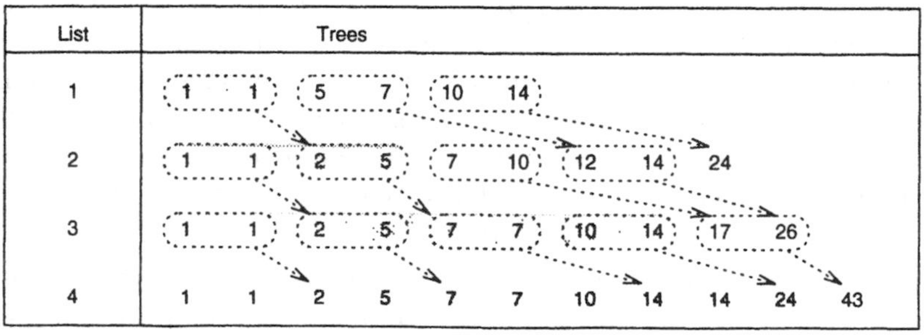

Fig. 1. Package-merge on the input weights $[1, 1, 5, 7, 10, 14]$ with $L = 4$

arising from the leaves of any tree rooted in the jth list is $2^{-(L-j+1)}$, so every package in the Lth list reduces K by 2^{-1}. That is, choosing $2n - 2$ trees from the Lth list reduces K from n to 1, guaranteeing that the code generated is a prefix code. Furthermore, the greedy manner in which the trees are constructed and chosen guarantees both that $l_i \leq L$ and that $B = \sum_{i=1}^{n} l_i p_i$ is minimised.

As described here, the package-merge method requires both $O(nL)$ time and $O(nL)$ space. Larmore and Hirschberg reduced the space requirement to $O(n)$ by implementing the method recursively and performing a controlled amount of reevaluation rather than storing intermediate results [3]. Our development, described in the next section, also starts with the space-inefficient package-merge and arrives at a third version that requires $O(L^2)$ auxiliary space.

Finally, note that it is possible to store the solution as a list $a = [a_j \mid j \in \{1 \ldots L\}]$, where a_j is the number of active leaves in the jth list, thereby avoiding the need to store the list l. In the example $a = [2, 3, 6, 6]$ when $L = 4$ and $a = [4, 6, 6]$ when $L = 3$. If conversion is required, a straightforward $O(n)$-time loop suffices to convert a into the corresponding list l. Boundary package-merge, described in the next section, produces a as its output.

4 Boundary Package-Merge

The package-merge algorithm develops trees in an exhaustive manner; that is, list $j - 1$ is completely created before construction of list j is commenced, and all of lists 1 through to $L - 1$ are fully instantiated before even the initial tree in list L is determined. The first key observation we make is that trees can also be built as a *demand-driven* or *lazy* process, beginning with the roots of the $2n - 2$ trees in the Lth list and adding the necessary children to complete them [6]. The roots must be created in increasing weight order, so for every package that appears in list L the weight of two items in list $L - 1$ must have been calculated. This in turn leads to the creation of items in list $L - 2$, and so on. Since there are no packages in the first list these cascading demands can always eventually be satisfied.

The basic operation in this demand-driven process is the appending of a tree to some list j, where $1 \leq j \leq L$. Each time this operation is performed it is necessary to determine whether the next item is a leaf or a package. The next leaf has weight p_{c+1}, where all leaves of weight p_k for $1 \leq k \leq c$ have already been included in list j; and the next package is of weight equal to the sum of the next two unused trees in list $j - 1$. As in the original package-merge method the candidate with the smaller weight is selected. Hence, when a tree is required in list j two trees must be instantiated in list $j - 1$ before the choice between package and leaf can be made. These two trees will be referred to as *lookahead* trees, as they are created to allow knowledge of the weight of the next package in list j, but the package they form may not become part of an active tree until some time in the future. When the two lookahead trees in list $j - 1$ are eventually used to form a package in list j two more lookahead trees must be created in list $j - 1$ so that the weight of the next package in list j can be determined.

Input	A set of L lists of trees, a list of n symbol weights, and the index j of the list in which an item is required. Suppose that list j currently contains c singleton trees.
Step 1	If $j = 1$ then create a singleton tree d of weight p_{c+1} and append it to list 1. Return.
Step 2	Let t_1 and t_2 be the two lookahead trees in list $j - 1$ and s be the sum of the weights of their roots. If s is larger than p_{c+1} then do Step 3a, otherwise do Step 3b.
Step 3a	Create a new leaf d with weight p_{c+1}. Append d to list j.
Step 3b	Create a new tree $d = s(t_1, t_2)$. Append d to list j. Invoke LazyPM twice with parameter $j-1$ to generate new lookahead trees in list $j - 1$.
Output	The same set of L lists, with one extra tree in list j and perhaps extra trees in lists 1 to $j - 1$.

Fig. 2. Algorithm LazyPM

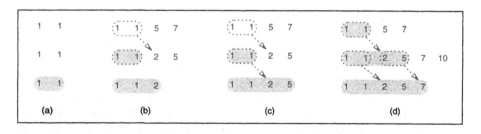

Fig. 3. Demand-driven list development on the weights $[1, 1, 5, 7, 10, 14]$ with $L = 3$ after: (a) two trees have been created; (b) three trees have been created; (c) four trees have been created; and (d) five trees have been created

Algorithm LazyPM in Figure 2 describes this process, accepting a list number as a parameter, and then generating the next tree in that list using the demand-driven method. When applied $2n - 2$ times with parameter L to a set of initialised lists, LazyPM produces all active trees. Initially, each list contains as lookahead trees two leaves with weight p_1 and p_2—these are the first two items in all of the L lists, since all of the weights are assumed to be positive and the first package in each list has weight $p_1 + p_2$.

Figure 3 shows the lists from the earlier example when evaluated using algorithm LazyPM with $L = 3$. Figure 3a shows the state of the lists after two trees in L have been processed, with Figures 3b, 3c, and 3d showing the lists after creation of the third, fourth and fifth trees in list L respectively.

The advantage of the demand-driven approach to list development is that all of the leaves in all of the constructed trees rooted in list L are definitely known to be active, as exactly $2n - 2$ trees are created in list L. This means that the leaves of each level L tree can be processed and included into a running solution as soon as that tree is formed, and then the space occupied by that tree freed for reuse. For example, in Figure 3d the first five trees rooted in list L (shown in the shaded region) are known to be active. These trees can be discarded and a set to $[2, 3, 3]$, where a is the active leaf list described earlier. Unfortunately, the peak space requirement is still $\Theta(nL)$. To see this, consider the case when $p_i = 1$, for all i. For large n and L, two lookahead trees in list $L - 1$ correspond to four outstanding trees in list $L - 2$; moreover, two more lookahead trees are required. These six trees correspond to 12 trees (plus two more lookahead trees) in list $L - 3$, and so on. Hence, for this example $\Theta(nL)$ space can still be required.

By the construction of the lists, however, we know that if the rightmost active leaf in list j has weight p_c, then $a_j = c$, and this is the second key observation that makes the new method possible. This observation suggests that building complete trees is wasteful, as a can be extracted knowing only the rightmost active item at each depth in each tree. Boundary package-merge exploits this observation by manipulating lists of one-dimensional *chains* rather than two-dimensional trees. Each chain corresponds to an equivalent tree of the same weight in the package-merge lists. A node in a chain is a *package node* if its weight is derived from the sum of two weights from the previous list, or a *leaf node* otherwise. A node in list j differs from a tree, however, in that it stores a single pointer into the list $j - 1$, and it also stores a count of the number of leaf nodes to its left in list j including itself. Let the pair $\langle w, c \rangle$ define a node, where w is the weight of the node and c is the leaf node count; and let each of the L lists be a list of chains of nodes.

Each chain forms a boundary between the nodes that will become active if the head of the chain becomes active—the nodes that were children of this tree in the two-dimensional package-merge—and those nodes that belong to other trees. Hence, if the head node in any chain becomes active it is known that all other nodes in chain are active. Furthermore, for each node in the chain that becomes active, say $\langle w_j, c_j \rangle$ in list j, it is known that the first c_j leaf nodes in list j have become active, and that any others remain of indeterminate status as lookahead nodes. The chain beginning at the $(2n - 2)$nd node in list L gives the result a by setting $a_j = c_j$ for each node $\langle w_j, c_j \rangle$ in the chain. For example, the final boundary chain for the lists in Figure 1 is $\langle 24, 6 \rangle (\langle 14, 6 \rangle (\langle 5, 3 \rangle (\langle 1, 2 \rangle)))$, giving $a = [2, 3, 6, 6]$.

The revised method is described in Figure 4. All L lists are initialised to contain the first two lookahead chains $\langle p_1, 1 \rangle$ and $\langle p_2, 2 \rangle$. The need for $2n - 2$ chains in list L again drives the process.

At each step there are two candidate chains to be considered for entry into list j. The first candidate is a singleton chain $\langle p_{c+1}, c + 1 \rangle$, where all leaf nodes of weight p_k for $k \leq c$ have already been added to list j. The second candidate is a package with weight equal to the sum of the weights of the head nodes in the two lookahead chains in list $j - 1$, and whose count is equal to the count of

Input	A set of L lists of chains, a list of n symbol weights, and the index j of the list in which an item is required. Suppose that $\langle w, c \rangle (y)$ is the last chain in list j, with weight w, count c, and tail y. That is, there are c singletons in list j.
Step 1	If $j = 1$ then create a singleton chain $\langle p_{c+1}, c + 1 \rangle$ and append it to list 1. Return.
Step 2	Let s be the sum of the weights of the heads of the two lookahead chains in list $j-1$. If s is larger than p_{c+1} then do Step 3a, otherwise do Step 3b.
Step 3a	Create a new chain $d = \langle p_{c+1}, c + 1 \rangle (y)$. Append d to list j.
Step 3b	Let z be the second of the two lookahead chains in list $j - 1$. Create a new chain $d = \langle s, c \rangle (z)$. Append d to list j. Invoke BOUNDARYPM twice with parameter $j - 1$.
Output	The same set of L lists, with one extra chain in list j and perhaps extra chains in lists 1 to $j - 1$.

Fig. 4. Algorithm BOUNDARYPM

the head of the chain currently at the end of list j.

Figure 5 shows how Steps 3a and 3b of BOUNDARYPM affect the list structure. Figure 5a shows the input to the algorithm, with the last chain in the Lth list labelled $\langle w, c \rangle (y)$ and the second lookahead chain in list $L - 1$ labelled z. Chain $\langle w, c \rangle (y)$ defines the chain of rightmost active nodes (the current solution boundary), so all nodes in and to the left of that chain are definitely in the active set. The next symbol weight that has not been used in list L is p_{c+1}. If $p_{c+1} < s$, where s is the sum of the weights of the nodes at the heads of the two lookahead chains in list $L - 1$, then a chain with head $\langle p_{c+1}, c + 1 \rangle$ and tail y is added to

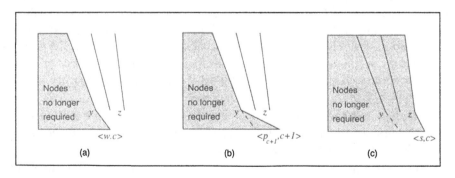

Fig. 5. Addition of a node to list L: (a) initial situation; (b) if Step 3a is used; and (c) if Step 3b is used

list L and the boundary shifts to the position shown in Figure 5b. On the other hand (Figure 5c), if $p_{c+1} \geq s$, then the chain added to list L has tail z and head $\langle s, c \rangle$. In this case no new leaves are added in list L, so the count is unchanged from the node $\langle w, c \rangle$; but all nodes in the chain are confirmed as active, so the tail of the chain is inherited from z, the second lookahead chain in list $L - 1$. As more nodes are created in list L the boundaries move further to the right; the final boundary after $2n - 2$ nodes have been created in list L defines the desired L-limited prefix code.

5 Analysis

We now examine the computational resources consumed during the execution of algorithm BOUNDARYPM, first considering the space required.

At any given point of time during the execution of the algorithm there are exactly two lookahead chains extant in each of the first $L - 1$ lists, and two chains are required in the Lth list—one current chain, and one new chain being constructed. Each extant chain contains just one node in each prior list. Moreover, chains might coalesce as they link through the lists. The total number M of nodes actually required at any given point in time is thus bounded above by $2L + 2(L - 1) + \cdots + 2 = 2 \sum_{i=1}^{L} i = L(L + 1)$.

Suppose that a pool of free nodes of size $2M$ is allocated at the start of the algorithm. Nodes are allocated at Steps 1 and 3 of Figure 4, but are nowhere deallocated. Indeed, it is not at all obvious how explicit deallocation can take place in any efficient manner. Fortunately, deallocation of nodes no longer required can be done economically by periodically performing a *garbage collection* step. Whenever the pool of available nodes is empty, the set of current chains is traversed and all nodes on them marked as "in-use". Then, the complete node pool is inspected sequentially and nodes not currently "in-use" collected onto a free list. Both of these steps require $O(M)$ time, and if the pool contains $2M$ nodes, the combined mark/collect process is guaranteed to release at least M nodes onto the free list. Thus, the amortized cost of each node allocation is $O(1)$, and the total memory required is $2M = O(L^2)$ nodes plus $O(L)$ space for the various list indexing arrays. Therefore, we have

Theorem 1. *The boundary package-merge algorithm described above requires $O(L^2)$ auxiliary space to construct a minimum-redundancy L-limited prefix code for an alphabet of n symbols.*

Let us now consider the running time of BOUNDARYPM. The two lemmas below follow directly from the description of the algorithm:

Lemma 2. *There are at most $2nL$ list items created by boundary package-merge.*

Proof: By induction, showing that each list contains fewer than $2n$ items. The first list contains n items, establishing a basis. Assume that list j has $i < 2n$ items. Then list $j + 1$ has $n + \lfloor i/2 \rfloor < 2n$ items. □

Lemma 3. *Each execution of algorithm* BOUNDARYPM *takes* $O(1)$ *time.*

Proof: If a pointer is maintained to the end of each list, nodes can be appended to the list in constant time. This array of pointers consumes $O(L)$ space, and does not impact the space bound of Lemma 1. Comparison, addition and assignment are all $O(1)$ operations by assumption, so Steps 1 and 2 take $O(1)$ time. Provided the pool of nodes is at least twice the size of the peak requirement, the cost of creating a new node is $O(1)$ amortized time, as discussed above. Steps 3a and 3b also require chain manipulations, but these involve a single pointer assignment each, so are also $O(1)$. Finally, note that the recursive calls at Step 3b are not counted, since they result in separate executions of BOUNDARYPM. □

Lemmas 2 and 3 yield

Theorem 4. *The boundary package-merge algorithm described above requires* $O(nL)$ *time to construct a minimum-redundancy L-limited prefix code for an alphabet of n symbols.*

6 Discussion

The boundary package-merge algorithm has been implemented and tested on the frequency distribution of the approximately 1,000,000 distinct words appearing in a corpus containing three gigabytes of English text. A minimum-redundancy code for this distribution averages 11.521 bits per word and has a maximum codelength of 29 bits; the boundary package merge method calculated an $L = 22$ length-limited code in approximately two minutes on a mid-range workstation using just a few kilobytes of auxiliary memory. In contrast, our implementation of Larmore and Hirschberg's algorithm requires over 60 megabytes of auxiliary memory [6] for that input data. Surprisingly, this heavily restricted code still attains 11.846 bits per symbol.

A number of other methods have been devised recently for calculating length-limited codes. Based upon a problem reduction due to Larmore and Przytycka [4], Schieber [7] has given an $O(n2^{O(\sqrt{\log L \log\log n})})$-time and $O(n)$-space algorithm for this problem. At the time of writing we know of no implementation of this method. Nevertheless, it provides another starting point from which the space-efficiency of the length-limited coding problem can be explored, and may yield algorithms of practical significance.

Moffat *et al.* [6] considered an alternative formulation in which each problem instance is described by a list $[(w_i, f_i) \mid i \in \{1 \dots r\}]$, where the pair (w_i, f_i) represents f_i repetitions of weight w_i, and $\sum_{i=1}^{r} f_i = n$. In this case $O(r \log(n/r))$ time and space suffices to generate a minimum-redundancy code, and $O(Lr \log(n/r))$ time and space suffices for the generation of a length-limited code.

Acknowledgement

This work was supported by the Australian Research Council.

References

1. T.C. Hu and K.C. Tan. Path length of binary search trees. *SIAM Journal of Applied Mathematics*, 22(2):225–234, March 1972.
2. D.A. Huffman. A method for the construction of minimum-redundancy codes. *Proc. Inst. Radio Engineers*, 40(9):1098–1101, September 1952.
3. L.L. Larmore and D.S. Hirschberg. A fast algorithm for optimal length-limited Huffman codes. *Journal of the ACM*, 37(3):464–473, July 1990.
4. L.L. Larmore and T.M. Przytycka. Constructing Huffman trees in parallel. *SIAM Journal on Computing*. To appear.
5. A. Moffat and J. Katajainen. In-place calculation of minimum-redundancy codes. In *Proc. Workshop on Algorithms and Data Structures*, Kingston University, Canada, August 1995. Springer-Verlag. To appear.
6. A. Moffat, A. Turpin, and J. Katajainen. Space-efficient construction of optimal prefix codes. In *Proc. IEEE Data Compression Conference*, pages 192–201, Snowbird, Utah, March 1995. IEEE Computer Society Press, Los Alamitos, California.
7. B. Schieber. Computing a minimum-weight k-link path in graphs with the concave Monge property. In *Proc. 6th Ann. Symp. Discrete Algorithms*, pages 405–411, San Francisco, California, January 1995. SIAM, Philadelphia, Pennsylvania.
8. J. van Leeuwen. On the construction of Huffman trees. In *Proc. 3rd International Colloquium on Automata, Languages, and Programming*, pages 382–410, Edinburgh University, Scotland, July 1976.
9. D.C. Van Voorhis. Constructing codes with bounded codeword lengths. *IEEE Transactions on Information Theory*, IT-20(2):288–290, March 1974.

Computing in linear time a chord from which a simple polygon is weakly internally visible

Binay K. Bhattacharya*
Department of Computer Science
The University of Newcastle
Callaghan, NSW 2308, Australia

Asish Mukhopadhyay
Department of CSE
IIT Kanpur, India

August 28, 1995

[Extended Abstract]

Abstract

A simple polygon is said to be weakly internally visible from a line segment lying inside it (chord, if the endpoints of the segment lie on the boundary of the polygon), if every point on the boundary of the polygon is visible from some point of this line segment. In this paper we describe a simple linear time algorithm for computing such a chord, when it exists. We also computed all non-redundant components of P which must be intersected by any chord of P from which P is *wiv*. The most interesting aspects of this algorithm are that we have been able to dispense with the complicated triangulation algorithm of Chazelle [2] as well as the algorithm in [7] for computing all shortest paths from a given vertex in a triangulated polygon. As a consequence of this, we can now use this algorithm to derive an optimal algorithm for triangulating a weakly internally visible polygon. Also these results can now be used in [3] to find a shortest line segment from which a polygon is weakly internally visible in optimal linear time.

1 Introduction

Visibility is an important sub-area of Computational Geometry. This notion has been extant in the mathematical literature long before it was introduced into Computational Geometry [11]. Research into the *computational aspects* of visibility perhaps started with the well-known art-gallery problem posed by Victor Klee [10]. The problem was to determine the minimum number of guards sufficient to cover the interior of an art gallery room.

---

*On leave from School of Computing Science, Simon Fraser University, Burnaby, BC, Canada, V5A1S6

A concrete example of a visibility problem is the following: Given a point (*viewpoint*) lying inside a simple polygon (*scene consists of this alone*), compute the part of the polygon visible (*notion of visibility : a point of the polygon is visible from the given point if these can be joined by a straight line segment which does not intersect the exterior of the polygon*) from this point [5]. If we replace the given point by a line segment lying inside the polygon we have a set of viewpoints instead. Define a point of the polygon to be visible from the line segment if it can be joined to a point on the line segment without intersecting the exterior of the polygon. If every point of the polygon is thus visible, it is said to be *weakly internally visible* (*wiv* from now on) from this line segment. This notion of weak and various other kinds of visibility were introduced into the Computational Geometry literature by Avis and Toussaint [1].

The problem we consider in this paper is to find a chord of a given polygon P from which it is *wiv*. Previous attempts on this problem, include an $O(n \log n)$ solution by Ke [9] and another one by Doh and Chwa [4]. Very recently Das and Narasimhan [3] presented an optimal algorithm to compute the shortest line segment from which a given triangulated polygon is *wiv*. A polygon can be triangulated in O(n) time [2]. However, the algorithm is not practical and its main impact is considered to be of theoretical nature only. Here we show how a *wiv* chord can be determined in linear time which does not require a triangulated simple polygon. We also show that if a given polygon is weakly internally visible (*wiv*), we can compute all non-redundant components of the polygon which must be intersected by any internally visible line segment. Once these information are known, the algorithm of [3] can be used to compute the shortest visible line segment in a non-triangulated polygon.

2   Geometric and other preliminaries

Let P be a *simple polygon* with n vertices. For any two points p and p' on the boundary of P, $bd(p, p')$ denotes the polygonal boundary of P from p to p', traversed in anti-clockwise order. $size(bd(p, p'))$ denotes the *size* of $bd(p, p')$.

Let $\overline{xy}$ be the line segment joining the points x and y. A *chord* of a polygon P is a line segment whose endpoints lie on the boundary of P. Two points of P are said to be *visible* if the line segment joining these does not intersect the exterior of P.

Let $SP(p, p')$ be the (Euclidean) *Shortest Path* between the two points p and p'. In the discussion that follows this section we will often talk about a weaker type of shortest path between two points p and p' on the boundary of P. The shortest path from p to p' is computed by just ignoring $bd(p', p)$. The notation we will use for such a path is $WSP(p, p')$.

Let $e = \overline{vv'}$ be an edge of P where v is a reflex vertex. When e is extended into the interior of P along the ray $\overrightarrow{v'v}$ to intersect its boundary, the edge gives rise to a chord of P, say $c(e)$. Of the two resulting subpolygons, the one that contains the convex vertex v' is a crucial one since any internal line segment from which P is weakly visible must necessarily intersect this subpolygon, which we call a *component* of P, say $P(e)$ (Fig. 1(a)). $c(e)$ bounds this component. A subpolygon of P is called *critical* if any line segment from which P is *wiv* must intersect the subpolygon. A critical subpolygon is shown in Fig. 1(b). The subpolygon is determined by $bd(x, y)$ and $\overline{xy}$. Clearly, each component of P is a critical subpolygon. A critical subpolygon is called *redundant* if it contains a component of P, otherwise, it is called *non-redundant*. All components generated by reflex edges (both endpoints are reflex vertices) are redundant. Clearly, redundant components can be ignored for our problem.

The following lemmas establish the importance of non-redundant components.

Lemma 2.1: A (non-convex) polygon P is *wiv* from a chord l iff every non-redundant component of P intersects l.

Lemma 2.2: $P(e)$ is non-redundant if and only if $P(e)$ is *wiv* from $c(e)$.

Lemma 2.2 implies that if $P(e)$ is not *wiv* from $c(e)$, then $P(e)$ contains at least one non-redundant component of P.

3 Computing either a *wiv* chord or a pair of disjoint non-redundant critical subpolygons

Let $e = \overline{vv'}$ be an edge of P where v is a reflex vertex and v' is a convex vertex. Let $P(e)$ be the component generated by e. Let $c(e)$ be its bounding chord.

If both P and $P - P(e)$ are visible from $c(e)$, as can be determined by running the algorithm of Avis and Toussaint [1], then we return $c(e)$ as a *wiv* chord. Otherwise assume that $P(e)$ is not visible from $c(e)$. Snce $P(e)$ is not *wiv* from $c(e)$, $P(e)$ contains a non-redundant component (Lemma 2.2). Let q_0, q_1, ..., q_m be the vertices of $P(e)$, where $c(e) = \overline{q_0 q_m}$. Let $e' = \overline{qq'}$, q a reflex vertex, be the edge which generates a component contained in $P(e)$. Let its bounding chord be $c(e') = \overline{qt}$. We call the component *Type A* if t is on $bd(q_0, q)$, otherwise, it is called *Type B*. We make use of this characterization to find a critical non-redundant subpolygon of P in two stages.

We first traverse the polygonal boundary of $P(e)$ in anti-clockwise order, dynamically maintaining the weaker type of shortest path from the vertex q_0 to a vertex q_{i-1}, up to which the search has progressed without finding a component-generating edge of the required type. This technique is similar to the one used in [6].

Let $q_{j_1}, q_{j_2}, \dots, q_{j_k}$ be the vertices on the weak shortest path from q_0 to q_{i-1}, $q_{j_1} = q_0$ and $q_{j_k} = q_{i-1}$. In trying to push the search to the next vertex q_i, the following cases arise:

Case 1: Both $q_{j_{k-1}} q_{j_k} q_i$ and $q_{i-2} q_{i-1} q_i$ constitute right turns.

Geometrically, the vertex q_i lies to the right of the ray $\overrightarrow{q_{j_{k-1}} q_{j_k}}$ (Fig. 2(a)). In this case, the shortest path up to q_i is found by simply augmenting the given shortest path with the vertex q_i, and we continue unless $i = m$.

Case 2: $q_{j_{k-1}} q_{j_k} q_i$ constitute a left turn.

Geometrically, the vertex q_i lies to the left of the ray $\overrightarrow{q_{j_{k-1}}q_{j_k}}$ (Fig. 2(b)). In this case, we find the vertex q_{j_t} of $WSP(q_0, q_{j-1})$, closest to q_i, such that $\overrightarrow{q_iq_{j_t}}$ is tangent to it. Hence, $WSP(q_0, q_j) = (q_{j_1}, q_{j_2}, ..., q_{j_t}, q_i)$. Again, we continue unless $i = m$.

Case 3: $q_{j_{k-1}}q_{j_k}q_i$ constitute a right turn and $q_{i-2}q_{i-1}q_i$ constitute a left turn.

In geometrical terms, the vertex q_i lies inside the pocket, determined by the last edge on the shortest path found so far and the subchain of the polygon $P(e)$ between the endpoints of this edge (Fig. 2(c)).

Let q_{last} be the last processed vertex (see Fig. 3). Let q_r be a reflex vertex of $bd(q_{last}, q_m)$ such that the line through q_{last} and q_r supports the part of the chain $bd(q_{last}, q_m)$ in the pocket that contains q_i. Consider the part of the line that contains q_r and is contained in P. In Figure 3 this is $\overline{y_1y_2}$. It is possible that $q_{last} = q_m$. In this case there does not exist any q_r and therefore, we take $\overline{y_1y_2}$ as $\overline{q_0q_m}$.

In either case the subpolygon, say Q, determined by $\overline{y_1y_2}$ and $bd(y_1, y_2)$ is critical. If Q is wiv from $\overline{y_1y_2}$, Q is a critical, non-redundant subpolygon of P (Lemma 2.2). If Q is not wiv from $\overline{y_1y_2}$, Q contains a component of P. But Q can not contain any Type A component, otherwise, Case 3 situation would arise before the vertex q_{last} is encountered.

We now repeat the process to determine a critical subpolygon , say R, in Q such that R does not contain any Type B component. Therefore, R is a critical, non-redundant subpolygon of Q and hence of P. Let $\overline{z_1z_2}$ be the bounding chord of R.

Next we test whether P is wiv from $\overline{z_1z_2}$. If it is so, we return this wiv chord and terminate the subroutine. If it is not, the other subpolygon of P, say $\overline{R}$, determined by the chord $\overline{z_1z_2}$, contains a component of P. We repeat the above process to determine a critical non-redundant subpolygon, say S, in $\overline{R}$. Thus we are able to determine two critical non-redundant subpolygons R and S which are disjoint.

Theorem 3.1: Two disjoint critical subpolygons of P, if exist, can be determined in $O(n)$ time.

4 Computing a chord of wiv

In this section, we assume that both the endpoints of the bounding chords of all components of P are known. Actually, only one endpoint of each bounding chord is known. We defer discussion on how to compute the other endpoint to the next section.

Let $\overline{ab}$ and $\overline{a'b'}$ be the bounding chords of two critical, non-redundant and disjoint subpolygons found in the previous section. Let $bd(a, b)$ and $bd(a', b')$ be the polygonal boundaries of the corresponding subpolygons, say P_{ab} and $P_{a'b'}$.

Any non-redundant component intersects exactly one or both the subpolygons P_{ab} and $P_{a'b'}$. Whether two components intersect or one contains the other can easily be determined in constant time by examining the endpoints of the polygonal boundaries of the components. If there exists a non-redundant component which intersects neither P_{ab} nor $P_{a'b'}$, P is not *wiv*.

We first consider the components which intersect exactly one of the polygonal chains $bd(a, b)$ and $bd(a', b')$. Consider a component whose chord is $\overline{xy}$, with the endpoint x on $bd(a, b)$ and the endpoint y not on $bd(a', b')$. In this case either the component contains $bd(a', b')$ entirely or does not contain $bd(a', b')$ at all. In the first case the component being considered is redundant and hence ignored. In the second case we determine the intersection of the component with $bd(a, b)$. Similarly we update $bd(a', b')$ for the components intersecting $bd(a', b')$ only without containing it.

Let a_r and b_r be the new positions of a and b respectively after we have processed all chords of the above type. Let $E_{ab} = bd(a_r, b_r)$. Similarly, let a'_r and b'_r be the new positions of a' and b'. Let $E_{a'b'} = bd(a'_r, b'_r)$. Any chord that has one endpoint on $E_{a,b}$ and the other endpoint on $E_{a',b'}$ will intersect all the components used to compute $E_{a,b}$ and $E_{a',b'}$. We next consider the components which intersect both the chains $bd(a, b)$ and $bd(a', b')$ without containing them. These components are called *cross-components* and the endpoints of their bounding chords lie on $bd(a, b)$ and $bd(a', b')$. Some of the cross-components may not intersect the chains $bd(a_r, b_r)$ or $bd(a'_r, b'_r)$. Any cross-component which contains either $bd(a_r, b_r)$ or $bd(a'_r, b'_r)$ is redundant and is ignored. If any cross-component does not intersect both $bd(a_r, b_r)$ and $bd(a'_r, b'_r)$, P can not be *wiv*. A cross-component which does not contain $bd(a_r, b_r)$ is used to update $bd(a'_r, b'_r)$. Similarly, a cross-component which does not contain $bd(a'_r, b'_r)$ is used to update $bd(a_r, b_r)$. We repeat this process till none of the points a_r, b_r, a'_r and b'_r can be updated any further. This entire process can be executed in linear time by ordering the updates carefully.

Unless P has been declared not *wiv*, the above process gives us the final polygonal chains $E_{a,b} = bd(a_f, b_f)$ and $E_{a',b'} = bd(a'_f, b'_f)$ (Fig. 4).

It is now easy to find a chord from which P is *wiv*, if one exists. We first find $WSP(b'_f, a_f)$. We also find $WSP(b_f, a'_f)$. If $WSP(b'_f, a_f)$ and $WSP(b_f, a'_f)$ are not linearly separable, clearly P is not *wiv*. Suppose that they are linearly separable. These convex paths along with the chains $E_{a,b}$ and $E_{a',b'}$, form a butterfly-shaped polygon. We join the cusps at b'_f and b_f. If this line segment does not intersect the paths, then we have our required chord. Otherwise, we compute a common tangent to the paths and find the intersections of the supporting line of this common tangent $E_{a,b}$ and $E_{a',b'}$. The polygon P is *wiv* from the chord bounded by these intersection points.

We thus claim that

Theorem 4.1: Given two disjoint non-redundant subpolygons of P, it is possible to compute an *wiv* chord in linear time.

5 Computing the endpoints of bounding chords

We first solve the following three problems. The endpoints of the chords of all the non-redundant components of P can be determined by the repeated applications of these problems.

Let Q be a subpolygon of P bounded by $\overline{xy}$. Let Q be wiv from $\overline{xy}$. Let $R = P - Q$. The three problems we first solve are:

Problem 1: Classify the components generated by the edges of Q into one of the following classes: (a) components that contain x only; (b) components that contain y only; and (c) components that contain both x and y.

Problem 2: Determine the bounding chords of the components generated by the edges of R and containing either x or y, but not both (i.e. the endpoints lie on $bd(x, y)$).

Problem 3: Determine the bounding chords of the components of Q containing both x and y.

Before solving these problems we first describe some useful properties which are the consequences of the fact that Q is wiv from $\overline{xy}$ [6]. Here Q is determined by $bd(x, y)$ of P and the chord $\overline{xy}$.

Fact 1. The shortest path from x to z, for any z on $bd(x, y)$, is always right-turning and the shortest path from y to z is always left-turning. This means that $WSP(x, z) = SP(x, z)$ and $WSP(z, y) = SP(z, y)$. When we draw these shortest paths we get what is called a funnel in the visibility literature.

Fact 2. The shortest path between any two arbitrary points of $bd(x, y)$ is convex.

Fact 3. Any subpolygon of P, contained in Q, is also wiv from its bounding chord.

Consider an edge $e = \overline{p_k p_{k+1}}$ of Q. Suppose we know the shortest paths from x to p_k and p_{k+1} and from y to p_{k+1} and p_k. Let the shortest paths from x to p_k and p_{k+1} be common up to some vertex, say v (Fig. 5). Let $SP(x, p_k) = (x, x_{i_1}, x_{i_2}, ..., v, x_{j_m}, x_{j_{m-1}}, ..., x_{j_1}, p_k)$ and $SP(x, p_{k+1}) = (x, x_{i_1}, x_{i_2}, ..., v, p_{k+1})$. If $m > 0$, we split the edge $\overline{p_k p_{k+1}}$ into subedges $\overline{p_{k,l} p_{k,l+1}}$, $l = 0, 1, ..., m$ such that the shortest path from x to any point p on the segment $\overline{p_{k,l} p_{k,l+1}}$ is $SP(x, p) = (x, x_{i_1}, x_{i_2}, ..., v, x_{j_m}, x_{j_{m-1}}, ..., x_{j_{l+1}}, p)$ (Fig. 5). Here $p_{k,0} = p_k$ and $p_{k,m+1} = p_{k+1}$. In fact for our purpose we need to know only the segments of these shortest paths incident on $p_{k,l}$, $l = 0, 1, 2, ..., m + 1$. We preprocess each edge of Q like this. Similarly we also preprocess each edge of Q using the shortest paths from y. Thus the edges of Q can be preprocessed in $O(size(bd(x, y)))$ time. The storage requirement is also $O(size(bd(x, y)))$.

Once all the edges of Q are preprocessed, it is possible to answer the following query in $O(1)$ time. Given a point z on a subedge $\overline{p_{k,l} p_{k,l+1}}$, and a ray, starting from z and towards the interior of Q, determine whether the ray intersects the chain $bd(x, z)$ or the chain $bd(z, y)$. We can do this for the chain $bd(x, z)$ by checking the segments $\overline{x_{j_l} p_{k,l}}$ and $\overline{x_{j_{l+1}} p_{k,l+1}}$. Clearly, if the ray does not intersect neither $bd(x, z)$ nor $bd(z, y)$, then it must intersect the chord $\overline{xy}$. Similar approach has also been used by Heffernan [8].

Therefore,

Lemma 5.1: Problem 1 can be solved in linear time.

Notice that the above arguments are still valid when the chord $\overline{xy}$ is replaced by the convex chain $WSP(x, y)$ and $bd(x, y)$ is wiv from $WSP(x, y)$.

5.1 The Solution of Problem 2

Here we assume that the edges of R whose corresponding components of P contain only either x or y but not both are known. We describe the algorithm for determining the endpoints of the bounding chords of the components which contain only y. The algorithm is similar for the other case. We assume here that the edges of Q are preprocessed as described in the previous section. Let $\overline{p_{i_1}p_{i_1+1}}$, $\overline{p_{i_2}p_{i_2+1}}$, ..., $\overline{p_{i_m}, p_{i_m+1}}$ be the anti-clockwise sequence of edges of R lying on $bd(y,x)$ whose corresponding components contain y.

For each component generating edge $\overline{p_{i_j}p_{i_j+1}}$ we need to determine the other endpoint of the bounding chord, say q_{i_j+1}. Suppose the components generated by the edges $\overline{p_{i_1}p_{i_1+1}}$, ..., $\overline{p_{i_{j-1}}p_{i_{j-1}+1}}$ are non-redundant. Clearly, we have so far visited the edges $bd(x, q_{i_{j-1}+1})$.

We now determine the bounding chord of the component corresponding to the edge $\overline{p_{i_j}p_{i_j+1}}$. Clearly, if the extension of the edge $\overline{p_{i_j}p_{i_j+1}}$ towards the interior of P does not intersect the chord $\overline{p_{i_{j-1}+1}q_{i_{j-1}+1}}$, the corresponding component is redundant as it contains the component generated by $\overline{p_{i_{j-1}}p_{i_{j-1}+1}}$ (Fig. 6). Suppose the extension does intersect $\overline{p_{i_{j-1}+1}q_{i_{j-1}+1}}$. We then start searching the edges of $bd(q_{i_{j-1}+1}, y)$ in anti-clockwise order, starting from $q_{i_{j-1}+1}$, till we find an edge that intersects the extension. Let s_{i_j+1} be the intersection point. Consider the ray r starting from s_{i_j+1} towards p_{i_j+1}. We can now determine in $O(1)$ time whether the ray r first intersects $bd(x, s_{i_j+1})$, $bd(s_{i_j+1}, y)$ or the segment $\overline{xy}$. Clearly, the ray can not intersect both $bd(x, s_{i_j+1})$ and $bd(s_{i_j+1}, y)$, otherwise, Q can not be wiv from $\overline{xy}$. It can be shown that

Lemma 5.1.1: If $SP(x, s_{i_j+1})$ intersects the ray r, the component generated by the edge $\overline{p_{i_j}p_{i_j+1}}$ is redundant and the computed endpoints of the bounding chords of the remaining non-redundant components containing y and generated by the edges of $bd(p_{i_j+1}, x)$, lie on $bd(s_{i_j+1}, y)$.

If r intersects $bd(s_{i_j+1}, y)$ first, $\overline{p_{i_j+1}s_{i_j+1}}$ is not the correct bounding chord of the component i.e. $q_{i_j+1} \neq s_{i_j+1}$. In this case we keep on examining the edges of $bd(s_{i_j+1}, y)$ starting from s_{i_j+1} for the correct endpoint of $c(\overline{p_{i_j}p_{i_j+1}})$. If the ray r intersects the chord $\overline{xy}$ first, $\overline{s_{i_j+1}p_{i_j+1}}$ is the bounding chord of the component generated by the edge $\overline{p_{i_j}p_{i_j+1}}$ i.e. $q_{i_j+1} = s_{i_j+1}$. Thus we never backtrack the edges already visited. Therefore

Lemma 5.1.2 Problem 2 can be solved in linear time.

5.2 The Solution of Problem 3

Applying techniques similar to the ones used in solving Problems 1 and 2 we can show that (details omitted)

Lemma 5.2.1: Problem 3 can be solved in linear time.

Here we show that the endpoints of the bounding chords of the non-redundant components can be computed easily when a wiv chord is known. Let $\overline{xy}$ be a chord of P from which P is wiv. Let P_1 and P_2 be the subpolygons of P determined by $\overline{xy}$. We are interested in computing the bounding chords of the components generated by the edges of P_1 and P_2. We do not compute the bounding chord of a component if it is found to be redundant. The formal description of the algorithm is given below:

Step 1: Classify the generating edges of P_1 and P_2 whose corresponding components contain only x; only y; and both x and y. (Let these sets of edges be $P_i(x)$, $P_i(y)$ and $P_i(xUy)$ respectively, $i = 1, 2$.)

Step 2: Determine the bounding chords of the components corresponding to the generating edges of $P_i(x)$, $i = 1, 2$.

Step 3: Determine the bounding chords of the components corresponding to the generating edges of $P_i(y)$, $i = 1, 2$.

Step 4: Determine the bounding chords of the components corresponding to the generating edges of $P_i(xUy)$, $i = 1, 2$.

These steps can easily be implemented in linear time by applying the algorithms designed for Problem 1, Problem 2 and Problem 3 with minor modifications.

Similarly, we can compute the bounding chords of the non-redundant components of P when the bounding chords of two disjoint critical non-redundant subpolygons are known. Therefore

Theorem 5.2.1: If a *wiv* chord or two disjoint critical non-redundant subpolygons of P are known, all the bounding chords of the non-redundant components of P can be determined in $O(n)$ time.

6   Conclusions

In this paper we presented an optimal algorithm to recognize a weakly internally visible polygon. The algorithm also outputs a chord from which it is weakly visible. The paper presented a simple $O(n)$ algorithm to compute a critical non-redundant subpolygon. We also described an optimal algorithm to compute all non-redundant components of a *wiv* polygon. The important features of this algorithm are that we have been able to dispense with the extremely complicated triangulation algorithm [2] as well as the shortest path algorithm [7]. The result of this paper can now be used in [3] to find a shortest line segment from which a non-triangulated polygon is weakly internally visible in optimal linear time.

References

[1] D. Avis and G. Toussaint. An optimal algorithm for determining the visibility of a polygon from an edge. *IEEE Trans. on Computers*, C-30(12):910–912, 1981.

[2] B. Chazelle. Triangulating a simple polygon in linear time. *Discrete and Computational Geometry*, 6:485–524, 1991.

[3] G. Das and G. Narasimhan OPtimal linear-time algorithm for the shortest illuninating line segment in a polygon. In*Proceedings of the Tenth Annual Symposium on Computational Geometry*, pages 259-266, 1994.

[4] J-I Doh and K-Y Chwa. An algorithm for determining visibility of a simple polygon from an internal line segment. *J. of Algorithms*, 14:139–168, 1993.

[5] H. El Gindy and D. Avis. A linear algorithm for computing the visbility polygon from a point. *J. of Algorithms*, 2:186–197, 1981.

[6] S. K. Ghosh, A. Maheshwari, S. P. Pal, S. Saluja, and C. E. Veni Madhavan. Computing the shortest path tree in a weak visibility polygon. In *Proceedings of the Twelvth Annual FST & TCS Conference*, pages 369–389, 1991.

[7] L. Guibas, J. Hershberger, D. Leven, M. Sharir, and R. E. Tarjan. Linear time algorithms for visibility and shortest path problems inside a triangulated simple polygon. *Algorithmica*, 2:209–233, 1987.

[8] P. Heffernan. An optimal algorithm for the two-guard problem. In *Proceedings of the Ninth Annual symposium on computational geometry*, pages 348–358, 1993.

[9] Y. Ke. *Polygon visibility algorithms for weak visibility and link distance problems*. PhD thesis, The John Hopkins University, Baltimore, Maryland, 1989.

[10] J. O'Rourke. *Art gallery therems and algorithms*. Oxford University Press, 1987.

[11] F. A. Valentine. Minimal sets of visibility. *Proceedings of the American Mathematical Society*, 4:917–921, 1953.

31

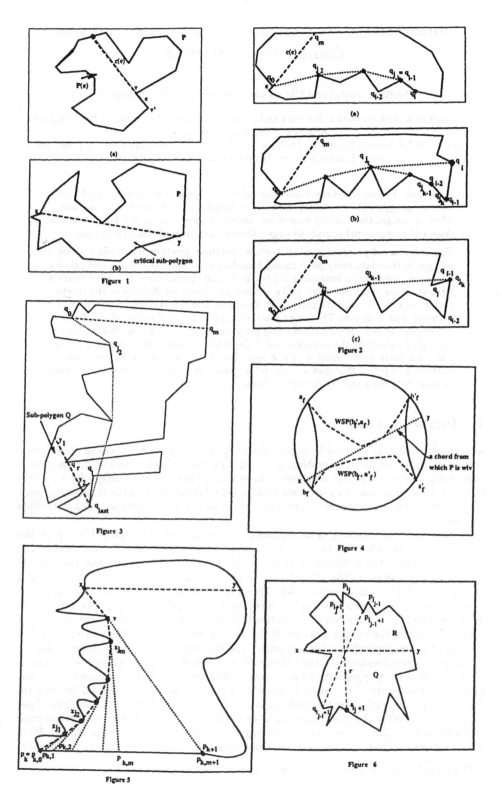

(a)

(b)
critical sub-polygon

Figure 1

(a)

(b)

(c)
Figure 2

Figure 3

Figure 4

Figure 5

Figure 6

Competitive Searching in Polygons—Beyond Generalised Streets*

Amitava Datta[1] and Christoph A. Hipke[2] and Sven Schuierer[2]

[1] Department of Mathematics, Statistics and Computing Science, University of New England, Armidale, NSW 2351, Australia.
[2] Institut für Informatik, Universität Freiburg, Am Flughafen 17, D-79110 Freiburg i. Br., FRG.
email: {hipke,schuiere}@informatik.uni-freiburg.de

Abstract. We consider a robot inside an unknown polygon P which has to find a path from a starting point s to a target point t. It is equipped with on-board cameras through which it can get the visibility map of its immediate surroundings. We define two new classes of polygons and provide strategies for searching these classes of polygons.

The first class of polygons is called *horizontal-vertical streets* or HV-streets and for a polygon in this class, every point on the boundary is visible from either a vertical or a horizontal line segment connecting the two polygonal chains from s to t. We provide a strategy under which the robot walks at most 14.5 times the distance of the shortest path from s to t to reach the point t. We also prove that this is an optimal strategy for searching such polygons. The second class of polygons is called *θ-generalized-streets* or *θ-G-streets* and every point on such a polygon is visible from at least one line segment at an angle $θ$ connecting the two polygonal chains between s and t. Here, $θ$ is an arbitrary but fixed angle with respect to the x axis. We provide a strategy for searching an isothetic polygon of this class and the robot travels at most 19.97 times the shortest distance between s and t under our strategy.

1 Introduction

A natural problem in robotics is to plan the path of a robot from a starting point s to a target point t. In the past this problem has mainly been considered when the environment is completely known in advance. However, a more natural and real life problem for a robot is to search for a target inside an unknown environment. We assume that the robot is equipped with some sensors, usually cameras, so that it can obtain the visibility map of its immediate surroundings.

The problem of searching in an unknown environment can be regarded as an *on-line* problem since the robot has to make its decisions based only on what it has seen so far. One way to judge the performance of the robot is to compare the distance travelled by the robot to the length of the shortest path from s to t. In other words, the robot's path is compared with that of an adversary's who knows the environment completely. This approach to analysing on-line algorithms was pioneered by Sleator and Tarjan [10]. The ratio of the distance travelled by the robot to the optimal distance from s to t is called the *competitive ratio* achieved by the robot's strategy. If this ratio is bounded by c, the strategy is called *c-competitive*. In particular, if c is a constant, the strategy is simply called *competitive*.

Lumelsky and Stepanov [7] study the problem of searching in an unknown environment when the robot is equipped with a tactile sensor. However, they do not analyse their strategies in the framework of competitive analysis. Papadimitriou and Yannakakis [9] were the first to study the path planning problem in the framework of competitive analysis. They consider the problem of a robot searching in an environment that is filled with n rectangular obstacles and are able to show that no strategy can obtain a better ratio than $\Omega(\sqrt{n})$. Blum et al. [3] present several deterministic and randomized algorithms when the environment has

---

* This research is supported by the DFG-Project "Diskrete Probleme", No. Ot 64/8-1.

rectangular as well as convex obstacles and their algorithms achieve a competitive ratio of $O(\sqrt{n})$. Blum and Chalasani consider the problem of improving the robot's performance if multiple trips are allowed from s to t [1]. Mei and Igarashi [8] improve upon some of the results obtained in Blum *et al.* [3] if the aspect ratio of the rectangles is bounded.

Even if we restrict ourselves to searching in a simple polygon, no strategy can achieve a competitive ratio that is better than $\Omega(n)$ [3].

Hence, more restricted polygon classes have been considered that allow constant competitive search strategies. Klein [5] was the first to consider the problem of searching for a target inside a class of polygons called **streets**. He assumes that both s and t are located on the polygon boundary and that the two chains from s to t in the clockwise and anticlockwise direction (called L- and R-chain, resp.) are mutually weakly visible. His strategy achieves a competitive ratio of < 5.73 and he proves a lower bound of $\sqrt{2}$ for the competitive ratio of searching in streets. Kleinberg [6] recently presented a strategy for searching streets which achieves a competitive ratio of < 2.83. His strategy achieves an optimal competitive ratio of $\sqrt{2}$ for searching isothetic streets. He also presented a strategy for searching in an unknown and arbitrary simple isothetic polygon. His strategy for this problem achieves a competitive ratio of $O(m)$, where m is the number of *essential cuts* in the polygon which can be as high as $O(n)$, when n is the number of vertices of the polygon.

Since for all practical purposes the distance travelled by the robot should not be too high compared to the optimal distance between s and t, it is important to explore classes of polygons which permit constant competitive search strategies. Recently, Datta and Icking [4] investigated a class of polygons called *generalized streets* or $\mathcal{G}$-streets. For a polygon in this class, every point on the polygon boundary is visible from at least one horizontal chord connecting the L- and R-chain. They call such a chord an *LR-chord*. This class of polygons is strictly larger than the class of streets. They provide a strategy for searching isothetic generalized streets and achieve an optimal competitive ratio of 9 in the L_1-metric and a ratio of $\sqrt{82}$ (~ 9.06) in the L_2-metric.

Since the polygon classes that have been investigated so far are severely restricted, the importance of finding larger classes that admit search strategies with a constant competitive ratio has been stressed in all papers on competitve searching in simple polygons [4, 5, 6].

In this paper, we present two new classes of polygons which permit search strategies with constant competitive ratio and which are strictly larger than isothetic generalized streets. We call a polygon of the first class a *horizontal-vertical street* or **HV-street**. Every point on the boundary of an HV-street is visible from either a horizontal or a vertical LR-chord. We prove a lower bound of 14.5 on the competitive ratio for searching in an HV-street in the L_1-metric and provide a strategy which achieves this competitive ratio.

A polygon of the second class is called a θ-$\mathcal{G}$-**street**. Every point on the boundary of such a polygon is visible from at least one LR-chord which has an inclination of θ with the x-axis. Here, θ is an arbitrary but fixed angle. We present a competitive strategy for searching a polygon in this class. Our strategy achieves a competitive ratio of 19.97 in the L_1-metric.

The rest of this paper is organized as follows. In Section 2 we present some definitions. A strategy for searching an HV-street is presented in Section 3. We prove that the strategy is correct and achieves a competitive ratio of 14.5 in the L_1-metric. Finally, in Section 4, we present a strategy for searching a θ-$\mathcal{G}$-street.

2 Preliminaries

Let P be a simple polygon in the plane with two distinguished points s and t on the boundary of P. We consider the problem that a robot has to find a path starting at s and ending at t. The clockwise (resp. counterclockwise) polygonal chain from s to t is called the *left chain* and denoted by L (resp. the *right chain* and denoted by R).

Definition 1. A line segment which intersects the interior but not the exterior of P and both of whose end points are on the boundary of P is called a **chord**.

If a chord c has one of its end points on L and the other end point on R, c is called an LR-chord.

In this paper, we introduce a new class of polygons called **horizontal-vertical streets** or **HV-streets** which can be easily shown to be strictly larger than the class of streets [5] or generalised streets [4].

Definition 2. A simple polygon is called a HV-street, if for every point p on $L \cup R$ (i.e., for every point on P), there exists either a horizontal or a vertical LR-chord c such that p is visible from at least one point on c.

2.1 Landmarks

In the following we consider a simple *isothetic* polygon P in the plane. We use a Cartesian coordinate system and the sides of the polygon P are parallel to either the x- or the y-axis. The notion of horizontal (resp. vertical) indicates a direction parallel to the x- (resp. y-) axis.

We assume for simplicity of exposition that no chord passes through more than one edge of the polygon. However, our strategy and its analysis can be modified easily in the general case. Notice that a chord divides the polygon into two or three regions depending on whether an edge e with two reflex vertices is incident to it or not.

The strategy we present explores the local neighborhood of LR-chords of P and searches for reflex vertices which give some indication about the location of t. We call such vertices *landmarks*.

Definition 3. Let c be an axis-parallel chord in P, l a reflex vertex of P, and e and e' the two edges incident to l such that e is parallel and e' orthogonal to c. Furthermore, let c_l be the maximal chord that contains e and let e' and c be contained in the same region of c_l (see Figure 1).

The vertex l is called a c-landmark if there is a line segment c' in P that is orthogonal to c and intersects both c and l and e'.

The vertex l is called a **generalized** c-landmark if there is a line segment c' in P that is orthogonal to c and intersects both c and c_l.

The edge e is called the **hidden edge** of l w.r.t. c and the chord c_l is called the **chord induced by** l.

Landmarks are important in our search since they may hide the target.

Lemma 4. *Let c be a maximal axis-parallel chord and P_c a region of c. If there is no landmark in P_c, then P_c is weakly visible from c.*

Definition 5. Let c be an axis-parallel chord of P and l a c-landmark. A chord c' which is orthogonal to c is called a **conjugate chord** of l if c' intersects c and the induced chord of l. A conjugate chord c' of l is said to be an **extreme conjugate** chord if there are no other conjugate chords c_1 and c_2 of l such that c' is between c_1 and c_2.

3 The Strategy for Searching an Unknown HV-street

In this section, we present and analyse our strategy for searching an unknown HV-street. First we explore some geometric properties of HV-streets.

The following property was proved in [4] for $\mathcal{G}$-streets and holds for any HV-street P as well.

Lemma 6. *A chord c is an LR-chord if and only if the points s and t are in two different regions of c. Moreover, every path from s to t in P intersects all LR-chords of P.*

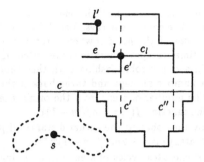

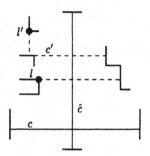

Fig. 1. A c-landmark l with hidden edge e, induced chord c_l, and extreme conjugate chords c' and c''. The reflex vertex l' is a generalized c-landmark.

Fig. 2. If $\hat{c}$ is a conjugate chord of c-landmark l and l' a $\hat{c}$-landmark, then the extreme conjugate chord c' of l' is an LR-chord.

In our strategy we move from one LR-chord to the next. If the robot is currently on a chord c, the region which contains the starting point s is called the *explored region of c* and denoted by $expl(c)$. The other region(s) are called the *unexplored regions of c* and their union is denoted by $unexpl(c)$. We only search in the unexplored regions of LR-chords.

For searching unknown HV-streets, extreme conjugate chords play an important role since extreme conjugate chords may be LR-chords. This can be decided by comparing the explored regions of two chords.

Definition 7. If c and c' are two chords in P, then c is said to **dominate** c' if the explored region of c is contained in the explored region of c'.

In Figure 1 the chord c' dominates the chord c''. The following lemma shows that domination is one criterion to recognize extreme conjugate chords as LR-chords.

Lemma 8. *Let c be an axis-parallel chord and l a c-landmark in the unexplored region of c with extreme conjugate chords c' and c''. If c' dominates c'', then c' is an LR-chord (see Figure 1).*

Sometimes it is necessary to explore parts of the polygon that are orthogonal to the search chord c of the robot.

Lemma 9. *Let c be an axis-parallel chord, l a c-landmark in the unexplored region of c, and $\hat{c}$ a conjugate chord of l. If l' is a $\hat{c}$-landmark such that both extreme conjugate chords of l' are in the unexplored region of c, then the extreme conjugate chord c' of l' that is closer to c is an LR-chord (see Figure 2).*

In our strategy, the robot repeatedly goes from a point p to the L_1 nearest point of another chord c induced by a landmark l which is visible from p. This method has been already used by Datta and Icking [4].

Lemma 10. *If c is an axis-parallel chord and l a c-landmark, then the robot can reach the L_1 nearest point on the chord induced by l by an xy-monotone path.*

3.1 The Strategy

Before describing our strategy in detail, we briefly discuss a method due to Baeza-Yates *et al.* [2] for searching a point in m concurrent rays. We call their method *multiway ray search*. Their model is the following. The robot is placed at the meeting point or origin of m concurrent rays and it has to find a point p which is situated in one of the rays. It is assumed that the robot has a fixed step size and the point p is situated in one of the rays at an integer multiple of steps away from the origin of the concurrent rays. The distance of the point p is unknown to the robot and it can only detect the point p when it reaches p. In [2], a deterministic strategy which achieves the optimal competitive ratio is presented. In this strategy, the robot visits the rays one by one in a round robin fashion until the point p is reached. In every ray, the robot goes a certain number of steps and turns back if the point p is not reached and explores the next ray. The number of steps from the origin the robot walks before the i-th turn is determined by the function $f(i) = (m/(m-1))^i$.

The worst case ratio of the distance traversed by the robot to the actual distance of p from the origin (i.e., the competitive ratio of this strategy) is $1 + 2m^m/(m-1)^{m-1}$.

For two concurrent rays (i.e., $m = 2$), the robot executes cycles of steps in increasing powers of 2 and the competitive ratio is 9. Similarly, for three concurrent rays, the robot executes cycles of steps in increasing powers of $\frac{3}{2}$ and the competitive ratio is 14.5. For exploring four rays the competive ratio is 19.97.

In our strategy, the robot searches sometimes in one, sometimes in two, and sometimes in three concurrent rays. When it searches in three rays, two of the rays are collinear and the initial direction of the third ray is perpendicular to these two rays.

When the robot searches on two concurrent rays, the two rays are either collinear or orthogonal to each other with the origin of the search in the middle. In the case of searching in three rays, we call the two collinear rays the *search chord of the robot*. The other ray is called the *orthogonal ray*.

The main purpose of the robot's strategy is to cross all the LR-chords of the polygon. So, in a generic step of our strategy, the robot moves from one LR-chord c to a new LR-chord c' in the unexplored region of c. We present our strategy below. We assume for ease of exposition that the starting point s is a convex vertex of the polygon. The other cases can be handled easily and we omit the details in this version.

If the robot is currently on chord c and it reached c first at the point p, then we call p the *arrival point* of the robot at c.

Procedure HV-search;
Input: the polygon P, the starting point s, and target point t;

Case 1 *The robot is at s.*

As mentioned earlier, we assume s to be a convex vertex. Suppose, the two edges incident to s are e' and e''. The robot performs a 2-way ray search on the maximal chords c' and c'' that are collinear with e' and e''. We set $c^{(old)} := \emptyset$.

The search stops if one extreme conjugate chord $\hat{c}$ of a c'- or c''-landmark has been found or t becomes visible. In the former we set $c := \hat{c}$ and $c^{(old)} := c'$ or $:= c''$, in the latter case we go to t by an xy-monotone path.

While t *is not reached*, execute the following cases repeatedly.

Case 2 *The arrival point p is not between two extreme conjugate chords of a c-landmark l.*

The robot starts either a 1-way or a 2-way ray search on c starting at p. It starts a 2-way ray search if the parts of c to both sides of the robot position belong to the unexplored region of $c^{(old)}$. It starts a 1-way ray search if only a part of c to one side of p belongs to the unexplored region of $c^{(old)}$.

The search stops if an extreme conjugate chord c' of a c-landmark is intersected or t becomes visible. In the former case the robot moves on c to the intersection point of c and c' and stops. We set $c^{(old)} := c$ and $c := c'$ and the case analysis starts again. In the latter case the robot moves toward t via an xy-monotone path.

Case 3 *The robot is between two extreme conjugate chords of a c-landmark l.*

Let c' and c'' be the extreme conjugate chords of l. The robot starts a either a 2-way or a 3-way ray search. As in Case 2 there are one or two rays collinear with c.

Moreover, another search ray orthogonal to c is started which may become an xy-monotone path in order to reach the induced chord c_l of l. By Lemma 10 the robot is able to construct such a path on-line. If there are other generalized c-landmarks visible from c_l, then the xy-monotone path is continued until either a c'- or c''-landmark is found or no generalized c-landmark above the current chord is visible anymore.

The search changes if one of the following conditions is met:

(i) An extreme conjugate chord d of a c-landmark (different from l) is intersected by one of the search paths on c. The search stops and we set set $c^{(old)} := c$ and $c := d$.

(ii) An extreme conjugate chord d of a c'- or c''-landmark is intersected by the orthogonal ray. The search stops and we set set $c := d$. $c^{(old)}$ remains unchanged.

(iii) The xy-monotone path reaches a point on c' or c'', say c'. The orthogonal search path now continues on c'. If the end point of c' is reached and there is no c'-landmark in the unexplored region of c', then the orthogonal search ray is stopped.

(iv) t becomes visible and the robot moves toward t via an xy-monotone path.

If an end point of c is reached in one of the Cases 2 or 3, then the search ray which has reached the end point is stopped.

end Procedure HV-search;

3.2 Analysis of the Strategy

Definition 11. The **path of the robot** is the path travelled by the robot taking only the search rays into account that led to new search chords.

In order to prove the correctness and competitiveness of Procedure HV-search we need to show that the path of the robot reaches t and that it is not too long. In order to show that it reaches t we make use of Lemma 6 and prove that all LR-chords are crossed.

Let the sequence of chords which constitute the search chords be numbered and denoted by (c_i). Similarly, the sequence $(c_i^{(old)})$ denotes the squence of old search chords that are orthogonal to c_i. The strategy is designed to move from one LR-chord to the next.

Lemma 12. *If l is a c_i-landmark in the unexplored region of $c_i^{(old)}$ such that both extreme conjugate chords are to one side of the arrival point, then one extreme conjugate chord dominates the other.*

Lemma 13. *If c_i is an LR-chord, then c_{i+1} is also an LR-chord.*

Furthermore, the strategy never crosses an LR-chord of a landmark twice and is, therefore, finite.

Lemma 14. *If c_i and c_j are two parallel chords with $i < j$, then c_i dominates c_j.*

The above lemma ensures that the robot progresses towards t and, finally, reaches it. In order to prove the competitiveness of our strategy we need some relation between the length of the path of the robot and the length of an L_1-shortest path. In the following we denote the arrival point at c_i by p_i. The following invariant holds during the execution of our strategy.

Lemma 15. *Either $c_i^{(old)}$ intersects c_i in the arrival point or the arrival point p_i of the robot at c_i is the closest point to s.*

Note that the latter can only occur if the robot follows an xy-monotone path in Case 3 of the strategy. In this case there is always a L_1-shortest path that includes the previous arrival point.

Lemma 16. *If q_i is the closest point to s and c_i is not intersected by $c_i^{(old)}$, then there is a L_1-shortest path from s to q_i that includes the arrival point p_{i-1} of c_{i-1}.*

We are now able to show that the path of the robot is, indeed, an L_1-shortest path.

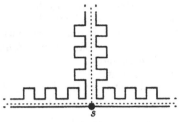

Fig. 3. A lower bound for searching an HV-street.

The point t can be placed in any one of the caves and the polygon still remains an HV-street. So, to search for t, the robot has to search for the projection of t on the three concurrent rays shown in the figure. Hence, searching for a target in an HV-street is as hard as searching for a point in 3 concurrent rays.

Lemma 17. *The path of the robot is an L_1-shortest path.*

Theorem 18. *The competitive ratio of Procedure HV-search is 14.5 in the L_1 metric and this is optimal.*

4   Isothetic θ-$\mathcal{G}$-streets

We now show that for a different class of isothetic polygons called θ-$\mathcal{G}$-streets competitive searching is also possible.

Definition 19. A chord inside the polygon P is called a θ-chord, if it is situated at an angle of θ with respect to the x-axis.

Definition 20. A simple polygon in the plane is called a θ-$\mathcal{G}$-**street** if for every boundary point $p \in L \cup R$, there exists an LR-θ-chord c such that p is visible from a point on c.

If we consider arbitrarily oriented polygons, then the class of $\mathcal{G}$-streets and the class of θ-$\mathcal{G}$-streets are isomorphic as every θ-$\mathcal{G}$-street can be rotated by $-\theta$ and we obtain a $\mathcal{G}$-street. However, if we consider isothetic polygons [4], then the situation changes considerably if non-axis parallel angles are allowed. Obviously, the class of isothetic θ-$\mathcal{G}$-streets, where $\theta \in [0, \pi)$ includes the class of isothetic $\mathcal{G}$-streets and is strictly larger.

Definition 21. Consider a reflex vertex l of the polygon P. Let e and e' be the edges incident to l such that e is parallel to the search direction. The vertex l is called a **forward landmark** (resp. **sideway landmark**), if it is visible from the search position p and p and e (resp. e') are in the same region of the maximal chord that contains e' (resp. e).
The chord **induced** by a forward (resp. sideway) landmark is the maximal chord that contains e' (resp. e). In Figure 4(a) l is a forward landmark and k is a sideway landmark.

4.1   The Strategy

Like our strategy for searching HV-streets the strategy for θ-$\mathcal{G}$-streets is also based on the multiway ray searching algorithm due to Baeza-Yates *et al*. In this case we have $m = 4$, which means we use a variation of this algorithm to competitively search a maximum of four rays or *search paths* simultaneously as described in Section 3.1. As opposed to the algorithm for HV-streets the ray search now continues through the whole process of the search, but the search paths are changed inbetween. That means search paths may be stopped and created. Points where a new search path is created are called *split points*.

All search paths are rectilinear. Every search path maintains a modified local version θ_0 of the angle θ. If the search direction is north or south, then $\theta_0 = \theta$, and if the search direction is west or east, then $\theta_0 = \theta + \pi/2$. There are three types of search paths, *main* search paths, and R- and L-search paths. R- and L-search paths are created by splitting a main search path.

We assume w.l.o.g. that at the beginning the robot is located on an edge with two convex vertices. The strategy starts by sending a main search path orthogonally away from the starting edge; there is no other search path at first. We now describe the strategy that pursues one search path.

Searching one direction

In the description of the strategy we assume for simplicity that the search direction is north.

The strategy continues a main search path until it reaches a chord c that is induced by a forward landmark l with respect to the search position p. It then identifies one of three cases and either proceed, split or turn. The same holds for L (resp. R)-search paths except that these paths only examine forward landmarks that are situated to the left (resp. to the right) of the search position.

If a search path turns or splits on the chord c_l induced by landmark l, then the new search path is orthogonal to original search direction does not examine any landmarks in the explored region of c_l before it has reached l.

We now describe the three different cases that can occur if the search path reaches a chord induced by a landmark l. To this end let C be the chain starting at l and ending at the projection of l in the search direction. For the case analysis we call a landmark *visible* if it has been visible at some point during the search process, not necessarily on the current search path.

Case 1 *There is no visible sideway landmark on chain C.*

(i) *On chain C there is a visible forward landmark l', whose induced chord does not intersect the projection of p in search direction. Furthermore, $0 < \theta_0 < \frac{\pi}{2}$ and l is to the left of p or $\frac{\pi}{2} < \theta_0 < \pi$ and l is to the right of p.*

In this case the search path turns into the direction of l. If it has been an L- or R-search path, then it continues as a main search path and the corresponding R- or L-search path is stopped.

(ii) *Case (i) does not hold.*

In this case the search is simply continued with the same search direction.

Case 2 *There is a visible sideway landmark on chain C and either $0 < \theta_0 < \frac{\pi}{2}$ and l is to the left of p or $\frac{\pi}{2} < \theta_0 < \pi$ and l is to the right of p.*

In this case the search path starts two main search paths by splitting to the side of the forward landmark. All other search paths are stopped.

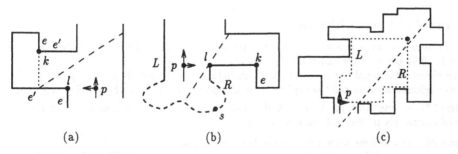

| | | |
| (a) | (b) | (c) |

Fig. 4. Cases 2, 3(ii) and 3(iii), the dashed line in (b) and (c) is the critical chord.

Before presenting the next case we have to introduce some definitions. The search path divides the part of the polygon seen so far into two regions. The *assumed L-part* of P includes every point on the boundary of P that has been to the left of the search position at some point on the search path and that was visible then. In the same way the *assumed R-part* is defined.

Consider the horizontal chord c_p induced by the search position. The θ-chords that pass through c_p can be ordered from left to right by the horizontal position of their intersection point with c_p.

Definition 22. If the forward landmark l is to the right (resp. to the left) of the search position, then the *critical chord* is defined as the rightmost (resp. leftmost) θ-chord that connects the assumed L-part (resp. R-part) of P to some point above the search position.

Note that the critical chord is the only θ-chord that touches both the assumed L- and R-part below the search direction.

Case 3 *There is a sideway landmark on chain C and either $0 < \theta_0 < \frac{\pi}{2}$ and l is to the right of p or $\frac{\pi}{2} < \theta_0 < \pi$ and l is to the left of p.*

We assume for ease of exposition that $0 < \theta_0 \le \frac{\pi}{2}$ and l is to the right of p. This case has to be subdivided further. Let k be the rightmost sideway landmark on chain C.

(i) *There is a forward landmark l' above l on the same side of the critical chord as p and the chord induced by l' does not intersect the critical chord.*

In this case the search path continues and no split occurs.

If there is no such landmark l', we examine the position of k:

(ii) *The critical chord is not visible from the hidden edge e of k (see Figure 4 (b)).*

Here the strategy starts two main search paths by splitting to the side of the forward landmark l. All other search paths are stopped.

(iii) *The critical chord is visible from e (see Figure 4(c)).*

In this case the action depends on the type of the search path. If it is a main search path, then the search path splits into an L- and R-search path. The L-search continues to north while the R-search path turns right. If the search path is already an L- or R-search path, then it just turns and no split occurs.

Let p be the search position and p' the projection of p in search direction. If after examining one of the previous cases there are no further forward landmarks above p and below p' for a search path it has to decide whether to stop or to turn. An L- or R-search path tries to continue an xy-monotone path, a main search path turns and looks for other forward landmarks.

If at any time during the search the target t becomes visible, then all other search paths are stopped and the search path goes to the target.

4.2 Proof of Competitiveness

In the following we give a short sketch of the proof that the strategy presented in the last section is 19.97-competitive under the L_1-metric.

The chord orthogonal to the search direction induced by the search position p divides P into two regions which we call the *explored* and *unexplored* region as in the previous section.

Lemma 23. *At any time in the search the path from p back to the starting points s is included completely in the explored region of P.*

Lemma 24. *At any time there are at most four search paths.*

The main part of the proof is then to show the following invariant. Here the cases have to be looked at in detail.

Invariant 25. *At any time during the search one of the search paths can be completed to an L_1-shortest path from s to t.*

Theorem 26. *The strategy is 19.97-competitive under the L_1-metric.*

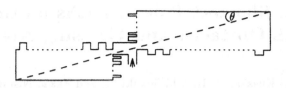

Fig. 5. A situation where the strategy uses four search paths. The target can be in any of the small caves.

5   Conclusions and Extensions

We have introduced two new classes of isothetic polygons, HV-**streets** and θ-$\mathcal{G}$-**streets**, that allow competitive strategies for a robot that starts at a vertex s and searches for a target t somewhere in the polygon only given the local visibility map. These classes are strictly larger than previously considered classes and the strategies we provide can be shown to be optimal. For HV-**streets** our strategy traverses a path that is at most 14.5 times as long as the L_1-shortest path from s to t. For θ-$\mathcal{G}$-**streets** the ratio is at most 19.97.

By putting the two results together, an infinite number of classes can be defined for which search strategies with a constant competitive ratio may exist. The definition is as follows. Let $\mathcal{O} = \{\theta_1, \ldots, \theta_k\}$ be a set of angles in $[0, \pi]$. A rectilinear polygon P is called an $\mathcal{O}$-*street* if, for any point p in P, there is a $\theta \in \mathcal{O}$ and a θ-oriented LR-chord c such that p sees a point of c. If the robot performs a search as described for θ-$\mathcal{G}$-streets for each θ in $\mathcal{O}$ in a round robin fashion until it has identified a new LR-chord, a search strategy with a competitive ratio that only depends on k seems possible.

If we leave the setting of isothetic polygons, no strategy has been found to search other polygons than streets. So an interesting open problem is to design strategies that allow a competitive search in arbitarily oriented polygon classes different from streets.

References

1. A. Blum and P. Chalasani. "An on-line algorithm for improving performance in navigation" *Proc. of 34th Annual IEEE Conference on Foundations of Computer Science*, (1993), pp. 2-11.
2. R. Baeza-Yates, J. Culberson and G. Rawlins. "Searching in the plane", *Information and Computation*, Vol. **106**, (1993), pp. 234-252.
3. A. Blum, P. Raghavan and B. Schieber. "Navigating in unfamiliar geometric terrain", *Proc. of 23rd Annual ACM Symp. on Theory of Computing*, (1991), pp. 494-504.
4. A. Datta and Ch. Icking. "Competitive searching in a generalized street", *Proc. of 10th Annual ACM Sypm. on Computational Geometry*, (1994), pp. 175-182.
5. R. Klein. "Walking an unknown street with bounded detour", *Computational Geometry: Theory and Applications* **1**, (1992), pp. 325-351.
6. J. Kleinberg. "On-line search in a simple polygon", *Proc. of 5th ACM-SIAM Symp. on Discrete Algorithms*, (1994), pp. 8-15.
7. V. J. Lumelsky and A. A. Stepanov. "Path-planning strategies of a point mobile automaton moving amidst unknown obstacles of arbitrary shape", *Algorithmica* **2**, (1987), pp. 403-430.
8. A. Mei and Y. Igarashi. "Efficient strategies for robot navigation in unknown environment" *Proc. of 21st International Colloquium on Automata, Languages and Programming*, (1994), to appear.
9. C. H. Papadimitriou and M. Yannakakis. "Shortest paths without a map", *Theoretical Computer Science* **84**, (1991), pp. 127-150.
10. D. D. Sleator and R. E. Tarjan. "Amortized efficiency of list update and paging rules", *Communications of the ACM* **28**, (1985), pp. 202-208.

Finding a Shortest Pair of Paths on the Plane with Obstacles and Crossing Areas

Yoshiyuki Kusakari *, Hitoshi Suzuki **, and Takao Nishizeki ***

Graduate School of Information Sciences,
Tohoku University, Sendai 980-77, Japan

Abstract. Given axis-parallel rectangular obstacles and crossing areas together with two pairs of terminals on the plane, our algorithm finds a shortest pair of rectilinear paths which connect the pairs of terminals and neither pass through any obstacle nor cross each other except in the crossing areas. The algorithm takes $O(n \log n)$ time and $O(n)$ space, where n is the total number of obstacles and crossing areas.

1 Introduction

The "shortest non-crossing path problem" often appears in the design of single-layer VLSI layouts, where "non-crossing paths" should not cross each other on the plane but may share common points or line segments unlike "disjoint paths." The problem asks to find non-crossing paths of the minimum total length, each connecting a specified terminal pair in a two-dimensional plane region without passing through any obstacles, *i.e.* build-in blocks. Lee *et al.* presented an algorithm to find shortest k non-crossing rectilinear paths in a plane region with n_o rectangular obstacles in time $O((k^2!)^k n \log n)$, where k is the number of terminal pairs [4]. Throughout the paper, we denote by n the size of an input for problems. Therefore $n = k + n_o$ for this problem. The algorithm does not seem practical when k is large. On the other hand, Takahashi *et al.* presented two efficient algorithms which find shortest k non-crossing paths in time $O(n \log n)$ for any k [7, 8]. One finds shortest k non-crossing rectilinear paths for the case when all terminals lie on the outer boundary of a plane region and the boundary of one of the rectangular obstacles [7]. The other finds shortest non-crossing paths in plane graphs in which all terminals lie on any two face boundaries [8].

On the other hand, the following problem often appear in the design of two-layer VLSI layouts: given two-layer routing regions called "crossing areas" in addition to obstacles on the plane, find a shortest pair of rectilinear paths which connect two pairs of terminals and neither pass through any obstacles nor cross each other except in the crossing areas. It should be noted that the found paths may cross each other in crossing areas. In this paper, we give an efficient algorithm to solve this problem in time $O(n \log n)$, where n is the total number

---

* kusakari@nishizeki.ecei.tohoku.ac.jp
** hs@nishizeki.ecei.tohoku.ac.jp
*** nishi@ecei.tohoku.ac.jp

of obstacles and crossing areas. The time complexity is best possible within a constant factor. This algorithm can be used for the rip-up and re-route problem of VLSI routing, in which each of unrouted wires together with a routed wire obstructing it are ripped up and two new re-route wires of the minimum total length will be found for the two pairs of terminals [5].

2 Preliminaries

The x-coordinate of a point $p \in \mathbb{R}^2$ is denoted by $x(p)$, and the y-coordinate by $y(p)$. The point p is often denoted by $(x(p), y(p))$. In this paper we consider only *rectilinear* paths $P \subseteq \mathbb{R}^2$. The *length* of P is denoted by $l(P)$. For two points p_1 and p_2 on P, $P[p_1, p_2]$ denotes the subpath of P connecting p_1 and p_2. All rectangles are assumed *axis-parallel*, that is, all edges of them are axis-parallel.

We next formally define a "crossing" of two paths P_a and P_b on $\mathbb{R}^2$. Let a and a' be the ends of path P_a, and let b and b' the ends of P_b. Clearly P_a and P_b do not cross each other if $P_a \cap P_b = \phi$. Thus we may assume that $P_a \cap P_b \neq \phi$. Let Q be a connected component (*i.e.* a path) of $P_a \cap P_b$, and let q and q' be the ends of path Q. Consider a component Q such that $\{q, q'\} \cap \{a, a', b, b'\} = \phi$. One may assume without loss of generality that a, q, q' and a' appear on P_a in this order, and that b, q, q' and b' appear on P_b in this order. Further one may assume that $P_a[q, a]$, $P_b[q, b]$ and $Q[q, q']$ appear clockwise around q in this order in a small circle with a center on q. Then P_a and P_b *cross* each other if $P_a[q', a']$, $P_b[q', b']$ and $Q[q', q]$ appear clockwise around q' in this order in a small circle with a center on q'. We call such a component Q a *crossing component*, and call any point in a crossing component a *crossing point*. We say that P_a and P_b do not cross each other if P_a and P_b cross each other at none of the components Q of $P_a \cap P_b$ such that $\{q, q'\} \cap \{a, a', b, b'\} = \phi$.

We assume that there are n_o obstacles and n_c crossing areas in the plane, and that all of them are rectangles and do not overlap each other. Thus $n = n_o + n_c$. The set of obstacles is denoted by $\mathcal{O} = \{O_1, O_2, \cdots, O_{n_o}\}$, and the set of crossing areas by $\mathcal{C} = \{C_1, C_2, \cdots, C_{n_c}\}$. In Fig. 1 obstacles are drawn in gray rectangles and crossing areas in white rectangles. The left-top, left-bottom, right-top, right-bottom vertices of a rectangle $R \in \mathcal{O} \cup \mathcal{C}$ are denoted by $lt(R)$, $lb(R)$, $rt(R)$ and $rb(R)$, respectively. The left, right, top, and bottom edges of R are denoted by $le(R)$, $re(R)$, $te(R)$, and $be(R)$, respectively. The boundary of R is denoted by $B(R)$. The *routing region* A is a subregion of $\mathbb{R}^2$ excluding the obstacles, that is, $A = \mathbb{R}^2 - \bigcup \{R - B(R) | R \in \mathcal{O}\}$. Thus A contains the boundaries of obstacles, and hence paths can pass through a boundary of two touching obstacles. We denote by $d(p_1, p_2)$ the L_1-distance from p_1 to p_2. We denoted by $d_A(p_1, p_2)$ the length of a shortest path connecting p_1 and p_2 in A.

We next precisely define the problem of this paper. Given two pairs (a, a') and (b, b') of terminals in routing region A, let path P_a connect a and a' in A, and let path P_b connect b and b' in A. We say that (P_a, P_b) is a *feasible pair of paths* (*in routing region A with crossing areas*) if every crossing component Q of P_a and P_b contains a point in a crossing area: $Q \cap (\bigcup_{i=1}^{n_c} C_i) \neq \phi$. Note that

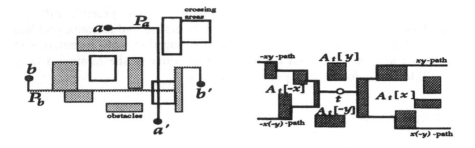

Fig. 1. A shortest pair of paths in a rout- **Fig. 2.** Subdivision of A into four regions.
ing region with crossing areas.

for each crossing component Q one can make such a pair of paths (wires) cross each other, that is, interchange layers at a crossing point $p \in Q \cap (\bigcup_{i=1}^{n_c} C_i)$ by putting "vias" on p in a crossing area, *i.e.*, a "two-layer routing region." Our problem is to find a feasible pair of P_a and P_b having the minimum total length, which we call an *optimal pair of paths (in routing region A with crossing areas)*. Fig. 1 illustrates such an optimal pair. We denote by $\mathcal{P}$ the set of all shortest pairs of paths, and by $l(\mathcal{P})$ the sum of lengths of paths P_a and P_b for a pair $(P_a, P_b) \in \mathcal{P}$. In this paper we will present an algorithm to find the length $l(\mathcal{P})$. It can be easily modified so that it actually finds a pair of paths $(P_a, P_b) \in \mathcal{P}$.

A feasible pair of paths may cross or may not cross each other. Any pair (P_a, P_b) of non-crossing paths is feasible. The set of shortest pairs of non-crossing paths in A is denoted by $\mathcal{P}^{nc}$. On the other hand, if a feasible pair (P_a, P_b) of paths cross each other, then P_a and P_b share a point in a crossing area: $P_a \cap P_b \cap (\bigcup_{i=1}^{n_c} C_i) \neq \phi$. We define that a *shortest pair of paths sharing points* is a shortest one among all feasible pairs of paths that share a point in a crossing area. The set of such pairs of paths is denoted by $\mathcal{P}^s$. Define lengths $l(\mathcal{P}^{nc})$ and $l(\mathcal{P}^s)$ similarly as $l(\mathcal{P})$. Then $l(\mathcal{P}) = \min\{l(\mathcal{P}^{nc}), l(\mathcal{P}^s)\}$ since $\mathcal{P} \subseteq \mathcal{P}^{nc} \cup \mathcal{P}^s$. Thus it suffices to find $l(\mathcal{P}^{nc})$ and $l(\mathcal{P}^s)$.

It is known that either (P_a, P_b') or (P_a', P_b) defined below is a shortest pair of non-crossing paths in A [4].

(1) P_a is an arbitrary shortest path connecting a and a' in A, and P_b' is a shortest path connecting b and b' in A without crossing P_a.

(2) P_b is an arbitrary shortest path connecting b and b' in A, and P_a' is a shortest path connecting a and a' in A without crossing P_b.

Both (P_a, P_b') and (P_a', P_b) can be found in time $O(n_o \log n_o)$, and hence $l(\mathcal{P}^{nc})$ can be found in time $O(n_o \log n_o)$ [4]. Thus we shall show that $l(\mathcal{P}^s)$ can be found in time $O(n \log n)$.

3 Shortest pair of paths sharing points

In this section we present an algorithm to find the length $l(\mathcal{P}^s)$. In Section 3.1 we define a set $\mathcal{P}_i^s$ of shortest pairs of paths sharing a point in C_i for each crossing area C_i, $1 \leq i \leq n_c$, define the length $l(\mathcal{P}_i^s)$ of such paths, and then prove that $l(\mathcal{P}^s) = \min\{l(\mathcal{P}_i^s)|1 \leq i \leq n_c\}$. In Section 3.2, as a preprocessing of our algorithm, we divide the routing region A into 25 subregions A_j, $1 \leq j \leq 25$, which are classified in eight groups. In Section 3.3, we present algorithms to find $l(\mathcal{P}_i^s)$; the algorithms depend on which group the subregion A_j containing C_i belongs to. Furthermore we show that all lengths $l(\mathcal{P}_i^s)$ can be found total in time $O(n \log n)$.

3.1 Definition and property of $\mathcal{P}_i^s$

For i, $1 \leq i \leq n_c$, we denote by $\mathcal{P}_i^s$ the set of shortest pairs of paths P_a and P_b in A that share a point in C_i, that is, $P_a \cap P_b \cap C_i \neq \phi$. Such a pair (P_a, P_b) may not be feasible because P_a and P_b may have a crossing component Q such that $Q \cap (\bigcup_{j=1}^{n_c} C_j) = \phi$. We define the length $l(\mathcal{P}_i^s)$ similarly as $l(\mathcal{P})$. Then $l(\mathcal{P}^s) \geq \min\{l(\mathcal{P}_i^s)|1 \leq i \leq n_c\}$ since all pairs in $\mathcal{P}^s$ are feasible but some pairs in $\mathcal{P}_i^s$ may not be feasible. However, this equation holds in equality:

Lemma 1. $l(\mathcal{P}^s) = \min\{l(\mathcal{P}_i^s)|1 \leq i \leq n_c\}$ □

Lemma 1 implies that one can determine $l(\mathcal{P}^s)$ by calculating $l(\mathcal{P}_i^s)$ for all i, $1 \leq i \leq n_c$.

3.2 Subdivision of the routing region

Our algorithm divides the routing region A into 25 subregions A_j, $1 \leq j \leq 25$, for each of which our algorithm calculates $\min\{l(\mathcal{P}_i^s)|C_i \subseteq A_j, 1 \leq i \leq n_c\}$. In this Section 3.2 we describe how to divide the routing region A.

Let L be a vertical line segment in A, and let x_L be the x-coordinate of L. We say that a point p in A is *visible from L in the x-direction* if $x(p) \geq x_L$, the foot of the horizontal line from p to L is on L, and the line intersects none of the obstacles. We similarly define the *visibility from L in the $(-x)$-direction* and the *visibility from horizontal line segment in the $(\pm y)$-direction*.

A path P in A is called *monotone in the x-direction* if the intersection of P and any vertical line is either a single point or a single line segment. Such a path is called an *x-path*. Similarly we define *$(-x)$-, y-, and $(-y)$-paths*.

We recursively define an *xy-path* $P \subseteq A$ starting at a point $t \in A$ as follows [1]:

(1) P is the semi-infinite horizontal line $y = y(t)$ for $x \geq x(t)$, and does not intersect any obstacle; or

(2) P follows the $+x$-direction to a vertical edge of an obstacle, then follows the $+y$-direction up to the upper end p of the edge, and then follows an xy-path starting at p.

We similarly define $x(-y)$-, $-xy$-, and $-x(-y)$-paths. These four paths, illustrated in Fig. 2, are called *prior paths starting at t*.

For a terminal $t \in \{a, a', b, b'\}$, the *x-region* $A_t[x]$ of A for t is the subregion of A which is bounded by the xy- and $x(-y)$-paths starting at t and contains the $+x$-semi-axis. We similarly define the $(-x)$-*region* $A_t[-x]$, *y-region* $A_t[y]$, and $(-y)$-*region* $A_t[-y]$, as illustrated in Fig. 2. One can easily observe the following lemma [1].

Lemma 2. *Let $t \in \{a, a', b, b'\}$ and $z \in \{\pm x, \pm y\}$. Then for any point p in $A_t[z]$, there exists a shortest path connecting t and p in A which is a z-path and is contained in $A_t[z]$.* □

The routing region A is divided into four subregions, $\pm x, \pm y$-regions, for each of the four terminals. Thus A is divided into $4^4 = 256$ subregions $A_j = A[z_1, z_2, z_3, z_4]$ $= A_a[z_1] \cap A_{a'}[z_2] \cap A_b[z_3] \cap A_{b'}[z_4]$, $1 \le j \le 256$, where $z_1, z_2, z_3, z_4 \in \{\pm x, \pm y\}$. However, one can easily know that at most twenty five A_j's are nonempty as follows. The four paths, each concatenating xy-path and $-x(-y)$-path starting at each of the four terminals, do not cross each other. Similarly the four paths, each concatenating $-xy$-path and $x(-y)$-path starting at each of the four terminals, do not cross each other. These two sets of four non-crossing paths divide A into at most $(4+1)^2 = 25$ nonempty subregions. For example, A in Fig. 3 is divided into 16 nonempty subregions $A_1, A_2, \cdots, A_{16}$.

Since one can find each prior path in time $O(n \log n)$ using $O(n)$ space [1], one can find the subdivision above in time $O(n \log n)$ using $O(n)$ space. It should be noted that a crossing area is not necessarily contained in one of the subregions. If a crossing area intersects two or more subregions, we divide it into several crossing areas by cutting it along prior paths on the boundaries of the subregions. The total number of crossing areas is $O(n_c)$ even if they are divided as above. Thus one may assume that each crossing area is completely contained in one of the 25 subregions.

If $(P_a, P_b) \in \mathcal{P}_i^s$, then P_a and P_b share a point p in the crossing area C_i. Therefore (P_a, P_b) is composed of four shortest paths connecting $p \in C_i$ and a, a', b, and b' and having the minimum total length. Thus we have

$$l(\mathcal{P}_i^s) = \min\{d_A(a, p) + d_A(a', p) + d_A(b, p) + d_A(b', p) | p \in C_i\}. \tag{1}$$

For a crossing area C_i, an *x-region* $A_{C_i}[x]$ is the subregion of A which is bounded by the xy-path starting at $rt(C_i)$, the $x(-y)$-path starting at $rb(C_i)$, and the right edge $re(C_i)$. We similarly define the $-x$-region $A_{C_i}[-x]$, y-region $A_{C_i}[y]$, $-y$-region $A_{C_i}[-y]$ for C_i. One can easily observe the following lemma.

Lemma 3. *Let $t \in \{a, a', b, b'\}$, and $z \in \{\pm x, \pm y\}$. Then $C_i \subseteq A_t[z]$ if and only if $t \in A_{C_i}[-z]$.*

By Lemma 3 and the symmetry of $\pm x, \pm y$ directions, the 25 subregions A_j's can be classified in the following eight groups depending on the relative location of the four terminals and crossing areas $C_i \subseteq A_j$.

Group 1: For every $C_i \subseteq A_j$, all the four terminals are in the same region $A_{C_i}[z]$ where $z \in \{\pm x, \pm y\}$, as illustrated in Fig. 4. (Subregions A_1, A_2, A_{10} and A_{16} in Fig. 3 belong to this group.)

Group 2: For every $C_i \subseteq A_j$, three terminals are in the same region $A_{C_i}[z]$ and the other terminal in the opposite region $A_{C_i}[-z]$.

Group 3: For every $C_i \subseteq A_j$, three terminals are in the same region $A_{C_i}[z]$ and the other terminal in an adjacent region $A_{C_i}[z']$ where $z' = \pm y$ if $z = \pm x$ and $z' = \pm x$ if $z = \pm y$.

Group 4: For every $C_i \subseteq A_j$, two terminals are in the same region $A_{C_i}[z]$ and the other two terminals in the opposite region $A_{C_i}[-z]$.

Group 5: For every $C_i \subseteq A_j$, two terminals are in the same region $A_{C_i}[z]$ and the other two terminals in one of the two adjacent regions $A_{C_i}[z']$.

Group 6: For every $C_i \subseteq A_j$, two terminals are in the same region $A_{C_i}[z]$, another in the opposite region $A_{C_i}[-z]$, and the other in an adjacent region $A_{C_i}[z']$.

Group 7: For every $C_i \subseteq A_j$, two terminals are in the same region $A_{C_i}[z]$, and exactly one of the other two terminals is in each of the two adjacent regions $A_{C_i}[\pm z']$.

Group 8: For every $C_i \subseteq A_j$, exactly one of the four terminals is in each of the four regions for C_i.

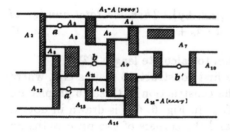

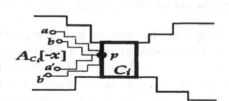

Fig. 3. Subdivision of A into 16 nonempty subregions $A_1, A_2, \cdots, A_{16}$.

Fig. 4. Group 1.

3.3 Algorithms

In this section we present algorithms to calculate $\min\{l(\mathcal{P}_i^s)|C_i \subseteq A_j, 1 \le i \le n_c\}$ for subregions A_j. The algorithms depend on which group A_j belongs to. However, we present only an algorithm for A_j belonging to Group 1 since one can easily modify the algorithm for other groups.

Let $t \in \{a, a', b, b'\}$, and let L be a horizontal line segment in A. A function $f_L^t : \{x | p = (x, y) \in L\} \to \mathbb{R}$ is defined as follows: $f_L^t(x) = d_A(t, p)$. The following lemma implies that the slope of this function takes values only ± 1, as illustrated in Fig. 5.

Lemma 4. *Let $t \in \{a, a', b, b'\}$. Let $z \in \{+y, -y\}$, and let L be a horizontal line segment in $A_t[z]$.*

(a) *Assume that, among terminal t and all vertices of obstacles in $A_t[z]$, exactly q points $v_1, v_2, \cdots, v_q$ are visible from L in the $(-z)$-direction, and that $x(v_1) \leq x(v_2) \leq \cdots \leq x(v_q)$. Then, for each i, $1 \leq i \leq q - 1$, there exists $x_i' \in \mathbb{R}$, $x(v_i) \leq x_i' \leq x(v_{i+1})$, such that*

$$\frac{d}{dx} f_L^t(x) = \begin{cases} +1 \text{ if } x(v_i) < x < x_i'; \\ -1 \text{ if } x_i' < x < x(v_{i+1}). \end{cases}$$

(b) *If $z = +y$, then the slope of $f_L^t(y)$ is $+1$. If $z = -y$, then the slope of $f_L^t(y)$ is -1.* □

A function $f_L^t(y)$ is similarly defined for a vertical line segment L. Then Lemma 4 holds also for a vertical line segment L.

Let $t \in \{a, a', b, b'\}$, $z \in \{\pm x, \pm y\}$, and let L be a vertical or horizontal line segment in $A_t[z]$. A point on L at which the slope of f_L^t changes is called a *bend point*. The function f_L^t can be represented in terms of the bend points and the slopes between consecutive bend points on L. Denote by E_{tz} the set of all edges of obstacles in $A_t[z]$, then the following lemma holds on the total number of bend points on edges $L_o \in E_{tz}$.

Lemma 5. *Let $t \in \{a, a', b, b'\}$ and $z \in \{\pm x, \pm y\}$. Then the total number of bend points on all edges $L_o \in E_{tz}$ is at most $4n_o + 1$.* □

Using these lemmas, we give an algorithm to calculate length $l(\mathcal{P}_i^s)$.

As a preprocessing of the algorithm, we first construct 16 graphs G_{tz}, where $t \in \{a, a', b, b'\}$ and $z \in \{\pm x, \pm y\}$. Since the construction of these 16 graphs are similar, we give only the construction of G_{ty}. In Fig. 6, G_{ty} is drawn in dotted lines. Add to G_{ty}, as vertices, terminal t and the corners of two prior paths on the boundary of $A_t[y]$, that is, xy- and $-xy$-path starting at t. Add to G_{ty}, as edges, line segments on the two paths connecting any two consecutive ones of these vertices. We add more vertices and edges to G_{ty} by means of the following plane sweep.

We move a horizontal sweep line L in the $+y$-direction from $y(t)$ with stopping on each of the horizontal edges of obstacles in $A_t[y]$. We keep data on L, and update the data whenever L stops on an edge. We use a *segment tree T_t* as the data structure [6]. $L \cap A_t[y]$ contains one or more horizontal line segments. Assume that, among terminal t, bend points and all vertices of obstacles in $A_t[y]$, exactly q points $v_1, v_2, \cdots, v_q$ are visible from line segments of $L \cap A_t[y]$ in the $(-y)$-direction, and that $y(v_1) \leq y(v_2) \leq \cdots \leq y(v_q)$. Then the segment tree T_t has exactly q leaves, say $l_1, l_2, \cdots, l_q$ from left to right. We store point v_h at leaf

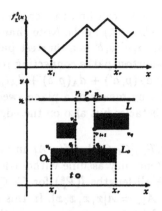

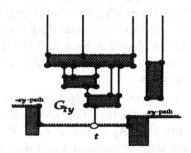

Fig. 5. Illustration for $f_L^t(x)$.   **Fig. 6.** Illustration for G_{ty}.

l_h for $1 \leq h \leq q$. Let $v_1', v_2', \cdots, v_q'$ be points on L having the same x-coordinate as $v_1, v_2, \cdots, v_q$. We store at leaf l_h, $1 \leq h \leq q$, the length of a shortest path from t to v_h'. We store at an internal node v of the segment tree T_t an interval $[i_1, i_2]$, where i_1 is the smallest x-coordinate of points stored at leaves which are the descendants of v, and i_2 is the largest one.

When sweep line L is set on the terminal t, we add a leaf storing the terminal t to the segment tree T_t.

When sweep line L stops on the bottom edge $be(O_k)$ of an obstacle $O_k \subseteq A_t[y]$, we add $lb(O_k)$ and $rb(O_k)$ to G_{ty} as vertices. Draw vertical line segments to $be(O_k)$ from points which are stored in T_t and whose x-coordinates are in interval $[x(lb(O_k)), x(rb(O_k))]$. Add to G_{ty} as vertices these intersection points on $be(O_k)$, and add to G_{ty} as edges these vertical line segments together with horizontal line segments on $be(O_k)$ connecting two consecutive ones of these vertices. After adding these vertices and edges to G_{ty}, we delete from T_t the leaves whose points have x-coordinates in interval $[x(lb(O_k), x(rb(O_k))]$. Furthermore we add to T_t two leaves storing $lb(O_k)$ and $rb(O_k)$.

On the other hand, when sweep line L stops on the top edge $te(O_k)$ of an obstacle $O_k \subseteq A_t[y]$, we find a bend point $p_k^* \in te(O_k)$ if there is. One can easily find $x(p_k^*)$ from $d_A(t, rt(O_k))$ and $d_A(t, lt(O_k))$. Add the three vertices p_k^*, $lt(O_k)$ and $rt(O_k)$ to G_{ty}, and add four line segments $le(O_k)$, $re(O_k)$, $[lt(O_k) - p_k^*]$, and $[p_k^* - rt(O_k)]$ to G_{ty} as edges. After adding these vertices and edges to G_{ty}, we insert to T_t three leaves storing p_k^*, $lt(O_k)$, and $rt(O_k)$.

When sweep line L stops on the topmost edge of obstacle $O_k \subseteq A_t[y]$, we execute the operation above, and add the infinite point to G_{ty} as a vertex. Add to G_{ty}, as the edges, line segments connecting the infinite point and the points stored at leaves of T_t. We thus complete the construction of G_{ty}.

We similarly find the length of the shortest path connecting t and each vertex of G_{ty} by the plane sweep.

Using 16 graphs G_{tz} found in the preprocessing, we next find lengths $l(\mathcal{P}_i^s)$ for all crossing areas C_i in subregion A_j, $1 \leq j \leq 25$. One can find the length $l(\mathcal{P}_i^s)$ by finding the point p shared by two paths for which (1) holds. Note that each of the four paths connecting p and four terminals a, a', b, b' is a shortest path in A. We denote by $d_A^4(p)$ the total length of four shortest paths connecting $p \in C_i$ and each of four terminals: $d_A^4(p) = d_A(p, a) + d_A(p, a') + d_A(p, b) + d_A(p, b')$. We find the shared point p and the length $l(\mathcal{P}_i^s)$ by means of a plane sweep in which sweep line L stops not only on edges of obstacles but also on the edges of crossing areas.

We then present an algorithm to find $\min\{l(\mathcal{P}_i^s)|C_i \subseteq A_j\}$ for A_j in Group 1. Algorithms for the other groups are similar. One may assume without loss of generality that $A_j = A[x, x, x, x]$. We compute all lengths $l(\mathcal{P}_i^s)$ for $C_i \subseteq A_j$, and take the minimum among them. In Fig. 3, $A_{10} = A[x, x, x, x]$. In this case, $a, a', b, b' \in A_{C_i}[-x]$ as illustrated in Fig. 4, and hence one may assume that p is on the left edge of each crossing area $C_i \subseteq A_j$. By Lemma 2, $P_a[a, p] \subseteq A_a[x]$, $P_a[a', p] \subseteq A_{a'}[x]$, $P_b[b, p] \subseteq A_b[x]$, and $P_b[b', p] \subseteq A_{b'}[x]$. Let x_{min} be the minimum x-coordinate of points in the subregion A_j. Then we initialize the sweep line L on the vertical line $x = x_{min}$, and move L to the $+x$-direction. We store on L the following segment tree T^*. Assume that, among all the top-left and bottom-left vertices of crossing areas in A_j and all vertices of four graphs G_{ax}, $G_{a'x}$, G_{bx} and $G_{b'x}$ in A_j, exactly q points $v_1, v_2, \cdots, v_q$ are visible from a vertical line segment of $L \cap A_j$ in the $(-x)$-direction, and that $y(v_1) \leq y(v_2) \leq \cdots \leq y(v_q)$. Then the segment tree T^* has exactly q leaves, say $l_1, l_2, \cdots, l_q$ from left to right, which store $v_1, v_2, \cdots, v_q$. Let v_h', $1 \leq h \leq q$, be the points on L having the same y-coordinate as v_h, then leaf l_h, $1 \leq h \leq q$, stores the length $d_A^4(v_h')$. Each internal node v of T^* stores leaf l_v and an interval $[i_1, i_2]$; i_1 is the minimum y-coordinate of points stored at the leaves l_h which are descendants of v, and i_2 is the maximum y-coordinate; l_v is the leaf storing the minimum $d_A^4(v_h')$ among all the leaves which are descendants of v.

When sweep line L stops on the left edge $le(O_k)$ of $O_k \subseteq A_j$, we delete some of the leaves l_h from T^*: if a leaf l_h of T^* stores a vertex v_h such that $y(v_h) \in [y(lb(O_k)), y(lt(O_k))]$, then we delete the leaf l_h from T^*. We then insert to T^* two leaves storing $lb(O_k)$ and $lt(O_k)$.

On the other hand, when sweep line L stops on the right edge $re(O_k)$ of $O_k \subseteq A_j$, we delete from T^* two leaves storing $lt(O_k)$ and $lb(O_k)$. Furthermore, we insert to T^* leaves storing vertices of four graphs G_{ax}, $G_{a'x}$, G_{bx}, $G_{b'x}$ on $re(O_k)$. These vertices v are not always contained in all of four graphs. For example, it happens that v is a vertex of G_{ax} but is not a vertex of $G_{a'x}$. In this case, $d_A(v, a)$ has already been computed, but $d_A(v, a')$ has not. However, $d_A(v, a')$ can be easily computed as follow. Let v' be the vertex of $G_{a'x}$ immediately above v on L, and let v'' be the vertex of $G_{a'x}$ immediately below v on L. From $d_A(v', a')$ and $d_A(v'', a')$ which have already been computed, one can easily compute $d_A(v, a')$. Similarly one can easily compute $d_A(v, b)$ and $d_A(v, b')$. Thus one can easily compute $d_A^4(v)$.

When sweep line L stops on the left edge $le(C_i)$ of a crossing area C_i, we insert $lb(C_i)$ and $lt(C_i)$ to T^* as leaves. We then find a leaf l_h of T^* such that the vertex v stored at l_h satisfies $y(v) \in [y(lb(C_i)), y(lt(C_i))]$ and $d_A^A(v_h')$ is the minimum among all these vertices v. The vertex v_h' is on the $le(O_k)$. The point p shared by the two paths that we wish to find is v_h', and $l(\mathcal{P}_i^s) = d_A^A(v_h')$.

In the plane sweep for A_j in Group 1, sweep line L does not stop on the right edge $re(C_i)$ of crossing areas C_i.

Finally, when sweep line L stops on the rightmost one of left edges $le(C_i)$ of $C_i \subseteq A_j$, we execute the operation above and complete the plane sweep of A_j in Group 1.

It is not difficult to show that our algorithm takes $O(n \log n)$ time and $O(n)$ space to find $l(\mathcal{P}^s)$.

4 Conclusion

In this paper we presented an efficient algorithm for finding a shortest pair of paths in a plane region with n_o rectangular obstacles and n_c rectangular crossing areas. In the algorithm we use a plane sweep and a segment tree tailored for our purpose. The algorithm runs in time $O(n \log n)$ where $n = n_o + n_c$. The complexity is best possible in a sense that finding a single shortest path between two points in a plane region with n_o rectangular obstacles requires time $\Omega(n_o \log n_o)$ [1].

It is rather straightforward to modify our sequential algorithm to an NC parallel algorithm which finds a shorest pair of paths in polylog time using a polynomial number of processors. Note that there are NC parallel algorithms for the shortest path problem [2, 3] and for the plane sweep [2].

References

1. P.J. de Rezend, D.T. Lee, and Y.F. Wu, *"Rectilinear shortest paths with rectangular barriers,"* Discrete and Computational Geometry, 4, pp. 41-53, 1989.

2. J. JáJá, An Introduction to Parallel Algorithms, Addison Wesley, Reading, MA, 1992.

3. P.N. Klein, "A linear processor polylog-time algorithm for shortest paths in planar graphs," Proc. of 34th Symp. on Found. of Comput. Sci., pp. 259-270, 1993.

4. D.T.Lee, C.F.Shem, C.D.Yang, and C.K.Wong, *"Non-crossing paths problems,"* Manuscript, Dept. of EECS, Northwestern Univ., 1991.

5. T. Ohtsuki(Editor), Layout Design and Verification, North-Holland, Amsterdam, 1986.

6. F.P. Preparata, and M.I. Shamos, Computational Geometry : An Introduction, Springer-Verlag, New York, 1985.

7. J. Takahashi, H. Suzuki, and T. Nishizeki, *"Finding shortest non-crossing rectilinear paths in plane regions,"* Proc. of ISAAC'93, Lect. Notes in Computer Science, Springer-Verlag, 762, pp. 98-107, 1993.

8. J. Takahashi, H. Suzuki, and T. Nishizeki, *"Shortest non-crossing paths in plane graphs,"* Algorithmica, to appear.

Logspace Verifiers, NC, and NP
(Extended Abstract)

Satyanarayana V. Lokam[1], Meena Mahajan[2], V Vinay[3]*

[1] Department of Computer Science, The University of Chicago
Chicago, IL 60637, U. S. A. . E-mail: satya@cs.uchicago.edu
[2] The Institute of Mathematical Sciences, C.I.T. Campus, Madras 600 113, India.
E-mail: meena@imsc.ernet.in
[3] Department of Computer Science and Automation, Indian Institute of Science,
Bangalore 560 012, India. E-mail: vinay@csa.iisc.ernet.in

Abstract. We explore the connection between public-coin interactive proof systems with logspace verifiers and $\mathcal{NC}$ using two different approaches. In the first approach, we describe an interactive proof system for accepting any language in $\mathcal{NC}$ after a logspace reduction, where the verifier is logspace-bounded and the protocol requires polylog time. These results are proved by describing $\mathcal{NC}$ computations as computations over arithmetic circuits using **maximum** and **average** gates, and then translating the arithmetic circuits into interactive proof systems in a natural way. In the second approach, we give a *characterization* of $\mathcal{NC}$ in terms of interactive proof systems where the verifier is logspace-bounded and runs in polylog time. The equivalent interactive proof systems work with error-correcting encodings of inputs, using the polylogarithmically checkable codes introduced in the context of transparent proofs.

We also characterize $\mathcal{NP}$ and $\mathcal{PSPACE}$ via public-coin interactive proof systems where the verifier is logspace-bounded, but has restricted access to auxiliary storage.

1 Introduction

An interactive proof system is a probabilistic game between an infinitely powerful prover and a randomized verifier. This model was introduced in [GMR85], and independently, in [Bab85](see also [BM88]). In [GMR85], the coin tosses of the verifier are hidden from the prover (this is called the *private coins* model), whereas in [Bab85, BM88], the coin tosses of the verifier are visible to the prover (this is called the *public coins* model). It was shown that for polynomial time verifiers both models have essentially the same computational power [GS86]. Generalizing the close connection between alternating Turing machines (ATMs) and interactive proof systems [LFKN92, Sha92, She92], Fortnow and Lund [FL93] give simulations between ATMs with simultaneous space and time bounds and

---

* Work done while the author was with the Institute of Mathematical Sciences, Madras 600 113, India.

interactive proof systems with space and time bounded verifiers. Condon, Lipton, Ladner [CL88, CL89, Con91, CL92], and Fortnow [For89, Ch.3], examined the power of logspace verifiers under many restrictions. In particular, it was shown [CL92] that with a logspace verifier, public coins can be strictly weaker than private coins, assuming $\mathcal{P} \neq \mathcal{NP}$. In this paper we extend some results of [For89, Ch.3], [FL93], and [CL92].

We study public-coin interactive proof systems with a space-bounded verifier. In Section 2, we define unbounded fanin arithmetic circuits with gates computing maximum and average. We establish a correspondence between *uniform* families of such circuits and public-coin interactive proof systems. The depth of the circuit equals the number of rounds while the size is exponential in the space of the verifier (Theorem 3). Similar circuits with bounded fanin have earlier been used in [For89, Ch.3], following intuition from [GS86], in the context of logspace verifiers. We use them in a more general setting. In particular, we consider *unbounded fanin* circuits so we can relate depth to rounds; due to non-associativity of the average operator, conversion from unbounded fanin to bounded fanin is not clear. We also ensure uniformity of our circuits. Both these differences are crucially used in our proofs to obtain results that properly generalize results from [For89, Ch.3].

In Section 3, we describe how to arithmetize computations of ATMs. A simple low-degree arithmetization, and a size-efficient arithmetization that requires the input to be present in an encoded form, are obtained (Theorems 4 and 5). These results build on techniques developed in [FL93]. Exploiting the fact that the verifier in our case has random access to encoded input, we are able to obtain arithmetizations of significantly smaller size than in [FL93].

In Section 4, we give public-coin interactive proof systems for $\mathcal{NC}$. Using the low-degree arithmetization of Section 3, we first show how to represent ATM computations as computations over the arithmetic circuits described in Section 2. Combining this with the results of Section 2, we get interactive proof systems for accepting any language in $\mathcal{NC}$ after a logspace reduction, where the verifier is logspace bounded and the protocol runs in polylog time (Theorem 6).

It was shown in [FL93] that every language in $\mathcal{NC}$ has a public-coin interactive proof system with a verifier using $O(\log n)$ space and $O(n \log^2 n)$ time. Our above-mentioned result identifies the polynomial-time bottleneck and isolates it in the logspace reduction, so that the rest of the protocol runs fast. Omitting the logspace reduction gives a polylog round interactive proof system for $\mathcal{NC}$.

In Section 5, we characterize $\mathcal{NC}$ exactly via interactive proof systems of polylog time and logspace complexity. Since a sublinear time verifier cannot even read the entire input, we supply the input in an *error-correcting encoding*, using the codes introduced in the context of transparent proofs [BFLS91] (see also [BF93]). Further, the encoding itself is still computable in $\mathcal{NC}$ and is probabilistically checkable in polylog time. Using this encoding, denoted γ, and the size-efficient arithmetizations of Section 3, we show that if a language L is in $\mathcal{NC}$ then $\gamma(L)$ is accepted by interactive proof systems with logspace verifier and polylog running time, where $\gamma(L)$ denotes the set of codewords corresponding to strings in L. Combining this with the reverse simulation from [FL93], we get the claimed equivalence of $\mathcal{NC}$ and interactive proof systems (Theorem 12).

In the last section of this paper, we show an interesting variant of the following result in [CL92]. Let $\mathcal{IPS}(a, b, c)$ denote the class of languages recognized by interactive proof systems in which a specifies public/private coins, b specifies the power of the verifier (polytime, logspace etc.,), and c gives the number of random bits used. In [CL92] the following result is proved: $\mathcal{NP} = \mathcal{IPS}(private, logspace, log\ random\ bits)$. *On the other hand*, $\mathcal{NLOG} \subseteq \mathcal{IPS}(public, logspace, log\ random\ bits) \subseteq \mathcal{LOGCFL}$.

We show that this weakening of the public coins model is offset by allowing the verifier access to an additional storage in the form of a pushdown stack. Thus, the verifier is a randomized $\mathcal{LOGCFL}$ machine. More precisely, we show (Theorem 13): $\mathcal{NP} = \mathcal{IPS}(public, logspace + PDA, log\ random\ bits)$. This strengthens another result in [CL92], namely $\mathcal{NP} = \mathcal{IPS}(public, polytime, log\ random\ bits)$, by replacing the polytime verifier by a $\mathcal{LOGCFL}$ verifier.

We also observe that by relaxing the pushdown stack to a readable stack (push-pop changes are restricted to the top, but stack contents can be read arbitrarily), we can capture $\mathcal{PSPACE}$. In fact, the verifier alternates between pushdown mode and scan mode (read internal contents) only once, and the stack is non-erasing (Theorem 14). Such a verifier is not known to be as powerful as $\mathcal{P}$ [JK89, VC90] and can potentially be weaker. Hence this is a conditional improvement on the standard poly-time verifier for $\mathcal{PSPACE}$ [LFKN92, Sha92, She92].

Notation: In the following, IPRDSP($r(n), s(n)$) and IPTISP($t(n), s(n)$) denote the sets of languages accepted by public-coin interactive proof systems using $r(n)$ rounds and $t(n)$ time, respectively, and with the verifier using $s(n)$ space. Also, ATISP($t(n), s(n)$) denotes the set of languages accepted by alternating Turing machines running in time $t(n)$ and space $s(n)$.

2 Arithmetic circuits and interactive proof systems

In this section, we consider a family of arithmetic circuits defined as follows. The circuit has two types of unbounded fanin gates: MAX and AVG. A MAX gate outputs the maximum of all its input values. An AVG gate returns the arithmetic mean of the values of all its inputs. In addition, there are special input gates called READ gates. These gates have one arithmetic-valued input i, one arithmetic-valued output o, and further one Boolean input b which is a circuit input. If $b = 1$, then $o = i$, otherwise $o = 0$. Thus at a READ gate, a circuit input can filter out an arithmetic value. We make the simplifying assumption that circuit inputs are read only through READ gates. All MAX and AVG gates have as inputs either the outputs of other gates or one of the arithmetic constants 0 or 1. The circuit returns a value at least 3/4 if the input is to be accepted, and a value at most 1/4 otherwise. (We do not consider circuits which return values in the range (1/4,3/4).) Such circuits, of size $S(n)$ and depth $D(n)$, form the family MAX,AVG($S(n), D(n)$). Unless otherwise stated, we will assume that the circuit families are uniform in the following sense: In a family of circuits of size $S(n)$, a description of the circuit can be generated by a DSPACE(log $S(n)$) machine, given input 1^n.

Apart from having unbounded fanin, our circuits differ from those of [For89, Ch.3] in that we use READ gates to read all circuit inputs. This permits the uniform constructibility of our circuits.

Our main result in this section shows that IPRDSP($r(n), s(n)$) is equivalent, *via* DSPACE($s(n)$)-reductions, to the class of languages accepted by uniform MAX,AVG -circuit families of size $2^{s(n)}$ and depth $r(n)$.

Lemma 1. *IPRDSP($r(n), s(n)$)* $\leq_{DSPACE(s(n))}$ *MAX,AVG($2^{s(n)}, r(n)$).*

Proof Sketch: The idea behind converting the interactive proof system to a circuit is as follows. Each configuration of the interactive proof system is a gate of the circuit. In the interactive proof system, when it is the prover's turn to respond, the prover chooses, out of all potential answers, that which maximizes the chances of convincing the verifier. So a prover's move from a configuration is captured by a MAX gate, whose children are configurations resulting from each potential prover response. Similarly, consider verifier's moves. We assume, w.l.o.g., that the verifier makes at most $s(n)$ coin tosses in a round, and they are all made at the beginning of the round. The coin toss sequence is captured by an unbounded fanin AVG gate. Deterministic computation by the verifier, as long as it does not refer to the input, can be "short-circuited" by the uniformity machine; the configuration c before the deterministic computation is a MAX gate with the configuration c' after the deterministic computation as its only input. However, if in the deterministic computation, the verifier accesses input bits, then this short-cutting is not possible, since the uniformity machine does not have access to input bits and so cannot compute c'. This is where the reduction steps in; for each pair of verifier configurations $\langle c, c' \rangle$, it outputs a bit $y_{c,c'}$, which is 1 iff the verifier can, from configuration c and using only deterministic moves (which may depend on the actual input), reach configuration c'. The uniformity machine can now construct a depth-two sub-circuit for the deterministic part of the verifier's computation: guess, through a MAX gate, the configuration c', and verify, by accessing the bit $y_{c,c'}$ through a READ gate, that the configuration is reachable. □

Lemma 2. *MAX,AVG($S(n), D(n)$)* $\subseteq$ *IPRDSP($D(n), \log(S(n))$).*

Theorem 3. *IPRDSP($r(n), s(n)$)* $\equiv_{DSPACE(s(n))}$ *MAX,AVG($2^{s(n)}, r(n)$).*

3   Arithmetizing the computations of ATMs

In this section we describe arithmetizations of ATISP($t(n), s(n)$) computations. To construct these, we borrow much of the technical machinery developed in [FL93]. However, the size of the arithmetizations in [FL93] is too high to obtain a characterization of $\mathcal{NC}$ (see Section 5). Exploiting the fact that the verifier in our case has random access to encoded input, we can obtain arithmetizations of significantly smaller size. In our proofs, we use the polynomial extension encoding scheme of [BFLS91]. Details of the constructions and proofs will appear in the full version.

Let $L \in \text{ATISP}(t(n), s(n))$ and let M be an ATM accepting L. We assume that M is in a normal form, described in [FL93]. Given an input x, we will construct an arithmetic expression whose value determines whether $x \in L$. Fix a finite field $\mathcal{F}$, a subset $\mathcal{H} \subset \mathcal{F}$ with $h := |\mathcal{H}|$, and a parameter t; the values $|\mathcal{F}|, h$, and, t will be specified later. In the following recursive definition for V_i, when $I_1, \ldots, I_t \in \mathcal{H}$, the t-tuple describes a configuration of M, the function Q defines the validity of M's move from I to I' after existential guess a followed by universal guess b, and R_x describes that M accepts in its final configuration.

$$
V_i(I_1, \ldots, I_t, x) = \begin{cases} R_x(I_1, \ldots, I_t) & \text{if } i = 0, \\ \coprod_{a=0}^{1} \prod_{b=0}^{1} \sum_{I' \in \mathcal{H}^t} Q(I_1, \ldots, I_t, I'_1, \ldots, I'_t, a, b) \cdot V_{i-1}(I'_1, \ldots, I'_t, x) \\ \qquad\qquad\qquad \text{otherwise.} \end{cases}
$$

(1)

where $\coprod_{x=0}^{1} p(x) := 1 - \prod_{x=0}^{1}(1 - p(x))$. For a fixed x, V_i maps $\mathcal{F}^t$ to $\mathcal{F}$.

Theorem 4. *Let $T := t(n)/2$, and V_i be defined by (1). If $I_{01} \ldots I_{0t} \in \mathcal{H}^t$ is the encoding of the initial configuration of M, then $V_T(I_{01}, \ldots, I_{0t}, x) = 1$ if M accepts x, and, 0 otherwise.*
Furthermore, the degree of each variable in Q and R_x is at most h.

The arithmetizations described above are of low degree but not of small size. Here *size* is the number of internal nodes in the "expression tree" of a formula. The *depth* of a formula is the length of the longest path from the root to a leaf in this tree. The formula size can be reduced if there is random access to encoded input. Accordingly, a different arithmetization is constructed below. The functions F and G_x have essentially the same meanings as Q and R_x above.

$$
\Phi_i(I_1, \ldots, I_t, x) = \begin{cases} G_x(I_1, \ldots, I_t) & \text{if } i = 0, \\ \coprod_{a=0}^{1} \prod_{b=0}^{1} \sum_{I' \in \mathcal{H}^t} F(I_1, \ldots, I_t, I'_1, \ldots, I'_t, a, b) \cdot \Phi_{i-1}(I'_1, \ldots, I'_t, x) \\ \qquad\qquad\qquad \text{otherwise.} \end{cases}
$$

(2)

Theorem 5. *Let $T := t(n)/2$ and Φ be defined by (2). If $I_{01} \ldots I_{0t} \in \mathcal{H}^t$ is the encoding of the initial configuration of M, then $\Phi_T(I_{01}, \ldots, I_{0t}, x) = 1$ if M accepts x, and, 0 otherwise.*
Furthermore, the degree of any variable in F is at most $O(h)$, $Size(F) = O(h^2 s^2(n))$, and $Depth(F) = O(\log s(n) + \log h)$; also, the degree of each variable in G_x is at most $O(\log n)$, $Size(G_x) = O(h^2 \log^2 n)$, and, $Depth(G_x) = O(\log h + \log \log n)$.

Some explanation as to the difference between (1) and (2) is in order here. In the interactive proof systems based on (1), the function $R_x(I_1, \ldots, I_t)$, for $I \in \mathcal{F}^t$, is evaluated, for a fixed x, by the reduction machine and is used to generate an input to the MAX, AVG circuit that is later simulated by an interactive proof system. On the other hand, in the interactive proof systems based on (2), the verifier is directly given random access to the polynomial extension code $\tilde{x}$ of x. The verifier uses the value of $\tilde{x}$ at a point in $\mathcal{F}^t$ to evaluate G_x. Since we want the verifier to run in polylog time, we need to ensure that the arithmetic expressions are small so they can be evaluated fast.

4 Fast logspace verifiers for $\mathcal{NC}$

In this section, we show a logspace reduction from $\mathcal{NC}$ languages to languages in IPTISP($polylog(n)$, $\log n$). Our main result is:

Theorem 6. $\mathcal{NC} \leq_{\mathcal{LOG}} IPTISP(log^{O(1)}(n), \log n)$.

First we observe that directly constructing an interactive proof system to verify the recurrence of Theorem 4 in the standard way, we get a protocol with polylog rounds, provided the parameter t in Theorem 4 is chosen carefully. Thus,

Theorem 7. $\mathcal{NC} \subseteq IPRDSP(log^{O(1)}(n), \log n)$.

However this construction will not directly give Theorem 6. For this, we first reduce $\mathcal{NC}$ to circuits over MAX and AVG gates. Then we follow the conversion of Lemma 2, and identify points where the reduction can be tightened to make the conversion more time-efficient.

Lemma 8. $\mathcal{NC}^k \leq_{\mathcal{LOG}} MAX,AVG(n^{O(1)}, \log^{k+1} n)$.

Proof Sketch: For simplicity of notation, let $(\mathcal{MAX}, \mathcal{AVG})^k$ denote the class MAX,AVG$(n^{O(1)}, \log^k n)$.

We use the ATM definition of $\mathcal{NC}$; $\mathcal{NC}^k = \text{ATISP}((\log n)^k, \log n)$. The ATM computation is arithmetized as in Theorem 4. The $(\mathcal{MAX}, \mathcal{AVG})^{k+1}$ circuit we construct will essentially simulate an LFKN-like protocol on recurrence (1) for verifying that $V_T(I_{01}, \ldots, I_{0t}, x) = 1$. Thus in the $(\mathcal{MAX}, \mathcal{AVG})^{k+1}$ circuit, gate labels have the following information: gate type (MAX or AVG), depth of polynomial (subscript i of formula V_i currently being checked), claimed value of polynomial currently being checked, values of variables which have been initialised, and in the case of an AVG gate, the coefficients of a univariate polynomial of degree h. The root gate of the MAX, AVG circuit has the information: gate type = MAX, depth = T, value = 1, and initialised variables $I_{01}, \ldots, I_{0t}$ where I_0 represents the initial ID of the ATM.

We only mention the points of difference from the standard protocol.

When the equality to be verified involves a computation of the function R, i.e. at a gate labelled $\langle MAX, 0, v, \alpha_1, \ldots, \alpha_t \rangle$ where the equality to be checked is $v = R(\alpha_1, \ldots, \alpha_t)$, the desired circuit connection is that the gate receives 1 if the equality is satisfied, and 0 otherwise. However to know what constant is to be fed to the gate, the uniformity machine will have to evaluate R, which requires access to the input bits. Since this is not possible, we again exploit the trick used in the proof of Lemma 1; push the difficult part of the computation into the reduction. Specifically, define $B(\alpha_1, \ldots, \alpha_t, v) = 1$ if $v = R(\alpha_1, \ldots, \alpha_t)$ and 0 otherwise. The reduction machine outputs the string of bits $B(\alpha_1, \ldots, \alpha_t, v)$ (one bit for each $\alpha_1, \ldots, \alpha_t, v \in \mathcal{F}$). These bits are the input bits for the $(\mathcal{MAX}, \mathcal{AVG})^{k+1}$ circuit we construct. Now the uniformity machine connects to the MAX gate a READ gate with the constant input 1 and the circuit input $B(\alpha_1, \ldots, \alpha_t, v)$.

We also need to choose the parameters t and $\mathcal{F}$. The polynomials checked at each stage are of degree h, so there is a potential error probability of $h/|\mathcal{F}|$. There are $T(t + 2)$ steps at which variables are randomly instantiated. So, to bound the overall error by $1/4$, $|\mathcal{F}| \in O(Tht)$ suffices.

The circuit size itself must be polynomial, so gate labels should be $O(\log n)$ bits wide. The labels require $O((t+h) \log |\mathcal{F}| + \log T)$ bits. Since $T \in O((\log n)^k)$, it can be ignored here. So we require $(t + h) \log |\mathcal{F}| \in O(\log n)$.

The two constraints are satisfied by choosing $t \in O(\frac{\log n}{\log \log n - \log \log \log n})$. This gives $w \in O(\log \log n - \log \log \log n)$ and $h \in O((\frac{\log n}{\log \log n})^{O(1)})$. Thus $|\mathcal{F}|$ is also polylog in n, and the depth of the circuit is $T(t + 2) \in O((\log n)^{k+1})$. $\square$

Sketch of Proof of Theorem 6: The circuits constructed in the above proof are uniform in the sense that a description of the circuit can be generated in space logarithmic in its size. There is another notion of uniformity that can be used for circuit classes. A class of circuits is said to be U_{DC}-uniform if the direct connection language L_{DC} of the circuit, comprising of descriptions of individual edges, can be *recognized* deterministically in time logarithmic in the circuit size [Ruz81]. It is known [Ruz81] that U_{DC}-uniform $\mathcal{NC}$ coincides with $\mathcal{NC}$.

The MAX, AVG circuit constructed in Lemma 8 is "almost" U_{DC}-uniform "almost everywhere". Let us qualify this. Most edges can be recognised in time polylog in circuit size, though not logarithmic. Hence "almost" uniform. An exception is recognising edges which involve a computation of the form $v' = v/Q(\alpha_1, \ldots, \alpha_{2t+2})$. Hence "almost everywhere". However, by modifying the proof in certain nontrivial ways (details appear in the full version), we can push the computation of these edges too into the logspace reduction. If this reduction is used in Lemma 8, and then Lemma 2 is applied, then the result follows. $\square$

5 Characterizing $\mathcal{NC}$ by $\mathcal{IPS}$

In this section, we give an exact characterization of $\mathcal{NC}$ in terms of interactive proof systems where the verifier is restricted to run in polylog time and logspace. To allow this situation, we will require that the verifier be given random access to *encoded* input. We justify this requirement by the fact that such an encoding of the input instance can be done in $\mathcal{NC}$. Furthermore, the verifier can probabilistically check in polylog time if what is presented to him is a valid encoding. Such an encoding scheme was utilized in [BFLS91] in the context of *transparent proofs*. This additional feature added to the earlier model of interactive proofs enables us to give an exact characterization of $\mathcal{NC}$. *Notice that the encoded input is provided to the verifier directly and is not part of the proof as in the model of transparent proofs.*

As in the previous section, we view $\mathcal{NC}$ computations as ATM computations. The arithmetization we use is described in Theorem 5. Using standard techniques, this results in an interactive proof system with a verifier of time-space complexity polynomially related to that of the ATM. Our protocol has the same overall structure as that of [FL93] except in places where the input is accessed. Hence, we do not explicitly describe this protocol; the reader is referred to [FL93] for details. We basically perform an LFKN-type protocol on recurrence (2) for Φ_i. Our protocol essentially differs from [FL93] when $i = 0$, the only place where the verifier needs access to the input. Our parameters (degrees of polynomials, size of the field, etc.,) are also slightly different.

In (2), if $i = 0$, then Φ expresses a **Read** move by M: $\Phi(y_1, \ldots, y_t) = G_x(y_1, \ldots, y_t)$. At this point in the protocol, by induction, the verifier has values $\alpha_1, \ldots, \alpha_t, \beta \in \mathcal{F}$, and he needs to check if $\beta = G_x(\alpha_1, \ldots, \alpha_t)$. Here the Verifier uses the arithmetic expression for G_x given by Theorem 5. During its evaluation, the verifier needs the value of the extension $\tilde{x}$ of x at a point $(c_1, \ldots, c_t) \in \mathcal{F}^t$. He gets this value by appealing to the self-correction features of the encoding of x. From [BFLS91, Theorem 4.6], we know that the verifier can get any such value in $polylog(n)$ time per point. This is the key to making our verifier run in $polylog(n)$ time.

An analysis of the protocol (keeping in mind the reduced size of G_x) yields:

Lemma 9. *Let $\epsilon > 0$ be any constant. Then, in the protocol above, the verifier runs in time $(t(n)s(n)\log n)^{O(1/\epsilon)}$ and in space $O(s(n)\log t(n)/\log\log n)$.*

Let $\gamma : \Sigma^* \longrightarrow \Sigma^*$ be the coding function given by [BFLS91, Theorem 4.6]. For a set $S \subseteq \Sigma^*$, let $\gamma(S)$ be the set of codewords corresponding to strings in S. Using Lemma 9 and the properties of γ, we get,

Theorem 10. *If $L \in ATISP(t(n), s(n))$,
then $\gamma(L) \in IPTISP((t(n)s(n)\log n)^{O(1/\epsilon)}, s(n)\log t(n)/\log\log n)$.*

Simulation in the other direction is given by the following theorem:

Theorem 11 (FL93). *$IPTISP(t(n), s(n)) \subseteq ATISP(t(n)\log t(n), s(n))$.*

Now, using Ruzzo's characterization [Ruz81] of $\mathcal{NC}$, we get our main theorem for this section:

Theorem 12. *Let L be any language and let γ be as above. Then*
$$L \text{ is in } \mathcal{NC} \text{ iff } \gamma(L) \text{ is in } IPTISP(\, polylog(n)\,, \log n).$$

Proof. From Theorem 10, if $L \in \mathcal{NC}$, then $\gamma(L) \in$ IPTISP($\,polylog(n)\,,\log n$). For the other direction: Given L, we can compute $\gamma(L)$ in $\mathcal{NC}$ since γ is computable in $\mathcal{NC}$. Now if $\gamma(L)$ is in IPTISP($\,polylog(n)\,,\log n$), then by Theorem 11, there is an ATM that recognizes $\gamma(L)$ and runs in time $polylog(n)$ and space $O(\log n)$, giving uniform $\mathcal{NC}$ circuits that recognize $\gamma(L)$. Combining these with the $\mathcal{NC}$ circuits that compute γ, we then get $\mathcal{NC}$ circuits to recognize L. □

6 AuxPDA logspace verifiers capture $\mathcal{NP}$

In this section, we give a new characterisation of $\mathcal{NP}$, in terms of an interactive proof system where the verifier is a $\mathcal{LOGCFL}$ machine with a restricted number of coin tosses. This strengthens some results of [CL92].

Theorem 13. $\mathcal{IPS}$ (public, logspace + PDA, log random bits) $= \mathcal{NP}$

Proof Sketch: $\supseteq$: This proof uses ideas similar to [CL92], except that the privacy of coin tosses is replaced by access to a pushdown store. An $\mathcal{NP}$ membership proof can be written as an array, and verification of the proof involves some local row-wise operations and some local column-wise operations. The prover sends the array twice, once in row-major order and once in column-major order. As the array is being sent, the verifier can do the local checks. But now the verifier must also check that the prover did not send two different arrays in the two phases. Here Lipton's fingerprinting technique [Lip90] is used; for each array, a small fingerprint which fits in log space is probabilistically generated. If the arrays are different, the fingerprints differ with high probability. For this to work, the coin tosses involved in fingerprinting the first array should not be known to the prover while he is sending the second array. So the verifier stacks the arrays on the pushdown, and does the fingerprinting while popping the stack. $\square$

The crucial idea used in the above proof is that parts of the verification can be deferred, and the data required for the deferred verification can be stored on the auxiliary storage. In the above case, deferring helped make public coins effectively private. It can also be used to bring down a polynomial time verification to a logspace verification, provided data is stacked in appropriate form.

In particular, considering the verifier in interactive proof system for $\mathcal{PSPACE}$ [LFKN92, Sha92, She92], we get the following result. The stack used by the verifier here is not strictly a pushdown; PUSH and POP are allowed only at top-of-stack, but internal stack contents can be read or scanned.

Theorem 14. $\mathcal{PSPACE}$ is exactly equal to the class of languages accepted by interactive proof systems where the verifier uses $\log n$ space and has access to a non-erasing stack. (Non-erasing means that there are no POP moves.) Further, the class of languages remains unchanged even if the verifier is restricted to change between pushdown mode and scan mode, on the stack, only once.

Acknowledgments

We are grateful to Jaikumar Radhakrishnan from the Tata Institute of Fundamental Research, Bombay, and to Laci Babai, Lance Fortnow, Ashish Naik, and Janos Simon from the University of Chicago for helpful comments on earlier versions of this work.

References

[Bab85]   L. Babai. Trading group theory for randomness. In *Proc. 17th STOC*, pp 421–429, 1985.

[BF93]　L. Babai and K. Friedl. On slightly superlinear transparent proofs. Technical Report CS-93-13, Department of Computer Science, University of Chicago, 1993.

[BFLS91]　L. Babai, L. Fortnow, L.A. Levin, and M. Szegedy. Checking computations in polylogarithmic time. In *Proc. 23rd STOC*, pp 21–31, 1991.

[BM88]　L. Babai and S. Moran. Arthur-Merlin games: a randomized proof system and a hierarchy of complexity classes. *J. Comp. Syst. Sci.*, 36:254–276, 1988.

[CL88]　A. Condon and R. Ladner. Probabilistic game automata. *J. Comp. Syst. Sci.*, 36(3):452–489, 1988.

[CL89]　A. Condon and R. Lipton. On the complexity of space-bounded interactive proofs. In *Proc. 30th FOCS*, pp 462–467, 1989.

[CL92]　A. Condon and R. Ladner. Interactive proof systems with polynomially bounded strategies. In *Proc. 7th Conference on Structure in Complexity Theory*, pp 282–294, 1992.

[Con91]　A. Condon. The complexity of the max word problem and the power of one-way interactive proof systems. In *Proc. 8th STACS*, pp 456–465, 1991. LNCS 480.

[FL93]　L. Fortnow and C. Lund. Interactive proof systems and alternating time-space complexity. *Theoretical Computer Science*, 113:55–73, 1993. also in Proc. 8th STACS 1991, LNCS 480.

[For89]　L. Fortnow. *Complexity-theoretic aspects of interactive proof systems.* PhD thesis, M. I. T., May 1989. Tech. Rep. MIT/LCS/TR-447.

[GMR85]　S. Goldwasser, S. Micali, and C. Rackoff. The knowledge complexity of interactive proof systems. In *Proc. 17th STOC*, pp 291–304, 1985. full version in SIAM J. Comput., Vol 18(1), pp 186–208.

[GS86]　S. Goldwasser and M. Sipser. Private coins versus public coins in interactive proof systems. In *Proc. 18th STOC*, 1986. also in *Advances in Computing Research 5: Randomness and Computation*, JAI Press, Greenwich, CT, 1989.

[JK89]　B. Jenner and B. Kersig. Characterizing the polynomial hierarchy by alternating auxiliary pushdown automata. *RAIRO Theoretical Informatics and Applications*, 23:93–99, 1989. also in Proc. STACS(1988), LNCS **294** 118–125.

[LFKN92]　C. Lund, L. Fortnow, H. Karloff, and N. Nisan. Algebraic methods for interactive proof systems. *J. ACM*, 39(4):859–868, October 1992. also in Proc. 31st FOCS 1990, pp 1–10.

[Lip90]　R. J. Lipton. Efficient checking of computations. In *Proc. 7th STACS*, pp 207–215, 1990. LNCS 415.

[Ruz81]　W.L. Ruzzo. On uniform circuit complexity. *J. Comput. Syst. Sci.*, 22:365–383, 1981.

[Sha92]　A. Shamir. IP = PSPACE. *J. ACM*, 39(4):869–877, 1992. also in Proc. 31st FOCS 1990, pp 11–15.

[She92]　A. Shen. IP = PSPACE: Simplified proof. *J. ACM*, 39(4):878–880, 1992.

[VC90]　V. Vinay and V. Chandru. The expressibility of nondeterministic auxiliary stack automata and its relation to treesize bounded alternating auxiliary pushdown automata. In *Proc. 10th FST & TCS*, pp 104–114, 1990. LNCS 472.

Structure in Average Case Complexity

Christoph Karg[*] and Rainer Schuler

Abteilung für Theoretische Informatik, Universität Ulm, D-89069 Ulm
{chkarg|schuler}@informatik.uni-ulm.de

Abstract. In 1990 Schapire gave an equivalent characterization of Levin's notion of functions, that are polynomial on average. This characterization gives a very smooth translation from worst case complexity to average case complexity of the notions for time and space complexity. We prove tight space and time hierarchy theorems and discuss the structure of deterministic and nondeterministic average case complexity classes.

1 Introduction

Despite having bad worst case behavior, many algorithms are frequently used in practice because they are efficient on the average. A well known example is the Simplex algorithm, a worst case exponential time algorithm for linear programming which performs well in practice, even better than worst case polynomial time algorithms for the same problem. It seems that the instances which cause the bad worst case complexity do not occur in practical applications. A similar example, within P, is the Quicksort algorithm. Even though the worst case is $O(n^2)$ for all (deterministic) implementations, the Quicksort algorithm is often used in practice, since its average case complexity is $O(n \log n)$. Thus, in some cases, the average case complexity of a problem is a better measure than its worst case complexity.

A general theory of average case complexity was introduced by Levin. He defined a robust notion of "functions are polynomial on average" with respect to a probability distribution on all instances [14]. Since then this notion has been considered by many researchers. In particular the notion of completeness under various reductions has been studied [15, 9, 2, 22, 16, 21]. The basic objects of average case complexity are randomized decision problems, pairs consisting of a decision (or search) problem and a probability distribution on the instances of the problem. An open problem, a generalization of the famous "P $\overset{?}{=}$ NP" question, is whether all sets in NP can be solved deterministically in polynomial time on average under all "natural" (or easy) probability distributions. Levin considered polynomial time computable distributions (as natural) [14, 15]. This notion seemed to be too restrictive and later the more general notion of polynomial time samplable distributions has been proposed in [2].

A first important connection between average case complexity and worst case complexity was established in [2]. Ben-David *et al.* show that if all problems in NP can be solved deterministically in polynomial time on average under every

---

[*] The research of this author was supported by the Deutsche Forschungsgemeinschaft, grant No. Schö 302/4-1.

polynomial time computable distribution, then deterministic linear-exponential time is equal to nondeterministic linear-exponential time. (This indicates that it is unlikely that all sets in NP are polynomial time solvable on average under polynomial time computable distributions.) However, the question whether the above assumption implies that the polynomial time hierarchy PH (or even P^{NP}) can be solved efficiently on average, is not clear [19].

As pointed out in [9], using Levin's definition it is reasonable to define the notion of "polynomial on average" and "linear on average". In this paper we generalize Levin's notion and define "g on average" for arbitrary functions g following a characterization of "polynomial on average" given in [17] (see also [20, 21, 11]). The time-bound g is here a two placed function, where the first argument is the length of the input and the second argument is the inverse of a probability weight. Now a function f (i.e. the time bound of a Turing machine) is said to be g on average if for every probability weight ε the probability over all x that $f(x)$ exceeds $g(|x|, 1/\varepsilon)$ is smaller than ε. Note that the notion of "polynomial on average" remains unchanged, but now it is possible to consider functions that are g on average but not $o(g)$ on average. This allows to incorporate results from "classical" average case complexity into the framework given by Levin.

We remark at this place, that different approaches to define "g on μ-average" have been suggested [16, 5]. However, these approaches define notions which are different to Levin's definition of "polynomial time on average".

The aim of our paper is to give a characterization of average case complexity classes that are similar in structure to worst case complexity and allow to precisely classify problems according to their average case complexity. The paper is organized as follows:

In **section 2** we recall the necessary notions and definitions of average case complexity and give a definition of "g on μ-average". In **section 3** we define deterministic time- and space-bounded average case complexity classes and give time and space hierarchy theorems, which are as tight as those, known from worst case complexity theory. It is possible to prove the hierarchy theorems w.r.t. an "easy" (polynomial-time computable) distribution.

In **section 4** we define nondeterministic time- and space-bounded average case complexity classes. We require that the time-bounds of nondeterministic Turing machines are exactly real time computable. That is, a function f is a time-bound if for all x, $0^{f(x)}$ can be computed by a deterministic transducer in time $O(f)$ (cf. [8]). We feel that this requirement is justified since otherwise nondeterministic polynomial time on average is not contained in any deterministic (on average) time-bounded class. A similar requirement is given for space-bounded computations.

Section 5 compares the structure of average case complexity classes with the structure in worst case complexity. We show that deterministic average polynomial time (AP) is equal to nondeterministic average polynomial time (ANP) if and only if P = NP. This equivalence is extended to exponential time classes. Furthermore we show a similar relation for nondeterministic polynomial time and deterministic exponential time. To prove these results we extend a proof from Wang and Belanger [22].

In **section 6** we study upward collapse properties of average case complexity classes. In worst case complexity theory it is known that if P = NP, then EXP = NEXP. We show, that the same relationship also holds in average case complexity. If AP is equal to ANP then deterministic average exponential-time (AEXP) is equal to nondeterministic average exponential time (ANEXP). Book showed, that there are tally sets in NP − P if and only if E ≠ NE [3]. We define a tally coding of randomized decision problems which allows us to extend this result to average case complexity.

2 Preliminaries

In this paper we use the standard notations and definitions of computational complexity theory (see for example [1]). For an introduction to average case complexity the reader is referred to [9].

Let $\Sigma = \{0, 1\}$ be fixed and Σ^* denote the set of all finite strings over Σ. For every $x \in \Sigma^*$, let $|x|$ denote the length of x. A set (language) is always a subset of Σ^*.

A total function μ from Σ^* to $[0, 1]$ is called a **probability function** (or **density function**), if $\sum_x \mu(x) = 1$. The **probability distribution** μ^* of μ is given by $\mu^*(x) = \sum_{y \leq x} \mu(y)$, for all x. We allow probability functions to converge to a constant $c \neq 1$. For a set $X \subseteq \Sigma^*$ let $\mu(X) = \sum_{x \in X} \mu(x)$.

For a Turing machine M let $M(x)$ denote the output of M on input x. M is said to accept x if $M(x) = 1$ and $L(M)$ denotes the set of strings accepted by M. A Turing machine M is said to accept a set (language) L if $L = L(M)$.

The **conditional probability function** μ_n of μ is defined as $\mu_n(x) = \mu(x)/\mu(\Sigma^n)$ if $|x| = n$ and $\mu(\Sigma^n) > 0$, and $\mu_n(x) = 0$ otherwise.

A distribution μ^* is **computable**, if there exists a transducer M, such that for every $x \in \Sigma^*$, $|\mu^*(x) - M(x, 0^k)| < 2^{-k}$ [13]. If M is polynomial time-bounded in $|x| + k$, then we say that μ is ptime-computable.

The **expected value** of a total function $f : \Sigma^* \to \mathbb{R}_0^+$ with respect to a probability function μ is $\mathrm{Exp}_\mu[f(x)] = \sum_{x \in \Sigma^+} \mu(x) \cdot f(x)$. Let A be a boolean predicate, i.e. a function from Σ^* to $\{0, 1\}$. The **probability** of A with respect to μ, denoted by $\mathrm{Prob}_\mu[A(x)]$, is equal to the expected value of A with respect to μ, i.e. $\mathrm{Prob}_\mu[A(x)] = \mathrm{Exp}_\mu[A(x)]$.

A **randomized decision problem** is a pair (D, μ) consisting of a recursive language D and a computable probability function μ. An **average case complexity class** is a set of randomized decision problems.

The foundation of average case complexity is Levin's notion of "functions are polynomial on average".

Definition 1 ([14]). A function f is **polynomial on average** with respect to a probability function μ (or short: polynomial on μ-average), if there exists a constant $0 < \varepsilon \leq 1$, such that $\sum_{x \in \Sigma^+} f(x)^\varepsilon \cdot \mu(x)/|x| < \infty$. If $\varepsilon = 1$ then f is called **linear on μ-average**.

The set of average polynomial functions has similar closure properties as the set of polynomials [14, 9]. In this paper we use an equivalent characterization of "polynomial on average" given by Schapire. In [17] he shows, that a function f is polynomial on μ-average (in the sense of Levin as defined above) if and only if there exists a polynomial p from $\mathbb{N} \times \mathbb{N}$ to $\mathbb{R}_0^+$, such that for all $m > 0$, $\text{Prob}_\mu[f(x) > p(|x|, m)] < 1/m$. Using Schapire's characterization it is possible to generalize the notion of "polynomial on average" to "g on average" for arbitrary functions g.

Definition 2 (g **on** μ-**average**). Let f be a total function from Σ^* to $\mathbb{R}_0^+$ and μ be a probability function. f is g on μ-average for a function g from $\mathbb{N} \times \mathbb{N}$ to $\mathbb{R}_0^+$, if for all $m > 0$ $\text{Prob}_\mu[f(x) > g(|x|, m)] < 1/m$.

The definition takes into account that the average case measure depends not only on the given instance x but also on the probability $\mu(x)$ of the occurrence of x. If the instance x does not appear, i. e. $\mu(x) = 0$, it has no affect on the average case analysis. Similar if for some string x, $\mu(x) > 0$ then $f(x) \leq g(|x|, \lceil 1/\mu(x) \rceil)$.

The following propositions are examples that the definition of "g on average" is reasonable to analyze the average case complexity of algorithms. First it is shown that if a function has uniform complexity (on strings of the same length), then worst case complexity is equal to average case complexity under any distribution.

Proposition 3. *Let* f, g *be functions from* $\mathbb{N}$ *to* $\mathbb{N}$ *and* $g'(n, m) = g(n) \cdot g(m)$.
- *If* $\forall x : f(x) = g(|x|)$ *then* f *is* g' *on* μ-*average for every* μ.
- *If* $\forall x : f(x) > g'(|x|, 1)$ *then* f *is not* g' *on* μ-*average for any* μ.

It is possible to show that Quicksort has average case complexity $(n \log n) \cdot (m \log m)$ under a standard distribution.

As input to Quicksort we consider permutations on an initial sequence of the natural numbers. For every $n_0 \in \mathbb{N}$ a standard distribution λ can be defined as $\lambda(x) = 1/((n - n_0)(n - n_0 + 1)n!)$ if x is a permutation of $(1, \ldots, n)$ and $n > n_0$ and $\lambda(x) = 0$ otherwise. Let $Q(x)$ denote the running time of the Quicksort algorithm on input x. Recall, that $\sum_{x \in \Sigma^n} Q(x) \lambda_n(x) = \Theta(n \log n)$.

Lemma 4. *Let* λ_n *denote the uniform distribution on permutations of* $(1, \ldots n)$. *There exist constants* c_1, c_2 *and* n_0, *such that for all* $n > n_0$
(1) $\forall m > 0 :$ $\text{Prob}_{\lambda_n}[Q(x) > c_1 \, m \, n \log n] < 1/m$,
(2) $\text{Prob}_{\lambda_n}[Q(x) > (n \log n)/c_2] > 1/c_2$.

Proposition 5. *Let* $g(n, m) = (n \log n) \cdot (m \log m)$. *The Quicksort algorithm is* $c \cdot g$ *on* λ-*average but not* g/c *on* λ-*average, for some constant* c *and a standard distribution* λ.

Definition 6. Let f by a function from $\mathbb{N}$ to $\mathbb{R}_0^+$. f is called
- linear on μ-average, if $\exists c, d \in \mathbb{R}_0^+$, such that f is $c \cdot n \cdot m + d$ on μ-average.
- polynomial on μ-average, if there exists a polynomial $p : \mathbb{N} \times \mathbb{N} \to \mathbb{R}_0^+$ such that f is p on μ-average.
- linear-exponential on μ-average, if $\exists c, d \in \mathbb{R}_0^+$, such that f is $2^{c \cdot n \cdot m + d}$ on μ-average.
- exponential on μ-average, if there exists a polynomial $p : \mathbb{N} \times \mathbb{N} \to \mathbb{R}_0^+$ such that f is 2^p on μ-average.

As mentioned above, the set of polynomial on average functions is closed under addition, multiplication and exponentiation with a constant [14, 9]. The same also holds for the set of linear-exponential (exponential, resp.) on average functions.

3 Deterministic average case complexity

3.1 Time-bounded computations. For a deterministic Turing machine M, let $time_M(x)$ denote the length of the computation path (the number of transition steps) of M on input x. If the computation of M terminates on every input then $time_M(x)$ is total. We consider only Turing machines that terminate on every input.

A deterministic Turing machine M is polynomial on μ-average time-bounded, if $time_M(x)$ is polynomial on μ-average. This leads to the definition of the class AP, the set of problems that are solvable in polynomial time on average [14, 9, 2, 22].

Definition 7 (AP). A randomized decision problem (D, μ) is in AP if D is accepted by a deterministic Turing machine, which is polynomial on μ-average time-bounded.

We remark here that there are sets D not in P that are polynomial time solvable on μ-average for every ptime-computable distribution μ, i.e. $(D, \mu) \in$ AP. In general a deterministic Turing machine M is g on μ-average time-bounded, if $time_M(x)$ is g on μ-average. This notion is used to define deterministic average case time-bounded classes.

Definition 8 (ADTIME(g)). Let g be a function from $\mathbb{N} \times \mathbb{N}$ to $\mathbb{N}$. The randomized decision problem (D, μ) is in the class ADTIME(g) if D is accepted by a deterministic Turing machine, which is g on μ-average time-bounded.

Note that the function g in the above definition is *not* the (worst case) time-bound of the Turing machine M. If μ is a positive density function and $g(|x|, h(|x|))$ is time-constructible, where $h(n) \geq \max_{|x|=n}(\lceil 1/\mu(x) \rceil)$, then for every D with $(D, \mu) \in$ ADTIME(g) it holds, that $D \in$ DTIME($g(|x|, h(|x|))$).

Proposition 9. AP $= \bigcup_{k>0}$ ADTIME($n^k m^k + k$).

Definition 10 (AE,AEXP). A randomized decision problem (D, μ) is in AE (AEXP, resp.) if D is accepted by a deterministic Turing machine, which is linear-exponential (exponential, resp.) on μ-average time-bounded.

We get AE $= \bigcup_{k>0}$ ADTIME($2^{k \cdot n \cdot m + k}$) and AEXP $= \bigcup_{k>0}$ ADTIME($2^{n^k m^k + k}$).

3.2 Space-bounded computations. Let $space_M(x)$ denote the number of tape cells used by the Turing machine M on input x. We say, the Turing machine M is g on μ-average space-bounded, if $space_M(x)$ is g on μ-average. This leads to the definition of ADSPACE(g).

Definition 11 (ADSPACE(g)). Let g be a total function from $\mathbb{N} \times \mathbb{N}$ to $\mathbb{N}$. The randomized decision problem (D, μ) is in the class ADSPACE(g) if D is accepted by a deterministic Turing machine, which is g on μ-average space-bounded.

We are interested especially in randomized decision problems which can be computed by polynomial on the average space-bounded Turing machines. We denote the set of these problems by APSPACE, formally defined as APSPACE $= \bigcup_{k>0} \text{ADSPACE}(n^k m^k + k)$.

3.3 Hierarchy theorems. One major result in computational complexity theory was the proof of tight time and space hierarchy theorems. In this section we give similar hierarchy theorems for average case complexity. We define a ptime-computable distribution μ and show that there exists a tight hierarchy of tally-sets under the distribution μ. A different approach was followed in [7]. Grape *et al.* show that there exists a hierarchy of languages that are hard to compute on all but polynomial many strings for each length.

Theorem 12 (Time Hierarchy Theorem). *Let f and g be total functions from $\mathbb{N} \times \mathbb{N}$ to $\mathbb{N}$. If $f(n, m) \geq \log n + \log m$ and $\lim_{n \to \infty} \frac{g(n,n) \log g(n,n)}{f(n,n)} = 0$ then $\text{ADTIME}(g) \subsetneq \text{ADTIME}(f)$.*

Proof. Define the function $h : \mathbb{N} \to \mathbb{N}$ as $h(1) = 4$ and $h(k) = 2^{h(k-1)}$, if $k > 1$. Let $S = \{0^{h(k)} \mid k \geq 1\}$. The density function μ over S is defined as $\mu(0^{h(1)}) = 7/8$ and $\mu(0^{h(k)}) = \sum_{i=h(k-1)}^{h(k)-1} 2^{-i}$. For all $x \notin \{0^{h(k)} \mid k \geq 1\}$, $\mu(x)$ is equal to 0. The functions h and μ provide the following property

$$1 - \sum_{i=1}^{k} \mu(0^{h(i)}) < \frac{1}{h(k)} < \mu(0^{h(k)}). \tag{1}$$

Claim 13. *Let $T \subseteq S$. If $T \in \text{DTIME}(f(n,n))$ then $(T, \mu) \in \text{ADTIME}(f)$.*

Proof. Let M be a $f(n,n)$ time-bounded deterministic Turing machine with $L(M) = T$. For an arbitrary integer $m > 0$ let $k_m = \max\{i \in \mathbb{N} \mid h(i) \leq m\}$. Then for all $i \leq k_m$ we get $time_M(0^{h(i)}) \leq f(h(i), h(i)) \leq f(h(i), m)$. We can estimate the probability, that M on input x needs more than $f(|x|, m)$ steps, as:

$$\text{Prob}_\mu[time_M(0^n) > f(n, m)] \leq 1 - \sum_{i=1}^{k} \mu(0^{h(i)}) < \frac{1}{h(k)} \leq \frac{1}{m}. \qquad \square$$

Claim 14. *Let $T \subseteq S$. If $T \notin \text{DTIME}(g(n,n))$ then $(T, \mu) \notin \text{ADTIME}(g)$.*

Claim 15. *For some $T \subseteq S$, $T \in \text{DTIME}(f(n,n)) - \text{DTIME}(g(n,n))$.* $\qquad \square$

The fact $f(n, n) = \omega(g(n, n) \log g(n, n))$ is only needed in Claim 15 to simulate an arbitrary k-tape machine by a fixed 2-tape machine. For details see [10].

Corollary 16. AP $\subsetneq$ AE $\subsetneq$ AEXP.

Corollary 17. *For every integer $k > 1$ and any $\varepsilon > 0$ it holds*

$$\text{ADTIME}(n^k m^k) \subsetneq \text{ADTIME}(n^{k+\varepsilon} m^{k+\varepsilon}).$$

The hierarchy theorem for space-bounded computations is as follows:

Theorem 18 (Space Hierarchy Theorem). *Let f and g be total functions from $\mathbb{N} \times \mathbb{N}$ to $\mathbb{N}$. If $\lim_{n \to \infty} \frac{g(n,n)}{f(n,n)} = 0$ then $\text{ADSPACE}(g) \subsetneq \text{ADSPACE}(f)$.*

4 Nondeterministic average case complexity

Let $time_N(x)$, the running time of the nondeterministic Turing machine N on input x, be the length of the shortest accepting computation path if N accepts x, and 1 otherwise. Similar, the space needed by the nondeterministic Turing machine N on input x, denoted by $space_N(x)$, is the smallest number of tape cells on any accepting computation path, if N accepts x, and 1 otherwise.

4.1 Computable boundaries. In analogue to [6] we say, a total function $f : \Sigma^* \rightarrow \mathbb{N}$ is exactly real time computable if there exist a constant c and a deterministic Turing machine which for every input x halts within $c \cdot f(x)$ steps with a string of length $f(x)$ on its output tape.

Note that f depends on x itself and not on the length of x as in Cook's definition of real time computable functions.

Let g be a total function from $\mathbb{N} \times \mathbb{N} \rightarrow \mathbb{R}_0^+$. A nondeterministic Turing machine N is g on the μ-average time-bounded, if there exists an exactly real time computable function f, such that
- $f(x)$ is g on μ-average and
- $time_N(x) \leq f(x)$ for all $x \in L(N)$.

If g is a polynomial, we say N is polynomial on μ-average time-bounded (similar exponential, linear-exponential, resp.). A similar definition was proposed in [8].

Forcing the real time computability of the time-bound enables us to simulate nondeterministic computations in a deterministic manner by using f as a clock and checking each possible computation path of N on input x for acceptance. For simplicity we identify $time_N(x)$ with $f(x)$ and assume without loss of generality, that every computation of N on input x halts after exactly $f(x)$ steps.

The term g on the μ-average space-bounded is defined similarly for nondeterministic space-bounded computations.

Definition 19 (ANTIME(g), ANSPACE(g)). Let g be a total function from $\mathbb{N} \times \mathbb{N}$ to $\mathbb{N}$. The randomized decision problem (D, μ) is in ANTIME(g) (ANSPACE(g)) if D is accepted by a nondeterministic Turing machine, which is g on μ-average time-bounded (space-bounded, resp.).

We define the following nondeterministic average case complexity classes: ANP $= \bigcup_{k>0}$ ANTIME($n^k m^k + k$), ANE $= \bigcup_{k>0}$ ANTIME($2^{k \cdot n \cdot m + k}$), ANEXP $= \bigcup_{k>0}$ ANTIME($2^{n^k m^k + k}$) and ANPSPACE $= \bigcup_{k>0}$ ANSPACE($n^k m^k + k$).

We note here that in contrast to worst case complexity the requirement that the time-bounds are exactly real time computable is restrictive. There exist a language D and a density function μ, such that D is accepted by a nondeterministic Turing machine where its running time is polynomial on μ-average, but there exists no nondeterministic machine for D, where its running time is bounded by a function, that is exactly real time computable and polynomial on μ-average.

Denote with AverageNP the set of randomized decision problems (D, μ), where D is accepted by a nondeterministic Turing machine N whose running time $time_N(x)$ is polynomial on μ-average (but not necessarily polynomial time-bounded on μ-average as defined above) [22]. This class contains problems, which are not computable by any nondeterministic polynomial on average time-bounded machine.

Theorem 20. ANP $\subsetneq$ AverageNP.

Furthermore it holds that AverageNP is not contained in ADTIME(g) for any fixed (recursive) function g.

4.2 Relations between average case measures. Next we state some relations between average case time and space complexity measures. It is not surprising that these relations are the same as in worst case complexity.

Lemma 21. *Let g be a total function from $\mathbb{N} \times \mathbb{N}$ to $\mathbb{N}$.*

(1) ADTIME(g) $\subseteq$ ADSPACE(g), ANTIME(g) $\subseteq$ ANSPACE(g).
(2) ADTIME(g) $=$ co-ADTIME(g), ADSPACE(g) $=$ co-ADSPACE(g).
(3) There exists a constant $c > 0$, such that ANTIME(g) $\subseteq$ ADTIME(c^g).
(4) There exists a constant $c > 0$, such that ADSPACE(g) $\subseteq$ ADTIME(c^g).
(5) ANSPACE(g) $\subseteq$ ADSPACE(g^2).
(6) If $g(n, m) \geq \log n + \log m$ then ANSPACE(g) $=$ co-ANSPACE(g).

Theorem 22. *(1) ANP $\subseteq$ AEXP.*
(2) APSPACE $=$ ANPSPACE $=$ co-ANPSPACE.
(3) APSPACE $\subseteq$ AEXP.

5 Relations to worst case complexity

Now we give some (fundamental) relations between worst case and average case complexity classes. We compare deterministic and nondeterministic average case complexity classes with their worst case counterparts. We show that a collapse of a deterministic and a nondeterministic class in the average case is equivalent to a collapse in the worst case.

Proposition 23. *Let $C \in \{P, E, EXP, NP, NE, NEXP, PSPACE\}$. Then for all $D \in C$ and for any computable μ it holds that $(D, \mu) \in AC$.*

It is shown in [22], that if P $\neq$ NP then AP $\neq$ ANP. Wang & Belanger use in their proof (polynomial) complexity cores, which exist for NP-sets, if P $\neq$ NP. We show the converse direction by using explicitly the exactly real time computability of nondeterministic time-bounds. Furthermore we extend these results to the exponential time average case classes.

In the proof we use generalized complexity cores, which are studied in [4]. Let C be a class of languages and for any language A let C_A be the set $\{C \in C \mid C \subseteq A\}$. A set H is a **complexity core** (or hard core) for A with respect to C if for every $C \in C_A$ the intersection $C \cap H$ is a finite set. If $H \subseteq A$, then H is called a **proper complexity core** for A with respect to C.

Theorem 24 (Theorem 2.10 in [4]). *Let C be a recursively enumerable class of recursive sets that is closed under finite union and finite variation. Any infinite recursive set not in C has an infinite recursive proper complexity core with respect to C.*

Under the assumption P $\neq$ NP there exists a proper complexity core for SAT (or any NP-complete language) with respect to P [4]. Assigning high probability to the elements of the core we can construct a randomized decision problem in ANP which is not in AP. The same argument holds for exponential and linear-exponential time.

Theorem 25. *(1)* $P = NP$ *if and only if* $AP = ANP$.
(2) $E = NE$ *if and only if* $AE = ANE$.
(3) $EXP = NEXP$ *if and only if* $AEXP = ANEXP$.

The next result links the assumption that ANP is included in AE to worst case complexity.

Theorem 26. $NP \subseteq E$ *if and only if* $ANP \subseteq AE$.

6 Upward collapse properties and tally sets

A well known structural result in worst case complexity is the upward collapse property, that is, a collapse of some smaller deterministic and nondeterministic time-bounded classes implies a collapse of the respectively (exponentially) larger time-bounded classes. In this section we show that the upward collapse properties also hold in average case complexity. In particular the implication "If $P = NP$ then $EXP = NEXP$" similarly holds in average case complexity.

Theorem 27. *(1)* *If* $AP = ANP$ *then* $AE = ANE$.
(2) *If* $AP = ANP$ *then* $AEXP = ANEXP$.

Another well known result is from Book. In [3] he shows that $E \neq NE$ if and only if there exists a tally set in $NP - P$. Here we give an average case analogue of this theorem. Interestingly this leads to weaker exponential time on average classes called APE in the deterministic and ANPE in the nondeterministic case. Recall that the average case time-bound is a function in the length of the input and the probability weight of the strings which exceed the time-bound. For example in the definition of AP, to achieve a probability weight smaller than $1/m$ the time-bound has to be a function polynomial in m and in the length n of the input. Similar, for AE the function is allowed to be linear-exponential in m and n. In the case of APE (and ANPE) the time-bound is linear-exponential in n but polynomial in m.

Definition 28 (APE, ANPE).

$$APE = \bigcup_{k>0} ADTIME(2^{k \cdot n} m^k + k).$$
$$ANPE = \bigcup_{k>0} ANTIME(2^{k \cdot n} m^k + k).$$

Using the hierarchy theorem, we get immediately $APE \subsetneq AE$.

Theorem 29. $E = NE$ *if and only if* $APE = ANPE$.

Corollary 30. $AE = ANE$ *if and only if* $APE = ANPE$.

The tally encoding of the randomized decision problem (D, μ) is a randomized decision problem (D', μ') where $D' = \{0^x \mid x \in D\}$ and $\mu'(y) = \mu(x)$ if $y = 0^x$ for some $x \in \Sigma^*$ and $\mu'(y) = 0$ otherwise. This definition is based on the assumption that the tally encoding 0^x of an instance x occurs with the same probability as x itself. So we can use transformations between tally and binary encoding without changing the underlying probability distribution. We denote the tally encoding of (D, μ) with $tally(D, \mu)$. We say, (D, μ) is a randomized tally decision problem, if $D \subseteq \{0\}^*$ and $\mu(x) = 0$ if $x \notin \{0\}^*$.

Lemma 31. *Let* (D, μ) *be any randomized decision problem.*
(1) $(D, \mu) \in$ APE *if and only if* $tally(D, \mu) \in$ AP.
(2) $(D, \mu) \in$ ANPE *if and only if* $tally(D, \mu) \in$ ANP.

Theorem 32. AE $\neq$ ANE *if and only if there exists a randomized tally decision problem in* ANP $-$ AP.

References

1. J.L. Balcázar, J. Díaz, and J. Gabarró. *Structural Complexity I*. Springer, 1988.
2. S. Ben-David, B. Chor, O. Goldreich, and M. Luby. On the theory of average case complexity. *JCSS*, 44(2):193–219, 1992.
3. R.V. Book. Tally languages and complexity classes. *Inf. and Control*, 26:281–287, 1974.
4. R.V. Book and D. Du. The existence and density of generalized complexity cores. *J. ACM*, 34(3):718–730, 1987.
5. J. Cai and A. Selman. Average time complexity classes. Technical Report TR95-019, ECCC Trier, 1995.
6. S.A. Cook. A hierarchy for nondeterministic time complexity. *JCSS*, 7:343–353, 1973.
7. M. Goldmann, P. Grape, and J. Håstad. On average time hierarchies. *IPL*, 49:15–20, 1994.
8. O. Goldreich. Towards a theory of average case complexity. Technical report, Technion Haifa, Israel, 1988.
9. Y. Gurevich. Average case complexity. *JCSS*, 42(3):346–398, 1991.
10. F. C. Hennie and R. E. Stearns. Two-tape simulation of multitape turing machines. *J. ACM*, 13(4):533–546, 1966.
11. R. Impagliazzo. A personal view of average-case complexity. In *Proc. 10th STRUCTURE*, pages 134–147, 1995.
12. C. Karg. Strukturfragen im Umfeld der Durchschnittskomplexität. Master's thesis, Universität Ulm, 1994.
13. K.I. Ko and H. Friedman. Computational complexity of real functions. *TCS*, 20:323–352, 1982.
14. L. Levin. Problems, complete in "average" instance. In *Proc. 16th STOC*, page 465, 1984.
15. L. Levin. Average case complete problems. *SIAM J. Comput.*, 15:285–286, 1986.
16. R. Reischuk and C. Schindelhauer. Precise average case complexity. In *Proc. 10th STACS*, 1993.
17. R.E. Schapire. The emerging theory of average case complexity. Technical Report 431, MIT, 1990.
18. R. Schuler. Some properties of sets tractable under every polynomial-time computable distribution. *IPL*, 1995.
19. R. Schuler and O. Watanabe. Towards average-case complexity analysis of NP optimization problems. In *Proc. 10th STRUCTURE*, pages 148–159, 1995.
20. R. Schuler and T. Yamakami. Structural average case complexity. In *Proc. 12th FST&TCS LNCS 652*, pages 128–139, 1992.
21. R. Schuler and T. Yamakami. Sets computable in polynomial time on average. In *Proc. 1st International Computing and Combinatorics Conference*, 1995.
22. J. Wang and J. Belanger. On average P vs. average NP. In K. Ambos-Spies, S. Homer, and U. Schöning, editors, *Complexity Theory—Current Research*. Cambridge University Press, 1993.

Some Geometric Lower Bounds

Hank Chien[1] * and William Steiger[2] ** ***

[1] Computer Science, Harvard University, Cambridge, MA
[2] Computer Science, Rutgers University, New Brunswick, NJ

Abstract. Dobkin and Lipton introduced the connected components argument to prove lower bounds in the linear decision tree model for membership problems, for example the element uniqueness problem. In this paper we apply the same idea to obtain lower bound statements for a variety of problems, each having the flavor of element uniqueness. In fact one of these problems is a parametric version of element uniqueness which asks, given n inputs $a_1, \ldots, a_n$ and a query $x \geq 0$, whether there is a pair of inputs satisfying $|a_i - a_j| = x$; the case $x = 0$ IS element uniqueness. Then we apply some of these results to establish the fact that "search can be easier than uniqueness"; specifically we give two examples (one is the planar ham-sandwich cut) where finding or constructing a geometric object - known to exist - is less complex than answering the question about whether that object is unique. Finally we apply some of these results, along with a reduction argument, to get a nontrivial lower bound for the complexity of the least median of squares regression problem in the plane.

1 Introduction and Summary

Given inputs $a_1, \ldots, a_n$, the element uniqueness question is to decide whether or not the inputs are distinct. This is really a *membership problem* asking whether $\underline{a} = (a_1, \ldots, a_n) \in S_n$, the set of points in R^n with distinct coordinates. Dobkin and Lipton [4] showed that every linear decision tree that solves the membership problem for a set S must have depth $\Omega(\log(C(S)))$, where $C(S)$ denotes the number of path-connected components of S. In the element uniqueness problem, $C(S_n)$ is easily seen to be at least $n!$, showing the obvious $O(n \log n)$ algorithm to be optimal. More importantly, it established a method of utilizing topological arguments in deducing lower bounds on the complexity of computation. This point was emphasized by Steele and Yao [15] who extended the Dobkin-Lipton lower bound for membership problems to algebraic decision trees of fixed degree and Ben-Or [1] was able to improve these bounds. Björner, Lovász, and Yao [2] used more refined topological properties to obtain lower bounds for membership

---

* Research Supported by a Research Experiences for Undergraduates (REU) supplement to the NSF center grant to DIMACS.
** Research Supported in Part by NSF grant CCR-9111491.
*** The author expresses gratitude to the NSF DIMACS Center at Rutgers.

problems whose sets are not covered by the previous methods, for example the k-equal-inputs problem.

In the present paper we prove lower bound statements for a variety of problems, each having the flavor of element uniqueness. Many of the familiar lower bounds in computational geometry (e.g. convex hull [5], slope selection [3]) depend on reduction to sorting, or to element uniqueness. Some of the results given here seem to require direct topological reasoning. Although we only use the simple connected components arguments of Dobkin and Lipton, we still think that these lower bounds are interesting in themselves and that they may be of some use. Because of space limitations, some of the details will be left for the final paper.

In Section 2 we establish some preliminary results that are either used later or are of independent interest. One is for a parametric version of element uniqueness, PEU. Given $a_1, \ldots, a_n$, and $x \geq 0$ the problem is to decide whether there is a pair $i \neq j$ for which $|a_i - a_j| = x$. Element uniqueness is the case $x = 0$ which has complexity $\Theta(n \log n)$. Notice that if $x \geq \max(a_i) - \min(a_i)$, then we may discover that fact and then answer the query in linear time. Similarly if x is greater than all but a constant number of the absolute differences. On the other hand in Section 2 we prove

Theorem 1 *Suppose that for some constant $c > 0$*

$$\#\{i < j : x \leq |a_i - a_j|\} \geq c\binom{n}{2}. \tag{1}$$

Then any decision tree for PEU must have depth $\Omega(n \log n)$.

It is interesting to ask about possible super-linear lower bounds $g(n)$ for PEU if c in (1) is replaced by a sequence $c_n \to 0$.

Section 2 also has lower bounds for two other comparison tasks, the k-gap ranking and selection problem, and the even/odd partition problem. In addition there are lower bounds for two problems about line arrangements. Given n lines $\ell_1, \ldots, \ell_n$ in general position in the plane, the jth level of the arrangement is the set of points λ_j on some line of the arrangement, and which have $j - 1$ lines above them (w.r.t. y-coordinate). A natural question that arises in many geometric algorithms is whether a given line ℓ_i is ever part of the jth level. We prove

Lemma 3: *The level membership query "$\ell_i \in \lambda_j$?" has complexity $\Omega(n \log n)$.*

Next, given a set S of n elements, we consider search problems that seek an element $x \in S$ which has property $\mathcal{P}$. It is known in advance in these problems that $|\{x \in S : x \text{ has } \mathcal{P}\}| \geq 1$. The uniqueness question asks whether $|\{x \in S : x \text{ has } \mathcal{P}\}| \leq 1$. We establish lower bounds for the uniqueness problem that exceed upper bounds on the search problem. This separation justifies the title of Section 3: "Search Can be Easier Than Uniqueness". Perhaps the most interesting example of this phenomenon concerns ham-sandwich cuts in R^2. Given n red points $R = \{r_1, \ldots, r_n\}$ and m blue points $B = \{b_1, \ldots, b_m\}$

in R^2 (w.l.o.g both m and n are taken as odd) a *ham-sandwich cut* is a pair r_i, b_j such that the line incident to them has equal numbers of red points on each side and also equal numbers of blue points on each side. The ham-sandwich theorem [5] states that the set H of ham-sandwich cuts in $R \times B$ is not empty. Lo and Steiger, by giving an optimal algorithm, proved that the search problem for $R \times B$ is linear.

Proposition 1 *(Lo and Steiger [8]) The complexity of finding a ham sandwich cut is $\Theta(m + n)$.*

In Section 3 we show that the uniqueness question is more difficult. Assuming w.l.o.g. that $m \leq n$

Theorem 2 *The complexity of deciding if $|H| > 1$ is $\Omega(n \log n)$.*

In fact even knowing $|H| > 1$ we can give an $\Omega(n \log n)$ lower bound for finding a second ham-sandwich cut. The last three statements all pertain to the unit cost RAM model where each arithmetic operation and each binary comparison has cost $= 1$.

Section 4 considers the least median of squares (LMS) regression problem. Let $P_i = (x_i, y_i)$, $i = 1, \ldots, n$, be n given data points in the plane. The (LMS) regression problem asks for the minimizer (m^*, b^*) of the function

$$g(m, b) = \text{median}\{|y_i - (mx_i + b)|, i = 1, \ldots, n\}; \qquad (2)$$

this function measures how well the points are fit by the line $y = mx + b$. The minimizer defines the line $y = m^*x + b^*$ that "best" fits the points according to the criterion in (2). Proposed by Rousseeuw [11], it is important in robust statistics because of its high (50%) breakdown point: Informally, in order to make an arbitrarily large change in (m^*, b^*), at least half of the data points must be perturbed.

Steele and Steiger [14] showed that least median of squares regression is really a discrete optimization problem and they gave an $O(n^3)$ algorithm to compute it. Souvaine and Steele [13] improved this to an $O(n^2)$ time, $O(n^2)$ space algorithm and finally, Edelsbrunner and Souvaine [7], using a clever adaptation of topological sweep, gave an $O(n^2)$ time, $O(n)$ space algorithm to compute the LMS line.

Despite the fact that (1) Edelsbrunner and Souvaine asked for a matching $\Omega(n^2)$ lower bound and (2) Rousseeuw and Leroy [12] actually conjectured that the Souvaine-Steele algorithm had optimal time complexity, no non-trivial lower bound was known. Here we can only prove

Theorem 3 *The RAM cost to find the LMS regression line for n points is $\Omega(n \log n)$.*

As we point out in Section 4 there are some compelling reasons to believe that $\Omega(n^{1+\varepsilon})$ steps are necessary to compute the LMS line.

2 Preliminary Results

In this section we prove some simple results which we think are interesting in their own right and which may be used later on. The first concerns the even/odd partition problem in which an input $\underline{a} = (a_1, \ldots, a_n)$ is given, $n = 2m$ (even). The task is to find (but not sort) the elements of odd rank. This problem arose in trying to prove Theorem 3.

If we knew the permutation π for which $a_{\pi_1} \leq \cdots \leq a_{\pi_n}$, then the set $O = \{a_{\pi_1}, a_{\pi_3}, \ldots, a_{2m-1}\}$ with odd ranks would give the partition we seek, so $O(n \log n)$ is clearly an upper bound. Since there are fewer than 2^n partitions, the information theory lower bound for this task is only linear. Nevertheless

Lemma 1. *Every decision tree for the even/odd partition problem has depth at least $\Omega(n \log n)$.*

Proof: Let $\underline{a} = (a_1, \ldots, a_n)$ be an input with distinct coordinates. Each node in a decision tree asks about the sign of a linear function of the a_i's. Therefore each leaf of the tree is reached by inputs that are in the same intersection of halfspaces (i.e., convex set). Each leaf of the tree returns a set of $n/2$ indices $1 \leq j_1 < j_2 < \cdots < j_{n/2} \leq n$ pointing to the elements of odd rank. We claim that inputs that are sorted by permutations $\pi \neq \rho$, both having $j_1, \ldots, j_{n/2}$ as their odd (or even) entries, *must* go to different leaves. If not, there are points $p \neq q$, respectively sorted (component-wise) by permutations π and ρ, that are in the same connected component of R^n. Let $\underline{\ell}(t) = p + t(q - p)$ be a point on the line joining p and q and let $\underline{\sigma}(t)$ be the permutation that sorts the coordinates of $\underline{\ell}(t)$. As t varies from 0 to 1, $\underline{\sigma}(t)$ changes from π to ρ via a sequence of at most $n^2/2$ transpositions. Immediately after the first one, say at t_1, $\underline{\ell}(t_1^+)$ is a point for which $j_1, \ldots, j_{n/2}$ is not the set of odd (or the even) entries of $\underline{\sigma}(t_1^+)$, and in fact there are infinitely many such points (e.g., if there is another transposition at $t_2 > t_1$, all points $\underline{\ell}(t)$, $t \in (t_1, t_2)$ have different even/odd partitions than $\underline{p}$). $\square$

Next we consider ranking and selection of k-gaps, problems that arose naturally in proving Theorem 3. Given $\underline{a} = (a_1, \ldots, a_n)$, a_i, a_j form a k-gap if $|\text{Rank}(a_i) - \text{Rank}(a_j)| = k$; its size is $\sigma_{ij} = |a_i - a_j|$. Any of the $n - k$ k-gaps can be found in linear time and clearly, all can be found and sorted in time $O(n \log n)$. If there are a linear number of them, this is the optimal complexity for ranking and selection:

Lemma 2. *If $k \leq cn$ for some constant $c < 1$, the depth of a decision tree to rank or select k-gaps is $\Omega(n \log n)$.*

Proof: First we show that deciding whether the smallest k-gap is unique is easier than four selections and some linear-time computations. An algorithm that selects the smallest k-gap returns integers i, j which signify that ρ_i and ρ_j, the ranks of a_i and a_j, satisfy $|\rho_i - \rho_j| = k$ and that the other $n - k - 1$ k-gaps have size at least $\sigma = |a_i - a_j|$. Consider $a_i < a_j$, the other possibility being similar. σ is unique if and only if (i) the minimum k-gap (if any) of $\{a_r : a_r \leq a_i\}$

is larger than σ; (ii) the minimum k-gap (if any) of $\{a_r : a_r \geq a_j\}$ is larger than σ; (iii) the minimum (k-1)-gap of $\{a_r : a_r \neq a_i$ and $\rho_i - k < \text{Rank}(a_r) < \rho_i + k\}$ is larger than σ; (iv) the minimum (k-1)-gap of $\{a_r : a_r \neq a_j$ and $\rho_j - k < \text{Rank}(a_r) < \rho_j + k\}$ is larger than σ. The four sets can be computed in linear time after which, the four selections are enough to answer the uniqueness query.

Now consider $n = 2k$ and $p = (p_1, \ldots, p_n)$ where $p_i = i$, $i = 1, \ldots, k+1$, and $p_{k+i} = k + i + \varepsilon$, $i = 2, \ldots, k$. The components are the inputs to a k-gap uniqueness algorithm. For each permutation $\pi = (\pi_1, \ldots, \pi_{k+1})$ of the first $k+1$ integers, the point $p_\pi = (p_{\pi_1}, \ldots, p_{\pi_{k+1}}, p_{k+2}, \ldots, p_n)$ is the center of an open ball whose points all go to the same leaf of any decision tree that answers the uniqueness question. It is a YES leaf because the coordinates of p_π have one k-gap of size $\sigma = k$ and the rest have size $k + \varepsilon$. Note that the largest (k-1)-gap has size $k - 1 + \varepsilon$.

Suppose $\pi \neq \pi'$ are two permutations with $1 = \pi_1 = \pi_1'$ and $k+1 = \pi_{k+1} = \pi_{k+1}'$. It is straightforward to verify that there must be a value of $t \in (0,1)$ for which $t p_\pi + (1-t) p_{\pi'}$ has two k-gaps of size σ. Therefore points p_π and $p_{\pi'}$ must belong to different connected components, which implies that the tree has depth $\Omega(n \log n)$. $\qquad\square$

Certainly if $k \geq n - c$ for some constant c, all k-gaps may be found and sorted in linear time. On the other hand, it is interesting to consider the case when $n - k = c_n \uparrow \infty$.

Now we turn to a parametric version of the element uniqueness problem, (PEU): decide whether, given inputs $a_1, \ldots, a_n$ and a query $x \geq 0$, there is a pair with $|a_i - a_j| = x$.

Theorem 1 *Suppose that for some constant $c > 0$, $\#\{i < j : x \leq |a_i - a_j|\} \geq c\binom{n}{2}$. Then any decision tree for PEU must have depth at least $\Omega(n \log n)$.*

Proof: (Reduction from set disjointness [1]). Let $S = \{s_1, \ldots, s_m\}$ and $T = \{t_1, \ldots, t_n\}$ be the inputs for the problem "$S \cap T = \phi$?", $n \leq m$. In time $O(m+n)$ we compute $\min S$, $\max S$, $\min T$, $\max T$, and the medians of both sets. We may assume $\min S < \min T$ or else, in time $O(m+n)$, discard all $t_i < \min S$; similarly, we suppose $\max S > \max T$. We can also assume that the median of S is less than the median of T. Let $\delta = \max S - \min S$, and define $T' = T + 2\delta$. We solve PEU for $A = S \cup T'$ using $x = 2\delta$. If a difference $|a_i - a_j| = x$, one of the elements was in S, the other in T'. Also $t_i' - s_j = 2\delta$ if and only if $t_i = s_j$, so [1] implies that PEU is $\Omega((m + n) \log n)$. Finally, because of the condition on the medians, x is smaller than at least $mn/4$ of the pairwise differences in A. If $mn/4 > c\binom{m+n}{2}$ for some $c \in (0,1)$ (see (1)), then $n > 2cm$ and the PEU lower bound for the set A of size $m + n$ is $\Omega((m + n) \log c(m + n))$. $\qquad\square$

It is interesting to consider the complexity of PEU when the function $c\binom{n}{2}$ in Theorem 1 is replaced by one which constrains x less strigently, for example $cn^2/\log n$. The present proof seems to give no clue.

The last two results of this section concern line arrangements in the plane. We are given $(m_1, b_1), \ldots, (m_n, b_n)$, the slopes and intercepts of n lines $\ell_1, \ldots, \ell_n$

in general position. The lines partition the plane into a complex of convex cells called the *arrangement*. The jth level of the arrangement, λ_j, is defined as the closure of the set of all points which lie on a unique line of the arrangement and which have exactly $j - 1$ lines above them. In geometric algorithms it is often necessary to test whether a certain line ℓ_i is part of the jth level. The level membership query asks if $\ell_i \in \lambda_j$. It is easy to see that if $\text{Rank}(m_i) = k$, the answer is YES for all $j \in [k, n - k + 1]$ and this may be decided in linear time. On the other hand if $j \notin [k, n - k + 1]$, the intersections of ℓ_i with each of the other lines may be computed, the x-coordinates sorted, and then we can compute every level that ℓ_i meets in linear time. In fact this algorithm is optimal.

Lemma 3. *Any RAM algorithm for the level membership query must make $\Omega(n \log n)$ comparisons in the worst case.*

Proof: Consider problems with $2n + 1$ inputs $(m_1, b_1), \ldots, (m_{2n+1}, b_{2n+1})$, regarded as a point in R^{4n+2}. Our "canonical" input will have slopes $m_i = 1$, $m_{n+i} = -1$, and intercepts $b_i = -i$, $b_{n+i} = i + \varepsilon$, $i = 1 \ldots, n$. Line ℓ_{2n+1} has equation $y = 0$ (see Fig. 1). The level membership query for line ℓ_{2n+1} and level λ_n ("is the x-axis ever above the median level?") has answer NO for $\varepsilon > 0$.

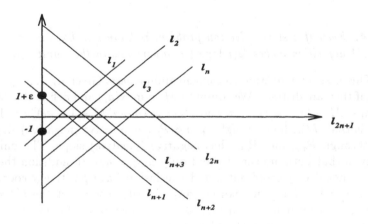

Fig. 1. *The canonical line arrangement.*

For any permutation π of the first n integers the point

$$P_\pi = (m_1, b_1, \ldots, m_n, b_n, m_{n+\pi_1}, b_{n+\pi_1}, \ldots, m_{n+\pi_n}, b_{n+\pi_n})$$

describes the canonical input, but with the last n line numbers permuted according to π; the answer is still NO. For permutations $\pi \neq \rho$ points P_π and P_ρ belong to different connected components. If not, for each $t \in (0, 1)$, the point $P(t) = tP_\pi + (1 - t)P_\rho$ is an input in the same component, by convexity. However as t varies from 0 to 1, lines forming levels λ_{n+1} through λ_{2n} will become permuted, starting with the ordering given by π and ending with the ordering given by ρ. A line that is a different levels at $t = 0$ and $t = 1$ will "move" at a constant rate between these positions, in a parallel fashion, as t varies through $(0, 1)$. Some line, say ℓ_{n+j} must move downwards. For some $t \leq 1/(n + 1)$ it will have decreased by exactly ε and will still not have crossed any other line as long as $\varepsilon < 1/(n + 1)$. This point, $P(t)$, is an input where the answer is YES, a contradiction. Therefore the tree must have $n!$ NO leaves. $\qquad \square$

The final geometric lower bound is for the slope selection problem [3]. n points $P_i = (x_i, y_i)$, $i = 1, \ldots, n$ are given, along with a rank $k, 1 \leq k \leq \binom{n}{2}$. It is required to compute that pair $i \neq j$ so the slope $(y_i - y_j)/(x_i - x_j)$ is the kth smallest amongst the slopes of the $\binom{n}{2}$ lines determined by the given points. The question was settled by

Proposition 2 *(Cole, Salowe, Steiger, Szemerédi [3]) The RAM complexity of slope selection is $\Theta(n \log n)$.*

The lower bound is a reduction to element uniqueness, and the upper bound, an elaborate algorithm. If we now consider a restricted class of inputs where the points are presented in order of increasing x-xoordinate (i.e., $x_1 < \cdots < x_n$), the optimality of the algorithm presented in [3] is open (because now element uniqueness of the x-coordinates may be checked in linear time). However we can prove

Lemma 4. *Even if a slope selection problem is known to have $x_1 < \cdots < x_n$, every RAM algorithm makes $\Omega(n \log n)$ comparisons in the worst case.*

Proof: This is still a reduction to element uniqueness. Given $a_1, \ldots, a_n$ we want to decide if they are distinct. We compute $M = \max(a_i)$ and define the $2n$ points $P_i = (i, a_i - M - 1)$, $i = 1, \ldots, n$ and $P_{n+i} = (n + i, M + 1 - a_i)$, $i = 1, \ldots, n$. The idea is that if the line through P_i and P_j has positive (negative) slope then the line through P_{n+i} and P_{n+j} has negative (positive) slope. The only other possibility is that both are zero (if and only if $a_i = a_j$). In addition the first n points have negative y-coordinates and the last n have positive y-coordinates. Since $x_i < x_j$ when $i < j$, we can use any algorithm to select the kth smallest slope determined by points with increasing x-coordinate. We use $k = n^2 - n$. The selected slope is negative if and only if the a_i are distinct. $\qquad \square$

3 Search Can be Easier Than Uniqueness

We study two search problems exemplifying the above heading. Both concern equi-partitioning of planar point sets. Given a set S of n points in the plane, a line ℓ is said to *bisect* S if both $|S \cap \ell^+| \leq n/2$ and $|S \cap \ell^-| \leq n/2$; i.e., each open halfspace has at most $n/2$ points of S. If n is odd, a bisecting line must be incident with a point of S. For the task of finding a bisecting line, we may assume n odd since otherwise, just delete any point p; a line that now bisects $S \backslash p$ (odd) also bisects S. In the *bisecting diameter* problem [6] S consists of n (unsorted) points $\theta_1, \ldots, \theta_n$ on the circumference of the unit circle. The task is to find a diameter d that bisects S. Let $H \subseteq S$ be the inputs θ_i such that the diameter d_i, incident with θ_i, bisects S. It is easy to prove (i) $|H| \geq 1$ (*existence*); (ii) if the inputs are sorted by radial angle ($O(n \log n)$), all $\theta_i \in H$ may be found in time $O(n)$ (*slow algorithm, cost* $= O(n \log n)$); (iii) a $\theta_i \in H$ can be found in $\Theta(n)$ (*fast search*). The uniqueness question is to decide if $|H| \leq 1$. The following result separates the search and uniqueness problems for bisecting diameters.

Lemma 5. *Every decision tree for the bisecting diameter uniqueness problem has depth at least* $\Omega(n \log n)$.

The proof depends on the lower bound for set equality [10] and is omitted.

The bisecting diameters problem may be somewhat artificial. The computation of a ham-sandwich cut however, is a partitioning problem which has some real applicability. In its dual version, we are given n red lines $R = \{r_1, \ldots, r_n\}$ and m blue lines $B = \{b_1, \ldots, b_m\}$ in general position in R^2. W.l.o.g. assume both $n = 2k + 1$ and $m = 2j + 1$ are odd and take $m \geq n$. It is required to find a *ham-sandwich cut*, namely a pair r_i, b_j such that the point $r_i \cap b_j$ has k red lines above and below and also j blue lines above and below. These points are the set $H = \rho_{k+1} \cap \beta_{j+1}$ of intersections of the red and blue median levels. The ham-sandwich theorem says that $|H| \geq 1$ and the Lo-Steiger theorem (Proposition 1) showed how to find an element of H in linear time. Here we prove

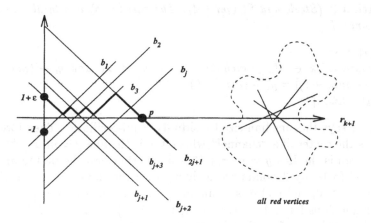

Fig. 2. *Uniqueness of a ham-sandwich cut.*

Theorem 2 *Given n red and m blue lines, a RAM algorithm to decide if $|H| > 1$ must make* $\Omega((m + n) \log(m + n))$ *comparisons in the worst case.*

Proof: (sketch, based on Lemma 3) Consider the $2k + 1$ red lines and $2j + 1$ blue lines shown in Figure 2. The blues have j positive slopes and $j + 1$ negative slopes, so the blue median level is the heavy line, ending with b_{2j+1}. There are k red lines with positive slopes and k with negative slopes. The x-axis is r_{k+1}, a red line and also the median level of the reds outside the vertical strip containing the red vertices. Once $p = r_{k+1} \cap b_{2j+1}$ is discovered and b_{2j+1} is discarded, the level membership query "$r_{k+1} \in \beta_{j+1}$?" for $b_1, \ldots, b_{2j}, r_{k+1}$ determines the existence of any other ham-sandwich cut. □

4 Least Median of Squares Regression

Given $P_1 = (x_1, y_1), \ldots, P_n = (x_n, y_n)$, n data points in general position in the plane, the least median of squares regression problem seeks a minimizer of the function

$$g(m, b) = \text{median}\{|y_i - (mx_i + b)|, i = 1, \ldots, n\}.$$

Steele and Steiger [14] studied the combinatorics of this optimization problem by first noting that the line $\ell_{m,b}$ with equation $y = mx + b$ partitions the n points into sets of small, median-sized, and big residuals.

$$S_{m,b} = \{i : |y_i - (mx_i + b)| < g(m, b)\}$$
$$M_{m,b} = \{i : |y_i - (mx_i + b)| = g(m, b)\}$$
$$B_{m,b} = \{i : |y_i - (mx_i + b)| > g(m, b)\}$$

They characterized local minima of g by proving

Proposition 3 *(Steele and Steiger [14]) The pair (m, b) is a local minimum of g if and only if*

1. $|M_{m,b}| = 3$.
2. *there are $i, j, k \in M_{m,b}$ such that $x_i < x_j < x_k$ and $y_i - (mx_i + b) = -[y_j - (mx_j + b)] = y_k - (mx_k + b)$.*
3. $|B_{m,b}| - |S_{m,b}| \geq 1$.

Condition 2 says that median-sized residuals "equi-oscillate". Therefore (i) the best line is the center of a "channel" whose top is the line $y = m^*x + b^* + \delta$ and whose bottom is the line $y = m^*x + b^* - \delta$; (ii) that half the points are within this channel (actually 3 points are incident with the top or bottom); (iii) that $g(m^*, b^*) = \delta > 0$; and (iv) δ is minimal. These properties are exploited by all efficient algorithms for LMS fitting, e.g., the $O(n^2)$ algorithm of [7]. The only nontrivial lower bound for this computational task is given by

Theorem 3 *Any RAM algorithm that computes the LMS line for n points must make $\Omega(n \log n)$ comparisons in the worst case.*

Proof: Consider a one parameter version of this problem, namely that of minimizing

$$f(b) = g(0, b) = \text{median}\{|y_i - b|, i = 1, \ldots, n\}. \tag{3}$$

It is easy to see (in the same spirit as Proposition 3) that the optimal b is the midpoint b^* of the interval of minimal width (say 2δ) which contains half the given y_i's (actually for some pair $i \neq j$, y_i is one of the endpoints and y_j is the other, so they form an $n/2$-gap). It is also clear that $\delta = f(b^*)$. Therefore if the y_i's were sorted, $\delta = \min(y_{j+n/2} - y_j, j = 1, \ldots, n/2)$, the smallest sized $n/2$-gap, and b^* is its midpoint. They could be found in linear time after sorting the $y's$, or in time $O(n \log n)$ overall. In fact, by Lemma 2, this is optimal. The proof is completed via a linear time reduction from minimizing f to minimizing g. □

In [14] it was also shown that g can have cn^2 local minima, $c > 1/16$. It is hard to imagine that a correct algorithm would not have to examine a fixed fraction of them. In view of Proposition 3 each local minimizer has the slope of the line joining a pair of the points. We are trying to reduce the problem of minimizing $h(n)$ inputs, $n = o(h(n))$ to a LMS regression problem for n points.

References

1. M. Ben-Or. "Lower Bounds for Algebraic Computation Trees". *Proc.* 15th *STOC*, (1983) 80-86.
2. A. Björner, L. Lovász, and A. Yao. "Linear Decision Trees: Volume estimates and Topological Bounds". *Proc.* 24th *STOC*, (1992) 170-177.
3. R. Cole, J. Salowe, W. Steiger, and E. Szemerédi. "An Optimal Time Algorithm for Slope Selection", *SIAM J. Comp. 18*, (1989) 792-810.
4. D. Dobkin and R. Lipton. "On the Complexity of Computations under Varying Sets of Primitives". *Lecture Notes in Computer Science 33*, 110-117, H. Bradhage, Ed., Springer-Verlag, 1975.
5. H. Edelsbrunner. *Algorithms in Combinatorial Geometry.* Springer-Verlag, Berlin, 1987.
6. H. Edelsbrunner. pers. com. 1986.
7. H. Edelsbrunner and D. Souvaine. "Computing Least Median of Squares Regression Lines and Guided Topological Sweep". *J. Amer. Statist. Assoc. 85* (1990) 115-119.
8. Chi-Yuan Lo and W. Steiger. "An Optimal-Time Algorithm for Ham-Sandwich Cuts in the Plane". *Second Canadian Conference on Computational Geometry,* (1990), 5-9.
9. Chi-Yuan Lo, J. Matoušek, and W. Steiger. "Algorithms for Ham-sandwich Cuts". *Discrete and Comp. Geom. 11*, (1994) 433-452
10. F.P. Preparata and Shamos, M.I. *Computational Geometry.* Springer-Verlag, New York, NY, 1985.
11. P. Rousseeuw. "Least Median of Squares Regression". *J. Amer. Statist. Assoc. 79* (1984) 871-880.
12. P. Rousseeuw and A. Leroy. *Robust Regression and Outlier Detection.* John Wiley, New York, 1987.
13. D. Souvaine and M. Steele. "Efficient Time and Space Algorithms for Least Median of Squares Regression". *J. Amer. Statist. Assoc. 82* (1987) 794-801.
14. M. Steele and W. Steiger. "Algorithms and Complexity for Least Median of Squares". Regression. *Discrete Applied Math. 14*, (1986) 93-100.
15. M. Steele and A. Yao. "Lower Bounds For Algebraic Decision Trees". *J. Algorithms 3*, (1982) 1-8.

The I/O-Complexity of Ordered Binary-Decision Diagram Manipulation

(Extended Abstract)

Lars Arge*

BRICS**, Department of Computer Science, University of Aarhus, Aarhus, Denmark

Abstract. We analyze the I/O-complexity of existing Ordered Binary-Decision Diagram manipulation algorithms and develop new efficient algorithms. We show that these algorithms are optimal in all realistic I/O-systems.

1 Introduction

Ordered Binary-Decision Diagrams (OBDDs) [6, 7] are the state-of-the-art data structure for boolean function manipulation and they have been successfully used to solve problems from numerous areas like e.g. digital-systems design, verification and testing, mathematical logic, concurrent system design and artificial intelligence [7]. There exist implementations of OBDD software packages on a number of sequential and parallel machines [4, 5, 14, 15]. Even though there exist very different sized OBDD representations of the same boolean function, OBDDs in real applications tend to be very large. In [4] for example OBDDs of Gigabyte size are manipulated in order to verify logic circuit designs, and researchers in this area would like to be able to manipulate orders of magnitude larger OBDDs. In such cases the Input/Output (I/O) communication becomes the bottleneck in the computation.

Until recently most research, both theoretical and practical, have concentrated on finding small OBDD-representations of boolean functions appearing in specific problems [5, 7, 13, 16], or on finding alternative succinct representations while maintaining the efficient manipulation algorithms [10]. Very recently however, researchers have begun to consider I/O-issues arising when the OBDDs get larger than the available internal memory, and experimental results show that very large speedups can be achieved with algorithms that try to minimize the access to external storage as much as possible [4, 15]. These speedups can be achieved because of the extremely large access time of external storage medias, such as disks, compared to the access time of internal memory. In the coming years we will be able to solve bigger and bigger problems due to the development of machines with larger and faster internal memory and due to increasing CPU

---

* This work was partially supported by the ESPRIT II Basic Research Actions Program of the EC under contract No. 7141 (project ALCOM II). Part of the work was done while a Visiting Scholar at Duke University. Email: large@daimi.aau.dk.

** Acronym for Basic Research in Computer Science, a Center of the Danish National Research Foundation.

speed. This will however just increase the significance of the I/O-bottleneck since the development of disk technology lack behind developments in CPU technology. At present, technological advances are increasing CPU speed at an annual rate of 40-60% while disk transfer rates are only increasing by 7-10% annually [17].

1.1 I/O-model and Previous Results

In this paper we will be working in the parallel I/O-model introduced by Aggarwal and Vitter [1] which models the I/O-system of many existing workstations. The model has the following parameters:

$$N = \# \text{ of elements in the problem instance}$$
$$M = \# \text{ of elements that can fit into main memory}$$
$$B = \# \text{ of elements per disk block}$$
$$D = \# \text{ of disks}$$

An I/O-operation in the model is the process of simultaneously reading or writing a block of data to or from each of the D disks [20]. The total amount of data transferred in one I/O is thus DB elements. The I/O-complexity of an algorithm is simply the number of I/Os it perform. Internal computation is free, and we always assume that the N elements initially are stored in the first $N/(DB)$ blocks on each disk. Typical values for workstations and file servers in production today are on the order of $10^6 \leq M \leq 10^8$, $B \simeq 10^3$ and $D \simeq 10^1$.

Early work on external-memory algorithms concentrated on sorting and permutation-related problems [1, 20]. More recently researchers have designed external memory algorithms for a number of problems in different areas. Most notably I/O-efficient algorithms have been developed for a large number of computational geometry [3, 12] and graph problems [9]. Also worth noticing in this context is [11] that addresses the problem of storing graphs in a paging-environment, but not the problem of performing computation on them, and [2] where a number of external (batched) dynamic data structures are developed. Finally, it is demonstrated in [8, 19] that the results obtained in the mentioned papers are not only of theoretical but also of great practical interest.

Some notes should be made about the typical bounds in the model. While $N/(DB)$ is the number of I/Os needed to read all the input, $\Theta(\frac{N}{DB} \log_{M/B} \frac{N}{B}) = \Theta(sort(N))$ is the number of I/Os needed to sort N elements [1]. Furthermore, the number of I/Os needed to rearrange N elements according to a given permutation is $\Theta(\min\{N/D, sort(N)\}) = \Theta(perm(N))$. Taking a closer look at these bounds for typical values of B, M, and D reveals that $\log_{M/B} \frac{N}{B}$ is less than 3 or 4 for all realistic values of N. This means that the sorting term in the permutation bound in all realistic cases will be smaller than N/D, such that $perm(N) = sort(N)$. The term in the bounds that really makes the difference in practice is the DB-term in the denominator. Normally we will be willing to accept an $\log_{M/B} \frac{N}{B}$ term in order to go from a bound like $O(N)$, where we generate a page-fault on every memory access, to a bound with a DB term in the denominator.

1.2 OBDDs and Previous Results

A branching program is a directed acyclic graph with one source (the root), where the sinks are labeled by boolean constants. The inner vertices are labeled by boolean variables and have two outgoing edges labeled with 0 and 1. If a vertex is labeled with x_i we say that it has index i. The evaluation of an input $a = (a_1, \ldots, a_n)$ starts at the root. At a vertex labeled x_i the outgoing edge with label a_i is chosen. The label of the sink reached this way equals $f(a)$ for the boolean function f represented by the branching program. An OBDD is a branching program for which an ordering of the variables in the vertices is fixed. For simplicity we assume that this ordering is the natural one, $x_1, \ldots, x_n$. If a vertex with label x_j is a successor of a vertex with label x_i, the condition $j > i$ has to be fulfilled. Note that an OBDD representing a boolean function of n variables can be of size 2^n, and that different variable orderings lead to representations of different size. There exist several algorithms (using heuristics) for choosing a variable-ordering that minimize the OBDD-representation of a given function [13, 16].

In [6] Bryant proved that for a given variable ordering and a given boolean function there is (up to isomorphism) exactly one OBDD — called the reduced OBDD — of minimal size. He also showed that the there are two fundamental operations on OBDDs — the *reduce* and the *apply* operation. The apply operation takes the OBDD-representation of two functions and computes the representation of the function formed by combining them with a binary operator. Finally, Bryant proved that using the following two reduction rules on a OBDD with at most one 0-sink and one 1-sink, until none of them are applicable anymore, yields the reduced OBDD: 1) If the two outgoing edges of vertex v lead to the same vertex w, then eliminate vertex v by letting all edges leading to v lead directly to w. 2) If two vertices v and w labeled with the same variable have the same 1-successor and the same 0-successor, then merge v and w into one vertex. Existing apply and reduce algorithms run in $O(|G|)$ and $O(|G_1| \cdot |G_2|)$ time, respectively [4, 6, 7, 14, 15, 18]. Here $|G|$ denote the size (the number of vertices) of the OBDD G.

Even though the I/O-system (the size of the internal memory) seems to be the primary limitation on the size of the OBDD problems one is able to solve practically today [4, 5, 6, 7, 15], it was only very recently that OBDD manipulation algorithms especially designed to minimize I/O was developed. In [4] and [15] it is realized that the traditional algorithms working in a depth-first or breadth-first manner on the involved OBDDs does not perform well when the OBDDs are too large to fit in internal memory, and new level-wise algorithms are developed.[3] The general idea in these algorithms is to store the OBDDs in a level-blocked manner,[4] and then try to access the vertices in a pattern that is as

---

[3] When we refer to depth-first, breadth-first and level-wise algorithms we refer to the way the apply algorithm traverse the OBDDs. All known reduce algorithms work in a level-wise manner.

[4] When we say that a OBDD is e.g. level-blocked we mean that the OBDD is stored in such a way that a given block only contain vertices from one level.

level-wise as possible. Previous algorithms did not explicitly block the OBDDs. In [4, 15] speed-ups of several hundreds compared to the "traditional" algorithms are reported using this idea.

In the rest of this paper we assume that an OBDD is stored as a number of vertices and that the edges are stored implicitly in these. We also assume that each vertex knows the index (level) of its two sons. This means that the fundamental unit is a vertex (e.g., an integer — we call it the *id* of the vertex) with an index, and a pointer and an index for each of the two sons. We assume that a pointer to a son is just the id of the son. The vertices also contain a few other fields used by the apply and reduce algorithms. Finally, we assume that a reduce operation is done as part of the apply operation after the actual apply operation is performed.

1.3 Our Results

In this paper we analyze the I/O-performance of the existing OBDD manipulation algorithms and develop new I/O-efficient algorithms. We show that all existing reduce and apply algorithms have a bad worst case performance, that is, they use $\Omega(|G|)$ and $\Omega(|G_1| \cdot |G_2|)$ I/Os in the worst case, respectively. We show that this is even true if "good" blockings are assumed — that is, if e.g. a depth-first blocking is assumed when analyzing the depth-first algorithm — and also true for the algorithms especially designed with I/O in mind. The main contribution however, is that we show that under some natural restrictions (fulfilled by all existing algorithms) $\Omega(perm(|G|))$ is a lower-bound on the number of I/Os needed to reduce an OBDD of size $|G|$. We show that this is the case even if we assume that the OBDD being reduced is depth-first, breadth-first or level-blocked, and even if we assume another intuitively good/optimal blocking which we will define later. Finally, we develop a new reduction algorithm that use $O(sort(|G|))$ I/Os, and a new apply algorithm that use $O(sort(|G_1| \cdot |G_2|))$ I/O-operations. As discussed in section 1.1 this is asymptotically optimal for all realistic I/O-systems. We believe that the developed algorithms are of practical interest due to relatively small constants in the asymptotic bounds.

2 The Reduce Operation

All the reduction algorithms [4, 6, 7, 14, 15] reported in the literature basically work in the same way. They all process the vertices level-wise from the sinks up to the root and assign a (new) unique integer label to each unique sub-OBDD root. Processing a level, under the assumption that all higher levels have already been processed, is done by looking at the new labels of the sons of the vertices on the level in question, checking if the reduction rules can be used, and assigning new labels to the vertices. While checking if reduction rule one can be used is easy, the algorithms either sort the vertices according to the labels of the sons, or maintain a (hash) table with an entry for each unique vertex generated so far, in order to check if reduction rule two can be used.

It is fairly easy to realize that all the existing algorithms perform badly in an I/O-environment, even assuming that the OBDDs are initially blocked in some "good" way. There are several reasons for this — for example the way some of the algorithms traverse the OBDD, or the "random" lookups in the table of unique vertices. The main reason however, is the visiting of the sons during the processing of a level of vertices. As there is no "nice" pattern in the way the sons are visited (mainly because a vertex can have large fan-in), the algorithms can in the worst case be forced to do an I/O each time a son is visited. The algorithms specifically developed with I/O in mind try to avoid some off all these page faults by visiting the sons in level order. This follows the general philosophy mentioned earlier that vertices should be visited level-wise. But still it is not hard to realize that also these algorithms could be forced to do an I/O every time a son is accessed, because there is no correlation between the order in which the sons on a given level are visited and the blocks they are stored in. In the full version of this paper a more complete analysis of the existing reduce algorithms is done, and it is shown that they all cause $\Omega(|G|)$ page-faults in the worst case when reducing an OBDD of size $|G|$.

2.1 I/O-Lower Bound on the Reduce Operation

In this section we will sketch the proof of the I/O-lower bound on the reduce operation. We prove the lower bound under the assumption that the reduce algorithm works like the known algorithms, that is, it works by assigning new labels to vertices and checking if one of the reduction rules can be used on vertices whose sons have already been assigned new labels. The precise assumption is that the new label is assigned to the original vertex, and that the sons are loaded into internal memory (if they are not there already) in order to obtain their new labels.

In [9] the *proximate neighbors problem* is defined as follows: We are given N elements in external memory, each with a key that is a positive integer $k \leq N/2$. For each possible value of k, exactly two elements have that key-value. The problem is to permute the elements such that elements with identical key-value are in the same block. We define a variant of the proximate neighbors problem called the *split proximate neighbors (SPN) problem*. This problem is defined similar to the proximate neighbors problem, except that we require that the keys of the first $N/2$ elements in external memory (and consequently also the last $N/2$ elements) are distinct. Following the I/O-lower bound proof on the proximate neighbors problem in [9] we can prove the following:

Lemma 1. *Solving the SPN problem require $\Omega(perm(N))$ I/Os in the worst case.*

Using Lemma 1 we can now prove the lower bound by reducing the SPN problem to the reduction problem. Given an SPN problem we create and block an unreduced OBDD as sketched in figure 1 and 2a). Figure 1 indicates how the SPN problem $8, 1, 2, 4, \cdots, 7, 6, 3, 5, 3, 8, 4, 7, \cdots, 1, 5, 3, 6$ is encoded in the highest levels of the OBDD. On top of the *base-blocks*, as we will call the blocks storing the highest levels of the OBDD, we build a complete tree. Figure 2a)

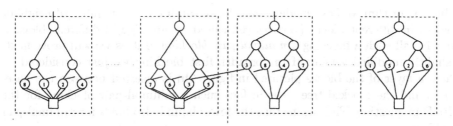

Fig. 1. Highest levels of the OBDD for a SPN problem $(B = 8)$.

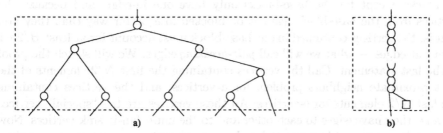

Fig. 2. a) One block of top blocking. b) Block used in proof of Lemma 4.

shows one block in the blocking of this tree. The blocking indicated on the figures is a breadth-first blocking of the OBDD. It is relatively easy to realize that if we construct the OBDD such that the id of every vertex containing an element from the first half of the SPN problem is the key of the element it holds, then the OBDD can be constructed and blocked in $O(|G|/(DB))$ I/Os. Now, whatever reduce algorithm we run on the produced OBDD, it will at some point have to hold both vertices with the same SPN problem key in main memory. This is due to our assumption about the reduce algorithm — the vertices containing elements from the last half of the SPN problem will at some point have to check the new label of their sons. This means that we can augment any reduce algorithm to solve the SPN problem.

In the full paper we give all the details in the above construction and show how we can also produce a level blocked or even depth-first blocked OBDD from an SPN problem in $O(|G|/(DB))$ I/Os. We also discuss precisely in which model of computation the construction works. As an algorithm for the SPN problem constructed in the above way use $O(|G|/(DB)) + O(reduce(|G|))$ I/Os, the following lower bound follows from Lemma 1.

Lemma 2. *Reducing an OBDD with $|G|$ vertices requires $\Omega(perm(|G|))$ I/Os in the worst case — even if the OBDD is depth-first, breadth-first, or level blocked.*

Even though Lemma 2 provides us with a lower bound on reducing an OBDD blocked in one of the natural ways for the different apply algorithms (and even in the way the OBDDs are actually blocked in the algorithms especially designed with I/O in mind), it does not mean that we could not be lucky and be able to reduce an OBDD in less I/Os, presuming that we had it blocked in some other clever way. Intuitively the best blocking strategy, under the assumptions on the reduce algorithm we have made, would be to minimize the number of neighbor vertices in different blocks (vertices connected by an edge in different blocks). We will call such a blocking a *minimal-pair blocking*. But as we will prove next, a slightly modified version of the breadth-first blocking we just considered — where a layer of the blocks pictured in figure 2b) is inserted between the base-blocks and the blocked tree — is in fact such a minimal-pair blocking for the OBDD in question. This means that the lower bound also holds for minimal-pair blockings.

Intuitively the blocking in figure 1 and 2 is a minimal-pair blocking because all blocks except for the base-blocks only have one in-edge, and because the vertices from the base-blocks cannot be blocked in a better way than they are, that is, the vertices contained in one base-block must account for at least $B/2+1$ in or out-edges — what we will call *pair-breaking* edges. We will sketch the proof of the last statement. Call the vertices containing the first $N/2$ elements of the split proximate neighbors problem for a-vertices, and the vertices containing the last $N/2$ elements for b-vertices. All these vertices are the "special" vertices because they have edges to each other and to the multi fan-in sink-vertices. Now call a sink-vertex that is connected with an a or b-vertex a c-vertex, and consider a block in an arbitrary blocking of the OBDD in figure 1 and 2.

Lemma 3. *Let K be a block containing a_1 a-vertices and their corresponding c-vertices and a_2 a-vertices without their corresponding c-vertices, together with b_1 b-vertices with their c-vertices and b_2 b-vertices without their c-vertices. Assume furthermore that a_1, a_2, b_1 and b_2 are all $\leq B/2$, and that at least one of the a_i's and one of the b_i's are non-zero. K has at least $a_1 + a_2 + b_1 + b_2 + k$ pair-breaking edges, where k is the number of a_1, a_2, b_1, b_2 that are non-zero.*

Sketch of proof: First we assume without loss of generality that all the a_1 a-vertices are from the same base-block. This is the best case as far as pair-breaking edges are concerned because we only need one c-vertex for all the a_1 a-vertices. We assume the same about the a_2 a-vertices and the b_1 and b_2 b-vertices, ending up with vertices from at most four base blocks.

We divide the proof in cases according to the value of a_1 and b_1. We illustrate the general idea in the proof with the case $a_1 \geq 1$ and $b_1 \geq 1$. In this case we need $a_1 + a_2 + b_1 + b_2 + 2$ of K's capacity of B vertices to hold the a and b-vertices and the two c-vertices. It can now be realized that the number of pair-breaking edges in a block consisting of *only* these vertices is $a_1 + 2(B/2 - a_1) + 3a_2 + b_1 + (B/2 - b_1) + 2b_2 + |(a_1 + a_2) - (b_1 + b_2)|$. Now we add vertices one by one in order to obtain the final block K. Because of the extra blocks (figure 2b) the number of pair-breaking edges can now only be reduced by one for each new vertex we add. The lemma then follows from the fact that $(a_1 + 2(B/2 - a_1) + 3a_2 + b_1 + (B/2 - b_1) + 2b_2 + |(a_1 + a_2) - (b_1 + b_2)|) - (B - (a_1 + a_2 + b_1 + b_2 + 2)) \geq a_1 + a_2 + b_1 + b_2 + k$.
□

In the full paper we show how Lemma 3 can be extended to the case where one of the variables is greater than $B/2$, and the case where all vertices in the block are of the same type, and that this leads to the following lemma:

Lemma 4. *The blocking of the OBDD showed in figure 1 and 2 is a minimal-pair blocking.*

To summarize, we have proved the following:

Theorem 5. *Reducing an OBDD with $|G|$ vertices - arbitrarily, level, depth-first, breadth-first or minimal-pair blocked - requires $\Omega(perm(|G|))$ I/Os in the worst case.*

2.2 I/O-Efficient Reduce Algorithm

Recall that the main problem with all the known apply algorithms with respect to I/O, is that when they process a level of the OBDD, they do a lot of I/Os in order to obtain the new labels of the sons of the vertices on the level in question. Our solution to this problem is simple — when a vertex is given a new label we "inform" all its immediate predecessors about it in a "lazy" way using a priority queue. Every time we give a vertex v a new label, we insert an element in the I/O-efficient priority-queue developed in [2] for each of the immediate predecessors of v. The elements in the priority-queue are ordered according to level and vertex-id of the "receiving" vertex, such that when we on a higher level want to know the new labels of sons of vertices on the level, we simply do deletemin operations on the queue until we have obtained all elements on the level in question. As the number of operations performed on the queue is linear in $|G|$, it follows from the $O((\log_{M/B} N/B)/(DB))$ amortized I/O-bound on the insert and deletemin operation proven in [2] that we overall use $O(sort(|G|))$ I/Os to manipulate the queue. In the full paper we will give all the details in the algorithm and prove the following:

Theorem 6. *An OBDD with $|G|$ vertices can be reduced in $O(sort(|G|))$ I/Os.*

As mentioned in the introduction, Theorem 5 and Theorem 6 together means that for all realistic I/O-systems we have developed an asymptotically optimal reduction algorithm.

3 The Apply Algorithm

The basic idea in all existing apply algorithms [4, 6, 7, 14, 15] is to use the recursive formula $f_1 \otimes f_2 = \overline{x_i} \cdot (f_1|_{x_i=0} \otimes f_2|_{x_i=0}) + x_i \cdot (f_1|_{x_i=1} \otimes f_2|_{x_i=1})$ to design a recursive algorithm. Here $f|_{x_i=b}$ denotes the function obtained from f when the argument x_i is replaced by the boolean constant b, and $\otimes$ is some binary operator. Using this formula Bryant [6] developed a recursive algorithm working in a depth-first manner on the involved OBDDs. In [14] and [4, 15] algorithms working in a breadth-first and in a level-wise manner are then developed. These algorithms work like Bryant's except that recursive calls (vertices who need to have their sons computed — we call them *requests*) are inserted in a global queue and in a queue for the level they should be computed on, respectively, and computed one at a time. In order to avoid generating the OBDD for a pair of sub-OBDDs more than once — which would result in exponential (in n) running time — all the algorithms use dynamic programming (maintain a table of size $|G_1| \cdot |G_2|$). Before doing a recursive call Bryant's algorithm tests if a simular call has been done before. Similarly the other algorithms test if a similar request already exists before inserting a new request in a queue.

It is relatively easy to realize that in order to hope for a good I/O-performance of the above sketched algorithms, we need to have the involved OBDDs blocked in a manner that corresponds to the algorithm that work on them — depth-first, breadth first, or level-wise. However, the algorithms still perform poorly mainly because there is no nice structure in the way the dynamic programming table is accessed — not even in the way one level of it is accessed. In the full paper we discuss this in detail and show that all the known algorithms in the worst case use $\Omega(|G_1| \cdot |G_2|)$ I/Os when trying to do an apply operation.[5]

3.1 I/O-Efficient Apply Algorithm

The main idea in our apply algorithm is to do the computation level-wise as in [4, 15], but to use a priority-queue to control the recursion. Using a priority-queue we do not need to have a queue for each level. In order to avoid the random lookups in the dynamic programming table, we do not check for duplicate requests when new requests are generated, but when they are about to be computed. This is done by ordering the priority-queue not only according to level, but secondary according to the id's of the nodes in the requests. In order to process one level of the requests we simply do deletemin operations on the priority-queue, and remove equal requests that will now appear next to each other in the order they are deleted from the queue. This way we get rid of the dynamic programming table. As in the case of the reduce algorithm we in total do a linear number of operations on the priority queue, and this and a careful analysis of the details in the algorithm leads to the following result:

Theorem 7. *The apply operation can be performed on two OBDDs G_1 and G_2 in $O(sort(|G_1| \cdot |G_2|))$ I/Os.*

Acknowledgments

The author would like to thank Allan Cheng for introducing him to OBDDs, Darren Vengroff and Jeff Vitter for inspiring discussions, and Peter Bro Miltersen for comments on earlier drafts of this paper.

---

[5] The natural question is of course why experiments with the level algorithms show so huge speedups compared to the traditional algorithms. The answer is partly that the traditional depth-first and breadth-first algorithms behave so poorly with respect to I/O that just considering I/O-issues, and actually try to block the OBDDs and access them in a "sequential" way, leads to large speedups. However, we believe that one major reason for the experimental success in [4] is that the OBDDs in the experiments roughly are of the size of the internal memory of the machines used. This means that one level of the OBDDs actually fits in internal memory. This again explains the good performance because the worst case behavior occurs when one level of the OBDD (or the dynamic programming table) does not fit in internal memory.

References

1. A. Aggarwal, J.S. Vitter: The Input/Output Complexity of Sorting and Related Problems. Communications of the ACM, 31 (9), 1988.
2. L. Arge: The Buffer Tree: A New Technique for Optimal I/O-Algorithms. In Proc. of 4th Workshop on Algorithms and Data Structures, 1995.
3. L. Arge, D.E. Vengroff, J.S. Vitter: External-Memory Algorithms for Processing Line Segments in Geographic Information Systems. In Proc. of 3rd Annual European Symposium on Algorithms, 1995.
4. P. Ashar, M. Cheong: Efficient Breadth-First Manipulation of Binary Decision Diagrams. In Proc. of 1994 IEEE International Conference on CAD.
5. S.K. Brace, R.L. Rudell, R.E. Bryant: Efficient Implementation of a BDD Package. In Proc. of 27'th ACM/IEEE Design Automation Conference, 1990.
6. R. Bryant: Graph-Based Algorithms for Boolean Function Manipulation. IEEE Transactions on computers, C-35 (8), 1986.
7. R. Bryant: Symbolic Boolean Manipulation with Ordered Binary-Decision Diagrams. ACM Computing Surveys, 24 (3), 1992.
8. Y.-J. Chiang: Experiments on the Practical I/O Efficiency of Geometric Algorithms: Distribution Sweep vs. Plane Sweep. In Proc. of 4th Workshop on Algorithms and Data Structures, 1995.
9. Y.-J. Chiang, M.T. Goodrich, E.F. Grove, R. Tamassia, D.E. Vengroff, J.S. Vitter: External-Memory Graph Algorithms. In Proc. of 6th ACM/SIAM Symposium on Discrete Algorithms, 1995.
10. J. Gergov, C. Meinel: Frontiers of Feasible and Probabilistic Feasible Boolean Manipulation with Branching Programs. In Proc. of 10th Symposium on Theoretical Aspects of Computer Science, LNCS 665, 1993.
11. M.T. Goodrich, M.H. Nodine, J.S. Vitter: Blocking for External Graph Searching. In Proc. of 1993 ACM Symposium on Principles of Database Systems.
12. M.T. Goodrich, J.-J. Tsay, D.E. Vengroff, J.S. Vitter: External-Memory Computational Geometry. In Proc. of 34th IEEE Foundations of Computer Science, 1993.
13. S. Malik, A.R. Wang, R.K. Brayton, A. Sangiovanni-Vincentelli: Logic Verification using Binary Decision Diagrams in a Logic Synthesis Environment. In Proc. of 1988 IEEE International Conference on CAD.
14. H. Ochi, N. Ishiura, S. Yajima: Breadth-First Manipulation of SBDD of Boolean Functions for Vector Processing. In Proc. of 28'th ACM/IEEE Design Automation Conference, 1991.
15. H. Ochi, K. Yasuoka, S. Yajima: Breadth-First manipulation of Very Large Binary-Decision Diagrams. In Proc. of 1993 IEEE International Conference on CAD.
16. R. Rudell: Dynamic Variable Ordering for Ordered Binary Decision Diagrams. In Proc. of 1993 IEEE International Conference on CAD.
17. C. Ruemmler, J. Wilkes: An introduction to disk drive modeling. IEEE Computer, 27 (3), 1994.
18. D. Sieling, I. Wegener: Reduction of OBDDs in linear time. Information Processing Letters, 48, 1993.
19. D.E. Vengroff, J.S. Vitter: I/O-Efficient Scientific Computation Using TPIE. In Proc. of 7th IEEE Symposium on Parallel and Distributed Processing, 1995.
20. J.S. Vitter, E.A.M Shrive: Algorithms for Parallel Memory I: Two-level Memories. Algoritmica, 12 (2), 1994.

Two Arc Disjoint Paths in Eulerian Digraphs

András Frank[†], Toshihide Ibaraki[‡] and Hiroshi Nagamochi[‡]

[†]Dept. of Computer Science, Mathematical Institute,
Eötvös University, Múseum körút 6-8,
Budapest VIII, Hungary 1088
[‡]Dept. of Applied Mathematics and Physics,
Kyoto University, Kyoto, Japan 606-01

Abstract. Let G be an Eulerian digraph, and $\{x_1, x_2\}, \{y_1, y_2\}$ be two pairs of vertices in G. An instance $(G; \{x_1, x_2\}, \{y_1, y_2\})$ is called feasible if it contains two arc-disjoint $x'x''$- and $y'y''$- paths, where $\{x', x''\} = \{x_1, x_2\}$ and $\{y', y''\} = \{y_1, y_2\}$. An $O(m + n \log n)$ time algorithm is presented to decide whether G is feasible, where n and m are the number of vertices and arcs in G, respectively. The algorithm is based on a structural characterization of minimal infeasible instances.

1 Introduction

Finding a set of edge-disjoint paths connecting pairs of specified vertices (called terminals) in a graph or a digraph is one of the classical and fundamental problems in graph theory, which has a wide variety of applications. A path between terminals s and t (or a directed path from s to t) is called an st-path. If the graph is undirected, an important result by Robertson and Seymour [6] says that edge-disjoint paths for k pairs $\{s_i, t_i\}$ of terminals, $i = 1, 2, \ldots, k$ can be obtained in polynomial time for a fixed k. In case of $k = 2$, a complete characterization of undirected graphs G that do not have edge-disjoint s_1, t_1- and s_2, t_2-paths is available (Seymour [7] and Thomassen [8]): Such G can be reduced to a graph G' that has a planar representation with the following properties:

(i) the four terminals have degree 2, and all other vertices are of degree 3, and

(ii) the terminals are located on the outer face in the order of s_1, s_2, t_1, t_2 (see Fig.1(a)).

Contrary to this, the arc-disjoint path problems in digraphs seem much more difficult. For example, the weak 2-linking problem (i.e., to decide whether there are arc-disjoint s_1, t_1- and s_2, t_2-paths) in a general digraph is known to be NP-complete [1]. However, if the digraph under consideration is Eulerian, the situation becomes slightly easier. For a given digraph $G = (V, E)$ with ordered terminal pairs (s_i, t_i), $i = 1, 2, \ldots, k$, call $H = (V, \{(t_i, s_i) \mid i = 1, 2, \ldots, k\})$ its demand digraph. The weak 2-linking problem in an Eulerian digraph $G + H$ is known to be polynomially solvable [2]. Furthermore, Ibaraki and Poljak [5] showed that the weak 3-linking problem for an Eulerian digraph $G + H$ can also be solved in polynomial time. It is based on the observations that weak 3-linking problem is equivalent to finding arc-disjoint $x_1 x_2$-, $x_2 x_3$- and $x_3 x_1$-paths in an Eulerian digraph with terminals x_1, x_2, x_3, and that the resulting problem is infeasible if and only if it is reducible to a 2-connected Eulerian digraph G', which has a planar representation (see Fig.1(b)) such that

(i) all terminals have degree 2, and all other vertices have degree 4, and

(ii) every face is a directed cycle, and all the terminals are located on the outer face in the order of x_3, x_2, x_1.

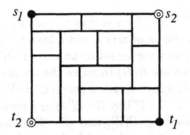

(a) Two edge-disjoint path problem for an undirected G.

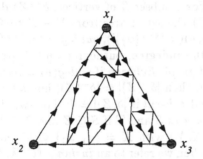

(b) Equivalent representation to the weak 3-linking problem for an Eulerian digraph $G+H$.

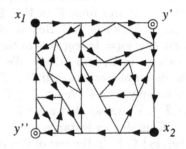

(c) Two arc-disjoint path problem for an Eulerian digraph G .

Figure 1: Examples of infeasible planar representations.

In this paper, we generalize the above result to the two arc-disjoint path problem in an Eulerian digraph G (but $G + H$ is not Eulerian), which asks to decide whether there are arc-disjoint paths connecting two *unordered* terminal pairs $\{x_1, x_2\}$ and $\{y_1, y_2\}$ (i.e., $x'x''$- and $y'y''$- paths, where either $x'x'' = x_1 x_2$ or $x_2 x_1$, and either $y'y'' = y_1 y_2$ or $y_2 y_1$). It can be shown that this problem includes the above weak 3-linking problem as a special case. We show that the problem can be solved in $O(m + n \log n)$ time, where m and n are the numbers of edges and vertices in G, respectively by deriving an analogue of the above structural characterization of infeasible instances: An Eulerian digraph G with four terminals x_1, x_2, y_1, y_2 is infeasible if and only if it is reducible to an Eulerian digraph G' that has a planar representation (see Fig.1(c)) such that

(i) all terminals have degree 2, and all other vertices have degree 4,

(ii) there is at most one cut vertex, and

(iii) every face is a directed cycle, and all terminals are located on the outer face in the order of x_1, y', x_2, y'', where $\{y', y''\} = \{y_1, y_2\}$.

The proof for this is, however, substantially different from that of [5].

2 Preliminaries

Let $G = (V, E)$ be a digraph. By a *path* or a *cycle*, we always imply a *directed* path or cycle, in which repetition of arcs is not allowed. Graph G, path P, or cycle C may be treated either as a vertex set or an arc set unless confusion arises. If it is necessary to specify, we use $E(G)$ to mean the arc set of a graph G, and $V(G)$ to mean the vertex set of a graph G. For a graph $G = (V, E)$ and an edge set $E' \subseteq E$, we denote the graph $(V, E - E')$ by $G - E'$. For a vertex set $Z \subset V$, the subgraph *induced* by Z is denoted by $G[Z]$, and $G - Z$ denotes the graph obtained by removing the vertices in Z together with the arcs incident to some vertices in Z. For a connected graph $G = (V, E)$, a vertex z is called a *cut vertex* if $G - \{z\}$ has more than one connected component. For a subset Z of vertices, $\delta^+(Z)$ denotes the set of arcs from Z to $V - Z$, $\delta^-(Z)$ the set of arcs from $V - Z$ to Z, and $\delta(Z) = \delta^+(Z) \cup \delta^-(Z)$. Denote by $indeg(v)(= |\delta^-(\{v\})|)$, $outdeg(v)(= |\delta^+(\{v\})|)$ and $deg(v)(= indeg(v) + outdeg(v))$ the indegree, outdegree and degree of a vertex v in G, respectively. We call a digraph *Eulerian* if $indeg(v) = outdeg(v)$ holds for all vertices v. If G is Eulerian, then $|\delta^+(Z)| = |\delta^-(Z)|$ holds for every nonempty $Z \subset V$. A set $Z \subset V$ is called a *k-cut* if $|\delta(Z)| = k$. For two disjoint $S, T \subset V$, we say that a cut $Z \subset V$ *separates* S and T if $S \subseteq Z$ and $T \subseteq V - Z$.

Consider an instance $(G = (V, E); X, Y)$ with $X = \{x_1, x_2\} \subseteq V$ and $Y = \{y_1, y_2\} \subseteq V$. In the rest of this paper, when we refer to an *instance* $(G; X, Y)$, we assume that G is Eulerian and is weakly connected (hence strongly connected since G is Eulerian). Each $t \in X \cup Y$ is called a *terminal*. We say that an instance $(G; X, Y)$ is *feasible* if it has two arc-disjoint $x'x''$- and $y'y''$- paths such that $\{x', x''\} = X$ and $\{y', y''\} = Y$; otherwise it is *infeasible*. We can easily show the next two lemmas.

Lemma 1. *Let P_X be an $x'x''$-path in $(G; X, Y)$ with $\{x', x''\} = X$. If y_1 and y_2 are connected in $G - E(P_X)$, then $(G; X, Y)$ is feasible.*

Proof: Since G is Eulerian, $G - E(P_X)$ has also an $x''x'$-path P'_X. Now $G_X = G - E(P_X) - E(P'_X)$ is a collection of connected components, each of which is Eulerian (possibly, a single vertex). If y_1 and y_2 are contained in the same connected component in G_X, then the instance is obviously feasible. Therefore, assume that y_1 and y_2 are contained in two distinct components H' and H'', respectively. Since y_1 and y_2 are connected in $G - E(P_X)$ but not connected in G_X, H' and H'' must contain vertices v' and v'' in $V(P'_X)$, respectively. Without loss generality, assume that P'_X visits v' before v''. Then, H' has a y_1v'-path, P'_X contains a $v'v''$-path, and H'' has a $v''y_2$-path. This implies that the instance is feasible. $\square$

Lemma 2. *If an instance* $(G; X, Y)$ *satisfies* $X \cap Y \neq \emptyset$, *then it is feasible.* □

In the following, therefore, we assume $X \cap Y = \emptyset$ for an instance $(G; X, Y)$. The following definition is necessary to derive a structural characterization of infeasible instances.

Definition 3. We say that an instance $(G; X, Y)$ with $X = \{x_1, x_2\}$ and $Y = \{y_1, y_2\}$ has an infeasible planar representation (IPR, for short) if the following conditions hold (see Fig.1(c)).
(i) All terminals have degree 2, and all other vertices have degree 4.
(ii) G has at most one cut vertex.
(iii) G is planar, and has a planar representation in which every face is a directed cycle (or equivalently, the arcs incident to a vertex are alternatively oriented out and in), and all the terminals lie on the outer face in the order of x_1, y', x_2, y'', where $\{y', y''\} = Y$. □

Lemma 4. *If an instance* $(G; X, Y)$ *has an IPR, it is infeasible.*
Proof: If an IPR has arc-disjoint $x'x''$- and $y'y''$-paths, where $\{x', x''\} = X$ and $\{y', y''\} = Y$, then these two paths must cross at some nonterminal vertex in the planar representation (since every terminal has degree 2, and is located on the boundary of the outer face). However, the two paths cannot cross at a nonterminal vertex, because the arcs incident to a vertex are alternatively oriented out and in. □

We say that an instance $(G; X, Y)$ is *minimal infeasible*, if it is infeasible but the instance $(G'; X, Y)$ obtained by contracting *any* arc becomes feasible. We want to show that the converse of Lemma 4 holds for such minimal infeasible instances. In case of $|V| = 6$, however, there is a minimal infeasible instance, as shown in Fig. 2, which is clearly infeasible but has no IPR. The following main theorem says that this is the only exception.

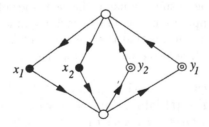

Figure 2: The minimal infeasible instance that has no IPR.

Theorem 5. *Let* $(G; X, Y)$ *be a minimal infeasible instance, and satisfy* $|V| \neq 6$. *Then* $(G; X, Y)$ *has an IPR.*
Proof: This is a consequence of Theorem 9, which will be stated in the next section. □

3 The Main Result

Given an instance $(G; X, Y)$, we check its infeasibility by trying to modify it to an instance with IPR, by applying the following reductions:

(1) Let Z be a 2-cut and $Z \cap (X \cup Y) = \emptyset$. Let u be the tail of the arc from $V - Z$ to Z, and v the head of the arc from Z to $V - Z$. Delete Z and add the arc (u, v).

(2) Let Z be a 2-cut, $|Z| \geq 2$, and Z contain exactly one terminal t. Then contract Z to t and delete the loops. (The resulting terminal t has degree 2.)

(3) Let Z be a 4-cut such that the subgraph induced by Z is connected, $|Z| \geq 2$, and $Z \cap (X \cup Y) = \emptyset$. Then contract Z into a single vertex and delete the loops.

A reducible cut Z in each of (1), (2) and (3) is called type-1, type-2 and type-3, respectively.

Lemma 6. *An instance $(G; X, Y)$ is feasible if and only if it is feasible after performing any of the reductions* (1), (2) *and* (3). $\qquad\square$

We say that an instance $(G; X, Y)$ is *reducible* if one of the above reductions (1)~(3) can be applied; otherwise *irreducible*.

Theorem 7. *Let $(G; X, Y)$ be a connected instance, and n and m be the numbers of vertices and edges in G, respectively. Then an irreducible instance $(G'; X, Y)$ of $(G; X, Y)$ can be found in $O(m + n \log n)$ time.*

Proof sketch: The following algorithm reduces all reducible cuts in a given instance $(G; X, Y)$.

1: Reducing all type-1 cuts. Contract four terminals x_1, x_2, y_1, y_2 into a single vertex t^* in G, and then find all 2-cuts in the resulting graph G'. This can be done in $O(m + n \log n)$ time by using the cactus graph that represents all minimum cuts [4]. Apply reduction (1) to each maximal 2-cut Z in G' such that $t^* \notin Z$ ("maximal" means that there is no 2-cut $Z' \subset Z$). This part can be done in $O(m + n)$ time since we can show that $Z_1 \cap Z_2 = \emptyset$ for any two such maximal 2-cuts Z_1, Z_2.

2: Reducing all type-2 cuts. For each terminal $t \in X \cup Y$, contract the other three terminals $X \cup Y - \{t\}$ into a single vertex $\bar{t}$, and find all 2-cuts in the resulting graph G' in $O(m + n \log n)$ time [4]. Apply reduction (2) to the maximal 2-cut Z in G' such that $\bar{t} \notin Z$, and apply reduction (2) to such Z. It can be shown that such maximal 2-cut Z is unique after step 1.

3: Reducing all type-3 cuts. Contract four terminals x_1, x_2, y_1, y_2 into a single vertex t^*, and find all 4-cuts in the resulting graph G' by constructing the cactus representation of all minimum cuts (in $O(m + n \log n)$ time [4]). Apply reduction (3) to each maximal 4-cut Z in G' such that $t^* \notin Z$. This part can be done in $O(m + n)$ time since we can show that $Z_1 \cap Z_2 = \emptyset$ for any two such maximal 4-cuts Z_1, Z_2 after steps 1 and 2. $\qquad\square$

Lemma 8. *Any minimal infeasible instance is irreducible.*

Proof: Obvious because any of the reductions (1), (2) and (3) can be performed by an adequate sequence of arc contractions, during which any feasible (resp., infeasible) instance never becomes infeasible (resp., feasible) by Lemma 6. □

Theorem 9. *Let $(G; X, Y)$ be an irreducible and infeasible instance, and satisfy $|V| \neq 6$. Then $(G; X, Y)$ has an IPR.* □

This is a slightly stronger statement than Theorem 5. If $|V| \leq 5$ or $|V| = 7$, its proof can be obtained by enumerating all cases. The general case with $|V| \geq 8$ will be discussed in the next section. Based on Theorems 7 and 9, we can test if a given instance $(G; X, Y)$ is feasible or not as follows. Given an instance $(G; X, Y)$, find an irreducible instance $(G'; X, Y)$ of $(G; X, Y)$ in $O(m + n \log n)$ time by Theorems 7. If G' has less than 7 vertices, its feasibility can be easily checked in $O(1)$ time (since we can show that any irreducible infeasible digraph G' with $|V| < 7$ has at most constant number of arcs). Otherwise test if $(G'; X, Y)$ has an IPR or not (this can be done in $O(m + n)$ time by using a fast planar drawing algorithm). If it has an IPR then it is infeasible; otherwise it is feasible. Therefore, we establish.

Theorem 10. *Let $(G; X, Y)$ be a given instance, and n and m be the numbers of vertices and edges in G, respectively. Then testing if $(G; X, Y)$ is feasible or not can be done in $O(m + n \log n)$ time.* □

4 Outline of the Proof

In this section, we outline the proof of Theorem 9.

Lemma 11. *An infeasible irreducible instance $(G = (V, E); X, Y)$ has the following properties:*

(i) *There is no 2-cut Z that separates X and Y.*
(ii) *Each terminal has degree 2.*
(iii) *Each nonterminal vertex has degree 4.*
(iv) *For any pair of vertices $u, v \in V$, there is at most one arc between them (i.e., at most one of (u, v) and (v, u) exists in E).*

Proof: (i) Assume that a 2-cut Z separates X and Y. Since G is connected and Eulerian, there are arc-disjoint $x_1 x_2$-path P_X and $x_2 x_1$-path P'_X. Clearly, one of them (say, P_X) contains no arcs from $\delta(Z)$. Similarly G has a $y' y''$-path P_Y, where $\{y', y''\} = Y$, such that $E(P_Y) \cap \delta(Z) = \emptyset$. These P_X and P_Y are clearly arc-disjoint in G, and hence $(G; X, Y)$ is feasible.

(ii) Assume $deg(x_1) \geq 4$ for a terminal $x_1 \in X$, without loss of generality. If G has two arc-disjoint $x_1 y_1$-path P_1 and $x_1 y_2$-path P_2, then $G - E(P_1) - E(P_2)$ has two arc-disjoint $y_1 x_1$-path P_3 and $y_2 x_1$-path P_4, since G is Eulerian. Let H be the connected component in $G - \bigcup_{1 \leq i \leq 4} E(P_i)$ that contains x_2. Since G is connected, H must contain a vertex z in $V(P_i)$ for some i. Assume $z \in$

$V(P_1) \cup V(P_2)$ (the case of $z \in V(P_3) \cup V(P_4)$ can be treated similarly). Then $E(P_1) \cup E(P_2) \cup E(H)$ contains a path P_X from x_1 to x_2 via z. However, y_1 and y_2 in $G - E(P_1) - E(P_2)$ is weakly connected by paths P_3 and P_4, and the instance would be feasible by Lemma 1, contradicting the assumption. Therefore, at least one of the above P_1 and P_2 does not exist, i.e., there must be a 2-cut W that separates $x_1 \in W$ and $Y \subseteq V - W$ by Menger's theorem. Since $deg(x_1) \geq 4$, we have $|W| \geq 2$. From (i) of this lemma, $x_2 \in V - W$ holds. This, however, implies that there is a reducible 2-cut W, a contradiction.

(iii) Omitted (see [3]). (iv) If there are multiple arcs (u, v) and (u, v) in E, then u and v are nonterminal vertices by (iii), and $deg(u) = deg(v) = 4$ holds. This means that $Z = \{u, v\}$ is a reducible 2- or 4-cut, a contradiction. Similarly if there are two arcs (u, v) and (u, v), it is easy to show that there is a reducible 2- or 4-cut $Z = \{u, v\}$, a contradiction. □

Lemma 12. *An irreducible instance $(G; X, Y)$ has at most two 2-cuts that separate $\{x_1, y'\}$ and $\{x_2, y''\}$, where $Y = \{y', y''\}$. If there are two such 2-cuts Z and Z', then G has a cut vertex z such that $Z = Z' \cup \{z\}$ or $Z = (V - Z') \cup \{z\}$. Conversely, any cut vertex is obtained in this manner.*
Proof: Omitted (see [3]). □

Now if Theorem 9 does not hold, there is a minimum counterexample $(G^*; X, Y)$; i.e., $(G^*; X, Y)$ is a irreducible infeasible instance with $n \neq 6$ vertices, but has no IPR, where G^* minimizes the number of vertices among all such instances. Clearly, $(G^*; X, Y)$ is irreducible by Lemma 8, and hence possesses the properties of Lemmas 11 and 12. The next lemma examines the cut vertex and 2-cuts in $(G^*; X, Y)$.

Lemma 13. *A minimum counterexample $(G^*; X, Y)$ has the following properties:*
(i) *There is no cut vertex.*
(ii) *There is no 2-cut Z such that $|Z \cap X| = |Z \cap Y| = 1$ and $Z - (X \cup Y) \neq \emptyset \neq (V - Z) - (X \cup Y)$.*
Proof: (i) Assume that $G^* = (V, E)$ has a cut vertex z, which has degree 4 by Lemma 11(iii) (note that no vertex with degree 2 is a cut vertex in a connected Eulerian digraph). Let Z' and Z'' be the vertex sets of the two components in $G^* - \{z\}$, and $(u', z), (u'', z), (z, v'), (z, v'') \in E$ be the four arcs incident to z. Since G^* is Eulerian, we can assume $u', v' \in Z'$ and $u'', v'' \in Z''$ without loss of generality. Clearly, each of $Z' \cup \{z\}$ and $Z'' \cup \{z\}$ is a 2-cut in G^*, and it must contain at least two terminals from the irreducibility. Hence, $|Z' \cap (X \cup Y)| = |Z'' \cap (X \cup Y)| = 2$. Since G^* is infeasible, $|Z' \cap X| = |Z' \cap Y| = 1$ by Lemma 11(i). Without loss of generality, assume that $Z' \cap (X \cup Y) = \{x_1, y_2\}$ and G^* has an Eulerian cycle which visits terminals in the order of y_2, x_1, y_1, x_2 (if there is an Eulerian cycle in the order of y_2, x_1, x_2, y_1, then the instance is feasible). We decompose instance $(G^*; X, Y)$ into $(G'; X', Y')$ and $(G''; X'', Y'')$ as follows. Let $G^*[Z']$ (resp., $G^*[Z'']$) denote the subgraph of G^* induced by Z' (resp., Z''), and G' (resp., G'') be the graph obtained by adding new vertices y'_1, x'_2

and new arcs $(u', y_1'), (y_1', x_2'), (x_2', v')$ (resp., new vertices y_2'', x_1'' and new arcs $(u'', y_2''), (y_2'', x_1''), (x_1'', v''))$ to $G^*[Z']$ (resp., $G^*[Z'']$). Regard $X' = \{x_1, x_2'\}, Y' = \{y_1', y_2\}, X'' = \{x_1'', x_2\}$ and $Y'' = \{y_1, y_2''\}$ as sets of new terminals. We show that both of the new instances $(G'; X', Y')$ and $(G''; X'', Y'')$ are irreducible. If $(G'; X', Y')$ has a reducible cut W, then W must separate y_1' and x_2' (otherwise W would be reducible in $(G^*; X, Y)$). Then W is a 2-cut such that $|W| \geq 2$ and $W \cap (X' \cup Y') = \{y_1'\}$ or $\{x_2'\}$. Since $deg(y_1') = deg(x_2') = 2$, $W - \{y_1'\}$ (or $W - \{x_2'\}$) is a 2-cut, which is reducible in $(G^*; X, Y)$, a contradiction. Similarly, irreducibility of $(G''; X'', Y'')$ follows. Note that $(G^*; X, Y)$ is infeasible only when both new instances $(G'; X', Y')$ and $(G''; X'', Y'')$ are infeasible. Each of these instances is smaller than G^*, and hence has an IPR. However, it is easy to see that G^* can have an IPR if both instances $(G'; X', Y')$ and $(G''; X'', Y'')$ have IPRs, a contradiction.

(ii) Analogously to (i), if there is such a 2-cut Z in $(G^*; X, Y)$, then we can decompose instance $(G^*; X, Y)$ into two instances based on Z and $V - Z$ so that each of new instances is an irreducible infeasible instance, and has an IPR (see [3] for detail). We then see that G^* has an IPR if these instances have IPRs, a contradiction. □

Let us introduce the following two operations.

Arc-splitting: Let w be a nonterminal vertex with four incident arcs (s_0, w), $(s_1, w), (w, s_2), (w, s_3)$, where s_0, s_1, s_2 and s_3 are all distinct. We say that arcs (s_0, w) and (w, s_2) are *split off* at w when the four arcs are replaced with new two arcs (s_0, s_2) and (s_1, s_3), eliminating w (see Fig.3).

Arc-hooking: Conversely, we say that two arcs $e = (s_0, s_2), e' = (s_1, s_3)$ are *hooked up* (with a new vertex w) when we replace these two arcs with the new arcs $(s_0, w), (w, s_2), (s_1, w)$ and (w, s_3) after introducing a new vertex w (see Fig.3).

Figure 3: Illustration of arc-splitting and arc-hooking.

Clearly, the instance resulting from an arc splitting (resp., arc hooking) remains infeasible (resp., feasible) if the original instance $(G; X, Y)$ is infeasible (resp., feasible). The proof of Theorem 9 consists of the following two major steps.

(A) Let us choose a nonterminal vertex w (with $deg(w) = 4$ by Lemma 11(iii)) adjacent to a terminal (say x_2) by an arc (x_2, w) in a minimum counterexample $(G^*; X, Y)$. Suppose that it is possible to split off two arcs at w so that the resulting instance $(G^*_w; X, Y)$ remains connected and irreducible, and has no cut vertex. We call such splitting (or the two arcs) *admissible*. This instance $(G^*_w; X, Y)$ is infeasible (from the infeasibility of $(G^*; X, Y)$), implying that $(G^*_w; X, Y)$ has an IPR, by the assumption on the minimality of G^*. It is shown that G^* always has an admissible splitting.

(B) Let $(G; X, Y)$ be an arbitrary irreducible infeasible instance that has an IPR, and choose the arc $e = (x_2, v)$ and any other arc e'. Then the instance $(G_{e,e'}; X, Y)$ obtained by hooking up e and e' satisfies one of the following properties (i)-(v):

(i) $(G_{e,e'}; X, Y)$ is reducible,

(ii) $(G_{e,e'}; X, Y)$ has an IPR,

(iii) $(G_{e,e'}; X, Y)$ is feasible,

(iv) $G_{e,e'}$ has a cut vertex,

(v) $(G_{e,e'}; X, Y)$ has a 2-cut $Z \subseteq V - \{w\}$ such that $|Z \cap X| = |Z \cap Y| = 1$ and $Z - (X \cup Y) \neq \emptyset \neq (V \cup \{w\} - Z) - (X \cup Y)$.

Note that (A) and (B) together prove Theorem 9. To see this, recall that G^* can be recovered from G^*_w by hooking up certain two arcs e and e'; i.e., $G^* = (G^*_w)_{e,e'}$. However, by the definition of G^* and Lemma 13, G^* satisfies none of (i)-(v), which is a contradiction. In other words, there exists no such counterexample $(G^*; X, Y)$; hence Theorem 9.

Now the main step (A) is proved in the next lemma.

Lemma 14. *Let w be the nonterminal vertex adjacent to a terminal x_2 by arc (x_2, w) in a minimum counterexample $(G^*; X, Y)$, and $(s_1, w), (w, s_2), (w, s_3)$ be other three arcs incident with w. Let $G^*_{s_2}$ (resp., $G^*_{s_3}$) denote the instance obtained from G^* by splitting arcs $(x_2, w), (w, s_2)$ (resp., $(x_2, w), (w, s_3)$) at w. Then one of these two splittings is admissible.*

Proof sketch: To show irreducibility of instance $(G^*_{s_2}; X, Y)$ (or $(G^*_{s_3}; X, Y)$), we first point out that the existence of a reducible 4-cuts in $(G^*_{s_2}; X, Y)$ (or $(G^*_{s_3}; X, Y)$) implies the existence of a certain type of 6-cut Z in $(G^*; X, Y)$ (note that $(G^*; X, Y)$ has no reducible 4-cut). By a further analysis (see [3] for detail), we can show that such a 6-cut Z, if any, satisfies $Z \subseteq V - (X \cup Y)$ and $|Z| \geq 4$. Then we show that any 6-cut $Z \subseteq V - (X \cup Y)$ with $|Z| \geq 4$ can be replaced with a 6-cut Z' with $|Z'| < |Z|$ so that (i) the resulting instance (say, $(H^*; X, Y)$) is infeasible if and only if so is $(G^*; X, Y)$, and (ii) $(G^*; X, Y)$ has an IPR if so does $(H^*; X, Y)$. By the minimality of G^*, this implies that $(G^*; X, Y)$ has no such 6-cut, and $(G^*_{s_2}; X, Y)$ or $(G^*_{s_3}; X, Y)$ does not have a reducible 4-cut. The other part of proof can be given by analogous reasoning. $\square$

Finally the main step (B) is proved in the following lemma.

Lemma 15. *Let $(G = (V, E); X, Y)$ be an irreducible instance that satisfies $|V| \geq 7$, have an IPR, and have no cut vertex. For arc $e = (x_2, u)$ with $x_2 \in X$*

and any arc $e' = (v, v') \in E$, let $(G_{e,e'}; X, Y)$ be the instance obtained by hooking up e and e' with a new vertex w. Then one of the properties (i)-(v), stated in the above step (B), holds.

Proof sketch: By analyzing the IPR of $(G; X, Y)$, it is rather easy to see what will happen after hooking up arcs $e = (x_2, u)$ and e'. Considering all possible cases separately, we can show that one of (i)-(v) holds in the resulting instance (see [3] for detail). $\square$

5 Discussion

For the arc-disjoint path problems $(G; X_i = \{s_i, t_i\}, i = 1, 2, \ldots, k)$ associated with Eulerian digraphs, different problem settings are conceivable depending upon the restrictions on G and the directions of the required paths: (i) either $G + H$ is Eulerian, where H is the demand digraph, or G itself is Eulerian, and (ii) either $s_i t_i$-paths are required for all i, or one of the $s_i t_i$- and $t_i s_i$-paths is required for each i. The result in [5] shows that $(G + H$ Eulerian, $s_i t_i$-path, $k = 3)$ can be solved in polynomial time, while our result here shows that $(G$ Eulerian, one of the $s_i t_i$- and $t_i s_i$-paths, $k = 2)$ can also be solved in polynomial time. By generalizing the proof in [1, 5], it is possible to prove that all types become NP-hard if k is considered as a part of input. Therefore, an interesting theoretical challenge will be to find out for each type the maximum constant k that permits a polynomial time algorithm, or to show that any constant k permits a polynomial time algorithm.

This paper is dedicated to the late Svatopluk Poljak, whose untimely death in a car accident is a great loss to our community.

References

1. S. Fortune, J. Hopcroft, and J. Wyllie, *The directed subgraph homeomorphism problem*, Theoret. Comput. Sci., 10, 1980, pp. 111-121.
2. A. Frank, *On connectivity properties of Eulerian digraphs*, in Graph Theory in Memory of G.A. Dirac, Annals of Discrete Mathematics 41, Noth-Holland, Amsterdam, 1988, pp. 179-194.
3. A. Frank, T. Ibaraki and H. Nagamochi, *Two arc disjoint paths in Eulerian digraphs*, Technical Report #95008, Kyoto University, Faculty of Engineering, Department of Applied Mathematics and Physics, 1995.
4. H. N. Gabow, *Applications of a poset representation to edge connectivity and graph rigidity*, Proc. 32nd IEEE Symp. Found. Comp. Sci., San Juan, Puerto Rico, 1991, pp. 812-821.
5. T. Ibaraki and S. Poljak, *Weak three-linking in Eulerian digraphs*, SIAM J. Disc. Math., 4, 1991, pp. 84-98.
6. N. Robertson and P.D. Seymour, *Graph minors XIII: The disjoint path problem*, 1986 (to appear in J. Combinatorial Theory B).
7. P.D. Seymour, *Disjoint paths in graphs*, Disc. Math., 29, 1980, pp. 239-309.
8. C. Thomassen, *2-linked graphs*, Europ. J. Combinatorics, 1, 1980, pp. 371-378.

Finding Dense Subgraphs

Yuichi Asahiro and Kazuo Iwama

Department of Computer Science and Communication Engineering
Kyushu University, Fukuoka 812, Japan
{asahiro, iwama}@csce.kyushu-u.ac.jp

Abstract. The dense subgraph problem (DSG) asks, given a graph G and two integers K_1 and K_2, whether there is a subgraph of G which has at most K_1 vertices and at least K_2 edges. When $K_2 = K_1(K_1 - 1)/2$, DSG is equivalent to well-known CLIQUE. The main purpose of this paper is to discuss the problem of finding *slightly* dense subgraphs. It is shown that DSG remains NP-complete for the set of instances (G, K_1, K_2) such that $K_1 \leq \frac{s}{2}$, $K_2 \leq K_1^{1+\epsilon}$ and $K_2 \leq \frac{e}{4}(1 + \frac{9}{20} + o(1))$, where s is the number of G's vertices and e is the number of G's edges. If the second restriction is removed, then the third restriction can be strengthened, i.e., DSG is NP-complete for $K_1 = \frac{s}{2}$ and $K_2 \leq \frac{e}{4}(1 + O(\frac{1}{\sqrt{s}}))$. The condition for K_2 is quite tight because the answer to DSG is always yes for $K_1 = \frac{s}{2}$ and $K_2 \leq \frac{e}{4}(1 - O(\frac{1}{s}))$. Furthermore there is a deterministic polynomial-time algorithm that finds a subgraph of this density.

1 Introduction

A graph is often associated with a relation over a set of people, i.e., a person corresponds to a vertex and an edge between two vertices shows some "good" relation between these two people. Thus a clique is a subset S of the people such that each person in S has the good relation with any other person in S. However, considering a clique in the real world, this condition seems to be too strong; it is more like a subset of the people in which there are relatively many pairs of well related persons. In terms of the graph, this is a subgraph that includes relatively many edges. In this paper, we discuss the complexity of finding such a subgraph, i.e., a *dense subgraph* of the given graph.

Formally, the *Dense Subgraph problem* (*DSG*) asks, given a graph G of s vertices and e edges and two integers K_1 and K_2, whether there is a subgraph of G which has at most K_1 vertices and at least K_2 edges. When $K_2 = K_1(K_1 - 1)/2$, DSG is equivalent to the well known CLIQUE problem and therefore the general DSG is NP-complete [Kar72]. So, the question is how this intractability changes if we restrict K_2, the degree of the density, into small values. As for K_1, we mainly discuss the case where $K_1 = s/2$ in this paper, which does not lose much generality.

Intuitively speaking, the problem of finding a subgraph of $s/2$ vertices and at least K_2 edges becomes harder as K_2 increases. Our main result shows this intuition is certainly true:

(1)DSG is NP-complete for the set of instances (G, K_1, K_2) such that $K_1 = s/2$ and $K_2 \leq \frac{e}{4}\left(1 + O\left(\frac{1}{\sqrt{s}}\right)\right)$.

(2)The answer to DSG is always yes if K_2 is a bit more restricted, i.e., $K_2 \leq \frac{e}{4}\left(1 - O\left(\frac{1}{s}\right)\right)$, and there is a polynomial time algorithm that always finds a subgraph of $s/2$ vertices and at least $\frac{e}{4}\left(1 - \left(\frac{1}{s-1}\right)\right)$ edges.

Thus the complexity of DSG exhibit so-called the *threshold behavior* which plays an important role in the complexity analysis. Actually there are many problems whose complexities jump at some point as the set of instances enlarges. For example, 2SAT is in P [EIS76] but 3SAT is NP-complete [Coo71]. Graph 3-Colorability is in P for graphs with maximum degree 3 but is NP-complete for graphs with maximum degree 4 [GJS76]. MAX CUT is in P for planer graphs [OD72, Had75] but is NP-complete for general ones [Kar72]. In those examples, however, one can recognize some kind of structural gap between the easy and hard classes, namely the threshold is somehow *structural*. By contrast, DSG's threshold is *nonstructural*, which exists just at one moment where the degree of required density increases continuously. Such nonstructural threshold must also exist in many problems but little have been known. For example consider K-CLIQUE: It is trivially in P if K is constant and is NP-complete for $K = s^\varepsilon$ for fixed $\varepsilon > 0$ [Kar72]. Thus the nonstructural threshold must exist between, but we have no more knowledge about that.

One can think of several applications of DSG. Among many others, we shall briefly mention its application to the security of the random test-instance generation which the same authors have been working on (the AIM project) [AIM95]. When generating test-instances for evaluating the performance of combinatorial algorithms empirically, one of our concerns is that the algorithms could be tuned so as to run fast especially for the benchmarks by exploiting the generation method of the benchmarks. We found that DSG is closely related to the security in this sense of the instance generators for the CNF satisfiability problem. The intractability of DSG is desirable for claiming the security of our test-instance generators (See [IM95] more about the security of instance generation).

2   Results

In this paper, a *graph* $G = (V, E)$ always means an undirected, simple graph. $|V|$ is denoted by s and $|E|$ by e. Also in this paper, a *subgraph* G' is determined only by a set $V' \subseteq V$ of vertices. Namely, G' is the so-called induced subgraph, or $G' = (V', E')$, where $E' = \{(v_1, v_2) \mid (v_1, v_2) \in E \text{ and } v_1, v_2 \in V'\}$. An instance of DSG is (G, K_1, K_2), where $G = (V, E)$ is a graph and K_1 and K_2 are integers such that $K_1 \leq |V|$ and $K_2 \leq |E|$. If we restrict the set of instances to

$$\{(G, K_1, K_2) \mid \text{for some integer } K, K_1 = K \text{ and } K_2 = K(K-1)/2\},$$

then it is equivalent to the K-CLIQUE problem. Thus K-CLIQUE is a special case of DSG, i.e., the problem of finding the most dense subgraphs.

In this paper, we are interested in the opposite direction, i.e., we wish to find a slightly dense subgraph. Intuitively, this problem seems to be easier, since less dense subgraphs more likely exist. For example if the instance is a random graph, a set of vertices which are taken from that of the largest degree to that of the K-th largest degree would probably be a nice solution. However, the worst-case complexity is high:

Theorem 1. *DSG is NP-complete for the set of instances* (G, K_1, K_2) *such that* $K_1 \leq \frac{s}{2}, K_2 \leq K_1^{1+\varepsilon}$ *and* $K_2 \leq \frac{e}{4}\left(1 + \frac{9}{20} + o(1)\right)$.

This is the most basic theorem, which can be extended to several different directions.

Theorem 2. *DSG is NP-complete even if* $K_1 = \frac{s}{2}$ *and* $K_2 \leq \frac{e}{4}\left(1 + O\left(\frac{1}{\sqrt{s}}\right)\right)$.

See Sect. 4 for proofs. Theorem 2 is fairly tight in terms of the relative degree of the density, since if $K_2 \leq \frac{e}{4}\left(1 - \frac{1}{s-1}\right)$ then it is not hard to prove that a subgraph of K_2 density always exists by using the probabilistic method[ASE92]. Furthermore, we can obtain such a subgraph:

Theorem 3. *There is a deterministic algorithm which finds a subgraph of $s/2$ vertices and at least* $\frac{e}{4}\left(1 - \frac{1}{s-1}\right)$ *edges.*

Two such algorithms, called GREEDY and GROUPING, are introduced in the next section. Note that if G is a random graph then the average number of edges included in $s/2$-vertex subgraph is also $\frac{e}{4}\left(1 - \frac{1}{s-1}\right)$. Theorem 3 says that we can select this dense subgraph for any particular graph. Recall that the restriction for K_2 in Theorem 2 is $\frac{e}{4}\left(1 + O\left(\frac{1}{\sqrt{s}}\right)\right)$. Thus $\frac{e}{4}$ is an important border: If the target is slightly larger than $\frac{e}{4}$, then the problem is hard for some graphs. The problem becomes easy for every graph if the target is slightly smaller than $\frac{e}{4}$.

Finally, DSG is also NP-complete for several other restricted subsets of instances (proofs are omitted):

Theorem 4. *DSG is NP-complete even if* $K_1 \leq |V|^{\varepsilon_1}$ *and* $K_2 \leq K_1^{1+\varepsilon_2}$ *for any (small)* $\varepsilon_1, \varepsilon_2 > 0$.

Theorem 5. *Under the same restriction as above, DSG is still NP-complete if the graph is* $|V|^{\varepsilon_3}$-*regular.*

3 Approximation Algorithms

In this section, we assume that $|V| = s$ is even. The following approximation algorithm, called GREEDY, is due to Tokuyama[Tok95]:

    Algorithm GREEDY

Step 1. Select a minimum-degree vertex v from G and remove it.

Step 2. Repeat step 1 until G has $s/2$ vertices. Then output G.

Proposition 1[Tok95]. *GREEDY outputs a subgraph of at least* $\frac{e}{4}\left(1 - \frac{1}{s-1}\right)$ *edges.*

The second approximation algorithm, called GROUPING, is based on the derandomization technique.

Algorithm GROUPING

Step 1. Let d_0 be the least even integer such that $d_0 \geq \log s$. Then the set V of G's vertices is partitioned into d_0 groups as evenly as possible, namely into (a) $d_0' = s - d_0 \left\lfloor \frac{s}{d_0} \right\rfloor$ groups of size $\left\lceil \frac{s}{d_0} \right\rceil$, and (b) $d_0 - d_0'$ groups of size $\left\lfloor \frac{s}{d_0} \right\rfloor$. The method of partitioning will be discussed later.

Step 2. Generate all $\left({}_{d_0}C_{d_0/2} \right)$ subsets including $d_0/2$ groups out of the whole d_0 groups. For each subset, let U be the set of vertices included in the subset and compute the number of edges between two vertices both in U. Now select U_0 for which the number of such edges is maximum.

Step 3. If $|U_0| \leq |G - U_0|$, then halt. Otherwise, if $|U_0| > |G - U_0|$, then we remove vertices from $|U_0|$ so that $|U_0|$ will be equal to $|G - U_0|$. Those removed vertices should be taken from the ones having as small degree as possible.

Proposition 2. *Let $e_{i,i}$ be the number of edges that connect two vertices both of which are in the ith group. Also, let $E = \sum_{i=1}^{d_0} e_{i,i}$. Then GROUPING outputs a subgraph of at least $X = \frac{e}{4} \left(1 - \frac{1}{d_0 - 1} + \frac{d_0}{d_0 - 1} \cdot \frac{E}{e} \right)$ edges (Proof is omitted).*

Thus GROUPING shows better performance (achieves larger X) when E is larger. If we set $E = e \cdot \frac{s - d_0}{d_0(s-1)}$, then $X = \frac{e}{4} \left(1 - \frac{1}{s-1} \right)$, i.e., the same performance as GREEDY. Note that this value for E is the same as the expected value for E when each group is randomly selected. If we let $E = 0$ then $X = \frac{e}{4} \left(1 - \frac{1}{\log s - 1} \right)$ which still can be written as $\frac{e}{4}(1 - o(1))$.

Thus we should construct the groups so that they will be dense in total. In other words, one can regard that GROUPING reduces the problem from obtaining dense $s/2$-vertex subgraphs into obtaining dense $s/\log s$-vertex subgraphs. Therefore it is not surprising that GROUPING works well for graphs for which dense groups are easily found. For example, consider the following graph of s vertices (See Fig. 5 for $s/2 \log s = 4$).

   (i) The vertices are divided into $\log s$ groups $V_1, V_2, \cdots, V_{\log s}$.

   (ii) Each V_i is further divided into two subgraphs X_i and Y_i of $\frac{s}{2 \log s}$ vertices.

   (iii) Edges exist between every pair (u, v) of vertices if u and $v \in X_i$ or $u \in X_i$ and $v \in Y_i$. (No edges among Y_i vertices.)

   (iv) Also edges exist between corresponding vertices of Y_i and Y_{i+1}.

To select each $X_i \cup Y_i$ as a single group seems to be the best, which is actually possible using simple heuristics, e.g., similar ones for obtaining maximal cliques. It turns out that GREEDY can achieve only 2/3 of the performance of GROUPING in this case.

4 Proof of Theorems 1 and 2

4.1 Outline

It is obvious that DSG is in NP. To prove its NP-hardness, we reduce 3SAT to DSG. An instance of 3SAT is a 3CNF formula $f = C_1 C_2 \cdots C_m$, where each *clause* C_i is in the form of $(x_i' + x_j' + x_k')$. It is known that 3SAT is still NP-complete if the instance meets the condition that every variable appears exactly six times in the formula. In this section, 3CNF formulas mean this type of formulas.

We first prove Theorem 1. In Sect. 4.2, we will describe how to transform a formula $f = C_1 C_2 \cdots C_m$ into a graph $G = (V, E)$ and two integers K_1 and K_2 in deterministic polynomial time. Also, it is shown that the two integers meet the condition of Theorem 1. In Sect. 4.3, it is proved that if f is satisfiable then there exists a subgraph of K_1 vertices and at least K_2 edges. Section 4.4 is for the opposite direction, i.e., if f is not satisfiable then any subgraph of K_1 vertices has less than K_2 edges. Section 4.5 is for the proof of Theorem 2.

4.2 Transformation from Formulas to Graphs

Recall that f consists of m clauses over n variables. The graph G includes two major portions, one, denoted by G_X, associated with the variables of f and the other, G_C, associated with the clauses of f. Also recall that each variable appears exactly six times and therefore f includes $6n$ literals and the number m of clauses is $2n$ (f is a 3CNF formula).

The most basic unit, $(G_{x,i}, G_{\overline{x,i}})$, of G_X is shown in Fig. 1. A circle denoted by $X_{i,j}$ or $\overline{X}_{i,j}$ is a complete graph of $3m$ vertices. $G_{x,i}$ is composed of t such complete graphs and the same for $G_{\overline{x,i}}$. As shown in Fig. 2, G_X includes un such units, where t is a polynomial in n and u is a constant, both of which will be determined later. (Thus the index i above changes from 1 to un.) Edges of G_X are placed as follows:

(1) Edges between every two vertices inside $X_{i,j}$ and $\overline{X}_{i,j}$ are drawn as mentioned above. The number of such edges is $3m(3m - 1)/2$ for one $X_{i,j}$ and $3m(3m - 1)tun$ in total.

(2) For any $j(1 \leq j \leq t)$, $3m$ vertices of $X_{i,j}$ are connected to $3m$ vertices of $X_{i,j+1}$ ($j+1 = 1$ when $j = t$) like a complete bipartite graph using $(3m)^2$ edges. The number of these edges is $9m^2 \cdot t \cdot 2 \cdot un = 18nm^2ut$.

(3) Edges between any pair of $G_{x,i}$ and $G_{x,j}$ ($i \neq j$): Unlike (2) above, $3mt$ vertices of $G_{x,i}$ are connected to $3mt$ vertices of $G_{x,j}$ using $3mt$ *parallel* edges. Exactly the same for between $G_{\overline{x,i}}$ and $G_{\overline{x,j}}$ ($i \neq j$) and for between $G_{x,i}$ and $G_{\overline{x,j}}$ ($i \neq j$). It should be noted that no edges are placed between $G_{x,i}$ and $G_{\overline{x,j}}$ if $i = j$. The number of these edges is $3mt \cdot (2un - 2) \cdot 2un/2 = 6n^2mu^2t - 6nmut$.

Thus, G_X has $6nmut$ vertices and

$$27nm^2ut + 6n^2mu^2t - 9nmut$$

edges in total. Recall that G_X corresponds to the n variables of the formula f. Actually, $G_{x,i}$ is associated with literal x_i and $G_{\overline{x,i}}$ with literal $\overline{x_i}$. Actually there are many more $G_{x,i}$ and $G_{\overline{x,i}}$ for $n + 1 \leq i \leq un$, which is just for the technical purpose.

The other main portion is G_C that is associated with the m clauses of f. See Fig. 3. G_C has $4m$ vertices, $v_{i,0}, v_{i,1}, v_{i,2}, v_{i,3}$, for $1 \leq i \leq m$. Index i shows the ith clause C_i of f. Among the four vertices, $v_{i,1}, v_{i,2}, v_{i,3}$ are associated with the three literals of C_i. $v_{i,0}$ is introduced again for the technical purpose.

(4) Edges are placed between every two vertices but between $v_{i,1}$ and $v_{i,2}$, between $v_{i,1}$ and $v_{i,3}$ and between $v_{i,2}$ and $v_{i,3}$. The number of these edges is $\{(4m-3)\cdot 3m + (4m-1)\cdot m\}/2 = 8m^2 - 5m$.

Finally, edges are placed between G_X and G_C as shown in Fig. 4.

(5) Suppose, for example, that $v_{i,2}$ of G_C corresponds to literal $\overline{x_k}$. Then we place un edges between $v_{i,2}$ and $G_{\overline{x,k}}$. Note that $G_{\overline{x,k}}$ contains a lot of vertices. So, when placing those edges, we have to follow the rule that there is at most one vertex in each $3m$-vertex complete graph (a circle in Fig. 1) which is connected to a vertex in G_C. This is possible by providing sufficiently many circles, or by selecting $t \geq 6un$. Then the number of these edges is $un \cdot 3m = 3unm$.

(6) As with (5), we place un edges from each $v_{i,0}$ to $G_{x,n+i}$. The above rule is exactly the same also. The number of edges is unm.

Thus we have obtained the graph $G = (V, E)$ such that

$$|V| = 6nmut + 4m,$$
$$|E| = 27nm^2ut + 6n^2mu^2t - 9nmut + 8m^2 - 5m + 4unm,$$

where $m = 2n$, t is a polynomial in n and u is a constant such that $t \geq 6un$.

As for K_1 and K_2, we select the following values:

$$K_1 = 3nmut + 2m \ (= |V|/2),$$
$$K_2 = \frac{27}{2}nm^2ut + \frac{3}{2}n^2mu^2t - 3nmut + 2nmu + 2m^2 - m.$$

Thus K_1 meets the condition of the theorem. As for K_2, let $u = 11$ and $t = n^c$. Then

$$K_1 = 66n^{c+2} + 4n,$$
$$K_2 = 957n^{c+3} - 66n^{c+2} + 52n^2 - 2n,$$
$$|E| = 2640n^{c+3} - 198n^{c+2} + 32n^2 + 78n.$$

Hence, if we choose a large constant as c, K_2 can be written as

$$K_2 = K_1^{1+\varepsilon_1}, \text{ and also as}$$
$$K_2 = \frac{|E|}{4}\left(1 + \frac{9}{20} + O\left(\frac{1}{|V|^{1-\varepsilon_2}}\right)\right).$$

Note that $\frac{9}{20}$ could become a smaller value by selecting a larger u, which appears to be desirable for us. However, for some reason mentioned in Sect. 4.4, we cannot select larger u than 11.

It is not hard to see that this construction of G can be carried out in deterministic polynomial time. In the following two subsections, we shall prove the correctness of the construction, i.e., $G = (V, E)$ has a subgraph of $K_1(= |V|/2)$ vertices and at least K_2 edges if and only if f is satisfiable.

4.3 Proof for If-Part

Suppose that f is satisfiable. Let $(x_1, x_2, x_3, \cdots) = (1, 0, 1, \cdots)$, where 1 (0) denotes true (false), be a satisfying truth assignment. Then we select one half vertices from the graph G as follows:

(1) From G_X, we select the vertices of $G_{x,1}, G_{\overline{x,2}}, G_{x,3}, \cdots$. Namely, if 1 is assigned to x_i in the above truth assignment, then we select $G_{x,i}$, otherwise we select $G_{\overline{x,i}}$.

(2) From G_C, we select all of the $v_{1,0}, v_{2,0}, v_{3,0}, \cdots, v_{m,0}$ and for each i, exactly one of $v_{i,1}, v_{i,2}$ and $v_{i,3}$ which corresponds to a literal being *true* by the assignment (if two or more such literals, any one of them).

As mentioned before, we surely selected $|V|/2$ vertices. The number of edges induced by these vertices is equal to the previously selected value for K_2. Thus, if f is satisfiable, we can select a subgraph meeting the condition.

4.4   Proof for Only-If Part

We will prove that if f is unsatisfiable, then any subgraph of G having K_1 vertices induces less than K_2 edges.

Lemma 1. *Consider only the G_C portion. If we wish to select $2m$ vertices from G_C that induces the maximum number of edges, we have to select $v_{1,0}, \cdots, v_{n,0}$ and exactly one from $v_{i,1}, v_{i,2}$ and $v_{i,3}$ for each $i(1 \leq i \leq m)$. This selection achieves $m(2m-1)$ edges (Proof is omitted).*

Lemma 2. *If we select $3nmut$ vertices from the G_X portion, i.e., a half vertices of G_X, the following selection gives us the maximum induced edges, whose number is $\frac{27}{2}nm^2ut + \frac{3}{2}n^2mu^2t - 3nmut$: Selecting either the whole $G_{x,i}$ or the whole $G_{\overline{x,i}}$ for each $1 \leq i \leq un$.*

Proof. We shall call this selection method *Selection A*. It is not hard to check that the number of edges by Selection A is equal to the above value. We shall consider the following three different cases for other ways of selection.

(Case 1.) The same as Selection A in that we never select a fractional part of $G_{x,i}$ or $G_{\overline{x,i}}$. In this case, a different way from Selection A means that we select both $G_{x,i}$ and $G_{\overline{x,i}}$. It is not hard to see that this selection induces less edges intuitively because there are no edges between $G_{x,i}$ and $G_{\overline{x,i}}$.

Before discussing the next case, we introduce the notation of "moving vertices". Let S be any set of $3nmut$ vertices of G_X. Then S can be written as

$$S = (S_A - S_{cut}) \cup (\overline{S_A} \cap S_{paste}),$$

where S_A is a set of the $3nmut$ vertices selected by Selection A, $\overline{S_A}$ is its complement, $S_{cut} \subseteq S_A$ and $S_{paste} \subseteq \overline{S_A}$. Since $|S| = |S_A| = |\overline{S_A}| = 3nmut$, $|S_{cut}|$ and $|S_{paste}|$ are the same. Namely, we can select S by first selecting some S_A by Selection A, then removing S_{cut} from S_A and the same number of vertices $(= S_{paste})$ are selected from $\overline{S_A}$. We shall say that S *can be selected by moving* $|S_{cut}|$ *vertices*. Now let us consider the second and third cases:

(Case 2.) The selection can be done by moving less than $3mt$ vertices. (It is not necessary to know how to determine these moving vertices.) Let k $(0 < k < 3mt)$ be the number of vertices that will move. Let S_A be the original selection by Selection A and let e^- be the number of edges in the subgraph that reduces by removing k vertices from S_A. Now the current subgraph has the edge set $S_A - S_{cut}$ $(|S_{cut}| = k)$. Then, by adding k vertices in $\overline{S_A}$, the number of edges of the subgraph increases. Let e^+ be that increasing number. We will show that $e^- > e^+$ no matter what the k vertices are.

Let G_A be the original subgraph induced by S_A. Then every vertex of G_A has the same degree, $d = 9m + un - 2$. Now consider any set $S_k \subseteq S_A$, $|S_k| = k$, and suppose that e_0 edges are induced by S_k. Then it turns out (proof is omitted) that

$$e^- = kd - e_0.$$

Hence this value becomes minimum when e_0 is maximum. Recall that $u = 11$. To make e_0 maximum, there are two different ways for $0 < k \leq 11n$.

(a)Take one vertex from each $G_{x,i}$ (or $G_{\overline{x,i}}$) of G_A for $1 \leq i \leq k$, so that these vertices are connected via parallel edges.

(b)Select some $G_{x,i}$ (or $G_{\overline{x,i}}$) of G_A and take k vertices, from $X_{i,j}$ and $X_{i,j+1}$.

In both cases the selected k vertices constitute a complete graph. So, in case (b), e^- can be written as

$$e^- = k(un - 1) + e_1 + e_0,$$

where $e_1 > 0$ is the number of edges between the k vertices and other vertices within $G_{x,i}$. Note that this value of e^- is exactly the same (although e_1 cannot be interpreted as above) in Case (a).

For $11n < k \leq 6m(= 12n)$, above (a) is no longer possible but a complete graph of k vertices can still be taken by the way of (b) above.

For $k > 6m$, it turns out that there is a unique way of maximizing e_0. That is, similarly to (b) above, to take the whole $X_{i,j}$ then the whole $X_{i,j+1}$, and so on until the number k is achieved (the last $X_{i,j+l}$ may be taken in part). e^- can be expressed exactly the same as previously. It should be noted that if u were too large, we can take a large complete graph for large k by method (a) above, this is not desirable for the discussion below and is the reason why we set $u = 11$.

Thus, when deleting the k vertices, these e^- edges are removed and the subgraph changes from G_A to, say, G'_A. After that the same number of vertices in $\overline{S_A}$ are added. One can see that the number of edges that increases by this addition is e_2 that is the number of the edges within those added vertices plus at most $k(un - 1)$ that is the number of edges between the added vertices and G'_A. Hence

$$e^+ \leq k(un - 1) + e_2.$$

Unlike e^-, we try to make this value maximum by maximizing e_2. A crucial point is that this e_2 cannot surpass the value e_0, since both are maximized numbers of edges within the k vertices. Now it follows that

$$e^- - e^+ \geq e_0 - e_2 + e_1 \geq e_1 > 0,$$

which is what we wanted to show.

(Case 3.) The selection can be done by moving at least $3mt$ vertices. We can combine Case 1 and Case 2 to show $e^- - e^+ > 0$ (details are omitted).

That concludes the proof of Lemma 2. ☐

Now we know that the selection described in Lemmas 1 and 2 makes the induced edges within G_X and G_C maximum under the condition that we take $3nmut$ vertices from G_X and $2m$ ones from G_C. We call this selection method *Selection B*. The rest of the proof is

(1) Selection B cannot achieve K_2 edges if f is not satisfiable.

(2) No other selections cannot either.

(1) is easy. To prove (2) we can use the moving-vertices principle introduced in the proof of Lemma 2. Again we need to check several cases, but the outline is exactly the same as before.

4.5 Proof of Theorem 2 (Sketch)

As shown in Fig. 6, G_1 and G_2 are added to graph $G = (V, E)$ that was constructed in Sect. 4.2. Let $|V| = N$ and $|E| = M$. G_1 and G_2 are both complete graphs of N^2 vertices. Between G_1 and G we place a complete bipartite connection. As for between G_1 and G_2, each vertex of G_1 is connected to $N^2 - N$ vertices of G_2 (and so that the degree of G_2's all vertices will be the same) in some straightforward way. The whole graph has $2N^2 + N$ vertices.

Now it turns out that if we wish to maximize the number of edges by taking $N^2 + N/2$ vertices, we have to take the whole G_1 and $N/2$ vertices from G. (If some vertices are taken from G_2 instead of from G, the number of edges decreases because of the way of connection between G_1 and G_2.) To maximize the edges within G, we can adopt the same selection as described in Sect. 4.1. Let us calculate the number e_0 of the whole edges and the number e_1 of the maximized edges in this subgraph:

$$e_0 = 2N^4 - N^2 + M,$$

$$e_1 = \frac{1}{2}N^4 + \frac{1}{2}N^3 - \frac{1}{2}N^2 + N^{1+\varepsilon}.$$

Hence $e_1 = \frac{e_0}{4}\left(1 + O\left(\frac{1}{N}\right)\right)$ and therefore we can select K_1 and K_2 that meet the condition of the theorem.

References

[AIM95] Y. Asahiro, K. Iwama and E. Miyano. Random Generation of Test Instances with Controlled Attributes. *DIMACS Series in Discrete Math. and Theor. Comput. Sci.*, 1995 (in press).

[ASE92] N. Alon, J. H. Spencer and P. Erdös. *The probabilistic method.* J.Wiley, 1992.

[Coo71] S. A. Cook. The complexity of theorem-proving procedures. In *Proc. 3rd Ann. ACM STOC*, pp.151-158, 1971.

[EIS76] S. Even, A. Itai and A. Shamir. On the complexity of timetable and multicommodity flow problems. *SIAM J. Comput.*, Vol.5, pp.691-703, 1976.

[GJS76] M. R. Garey, D. S. Johnson and L. Stockmeyer. Some simplified NP-complete graph problems, *Theor. Comput. Sci.* Vol.1, pp.237-267, 1976.

[Had75] F. O. Hadlock. Finding a maximum cut of a planar graph in polynomial time. *SIAM J. Comput.*, Vol.4, pp.221-225, 1975.

[IM95] K. Iwama and E. Miyano. Intractability of read-once resolution. In *Proc. 10th IEEE Structure in Complexity Conference*, 1995.

[Kar72] R. M. Karp. Reducibility among combinatorial problems. *Complexity of Computer Computations*, Plenum Press, N.Y., pp.85-103, 1972.

[OD72] G. I. Orlova and Y. G. Dorfman. Finding the maximum cut in a graph. *Engrg. Cybernetics*, Vol.10, pp.502-506, 1972.

[Tok95] T. Tokuyama, personal communication, 1995.

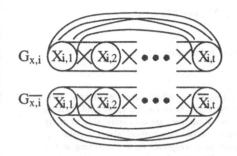

Fig. 1. $G_{x,i}$ and $G_{\overline{x,i}}$

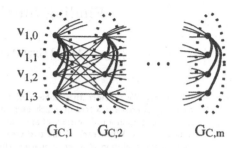

Fig. 3. G_C

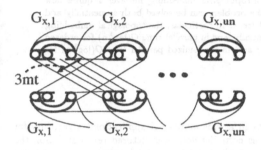

Fig. 2. G_X

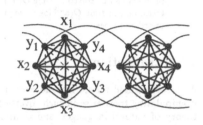

Fig. 5. A graph for which GROUPING is 3/2 better than GREEDY

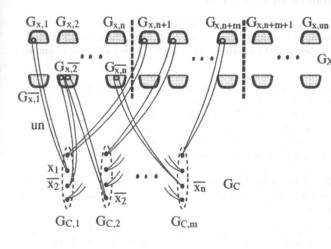

Fig. 4. The Whole G

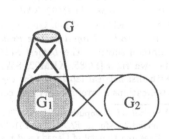

Fig. 6. Proof of Theorem 2

Finding Smallest Supertrees

Arvind Gupta* and Naomi Nishimura**

Abstract. We consider the problem of constructing the smallest *supertree* of two trees, where a supertree is a tree in which each of the two input trees can be embedded. Due to the use of trees in a wide variety of application areas, supertrees are also applicable in many different ways. When each of the two trees contains partial information about a data set, such as the evolution of a set of species, the supertree corresponds to a structuring of the data in a manner consistent with both original trees. The size of a supertree of two trees can also be used to measure the similarity between two arrangements of data, whether images, documents, or RNA secondary structures.

When the embedding relation is subgraph isomorphism, the problem reduces to that of finding the largest common subtree. The case of topological embedding, however, requires new techniques and algorithms. We show how the problem can be solved both sequentially and in parallel, for both ordered and unordered trees. In particular, the topological embedding problem can be solved in time $O(n^2)$ sequentially and in parallel time $O(\log^3 n)$ for ordered trees, and in time $O(n^{2.5} \log n)$ sequentially and in randomized parallel time $O(\log^3 n)$ for unordered trees.

1 Introduction

As trees are used for storing and accessing data in a wide number of applications, algorithms for manipulating trees arise in such diverse fields as compiler design, structured text databases, the theory of natural languages, and computational biology. Of particular interest are methods for combining or comparing the data associated with a pair of trees.

The *Smallest Supertree Problem* is one of several classes of problems concerning the relationships between two trees with respect to various embedding relations, such as subgraph isomorphism and topological embedding. For a particular embedding relation, the following classes of problems can be defined for either ordered or unordered input trees T and T':

Embeddable Tree Problem: Determine whether T is embeddable in T'.
Largest Subtree Problem: Determine the largest tree S such that S embeds in both T and T'.
Smallest Supertree Problem: Determine the smallest tree S such that both T and T' embed in S.

We can view the three classes of problems as tree, subtree (or subgraph), and supertree problems, listed in order of attention given by previous researchers to date.

As a problem on ordered trees, the Embeddable Tree Problem has been approached as pattern matching [Kos89, DGM90, KM91]; there has also been work done on the unordered case [Mat78, Chu87, LK89, GKMS90, GN95b]. For many embedding relations, the *Embeddable Tree Problem* reduces to the *Largest Subtree Problem* since T is embeddable in T' if and only if T is the largest common subtree. The *Largest Subtree Problem* has been studied extensively, due to interest in its own right [FG85, KKM92, SW, FT94a, FT94b, KA94, GN95a]; for T and T' general trees with distinct leaf labels and for the embedding relation topological embedding, this problem has applications in evolutionary tree comparison.

---
* School of Computing Science, Simon Fraser University, Burnaby, BC, Canada, V5A 1S6. email: arvind@cs.sfu.ca, FAX (604) 291-3045. Research supported by the Natural Sciences and Engineering Research Council of Canada and the Advanced Systems Institute.
** Department of Computer Science, University of Waterloo, Waterloo, Ontario, Canada, N2L 3G1. email: nishi@plg.uwaterloo.ca, FAX (519) 885-1208. Research supported by the Natural Sciences and Engineering Research Council of Canada and the Information Technology Research Centre.

In this paper, we consider the *Smallest Supertree Problem* for subgraph isomorphism and topological embedding, with both ordered and unordered trees as inputs. In certain cases, we will show the relationships between supertree and subtree problems; in others, we will develop new techniques.

A supertree is of interest in the context of editing, image clustering, genetics, and chemical structure analysis, as it gives a measure of the similarity of trees [JWZ94], and in the context of computational biology, as it gives a way to form an evolutionary tree [War93]. In the case of the paper by Jiang, Wang, and Zhang, the problem solved is that of ordered minor containment (a generalization of topological embedding); in the case of the paper by Warnow, the problem is that of minor containment where leaves have distinct labels and are constrained to map to each other.

Section 2 contains definitions and notation. In Section 3 we discuss the smallest supertree problem under subgraph isomorphism. We present sequential and parallel algorithms for unordered trees under topological embedding in Sections 4 and 5, and sequential and parallel algorithms for ordered trees under topological embedding in Section 6. Finally, in Section 7, we discuss directions for further work.

2 Preliminaries

The algorithms in this paper all apply to finite rooted trees. We will use the notation $V(T), E(T), |T|$, and $root(T)$, respectively, to refer to the node set, edge set, size (number of nodes), and root of a tree T. For nodes u and v in T, we will denote by $path(u, v)$ the (nodes of the) unique path from u to v in T. We consider both *ordered trees* (also called planar-planted trees in the literature), in which there is an order imposed on the children of each node, and *unordered trees*, in which there is no such order.

In our parallel algorithms, we will process the trees making use of a representation known as a *Brent tree*, defined below. To avoid confusion, we will refer to the *nodes* of a tree T using the Roman alphabet, and to the *vertices* of a Brent tree using the Greek alphabet. Moreover, we will distinguish between an arbitrary connected *subgraph* of T and a *subtree* of T, T_v, consisting of v and all its descendants. When we form a subgraph by removing one subtree from another, we will use $T \Upsilon T_v$ to denote the subgraph obtained by removing from T all proper descendants of v. In such a situation, we refer to v as a *scar*, and to $T \Upsilon T_v$ as a *scarred subtree*.

The embedding relation of topological embedding consists of the mapping of edges in one tree to paths in another. More formally, we define topological embedding as follows [GN95a]:

Definition 1. A tree T is *topologically embeddable* in a tree S, $T \leq_e S$, if there is a one-to-one function $\phi : V(T) \rightarrow V(S)$ such that for any $v, w, x \in V(T)$ the following properties hold. If w is a child of v, then $\phi(w)$ is a descendant of $\phi(v)$. If w and x are distinct children of v, then the path from $\phi(v)$ to $\phi(w)$ and the path from $\phi(v)$ to $\phi(x)$ have exactly the node $\phi(v)$ in common. Equivalently, we will say that there is a *topological embedding* of T in S and that ϕ is a topological embedding of T in S. The topological embedding is a *root-to-root topological embedding* if $\phi(root(T)) = root(S)$.

To process trees in parallel, we make use of a technique based on the results of Brent [Bre74] to divide a tree repeatedly into smaller and smaller subgraphs. The proofs of these lemmas, and all other omitted lemmas, can be found in the full paper [GN95c].

Lemma 2. *For T a tree with at least two nodes, there is a unique node v of T with children $c_0, \ldots, c_{k-1}$ such that $|T \Upsilon T_v| \leq \frac{|T|}{2}$ and $|T_{c_i}| \leq \frac{|T|}{2}$, for all $0 \leq i \leq k$.*

Lemma 3. *For T a tree with more than two nodes and l a leaf of T, there is a unique ancestor v of l such that if c is the child of v for which $l \in T_c$, then $|T \Upsilon T_v| \leq \frac{|T|}{2}$ and $|T_c| \leq \frac{|T|}{2}$.*

We can reduce T to constant size pieces by a logarithmic number of applications of the two lemmas. Seen as a two-step operation, the alternating steps applied are those of a *Brent break* (the dividing of a tree T into pieces T_v and $T \backslash T_v$), and a *child break* (the dividing of a tree into subtrees rooted at the children of the root). The application of a Brent break followed by a child break (*Brent restructuring*) results in subgraphs containing disjoint sets of nodes. We store all of these subgraphs and their interrelations in a tree known as the *Brent tree of T*, which can be formed in $O(\log^2 |T|)$ time using an $O(|T|^4)$-processor CREW PRAM [GN95b]. A vertex α in the Brent tree is labelled by a subgraph of T, and has edges to vertices representing smaller subgraphs obtained by a restructuring step.

The reader is referred to the bibliography for various papers in which this technique is applied [GN94, GN95a, GN95b], and to one paper in particular for a detailed discussion of its use [GN95b].

3 Subgraph isomorphism

For the embedding relation of subgraph isomorphism, the *Smallest Supertree Problem* reduces to the *Largest Subtree Problem*. To see this, consider a tree S such that both T and T' are subgraphs of S. Let $S[T]$ be the subgraph of S that is isomorphic to T and let $S[T']$ be the subgraph of S that is isomorphic to T'. We now consider the interrelation between T and T' in terms of $S[T]$ and $S[T']$.

The size of S is at most the sum of the sizes of T and T'; equality can only occur if T or T' is the empty tree. If the intersection R of $S[T]$ and $S[T']$ is nonempty, then R is itself a connected subgraph of S (and therefore of T and T'). The size of S will be the sum of $|R|$, $|S[T] \backslash R|$, and $|S[T'] \backslash R|$. Intuitively, to make S as small as possible, we would seem to require R to be as large as possible. In particular, R must be the largest common subtree of $S[T]$ and $S[T']$, or, equivalently, the largest common subtree of T and T'. From this we can conclude:

Theorem 4. *For any trees T and T' with a largest subtree of size L, the size of any smallest supertree of T and T' is $|T| + |T'| - L$.*

The known algorithms for the *Largest Subtree Problem* determine the largest common subtrees for all pairs of nodes in T and T'. We can thus obtain the following results, preserving the complexity of the algorithms for the *Largest Subtree Problem* with respect to subgraph isomorphism, for both ordered trees [ZS89] and unordered trees [GN95a].

Theorem 5. *For T and T' ordered trees, the* Smallest Supertree Problem *can be solved sequentially in time $O(|T| \cdot |T'| \cdot \min\{dp(T), lv(T)\} \cdot \min\{dp(T'), lv(T')\})$ and in parallel in time $O(|T| + |T'|)$, using $O(\min\{|T|, |T'|\} \cdot lv(T) \cdot lv(T'))$ processors for the problem of subgraph isomorphism where $dp(T)$ and $lv(T)$ are the depth and number of leaves in T, respectively.*

Theorem 6. *For T and T' unordered trees, $n = |T| + |T'|$, the* Smallest Supertree Problem *can be solved sequentially in time $O(n^{2.5} \log n)$ and in randomized parallel time $O(\log^3 n)$ using $O(n^{6.5})$ processors for the problem of subgraph isomorphism.*

4 Sequential topological embedding for unordered trees

Solutions to the *Smallest Supertree Problem*, like solutions to the *Embeddable Tree Problem* and the *Largest Subtree Problem*, make use of a dynamic-programming approach introduced by Matula and Reyner [Mat78, Rey77, VR89]. The basic paradigm for these algorithms is that of working up from the leaves to the root in each of the two trees, at each stage determining the size of the best solution to the original problem on smaller inputs, namely on trees T_a and T'_u. Determining the solution to the smaller problem, given solutions to yet smaller problems (in particular, solutions to problems where inputs are subtrees rooted at children of a and u), typically involves solving a bipartite matching problem. In the bipartite graph $G(X, Y, E)$, X will represent the children of a, Y will represent the children of u, and E will represent the known solutions to the problems involving pairs of children, one in T and one in T'.

The solutions to the various problems vary in the type of matching problem required, and the exact way in which the trees T and T' are processed from leaves to root. In the matchings to be computed in this paper, an edge weight between two nodes b and v will be the size of the smallest supertree of the trees T_b and T'_v. It may happen that parts of T and T' share no common node in the smallest supertree. To capture this possibility, adjacent to each node b of T we will introduce a "dummy" node of weight $|T_b|$. Choosing this edge is equivalent to choosing not to overlap T_b with any nodes of T'.

We use $MWM(\{b_1, \ldots, b_k\}, \{v_1, \ldots, v_\ell\})$ to denote the minimum weight perfect matching in the graph $G = (X, Y, E)$, where $X = \{b_1, b_2, \ldots, b_k, n_1, \ldots, n_\ell\}$, $Y = \{v_1, v_2, \ldots, v_\ell, m_1, \ldots, m_k\}$, and the edge set is specified as follows: for each i and j, the edge (b_i, v_j) of weight $S(b_i, v_j)$; for each i, the edge (b_i, m_i) of weight $S(b_i, \emptyset)$; for each j, the edge (n_j, v_j) of weight $S(\emptyset, v_j)$; and for each i and j, the edge (n_i, m_j) of weight 0.

All of our results deal with the problem of finding the size of the smallest supertree. Our algorithms are easily modified to also compute a smallest supertree: as well as keeping the size of the smallest supertree at each level, we keep track of the structure of such a supertree.

In order to determine the size of the smallest supertree of T and T', we will determine for each $a \in T$ and for each $u \in T'$ the size of the smallest tree S such that $T_a \leq_e S$ and $T'_u \leq_e S$. We denote this quantity by $S(T_a, T'_u)$, or $S(a, u)$ when T and T' are clear from context. In addition, we define $S(\emptyset, u)$ to be $|T'_u|$ and $S(a, \emptyset)$ to be $|T_a|$.

To determine $S(a, u)$, we note that in a smallest supertree of T_a and T'_u, one of the following cases must hold, for $b_1, \ldots, b_k$ the children of a and $v_1, \ldots, v_\ell$ the children of u:

1. for some p, $1 \leq p \leq \ell$, a subgraph of the supertree is formed from T_a and T'_{v_p}, with u, $T'_{v_1}, \ldots, T'_{v_{p-1}}, T'_{v_{p+1}}, \ldots, T'_{v_\ell}$ added in;
2. for some q, $1 \leq q \leq k$, a subgraph of the supertree is formed from T_{b_q} and T'_u, with a, $T_{b_1}, \ldots, T_{b_{q-1}}, T_{b_{q+1}}, \ldots, T_{b_k}$ added in; or
3. there is a node representing both a and u and subgraphs formed from some subtrees T_{b_j} and T'_{v_i}, for various values of j and i, with subtrees rooted at the remaining children of a and u added in.

To determine $S(a, u)$, given $S(b_j, u)$, $S(a, v_i)$, and $S(b_j, v_i)$, we determine the minimum of the following quantities (associated with the three cases above):

$$M_1 = \min\{S(a, v_p) + 1 + \sum_{i \neq p} S(\emptyset, v_i) \mid 1 \leq p \leq \ell\}$$
$$M_2 = \min\{S(b_q, u) + 1 + \sum_{j \neq q} S(b_j, \emptyset) \mid 1 \leq q \leq k\}$$
$$M_3 = MWM(\{b_1, \ldots, b_k\}, \{v_1, \ldots, v_\ell\}) + 1$$

It is clear that all quantities required for the computation of M_1 and M_2 have been computed in an earlier stage of dynamic programming. Therefore, the running time is dominated by the cost of finding a minimum weight perfect matching; such matchings can be found in time $O(n^{.5} m \log(nN))$ on an n-node, m-edge bipartite graph with weights of maximum size N [GT89]. Since the maximum edge weight in our graphs is $n = |T| + |T'|$, and since a matching is required for each a, u pair, the total time necessary is in:

$$O(\sum_{a \in V(T)} \sum_{u \in V(T')} (|\text{children of } a| + |\text{children of } u|)^{.5}(|\text{children of } a| \cdot |\text{children of } u|)(\log n))$$

$$\subseteq O(\sum_{a \in V(T)} \sum_{u \in V(T')} n^{.5}(|\text{children of } a| \cdot |\text{children of } u|)(\log n))$$

$$\subseteq O(n^{.5} \log n \sum_{a \in V(T)} |\text{children of } a| \sum_{u \in V(T')} |\text{children of } u|)$$

$$\subseteq O(n^{2.5} \log n).$$

5 Parallel topological embedding for unordered trees

We begin by extending the definition of S to handle scarred trees.

Definition 7. For $a, s \in V(T)$ with a an ancestor of s and $u, y \in V(T')$ with u an ancestor of y, define $S(a \curlyvee s, u \curlyvee y)$ to be the size of the smallest tree S such that: (1) S is a smallest supertree of $T_a \curlyvee T_s$ and $T'_u \curlyvee T'_y$; and (2) there is a distinguished node d of S and a topological embedding ϕ from $T_a \curlyvee T_s$ to S such that $\phi(s) = d$ and a topological embedding ϕ' from $T'_u \curlyvee T'_y$ to S such that $\phi'(y) = d$. Furthermore, we define $S(a, u \curlyvee y)$ to be $S(T_a, T'_u \setminus T'_y)$ and $S(a \curlyvee s, u)$ to be $S(T_a \setminus T_s, T'_u)$.

The next two lemmas give insight into the structure of the supertree.

Lemma 8. For $a \in V(T)$ and $u \in V(T')$, let S be a smallest supertree of T_a and T'_u. Let ϕ and ϕ' be the topological embeddings from T_a into S and T'_u into S, respectively. Then either $\phi(a) = \phi'(u)$, $\phi(a)$ is a descendant of $\phi'(u)$, or $\phi(a)$ is an ancestor of $\phi'(u)$.

Proof. If neither $\phi(a)$ nor $\phi'(u)$ is the root of S, it would be possible to contract any edge from the root of S to one of its children, thereby obtaining a smaller supertree and contradicting the minimality of S. ∎

Lemma 9. For $a, u, S, \phi,$ and ϕ' as in Lemma 8, suppose that $\phi(a)$ is an ancestor of $\phi'(u)$. Then, there is some descendant c of a such that $\phi(c) = \phi'(u)$. Furthermore, S is isomorphic to the tree obtained from T_a by replacing T_c by $S_{\phi(c)}$.

Proof. Suppose that for every descendant c of a, $\phi(c) \neq \phi'(u)$. Since $\phi(a)$ is an ancestor of $\phi'(u)$, there exists at least one descendant of a which maps to an ancestor of $\phi'(u)$; otherwise we would be able to form a smaller supertree by contracting an edge on the path from $\phi(a)$ to $\phi'(u)$. In particular, let b be the descendant of a such that $\phi(b)$ is an ancestor of $\phi'(u)$, and such that no descendant of b maps to an ancestor of $\phi'(u)$. Now consider the tree S' formed by contracting the edges on the path from $\phi(b)$ to $\phi'(u)$ in S. Since no nodes of T_a map to nodes on the contracted path, clearly T_a can be topologically embedded in S'. Similarly, we can show that T'_u can be topologically embedded in S', contradicting the minimality of S.

To prove the second part of the lemma, it suffices to show that $S \curlyvee S_{\phi(c)}$ is isomorphic to $T_a \curlyvee T_c$. Since S is a smallest supertree, clearly each node in S must be in the image of ϕ, ϕ', or both; any node violating this condition could be removed to form a smaller supertree. Consequently, we can conclude that for every node $k \in V(S \curlyvee S_{\phi(c)})$, $k = \phi(d)$ for some $d \in V(T_a \curlyvee T_c)$, and therefore $S \curlyvee S_{\phi(c)}$ is a subgraph of $T_a \curlyvee T_c$. Conversely, for every node d of $T_a \curlyvee T_c$, $\phi(d) \in S \curlyvee S_{\phi(c)}$ since c and $\phi(c)$ are cut nodes in their respective trees. ∎

As a starting point for discussing the parallel algorithm for this problem, we note that a naive parallelization of the sequential algorithm described in Section 4 would be superquadratic, as both T and T' could have depth $O(n)$. To reduce the running time, we restructure T' using Brent restructuring to form the tree $\mathcal{B}_{T'}$ of depth $O(\log n)$. We then work up level by level in $\mathcal{B}_{T'}$ from the leaves to the root. At a vertex α of $\mathcal{B}_{T'}$ labelled by the tree $T'_u \curlyvee T'_y$, we compute the smallest common subtree of $T'_u \curlyvee T'_y$ and every subtree $T_a \curlyvee T_s$ of T. In particular, we determine the two quantities $S(a, u \curlyvee y)$ and $S(a \curlyvee s, u \curlyvee y)$. Upon termination, we can easily extract $S(T, T')$. It now remains to show how to compute $S(a, u \curlyvee y)$ and $S(a \curlyvee s, u \curlyvee y)$ in the cases in which α is a Brent break and α is a child break. We present in detail only those cases in which the label of vertex α under consideration contains a scar; the unscarred case is very similar.

5.1 Case 1: α is a Brent break

For any Brent break vertex α, we can assume that the children of α in $\mathcal{B}_{T'}$ are labelled by $T'_u \curlyvee T'_x$ and $T'_x \curlyvee T'_y$ where $x \in path(u, y)$. Because of the order of evaluation of values of S, when we reach α in $\mathcal{B}_{T'}$, we will have computed, for every triple $a, t,$ and s of nodes in T, $t \in path(a, s)$, the quantities $S(a\curlyvee t, u\curlyvee x)$, $S(t\curlyvee s, x\curlyvee y)$, $S(a, u\curlyvee x)$, and $S(a, x\curlyvee y)$. In the course of the algorithm, we must compute $S(a\curlyvee s, u\curlyvee y)$ for every pair of nodes a and s of T with a an ancestor of s and compute $S(a, u\curlyvee y)$ for every node a of T.

Lemma 10. *For a, s, u, x, and y defined as above,*

$$S(a\curlyvee s, u\curlyvee y) = \min_{t\in path(a,s)}\{S(a\curlyvee t, u\curlyvee x) + S(t\curlyvee s, x\curlyvee y) - 1\}$$

and

$$S(a, u\curlyvee y) = \min\{S(a, u\curlyvee x) + S(\emptyset, x\curlyvee y) - 1, \min_{t\in V(T_s)}\{S(a\curlyvee t, u\curlyvee x) + S(t, x\curlyvee y) - 1\}\}.$$

Proof. For S a smallest supertree of $T_a \curlyvee T_s$ and $T_u' \curlyvee T_y'$ with topological embeddings ϕ and ϕ' respectively, we first show that $\phi(a)$, $\phi'(u)$, $\phi'(x)$, and $\phi'(y) = \phi(s)$ all occur on a common root-leaf path in S. Since u, x and y are on a common root-leaf path in T', $\phi'(u), \phi'(x)$ and $\phi'(y)$ are also on a common root-leaf path in S. Finally, since $\phi(s) = \phi'(y)$, $\phi(a) \in path(root(S), \phi'(y))$.

To prove the first part of the lemma, it will suffice to find a node $t \in path(a, s)$ such that $S(a\curlyvee s, u\curlyvee y) = S(a\curlyvee t, u\curlyvee x) + S(t\curlyvee s, x\curlyvee y) - 1$. If $\phi(a) \in V(S_{\phi'(x)})$, then all of $T_a \curlyvee T_s$ must map to $S_{\phi'(x)}$, and by the same argument as in Lemma 9, S consists of T_u' with the subtree T_x' replaced by $S_{\phi'(x)}$. For $t = a$, we then have $S(a\curlyvee s, u\curlyvee y) = |S| = |T_u'\curlyvee T_x'| + |S_{\phi'(x)}| = S(a\curlyvee a, u\curlyvee x) + S(a\curlyvee s, x\curlyvee y) - 1$, as needed. When $\phi(a) \notin V(S_{\phi'(x)})$, we can choose as t the unique node in $path(a, s)$ such that $\phi(t)$ is in $V(S_{\phi'(x)})$ but the parent of t is not in $V(S_{\phi'(x)})$. We then have $S(a\curlyvee s, u\curlyvee y) = |S| = |S\curlyvee S_{\phi'(x)}| + |S_{\phi'(x)}| = S(a\curlyvee t, u\curlyvee x) + S(t\curlyvee s, x\curlyvee y) - 1$, as needed.

The proof of the second part of the lemma is similar. ∎

5.2 Case 2: α is a child break

We now consider a vertex α in $\mathcal{B}_{T'}$ with children $\beta_1, \ldots, \beta_\ell$. The label of α is a tree $T_u'\curlyvee T_y'$, with $v_1, \ldots, v_\ell$ the children of u and the scar y a descendant of v_p. Then the label of β_p is the tree $T_{v_p}'\curlyvee T_y'$, and for all $j \neq p$, the label of β_j is the tree T_{v_j}'.

As in Case 1, we conclude that certain values have been computed when we reach α. In particular, for every pair of nodes a, s with $b_1, \ldots, b_k$ children of a and s a descendant of b_q, the following quantities have been computed:

1. $S(a, v_j)$, $1 \leq j \leq \ell, j \neq p$;
2. $S(a, v_p\curlyvee y)$;
3. $S(a\curlyvee s, v_p\curlyvee y)$;
4. $S(b_i, u\curlyvee y)$, $1 \leq i \leq k, i \neq q$;
5. $S(b_q\curlyvee s, u\curlyvee y)$;
6. $S(b_i, v_j)$, $1 \leq i \leq k, 1 \leq j \leq \ell, i \neq q, j \neq p$;
7. $S(b_i, v_p\curlyvee y)$, $1 \leq i \leq k, i \neq q$; and
8. $S(b_q\curlyvee s, v_p\curlyvee y)$.

To execute the algorithm, we must compute $S(a\curlyvee s, u\curlyvee y)$ and $S(a, u\curlyvee y)$ for every pair of nodes a and s of T with a an ancestor of s. We require the next lemma for computing $S(a\curlyvee s, u\curlyvee y)$.

Lemma 11. *For $a, s, u, y, b_1, \ldots, b_k, v_1, \ldots, v_\ell, p$, and q as defined above, let*

$M_1 = S(a\curlyvee s, v_p\curlyvee y) + S(\emptyset, u\curlyvee v_p) - 1;$

$M_2 = S(b_q\curlyvee s, u\curlyvee y) + S(a\curlyvee b_q, \emptyset) - 1;$ and

$M_3 = MWM(\{b_1, \ldots, b_{q-1}, b_{q+1}, \ldots, b_k\}, \{v_1, \ldots, v_{p-1}, v_{p+1}, \ldots, v_\ell\}) + S(b_q\curlyvee s, v_p\curlyvee y) + 1.$

Then, $S(a\curlyvee s, u\curlyvee y) = \min\{M_1, M_2, M_3\}$.

Proof. Let ϕ and ϕ' be the required topological embeddings from $T_a \Upsilon T_s$ and $T_u' \Upsilon T_y'$ into a smallest common supertree S such that $\phi(s) = \phi'(y)$. Then, by Lemma 8, either $\phi(a) = root(S)$, $\phi'(u) = root(S)$ or both; we consider each case in turn.

If $\phi(a) = root(S)$ and $\phi'(u) \neq root(S)$, then $\phi'(u) \in S_{\phi(b_q)}$. Hence, the image of $T_u' \Upsilon T_y'$ is contained entirely in $S_{\phi(b_q)}$ and $S_{\phi(b_q)}$ is a smallest supertree of $T_{b_q} \Upsilon T_s$ and $T_u' \Upsilon T_y'$. Moreover, the subtrees of T rooted at the children of a (with the exception of T_{b_q}) are all contained in $S \Upsilon S_{\phi(b_q)}$. Thus, since b_q is counted twice, $S(a \Upsilon s, u \Upsilon y) = M_2$. Similarly, we can show that if $\phi(a) \neq root(S)$ and $\phi'(u) = root(S)$ then $S(a \Upsilon s, u \Upsilon y) = M_1$.

If $\phi(a) = \phi'(u)$, then since $\phi(u) = \phi'(y)$ there must be a child f of S such that $\phi(b_q), \phi'(v_p) \in S_f$ and no other node of either $T_a \Upsilon T_{b_q}$ or $T_u' \Upsilon T_{v_p}'$ maps to S_f. Therefore, S_f is a smallest supertree of $T_{b_q} \Upsilon T_s$ and $T_{v_p}' \Upsilon T_y'$. But then the size of $S \Upsilon S_f$ is $MWM(\{b_1, \ldots, b_k\} \backslash \{b_q\}, \{v_1, \ldots, v_\ell\} \backslash \{v_p\})$ and it follows that $S(a \Upsilon s, u \Upsilon y) = M_3$. ∎

Due to its similarity to the preceding lemma, the following lemma is stated without proof.

Lemma 12. *For $a, u, y, b_1, \ldots, b_k, v_1, \ldots, v_\ell, p,$ and q as defined above, let*

$$M_1 = \min_{1 \leq j \leq \ell, j \neq p} \{S(a, v_j) + S(\emptyset, u \Upsilon v_j) - 1\};$$

$$M_2 = \min_{1 \leq i \leq k, i \neq q} \{S(b_i, u \Upsilon y) + S(a \Upsilon b_i, \emptyset) - 1\};$$

$$M_3 = S(a, v_p \Upsilon y) + S(\emptyset, u \Upsilon v_p) - 1; \text{ and}$$

$$M_4 = MWM(\{b_1, \ldots, b_k\}, \{v_1, \ldots, v_\ell\} \backslash \{v_p\}) + S(\emptyset, v_p \Upsilon y) + 1.$$

Then, $S(a, u \Upsilon y) = \min\{M_1, M_2, M_3, M_4\}$.

5.3 Handling T

The depth of T remains a hurdle to a fast parallel algorithm. We now show that when processing a vertex α of $\mathcal{B}_{T'}$ labelled by the tree $T_u' \Upsilon T_y'$, we can compute $S(a \Upsilon s, u \Upsilon y)$ and $S(a, u \Upsilon y)$ in parallel for all node pairs a and s in T. When α is a Brent break, this is possible by using the same procedure as in Section 5.1. The case of a child break, however, is more complicated. In particular, since we are handling all pairs a and s in T simultaneously, we will not necessarily know the quantity $S(b_q \Upsilon s, u \Upsilon y)$ before $S(a \Upsilon s, u \Upsilon y)$ is computed, and hence we will not be able to compute M_2 in Lemma 11. Similarly, we will not know $S(b_i, u \Upsilon y)$, $1 \leq i \leq k, i \neq q$ before we determine $S(a, u \Upsilon y)$ and therefore we will not be able to compute M_2 in Lemma 12.

However, notice that in both these lemmas, we can compute M_1 and M_3 (and in Lemma 12 M_4) directly for all descendants c of a. In particular, to compute M_2 in Lemma 11, consider a smallest supertree S of $T_a \Upsilon T_s$ and $T_u' \Upsilon T_y'$ such that ϕ and ϕ' are topological embeddings of $T_a \Upsilon T_s$ and $T_u' \Upsilon T_y'$ into S respectively such that $\phi(a) = root(S)$ and $\phi'(u) = root(S)$. Then by Lemma 8, there is a descendant c of a, $c \in path(a, s)$, such that $\phi(c) = \phi'(u)$. Let $d_1, \ldots, d_{k'}$ be the children of c with s a descendant of d_m. Then,

$$S(c \Upsilon s, u \Upsilon y) = MWM(\{d_1, \ldots, d_{k'}\} \backslash \{d_m\}, \{v_1, \ldots, v_\ell\} \backslash \{v_q\}) + S(d_m \Upsilon s, v_p \Upsilon y) + 1.$$

Although we do not know which node of $path(a, s)$ corresponds to c, we can try all possibilities in parallel. Then M_2 will be $\min\{S(c \Upsilon s, u \Upsilon y) + S(a \Upsilon c, \emptyset) - 1 \mid c \in path(a, s)\}$.

Similarly, we can compute the value of M_2 in Lemma 12 as follows:

$$M_2 = \min_{c \in V(T_a), c \neq a} \{S(c, u \Upsilon y) + S(a \Upsilon c, \emptyset) - 1\}$$

$$= \min_{c \in V(T_a), c \neq a} \{\{MWM(\{d_1, \ldots, d_{k'}\}, \{v_1, \ldots, v_\ell\}) \mid d_1, \ldots, d_{k'} \text{ children of } c\} + S(a \Upsilon c, \emptyset)\}.$$

Theorem 13. *For T and T' unordered trees, $n = |T|+|T'|$, the* Smallest Supertree Problem *can be solved in randomized $O(\log^3 n)$ parallel time using $O(n^{7.5})$ processors for the problem of topological embedding.*

Proof. The correctness of the algorithm follows from the proceeding discussion; we show that the algorithm meets the resource constraints in the theorem. Since the Brent tree of T' can be created in time $O(\log^2 n)$ with $O(n^4)$ processors [GN95b], it will suffice to show that each of the $O(\log n)$ levels of the Brent tree can be processed in time $O(\log^2 n)$ using $O(n^{7.5})$ processors. We give details of the calculations for handling a child break using Lemma 11; the remaining cases require fewer resources.

For a particular vertex α, and a particular pair of nodes a and s in T, M_1 can be determined using table look-up. M_3 can be determined using a minimum weight perfect matching, which can be set up and solved in randomized time $O(\log^2 n)$ with $O(n^{5.5})$ processors [MVV87]. To calculate M_2 requires solving matching problems on all $O(n)$ nodes c that are proper descendants of a, requiring a total of $O(n^{6.5})$ processors. The final processor count is derived from the fact that there are $O(n^2)$ choices for the nodes a and s in T and $O(n)$ vertices at a level of the Brent tree, reduced to $O(n^{7.5})$ by reusing matching information (since the matchings to compute M_2 are the same as those required for M_3). $\blacksquare$

6 Topological embedding for ordered trees

The main distinction between the algorithms for unordered and ordered trees is the type of matching problem that must be solved. The dynamic programming formulation given in Section 4 and the Brent restructuring used in Section 5 are the same for ordered and unordered trees. In the remainder of this section we describe the new matching problem and sequential and parallel algorithms for finding supertrees of ordered trees.

When processing nodes u and a, the ordering among the children of u and a is significant; we require a method of matching the children so that the order is preserved. In essence, we would like to solve a matching problem on the graph $G = (X, Y, E)$ where $X = \{b_1, \ldots b_k\}$, $Y = \{v_1, \ldots, v_\ell\}$, and E consists of edges (b_j, v_i) with weights $S(b_j, v_i)$. A solution to this problem would be the smallest cost matching in which no edges cross and the cost of a solution is the sum of the edge weights plus the sum over all unmatched nodes b_j of $S(b_j, \emptyset)$, plus the sum over all unmatched nodes v_i of $S(\emptyset, v_i)$. For $C[i, j]$ denoting the weight of this matching, the problem can be formulated in terms of dynamic programming in a manner similar to that of Jiang, Wang, and Zhang in their work on tree alignment [JWZ94]:

$$C[i, j] = \min\{C[i-1, j-1] + S(b_i, v_j), C[i-1, j] + S(b_i, \emptyset), C[i, j-1] + S(\emptyset, v_j)\}$$

Since $C[i, j]$ can be computed from $C[i-1, j-1]$, $C[i-1, j]$, and $C[i, j-1]$ in constant time, we can compute $C[k, \ell]$ sequentially in time $O(k \cdot \ell)$, resulting in the following theorem.

Theorem 14. *For T and T' ordered trees, $n = |T| + |T'|$, the* Smallest Supertree Problem *can be solved sequentially in $O(n^2)$ time.*

For the parallel algorithm, we create an edge-weighted graph with a node for each pair (i, j), and directed edges from (i, j) to $(i-1, j)$ of weight $S(b_i, \emptyset)$, to $(i, j-1)$ of weight $S(\emptyset, v_j)$, and to $(i-1, j-1)$ of weight $S(b_i, v_j)$. By the results in Section 5, all edge weights will have been computed by the time this problem is to be solved. Then, finding a minimum weight matching corresponds to finding the shortest path in the graph, a problem that can be solved in time $O(\log^2 n)$ using $O(n^3)$ processors [GM88]. Using an analysis similar to that in Theorem 13, we obtain the following:

Theorem 15. *For T and T' ordered trees, $n = |T| + |T'|$, the* Smallest Supertree Problem *can be solved in parallel time $O(\log^3 n)$ using $O(n^8)$ processors for the problem of topological embedding.*

As a corollary to this result, we can improve the time bound on the Zhang and Shasha parallel algorithm for finding the smallest supertree for two ordered trees under subgraph isomorphism:

Corollary 16. *For T and T' ordered trees, the* Smallest Supertree Problem *can be solved in parallel time $O(\log^3 n)$ using $O(n^8)$ processors for the problem of subgraph isomorphism.*

7 Directions for further research

Although all the algorithms in this paper take as input unlabelled trees, it is straightforward to extend them to handle labels. The known algorithms for computing smallest supertrees for minor containment are for ordered, labelled trees [JWZ94] and unordered, leaf-labelled trees [War93]; it would be interesting to consider other labelling schemes as well as other embeddings.

The work on the *Smallest Supertree Problem*, and most of the work on the *Largest Subtree Problem* has concentrated on two-input versions of the problem. Keselman and Amir have shown that the three-input version of the *Largest Subtree Problem* is NP-complete for subgraph isomorphism [KA94]. There are no known such results for the *Smallest Supertree Problem*.

Acknowledgements

We would like to thank Jianghai Fu for bringing to our attention the work of Jiang, Wang, and Zhang and the subject of supertrees in general, Prabhakar Ragde for fine-tuning of our analysis, and an anonymous referee for careful reading of our paper.

References

[Bre74] R. Brent, The parallel evaluation of general arithmetic expressions, *Journal of the ACM* **21**, 2 (1974), pp. 201–206.

[Chu87] M. J. Chung, $O(n^{2.5})$ time algorithms for the subgraph homeomorphism problem on trees, *Journal of Algorithms* **8**, (1987), pp. 106–112.

[DGM90] M. Dubiner, Z. Galil, and E. Magen, Faster tree pattern matching, *Proceedings of the 31st Annual Symposium on Foundations of Computer Science*, pp. 145–150, 1990.

[FT94a] M. Farach and M. Thorup, Fast comparison of evolutionary trees, *Proceedings of the Fifth Annual ACM-SIAM Symposium on Discrete Algorithms*, pp. 481–488, 1994.

[FT94b] M. Farach and M. Thorup, Optimal evolutionary tree comparison by sparse dynamic programming, *Proceedings of the 35th Annual Symposium on Foundations of Computer Science*, pp. 770-779, 1994.

[FG85] C.R. Finden and A.D. Gordon, Obtaining common pruned trees, *Journal of Classification* **2**, (1985), pp. 255–276.

[GT89] H. Gabow and R. Tarjan, Faster scaling algorithms for network problems, *SIAM Journal on Computing* **18**, 5 (1989), pp. 1013–1036.

[GM88] H. Gazit and G. L. Miller, An improved parallel algorithm that computes the BFS numbering of a directed graph, *Information Processing Letters* **28**, (1988), pp. 61–65.

[GKMS90] P. Gibbons, R. Karp, G. Miller, and D. Soroker, Subtree isomorphism is in random NC, *Discrete Applied Mathematics* **29** (1990), pp. 35–62.

[GN95a] A. Gupta and N. Nishimura, Finding largest common embeddable subtrees, *Proceedings of the Twelfth Annual Symposium on Theoretical Aspects of Computer Science*, pp. 397–408, 1995.

[GN95c] A. Gupta and N. Nishimura, Finding largest subtrees and smallest supertrees, Technical Report CS-95-30, Department of Computer Science, University of Waterloo, July 1995.

[GN95b] A. Gupta and N. Nishimura, The parallel complexity of tree embedding problems, *Journal of Algorithms* **18**, 1 (1995), pp. 176–200.

[GN94] A. Gupta and N. Nishimura, Sequential and parallel algorithms for embedding problems on classes of partial k-trees, *Proceedings of the Fourth Scandinavian Workshop on Algorithm Theory*, pp. 172–182, 1994.

[JWZ94] T. Jiang, L. Wang, and K. Zhang, Alignment of trees – an alternative to tree edit, *Combinatorial Pattern Matching*, pp. 75–86, 1994.

[KA94] D. Keselman and A. Amir, Maximum Agreement Subtree in a Set of Evolutionary Trees - Metrics and Efficient Algorithms, *Proceedings of the 35th Annual Symposium on Foundations of Computer Science*, pp. 758-769, 1994.

[KM91] P. Kilpeläinen and H. Mannila, Ordered and unordered tree inclusion, to appear in *SIAM Journal on Computing*, preliminary version appeared as The Tree Inclusion Problem, *TAP-SOFT'91, Proceedings of the International Joint Colloquium on Trees in Algebra and Programming (CAAP 91)*, pp. 202-214, 1991.

[Kos89] S. R. Kosaraju, Efficient tree pattern matching, *Proceedings of the 30th Annual Symposium on Foundations of Computer Science*, pp. 178-183, 1989.

[KKM92] E. Kubicka, G. Kubicki, and F.R. McMorris, On agreement subtrees of 2 binary trees, *Congressus-Numerantium* **88** (1992), pp. 217-224.

[LK89] A. Lingas and M. Karpinski, Subtree isomorphism is NC reducible to bipartite perfect matching, *Information Processing Letters* **30** (1989), pp. 27-32.

[Mat78] D. Matula, Subtree isomorphism in $O(n^{5/2})$, *Annals of Discrete Mathematics* **2** (1978), pp. 91-106.

[MVV87] K. Mulmuley, U. Vazirani, and V. Vazirani, Matching is as easy as matrix inversion, *Proceedings of the 19th Annual ACM Symposium on the Theory of Computing*, pp. 345-354, 1987.

[Rey77] S. W. Reyner, An analysis of a good algorithm for the subtree problem, *SIAM Journal on Computing* **6**, 4, (1977), pp. 730-732.

[SW] M. Steel and T. Warnow, Kaikoura tree theorems: Computing the maximum agreement subtrees. Submitted for publication.

[VR89] R. M. Verma and S. W. Reyner, An analysis of a good algorithm for the subtree problem, corrected, *SIAM Journal on Computing* **18**, 5 (1989), pp. 906-908.

[War93] T. Warnow, Tree compatibility and inferring evolutionary history, *Proceedings of the Fourth Annual ACM-SIAM Symposium on Discrete Algorithms*, pp. 382-391, 1993.

[ZS89] K. Zhang and D. Shasha, Simple fast algorithms for the editing distance between trees and related problems, *SIAM Journal on Computing* **18**, 6 (1989), pp. 1245-1262.

Weighted Domination on Cocomparability Graphs

Maw-Shang Chang *

Department of Computer Science and Information Engineering
National Chung Cheng University
Min-Hsiun, Chiayi 621, Taiwan, Republic of China
Email: mschang@cs.ccu.edu.tw

Abstract. It is shown in this paper that the weighted domination problem and its two variants, the weighted connected domination and weighted total domination problems are NP-complete on cocomparability graphs when arbitrary integer vertex weights are allowed and all of them can be solved in polynomial time if vertex weights are integers and less than or equal to a constant c. Besides, an $O(|V|^2)$ algorithm is given for the weighted independent perfect domination problem of a cocomparability graph $G = (V, E)$.

1 Introduction

Many NP-complete graph problems become tractable when restricted to special subclasses of perfect graphs. This motivates the search for larger classes for which the problem is still tractable. Interval and permutation graphs are two well-known graph classes that admit many polynomial time algorithms for NP-complete graph problems. However, if we generalize both in a natural way to chordal and comparability graphs respectively, many problems become NP-complete. Recently, attention has been drawn to cocomparability graphs, a class of graph that properly contains interval and permutation graphs. It seems very likely that problems solvable in polynomial time on interval and permutation graphs remain tractable on cocomparability graphs. For example, hamiltonian cycle and hamiltonian path problems were shown solvable in polynomial time [6, 7, 8]. The domination problem and its variants, independent, connected and total domination problems are solvable too, except that finding a minimum cardinality dominating clique is NP-hard [11]. In addition, the weighted independent perfect domination problem is solvable in polynomial time [3].

We note that the domination, connected and total domination problems are NP-complete on chordal graphs. The independent domination problem is solvable in linear time on chordal graphs [9]. But it becomes NP-complete when arbitrary integer vertex weights are allowed [2]. On the other hand, the weighted independent domination problem is solvable in polynomial time on cocomparability graphs even arbitrary vertex weights are allowed [11]. Whether the weighted

---

* Supported partly by the National Science Council of the Republic of China under grant NSC 85-2121-M-194-020.

domination, connected and total domination problems are solvable in polynomial time on cocomparability graphs was left unresolved in [11]. One is tempted to conjecture that they are solvable in polynomial time. Our studies showed the conjecture was not correct. In subsection 2.1, we prove that the weighted domination, connected and total domination problems are NP-complete on cobipartite graphs when arbitrary integer vertex weights are allowed. Note that cobipartite graphs are properly contained in cocomparability graphs. In subsection 2.2, we give a polynomial time algorithm for the weighted domination on cocomparability graphs with bounded integer vertex weights, i.e. the weight of every vertex is an integer and is less than or equal to a constant c. The algorithm can be easily modified to solve the weighted connected domination and total domination problems on cocomparability graphs with bounded integer vertex weights.

The running times of the algorithms given by Chang, Rangan and Coorg [3] for the weighted independent perfect domination problem are $O(|V||E|)$ and $O(|V|^{2.37})$, respectively. The algorithms suffer from determining whether two vertices have a neighbor in common for every pair of non-adjacent vertices. In section 3, we propose an $O(|V|^2)$ algorithm that avoids this bottleneck of their algorithms.

2 Weighted Domination

A *comparability graph* is a graph $G = (V, E)$ whose vertex set has a *transitive ordering*, i.e., an ordering of V into $1, 2, \ldots, n$ such that $i < j < k, (i, j) \in E$, and $(j, k) \in E$ imply $(i, k) \in E$. There is an $O(|V|^{2.37})$ algorithm [15] to test if a graph is a comparability graph. In the case of a positive answer, an algorithm produces a transitive ordering in $O(|V| + |E| \log |V|)$ time [12]. A *cocomparability graph* is the complement of a comparability graph, or, equivalently, if its vertex set has a *cocomparability ordering*, which is an ordering of V into $1, 2, \ldots, n$ such that $i < j < k$ and $(i, k) \in E$ imply $(i, j) \in E$ or $(j, k) \in E$. Through out the paper, we assume that vertices are numbered in cocomparability ordering with $1, 2, \ldots, n$. A *dominating set* of a graph $G = (V, E)$ is a subset D of V such that every vertex not in D is adjacent to at least one vertex in D. The *domination problem* is to find a minimum dominating set of the given graph. Suppose that every vertex $v \in V$ is associated with a weight, denoted by $w(v)$. The *weighted domination problem* involves finding a dominating set D of the given graph such that its weight $w(D) = \sum\{w(v) : v \in D\}$ is minimum. The domination problem is just the weighted domination problem with $w(v) = 1$ for each vertex v. A dominating set D is *independent, connected*, or *total* if the subgraph induced by D has no edge, is connected, or has no isolated vertex, respectively. A set D of vertices is called a dominating clique if D is a dominating set and the subgraph induced by D is complete. A lot of work has been done to clarify the algorithmic complexity of these problems when restricted to special classes of graphs. For an overview see [5].

2.1 The NP-complete Results

In this subsection we prove that all of the weighted domination, connected domination and total domination problems are NP-complete for cocomparability graphs when arbitrary integer vertex weights are allowed. These problems can be formulated as follows:

WEIGHTED DOMINATION (CONNECTED, TOTAL DOMINATION) PROBLEM

INSTANCE: Graph $G = (V, E)$, each vertex $v \in V$ is associated with a weight $w(v)$; a number W, $W \leq \sum\{w(v) : v \in V\}$.

QUESTION: Is there a subset $D \subseteq V$ with $w(D) \leq W$ such that D is a dominating set (a connected dominating set or a total dominating set) of G?

The reduction of the proof is from the hitting set problem (See [SP8] of [10]).

HITTING SET PROBLEM

INSTANCE: Collection C of subsets of a finite set S, positive integer $K \leq |S|$.

QUESTION: Is there a subset $S' \subseteq S$ with $|S'| \leq K$ such that S' contains at least one element of each subset in C?

Theorem 1. *The weighted domination, connected domination and total domination problems on cocomparability graphs are NP-complete.*

Proof. Clearly, the weighted domination, connected domination, total domination problems are NP problems. We transform the hitting set problem to these problems. For an instance of the hitting set problem we construct graph $G = (V, E_{SS} \cup E_{SC} \cup E_{CC})$ where $V = S \cup C$, $E_{SS} = \{(s_i, s_j) : s_i, s_j \in S, i \neq j\}$, $E_{SC} = \{(s_i, C_i) : s_i \in S, C_i \in C, s_i \in C_i\}$, and $E_{CC} = \{(C_i, C_j) : C_i, C_j \in C, i \neq j\}$. The weight of each vertex in V is as follows: $w(s_i) = 1$ for all $s_i \in S$ and $w(C_i) = |S| + 1$ for all $C_i \in C$. It is straightforward to verify that G is a cocomparability graph since both S and C are complete subgraphs in G and G can be considered as the complement of a bipartite graph with two vertex sets S and C.

If the hitting set problem instance has a hitting set D of size at most K, then D is a dominating set of G too. Apparently, the induced subgraph of D is connected and thus has no isolated vertices. In other words, we have a dominating set, a connected dominating set and a total dominating set of weight K. On the other hand, suppose G has a dominating set D of weight K where $K \leq |S|$. Since the weight of any vertex in C is greater than $|S|$, $D \subseteq S$ and hence the induced subgraph of D is connected and thus has no isolated vertices. Since D dominates C, each $C_i \in C$ has at least one element in D. Hence D is a hitting set of size K of the hitting set problem instance. Q.E.D.

2.2 Polynomial Algorithms

In this subsection, we give a polynomial time algorithm for the weighted domination problem on cocomparability graphs where vertex weights are bounded integers. Manacher and Mankus showed that any algorithm for finding a minimum weighted dominating set for nonnegative weights can be extended to handle negative weights without loss of efficiency [13]. Thus, we assume that $G = (V, E)$ is a cocomparability graph such that $w(v)$ is an integer and $0 \leq w(v) \leq c$ for every $v \in V$ where c is a constant. For technical reason, we add two isolated vertices, 0 and $n + 1$ with $w(0) = w(n + 1) = 0$ to G to obtain a new cocomparability graph G_a with a cocomparability ordering $0, 1, 2, \ldots, n, n+1$. Note that D is a dominating set of G if and only if $D \cup \{0, n + 1\}$ is a dominating set of G_a. For simplicity, we assume that G is the new graph added with 0 and $n + 1$. We need some notation.

- For $0 \leq i \leq n + 1$, $N(i) = \{j : j \in V, (i, j) \in E\}$, $N[i] = N(i) \cup \{i\}$, $N^+[i] = \{j : j \in N[i], i \leq j\}$, and $N^-[i] = \{j : j \in N[i], j \leq i\}$.
- For $0 \leq i < j \leq n + 1$, let $N[i, j] = \{k : k \in N[i] \cup N[j], i \leq k \leq j\}$ and $B_{i,j} = \{k : k \in V, i < k < j\}$, respectively.
- For a subset X of V, $N[X] = \bigcup_{i \in X} N[i]$, $ZW(X) = \{i : i \in X, w(i) = 0\}$, $\min(X)$ and $\max(X)$ denote the leftmost and rightmost vertices of X in cocomparability ordering, respectively. And $G[X]$ denotes the induced subgraph of X in graph G.
- For a vertex $v \in V$ (resp. a set X of vertices of V) and a subset $D \subseteq V$, a vertex in $N[v] - N[D - \{v\}]$ (resp. $N[X] - N[D - X]$) is called a *private neighbor of v (resp. X) with respect to D*. We may say that a vertex $u \in N[v] - N[D - \{v\}]$ (resp. $u \in N[X] - N[D - X]$) is a private neighbor of v (resp. X) if D is understood without ambiguity.

The following lemma can be proved easily.

Lemma 2. *Suppose S is a set of vertices of cocomparability graph G where $G[S]$ is connected. Then, S dominates the vertex set $\{\min(S), \min(S)+1, \ldots, \max(S)\}$.*

Let D be a minimum weighted dominating set of cocomparability graph $G = (V, E)$ where $G[D]$ has $q+2$ connected components, $\{0\} \equiv C_0, C_1, \ldots, C_q, C_{q+1} \equiv \{n + 1\}$, sorted in increasing order of $\max(C_r)$'s for $0 \leq r \leq q + 1$. For each component C_r, $0 < r < q + 1$, define $lpr(C_r)$ and $rpr(C_r)$ to be the minimum and maximum private neighbor of C_r with respect to D respectively. Let $S(C_r)$ be a shortest path from $\min(C_r)$ to $\max(C_r)$ in the induced subgraph $G[C_r]$ of C_r. Let $L(C_r)$ (resp. $R(C_r)$) be the set of vertices of $C_r - S(C_r)$ that has a private neighbor v with respect to D such that $v < \min(C_r)$ (resp. $\max(C_r) < v$). Note that $D \cup ZW(V)$ is also a minimum weighted dominating set. Thus, there exists a minimum weighted dominating set D such that $ZW(V) \subseteq D$.

Lemma 3. *Suppose D is a minimum weighted dominating set of cocomparability graph G where $ZW(V) \subseteq D$ and $G[D]$ has $q + 2$, $q > 0$, connected components, $C_0 \equiv \{0\}, C_1, C_2, \ldots, C_q, C_{q+1} \equiv \{n+1\}$, sorted in increasing order of $\max(C_r)$'s for $0 \leq r \leq q + 1$. For $0 < r \leq q + 1$,*
(1) $\max(C_{r-1}) < \min(C_r)$;
(2) $L(C_r) \subseteq N^+[\min(C_r)]$ and $R(C_r) \subseteq N^-[\max(C_r)]$;
(3) $(C_{r-1} \cap N^-[i]) \cup (C_r \cap N^+[j])$ dominates $B_{i,j}$ where $i = \max(C_{r-1})$ and $j = \min(C_r)$;

(4) $(D - C_r) \cup S(C_r) \cup \{lpr(C_r), u, rpr(C_r), v\}$ is a dominating set of G where $u \in C_r$, $v \in C_r$, $(lpr(C_r), u) \in E$, and $(rpr(C_r), v) \in E$;

(5) $(D - C_r) \cup S(C_r) \cup (C_r \cap N^+[\min(C_r)]) \cup (C_r \cap N^-[\max(C_r)])$ is a dominating set of G;

(6) if $|S(C_r)| \leq 3$, then $w(C_r) \leq 7c$;

(7) if $|S(C_r)| > 3$, then $(C_r \cap N^+[\min(C_r)]) \cap (C_r \cap N^-[\max(C_r)]) = \emptyset$;

(8) if $|S(C_r)| > 3$, then $w(C_r \cap N^+[\min(C_r)]) \leq 4c$ and $w(C_r \cap N^-[\max(C_r)]) \leq 4c$;

Proof. Omitted.   Q.E.D.

Immediately following statement (5) of Lemma 3, we have the following lemma.

Lemma 4. *Given a cocomparability graph G, there exists a minimum weighted dominating set D such that $C = S(C) \cup (C_r \cap N^+[\min(C_r)]) \cup (C_r \cap N^-[\max(C_r)])$ for every connected component C of $G[D]$.*

Our algorithm solves the problem by constructing a weighted directed graph $G'' = (V'', E'')$ such that a minimum weighted dominating set D of cocomparability graph $G = (V, E)$, satisfying the property given in Lemma 4, corresponds to a minimum weighted path of G''. Note that we refer to an element of V'' as a node and an element of V as a vertex for clarity. Each node of G'' corresponds to a set of vertices of G. The basic ideas for constructing G'' are as follows. For each vertex $i \in V$, we construct two nodes, v_i and $Z1(i)$, and two sets, $XL(i)$ and $XR(i)$, of nodes:

$XL(i) = \{X : i \in X, X \subseteq N^+[i], ZW(N^+[i]) \subseteq X, w(X) \leq 4c\}$ and
$XR(i) = \{X : i \in X, X \subseteq N^-[i], ZW(N^-[i]) \subseteq X, w(X) \leq 4c\}$.

For each $0 < i < j < n+1$ where $\max(ZW(N^+[i])) \leq j$ and $i \leq \min(ZW(N^-[j]))$, we construct a set $Z2(i, j)$ of nodes:

$Z2(i, j) = \{Z : Z \subseteq N[i, j], w(Z) \leq 7c, ZW(N^+[i]), ZW(N^-[i]) \subseteq Z, N^+[i] \cap N^-[j] \cap Z \neq \emptyset\}$.

Let $V'' = V' \cup XR \cup XL \cup Z1 \cup Z2$ where $XR = \bigcup_{1 \leq i \leq n} XR(i)$, $XL = \bigcup_{1 \leq i \leq n} XL(i)$, $Z1 = \{Z1(i) : 0 \leq i \leq n + 1\}$, $Z2 = \bigcup_{1 \leq i < j \leq n} Z2(i, j)$, and $V' = \{v_i : 0 \leq i \leq n + 1\}$. Clearly, $|V''| = O(|V|^{7c})$.

Next, we show how to construct edges between two nodes of V''. Let the edge of G'' directed from node X to node Y be denoted by $\langle X, Y \rangle$. For each edge (i, j) in G, we construct two directed edges $\langle v_i, v_j \rangle$ and $\langle v_j, v_i \rangle$. Let $E_d = \{\langle v_i, v_j \rangle, \langle v_j, v_i \rangle : v_i, v_j \in V', (i, j) \in E\}$. For each $X \in XL(i)$ and $v_j \in V'$ where $0 < i < j \leq n$, $j \notin N^+(i)$, and $N(j) \cap X \neq \emptyset$, we construct a directed edge $\langle X, v_j \rangle$. Let $E_L = \{\langle X, v_j \rangle : v_j \in V', X \in XL(i), 0 < i < j \leq n, j \notin N(i), N(j) \cap X \neq \emptyset\}$. By symmetry, we construct the set E_R of directed edges in G'' where $E_R = \{\langle v_i, X \rangle : v_i \in V', X \in XR(j), 0 < i < j \leq n, i \notin N(j), N(i) \cap X \neq \emptyset\}$. For $X_1 \in XL(i)$ and $X_2 \in XR(j)$ where $0 < i < j < n + 1$, $(i, j) \notin E$, $X_1 \cap X_2 = \emptyset$, $\max(X_1) < j$, $i < \min(X_2)$, and there exists a vertex $h \in X_1$ and a vertex $k \in X_2$ such that $(h, k) \in E$, we construct a directed edge $\langle X_1, X_2 \rangle$. Let $E_{LR} = \{\langle X_1, X_2 \rangle : X_1 \in XL(i), X_2 \in XR(j), 0 < i < j < n + 1, (i, j) \notin E,$

$X_1 \cap X_2 = \emptyset$, $\max(X_1) < j$, $i < \min(X_2)$, and there exists a vertex $h \in X_1$ and a vertex $k \in X_2$ such that $(h, k) \in E\}$. For $X_1 \in XR(i)$ and $X_2 \in XL(j)$ where $0 < i < j < n + 1$, and $X_1 \cup X_2$ dominates $B_{i,j}$, we construct a directed edge $\langle X_1, X_2 \rangle$. Let $E_{RL} = \{\langle X_1, X_2 \rangle : X_1 \in XR(i), X_2 \in XL(j), 0 < i < j < n+1, X_1 \cup X_2 \text{ dominates } B_{i,j}\}$. Other sets of directed edges of G'' are E_{RZ}, E_{ZL}, and E_Z. An edge of E_{RZ} is directed from a node of XR to a node of $Z1 \cup Z2$. An edge of E_{ZL} is directed from a node of $Z1 \cup Z2$ to a node of XL. An edge of E_Z is directed from a node of $Z1 \cup Z2$ to another node of $Z1 \cup Z2$. They are formally defined in the following:
$E_{RZ} = \{\langle X, Z \rangle : X \in XR, Z \in Z1 \cup Z2, \max(X) < \min(Z), X \cup Z \text{ domi-}$ nates $B_{\max(X),\min(Z)}\}$, $E_{ZL} = \{\langle Z, X \rangle : X \in XL, Z \in Z1 \cup Z2, \max(Z) < \min(X), X \cup Z \text{ dominates } B_{\max(Z),\min(X)}\}$, and $E_Z = \{\langle Z_1, Z_2 \rangle : Z_1, Z_2 \in Z1 \cup Z2, \max(Z_1) < \min(Z_2), Z_1 \cup Z_2 \text{ dominates } B_{\max(Z_1),\min(Z_2)}\}$.
Finally, we let $E'' = E_d \cup E_L \cup E_R \cup E_{LR} \cup E_{RL} \cup E_{RZ} \cup E_{ZL} \cup E_Z$.

Each node $q \in V''$ corresponds to a set of vertices in V, denoted by $V(q)$. Define the weight $w(q)$ of each node $q \in V''$ to be the total weight of vertices in $V(q)$, i.e. $w(q) = w(V(q))$. For a set Q of nodes of V'', define $V(Q) = \bigcup_{q \in Q} V(q)$ and $w(Q) = \sum_{q \in Q} w(q)$. For simplicity, we use q to denote $V(q)$ for a node q of G'' if there is no ambiguity.

Let D be a minimum weighted dominating set satisfying the condition given in Lemma 4 and C be a connected component of $G[D]$ where $i = \min(C)$ and $j = \max(C)$. If $C = \{i\}$, then C corresponds to node $Z1(i)$. If $1 < |S(C)| \leq 3$, then it corresponds to a node of $Z2(i, j)$ since $w(C) \leq 7$. If $|S(C)| = 4$, then C can be partitioned into two parts such that one part is $C \cap N^+[i]$ which corresponds to a node X of $XL(i)$, the other part is $C \cap N^-[j]$ which corresponds to a node Y of $XR(j)$, and there exists an edge $\langle X, Y \rangle \in E_{LR}$. If $|S(C)| > 4$, then C can be partitioned into three parts such that one part is $C \cap N^+[i]$ which corresponds to a node X of $XL(i)$, the second part is $C \cap N^-[j]$ which corresponds to a node Y of $XR(j)$, the third part is $S(C) - (X \cup Y)$ which corresponds to a path from node v_h to node v_k where $S(C) = i \to i' \to h \to \cdots \to k \to j' \to j$, and there exist edges $\langle X, v_h \rangle \in E_L$, $\langle v_k, Y \rangle \in E_R$. In other words, if $|S(C)| > 4$, then C corresponds to a path from node X to node Y. Suppose $G[D]$ has $q + 2$, $q > 0$, connected components, $C_0 \equiv \{0\}, C_1, C_2, \ldots, C_q, C_{q+1} \equiv \{n+1\}$, sorted in increasing order of $\max(C_r)$'s for $0 \leq r \leq q + 1$. Clearly C_0 and C_{q+1} correspond to nodes $Z1(0)$ and $Z1(n+1)$, respectively. Consider any two consecutive connected components C_{r-1} and C_r where $1 \leq r \leq q + 1$. If both $|S(C_{r-1})| \leq 3$ and $|S(C_r)| \leq 3$, then there is a directed edge from the node corresponding to C_{r-1} to the node corresponding to C_r (See the definition of E_Z). If $|S(C_{r-1})| \leq 3$ and $|S(C_r)| > 3$, then there is a directed edge from the node corresponding to C_{r-1} to the node corresponding to $C_r \cap N^+[\min(C_r)]$ (See the definition of E_{ZL}). If $|S(C_{r-1})| > 3$ and $|S(C_r)| \leq 3$, then there is a directed edge from the node corresponding to $C_{r-1} \cap N^-[\max(C_{r-1})]$ to the node corresponding to C_r (See the definition of E_{RZ}). If both $|S(C_{r-1})| > 3$ and $|S(C_r)| > 3$, then there is a directed edge from the node corresponding to $C_{r-1} \cap N^-[\max(C_{r-1})]$ to the node corresponding to $C_r \cap N^+[\min(C_r)]$ (See the definition of E_{RL}). Following the above observations, we have the following lemma.

Lemma 5. *For a minimum weighted dominating set D of G, satisfying the property given in Lemma 4, there exists a path P from node $Z1(0)$ to node $Z1(n+1)$ in G" where $w(P) = w(D)$.*

Lemma 6. *If P is a directed path from node $Z1(0)$ to node $Z1(n+1)$ in G", then $V(P)$ is a dominating set of G.*

Proof. Omitted. Q.E.D.

Theorem 7. *A minimum weighted dominating set of a cocomparability graph G, where vertex weights are bounded integers, can be found in polynomial time.*

Proof. Immediately following Lemma 5 and 6, we can find a minimum weighted dominating set of G by finding a minimum weighted path from node $Z1(0)$ to node $Z1(n+1)$ in G". Since the number of nodes of graph G" is $O(|V|^{7c})$ and all weights of nodes are nonnegative, it can be implemented to run in polynomial time. Q.E.D.

The above algorithm can be slightly modified to solve the weighted total domination and weighted connected domination problems in polynomial time when restricted to cocomparability graphs with bounded integer vertex weights. We omit the proof of the following theorem.

Theorem 8. *A minimum weighted total dominating set and a minimum weighted connected dominating set of a cocomparability graph G, where vertex weights are bounded integers, can be found in polynomial time.*

3 Weighted Independent Perfect Domination

A *perfect dominating set* of a graph $G = (V, E)$ is a subset D of V such that every vertex not in D is adjacent to *exactly one* vertex in D. The *perfect domination problem* involves finding a minimum perfect dominating set of the given graph. Suppose that every vertex $v \in V$ is associated with a weight $w(v)$ and every edge $e \in E$ has a weight $w(e)$. The *weighted perfect domination problem* involves finding a perfect dominating set D such that its weight

$$w_e(D) = \sum \{w(v) : v \in D\} + \sum \{w(u,v) : u \notin D, v \in D, \text{ and } (u,v) \in E\}$$

is minimum. The perfect domination problem is just the weighted perfect domination problem with $w(v) = 1$ for each vertex v and $w(e) = 0$ for each edge e. Yen and Lee [16] proved that the perfect domination problem is NP-complete for bipartite graphs and chordal graphs. Yen and Lee [17] also considered the following variants of perfect domination. A perfect dominating set D is *independent, connected,* or *total* if the subgraph induced by D has no edge, is connected, or has no isolated vertex, respectively. They gave NP-complete results of these variants on bipartite graphs and chordal graphs, except for the connected domination problem on chordal graphs. On the other hand, Chang and Liu [4] gave a linear time algorithm for the weighted connected perfect domination problem on chordal graphs by using clique-tree structures of chordal graphs. Not all graphs contain an independent perfect dominating set. An independent perfect dominating set of a graph G is also a minimum dominating set of G [1]. In this section, we give an $O(|V|^2)$ algorithm for the weighted independent perfect domination problem for a cocomparability graph $G = (V, E)$. This problem was first studied by Chang, Rangan and Coorg [3]. Define $\overline{w}(v) = w(v) + \sum\{w(u, v) : (u, v) \in E\}$. Then, for an independent perfect domination set D, $w_e(D) = \sum\{\overline{w}(v) : v \in D\}$. Note that $\overline{w}(v)$ for all $v \in V$ can be computed in $O(|E|)$ time. Thus, for solving the weighted independent perfect domination problem, without loss of generality, we may assume that $w(e) = 0$ for all $e \in E$ [3]. Let $G = (V, E)$ be a cocomparability graph with a given cocomparability ordering. Note that D is an independent perfect dominating set of G if and only if $D \cup \{0, n + 1\}$ is an independent perfect dominating set of G_a. For technical reasons, in this section we also assume that G has been added two isolated vertices 0 and $n + 1$ with $w(0) = w(n + 1) = 0$ and a cocomparability ordering $0, 1, 2, \ldots, n, n + 1$. For convenience we need the following notation, where v is a vertex. Some of them were defined in [3].

- For $0 \le v \le n + 1$, $high(v) = \max(N[v])$, $low(v) = \min(N[v])$, $d^+(v) = |N^+[v]|$, $d^-(v) = |N^-[v]|$, $s^-(v) = \max(\{0, 1, 2, \ldots, v\} - N[v])$, $s^+(v) = \min(\{v, v + 1, \ldots, n, n + 1\} - N[v])$.
- For a vertex v, $high(v) > s^+(v)$, R_v is an $(n+2)$-dimensional vector defined as follows: $R_v(u) = 1$ if $s^+(v) < u \le high(v)$ and $u \in N[v]$; $R_v(u) = 0$ otherwise.
- For a vertex v, $low(v) < s^-(v)$, L_v is an $(n+2)$-dimensional vector defined as follows: $L_v(u) = 1$ if $low(v) < u \le s^-(v)$ and $u \notin N[v]$; $L_v(u) = 0$ otherwise.

The following lemma was proved in [3].

Lemma 9. $D = \{0 \equiv v_0 < v_1 < v_2 < \ldots < v_r < v_{r+1} \equiv n + 1\}$ is an independent perfect dominating set of a cocomparability graph G if and only if the following three conditions hold for all $1 \le i \le r + 1$:
(1) $high(v_{i-1}) < v_i$,
(2) $v_{i-1} < low(v_i)$, and
(3) $\{x : x \in V, v_{i-1} \le x \le v_i\}$ is the disjoint union of $N^+[v_{i-1}]$ and $N^-[v_i]$.

Based upon the above lemma, $O(|V||E|)$ and $O(|V|^{2.37})$ time algorithms were proposed by Chang, Rangan and Coorg [3]. The bottleneck of their algorithms is to check whether $N[u] \cap N[v] = \emptyset$ for each pair of non-adjacent vertices u and v. Lemma 9 can be written in the following form, which is more useful in designing an efficient algorithm.

Lemma 10. $D = \{0 \equiv v_0 < v_1 < v_2 < \ldots < v_r < v_{r+1} \equiv n + 1\}$ is an independent perfect dominating set of a cocomparability graph G if and only if, for all $1 \le i \le r + 1$, one of the following two conditions holds.

(1) $s^+(v_{i-1}) = \text{high}(v_{i-1}) + 1 = \text{low}(v_i)$ *and* $s^-(v_i) = \text{low}(v_i) - 1 = \text{high}(v_{i-1})$,
or
(2) $v_{i-1} < s^+(v_{i-1}) = \text{low}(v_i) < \text{high}(v_{i-1}) = s^-(v_i) < v_i$ *and* $R_{v_{i-1}} = L_v$.

Working from Lemma 10, we design the following algorithm.

Algorithm WIPD. Find a weighted independent perfect dominating set of a cocomparability graph.
Input. A cocomparability graph $G = (V, E)$ with a cocomparability ordering $0, 1, \cdots, n, n+1$, in which each vertex v is associated with a weight $w(v)$.
Output. A minimum weighted independent perfect dominating set D of G.
Method.
1.   weight$(0) \leftarrow 0$;
2.   **for** $v = 1$ **to** $n + 1$ **do**
3.   weight$(v) \leftarrow \infty$;
4.   **for all** $u < v$ satisfying either
          (C1) $s^+(u) = \text{high}(u) + 1 = \text{low}(v)$ *and* $s^-(v) = \text{low}(v) - 1$ or
          (C2) $u < s^+(u) = \text{low}(v) < \text{high}(u) = s^-(v) < v$ *and* $R_u \equiv L_v$.
          do
5.   **if** (weight$(u) + w(v) <$ weight(v))
6.   **then** { weight$(v) \leftarrow$ weight$(u) + w(v)$; previous$(v) \leftarrow u$; }
          end do;
       end do;
7. $D \leftarrow \emptyset$;
8. $v \leftarrow$ previous$(n + 1)$;
9.   **while** ($v \neq 0$) **do** $\{D \leftarrow D \cup \{v\}; \ v \leftarrow$ previous(v); }

The bottleneck of this algorithm is determining whether (C2) holds. Whether (C2) holds can be checked in time $O(|V|)$ for a pair of vertices u and v by a straightforward implementation. This leads to an $O(|V|^3)$ time algorithm. Note that low(v), high(v), $s^+(v)$ and $s^-(v)$ of a vertex $v \in V$ can be computed in time $O(|V|)$. Also, we note that there are at most $O(|V|)$ L vectors and R vectors. Each vectors can be constructed in $O(|V|)$ time if low(v), high(v), $s^+(v)$ and $s^-(v)$ are available. These vectors can be sorted in lexicographic ordering in $O(|V|^2)$ time by using a radix sorting algorithm (see p. 115 of [14]). The key idea in the sorting step is that vectors should not be moved away from their initial place. This can be done by using pointers. Only pointers are moved during sorting steps. After sorting, the vectors with the same value will be in consecutive positions in the sorted sequence. By comparing every two vectors adjacent in the sorted sequence, in $O(|V|^2)$ time we can partition vectors into lists such that vectors in the same list are all equal, and two vectors from two different lists are not equal. Then, for a vertex v we can find all vertices u such that $u < s^+(u) = \text{low}(v) < \text{high}(u) = s^-(v) < v$ *and* $R_u = L_v$ in $O(|V|)$ time. In other words, lines 4, 5, and 6 of Algorithm WIPD can be implemented in $O(|V|)$ time. Thus, we have the following theorem:

Theorem 11. *Given a weighted cocomparability graph G with a cocomparability ordering of the vertices, a minimum weighted independent perfect dominating set of G can be found in $O(|V|^2)$ time and space.*

References

1. D. W. Bange, A. E. Barkauskas, and P. T. Slater, Efficient dominating sets in graphs, *Applications of Discrete Mathematics*, R. D. Ringeisen and F. S. Roberts, eds., SIAM, Philad. (1988)189-199.
2. G. J. Chang, *Private communication*.
3. G. J. Chang, C. Pandu Rangan and S. R. Coorg, Weighted independent perfect domination on cocomparability graphs, *Lecture Notes in Computer Science*, Vol. 766, Springer-Verlag, (1993) 506-514.
4. M. S. Chang and Y. C. Liu, Polynomial algorithms for the weighted perfect domination problems on chordal graphs and split graphs, *Information Process. Lett.* 48 (1993) 205-210.
5. D. G. Corneil and L. Stewart, Dominating sets in perfect graphs, *Discrete Math.* 86 (1990) 145-164.
6. P. Damaschke, J. S. Deogun, D. Kratsch and G. Steiner, Finding Hamiltonian paths in cocomparability graphs using bump number algorithm, *Order* 8 (1992) 383-391.
7. J. S. Deogun and G. Steiner, Hamiltonian cycle is polynomial on cocomparability graphs, *Discrete Appl. Math.* 39 (1992) 165-172.
8. J. S. Deogun and G. Steiner, Polynomial algorithms for Hamiltonian cycle in cocomparability graphs, *SIAM J. Comput.*, to appear.
9. M. Farber, Independent domination in chordal graphs, *Operations Research Lett.* 4 (1982) 134-138.
10. M. R. Garey and D. S. Johnson, *Computers and Intractability: A guide to the theory of NP-completeness*, (1979) W. H. Freeman, New York.
11. D. Kratsch and L. Stewart, Domination on cocomparability graphs, *SIAM J. Disc. Math.* 6 (1993), 400-417.
12. R. M. McConnell and J. P. Spinrad, *Linear time modular decomposition and efficient transitive orientation of comparability graphs*, Proc. of the fifth annual ACM-SIAM symposium on discrete algorithms (1994) 536-545.
13. G. K. Manacher and T. A. Mankus, Incorporating negative-weight vertices in certain vertex-search graph algorithms, *Information Process. Lett.*, 42 (1992) 293-294.
14. R. Sedgewick, *Algorithms*, Addison-Wesley Publishing Company, Massachusetts (1983).
15. J. Spinrad, On comparability and permutation graphs, *SIAM J. Comput.* 14 (1985) 658-670.
16. C. C. Yen and R. C. T. Lee, The weighted perfect domination problem, *Inform. Processing Letters* 35 (1990) 295-299.
17. C. C. Yen and R. C. T. Lee, The weighted perfect domination problem and its variants, manuscript.

The Parallel Complexity of Approximating the High Degree Subgraph Problem

(Extended Abstract)

A.E. Andreev[1], A. Clementi[2], P. Crescenzi[3],
E. Dahlhaus[4], S. de Agostino[3], J.D.P. Rolim[2]

[1] University of Moscow
[2] University of Geneva
[3] University of Rome
[4] University of Sydney

Abstract. The HIGH DEGREE SUBGRAPH problem is to find a subgraph H of a graph G such that the minimum degree of H is as large as possible. This problem is known to be P-hard so that parallel approximation algorithms are very important for it. Our first goal is to determine how effectively the approximation algorithm based on a well-known extremal graph result parallelizes. In particular, we show that two natural decision problems associated with this algorithm are P-complete: these results suggest that the parallel implementation of the algorithm itself requires more sophisticated techniques. Successively, we study the HIGH DEGREE SUBGRAPH problem for random graphs with *any* edge probability function and we provide different parallel approximation algorithms depending on the type of this function.

1 Introduction

The HIGH DEGREE SUBGRAPH problem is to find a subgraph H of a graph G such that the minimum degree of H is as large as possible. This problem has been previously studied in [2] to develop techniques to approximate P-complete problems and in [9] to identify strictly related problems that have different complexities.

1.1 Previous Results

The HIGH DEGREE SUBGRAPH problem is to find a subgraph H of a graph G such that problem which is based on the fact that, for any integer $d > 0$, the largest subgraph G_d of G whose minimum degree is at least d can be found by repeatedly deleting nodes of degree less than d until all remaining nodes have degree at least d. In [2] it is shown that, for any $d \geq 3$, deciding whether G_d is empty is a P-complete problem. As a consequence, the HIGH DEGREE SUBGRAPH problem cannot be solved efficiently in parallel (unless NC is equal to P) and parallel approximation algorithms are very important for this problem. In [2] a parallel approximation algorithm for the "non-constructive" version of the HIGH DEGREE SUBGRAPH problem is given. This version consists of finding the largest d such that G_d is not empty and the algorithm is based on the following extremal graph result [8].

Lemma 1. *If a graph G has n vertices and m edges, then $G_{\lfloor m/n \rfloor}$ is not empty.*

The proof of the above lemma can be derived either by the node-deletion procedure previously described or by the following edge-contraction technique. Recall that "contracting" an edge (u, v) on the vertex u means performing the following operations: (1) delete the edge (u, v), (2) connect u with all neighbors of v which are not already connected to u, and (3) delete v and all its incident edges (see the following picture).

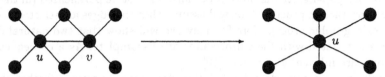

The edge-contraction technique finds a "dense subgraph", i.e., a subgraph whose degree is at least $\lfloor m/n \rfloor$ by first selecting a vertex u in V and then iteratively applying the following step as far as it is possible.

If an edge (u, v) exists such that contracting (u, v) on u yields a graph G' satisfying

$$\frac{|E(G')|}{|V(G')|} \geq \frac{m}{n}$$

where $V(\cdot)$ and $E(\cdot)$ denote the set of vertices and the set of edges of a graph, respectively, then contract (u, v) on u.

Let us denote as G_0 the graph obtained at the end of this procedure. It is easy to prove that if G is connected then the subgraph H of G induced by the neighbors of u in G_0 has degree greater than or equal to $\lfloor m/n \rfloor$ [17] (observe that if G is not connected then it suffices to choose u within the connected component of G whose edge-node ratio is maximum).

The parallel approximation algorithm for the non-constructive version of the HIGH DEGREE SUBGRAPH problem runs in parallel, for any $d \in [1, |V| - 1]$, a node-deletion procedure almost identical to that described above: it repeatedly deletes all nodes of degree less than d until at least one-half of the remaining nodes have degree at least d. Clearly, at each step the number of nodes is halved so that the number of steps is at most $\log n$. Moreover, if the graph G' returned by the procedure is not empty then it contains at least $\frac{1}{4}dn'$ edges where n' denotes the number of nodes in G'. From Lemma 1 it then follows that $|G_{\lfloor d/4 \rfloor}| \geq |G'_{\lfloor d/4 \rfloor}| > 0$. Let $d_{\max}$ be the largest value of d for which the procedure returns a non-empty graph and let d^* be the largest d such that G_d is not empty. Then

$$\frac{d^*}{\lfloor d_{\max}/4 \rfloor} \leq 4,$$

that is, the value $\lfloor d_{\max}/4 \rfloor$ is 4-approximate. Observe that by requiring that the node-deletion procedure runs until at least $1 - \epsilon$ of the remaining nodes have degree at least d, we obtain a $2/(1 - \epsilon)$-approximate value for any arbitrary small $\epsilon > 0$ (clearly, as smaller is ϵ as larger is the number of steps required by the procedure).

1.2 Our Results

As it should be clear from the above discussion, the non-constructive parallel approximation algorithm would turn into a constructive one if the proof of Lemma 1 could be efficiently parallelized. As already observed, this proof can be derived by either a node-deletion procedure or by an edge-deletion procedure. The first two results of this paper state that both procedures cannot be parallelized (in the case of the edge-contraction procedure, we will assume that the edges are selected according to the lexicographic order). More formally, we will show that two natural decision problems associated with these procedures are P-complete. As a consequence, the algorithms are *inherently sequential* [10].

Clearly, these results do not settle the main open question whether the HIGH DEGREE SUBGRAPH problem is approximable in parallel. However, they suggest that the two algorithms are not easy to be parallelized and more sophisticated (or completely different) techniques have to be found in order to develop a parallel approximation algorithm. The results also emphasize the fact that while the HIGH DEGREE SUBGRAPH problem is easy to be approximated in a non-constructive way, the design of a constructive parallel approximation algorithm seems to be an harder task thus providing a clear motivation for studying this problem in a more relaxed "framework". We refer here to the analysis of the average-case complexity of the HIGH DEGREE SUBGRAPH problem. As a second step towards understanding the complexity of approximating this problem, we indeed show parallel approximation algorithms for this problem running in poly-logarithmic *expected* time and using a polynomial number of processors. As probability distribution on the input space, we will make use of the standard model of *random graphs* $\mathcal{G}(n, P(edge) = p(n))$ [4], where edges are chosen independently and with an arbitrary *edge probability function* $p(n)$, and we will provide different parallel algorithmic solutions depending on the type of this function. Recent results on parallel algorithms for random graphs can be found in [7, 13, 16]. In particular, whenever $p(n) = k/n$ with $4 \leq k < \log^2 n$, that is, a family of *sparse* graphs is generated, we present an approximation parallel algorithm running in $O(\log n \log \log n \log k)$ expected time with nk number of processors which gives, when k is constant, a subgraph H having minimum degree not smaller than $d^*/6 - o(1)$, where d^* denotes the measure of an optimum solution. Furthermore, when $k = k(n) \to \infty$, then the subgraph H generated by the algorithm has minimum degree not smaller than $d^*/2 - o(1)$. The interest in sparse random graphs lies in the fact that they exhibit a rather irregular and rich structure [1, 4]. Consequently, the corresponding parallel algorithm and its complexity analysis requires non trivial probabilistic arguments. The algorithm is based on a more sophisticated version of the node-deletion procedure. On the other hand, we will see that, if $p(n) < \frac{4}{n}$ or $p(n) = \Omega(\frac{\log^2 n}{n})$, the problem can be trivially approximated in parallel.

Due to the lack of space, the proofs of our results are not contained in this extended abstract and will appear in the full version of the paper.

1.3 Further Discussions

The node-deletion procedure and the related density technique, introduced in [2], have been successively used to prove that both the *maximum edge connectivity sub-*

graph problem and the *maximum vertex connectivity subgraph* problem are approximable in parallel [14]. However, for reasons equivalent to those observed in this paper, the corresponding algorithms do not give constructive solutions. An interesting open question thus consists in deriving constructive parallel approximation algorithms for these problems.

1.4 Preliminaries

Background material on NC, P. P-completeness and the importance of these notions can be found in [11]. Preliminary information about parallel algorithms are contained in [12]. Discussion concerning parallel approximation algorithms for P-complete and NP-complete problems can be found in [6]. We adopt the following terminology. For any set S, $|S|$ denotes the cardinality of S. For an input graph $G = (V, E)$ we assume the vertices in V are specified in a particular order. This can be thought of as a vertex numbering. We assume the input graphs are not directed, are not multigraphs and contain no self-loops. Given a graph $G = (V, E)$ and a subset of nodes $V' \subseteq V$, $G' = (V', E')$ is the induced subgraph of G where $E' = \{(u, v) \in E : u, v \in V'\}$. For any graph G, $V(G)$ and $E(G)$ will denote the set of vertices and the set of edges of G, respectively. Given a graph $G = (V, E)$ and a subset $H \subseteq V$, we denote by $d_H(v)$ the degree of a vertex v restricted to H (if $H = V$ then we will simply write $d(v)$). We also define the following two functions: given a graph G, $d^*(G) = \max\{d(H) : H \subseteq V\}$ where $d(H) = \min\{d_H(v) : v \in H\}$. Moreover, for any integer d, G_d denotes the (unique) largest subset $H \subseteq V$ such that $d(H) \geq d$.

2 The P-completeness of the Node-Deletion Technique

In this section we formally describe the node-deletion procedure needed to prove Lemma 1. The input/output specifications of this procedure are as follows.

Input: A graph G with n nodes and m vertices.
Output: An induced subgraph G' such that the minimum degree of G' is at least $\lfloor m/n \rfloor$.

The algorithm is shown below (in the following description, C denotes the set of vertices that are deleted).

```
algorithm node-delete;
begin
    C := ∅;
    while a node v exists such that d(v) < ⌊m/n⌋ do
    begin
        G := the graph obtained by deleting v;
        C := C ∪ {v};
    end;
end.
```

In order to measure the difficulty of parallelizing this procedure, we will examine the parallel complexity of a natural decision problem that incorporates the algorithm. This approach has been previously adopted in [10, 3, 5]. The decision problem is defined as follows.

Problem: NODE DELETE.
Istance: A graph G and a set of vertices H.
Question: Is H contained in the node set C computed by the node-delete algorithm with input G?

Theorem 2. *The* NODE DELETE *problem is P-complete even if restricted to connected graphs.*

3 The P-completeness of the Edge-Contraction Technique

In this section we formally describe the edge-contraction procedure. The input/output specifications of this procedure are as follows.

Input: A graph G with a numbering on the vertices $v_1, \ldots, v_n$.
Output: An induced subgraph G' such that the minimum degree of G' is at least $\lfloor m/n \rfloor$ where m denotes the number of edges of G.

As stated in the introduction, the algorithm repeatedly performs the contraction of an edge incident to a specific node until the resulting graph has an edge-node ratio less than m/n. In the description of the procedure, we will assume that the graph is connected (otherwise, we can first look for the connected component whose edge-node ratio is maximum), that the chosen node is v_1, and that the edges are selected according to the lexicographic order (i.e., (v_1, v_j) is considered before (v_1, v_k) for any j and k with $1 < j < k$). The algorithm is then shown below (in the following description, C denotes the set of vertices that are "eaten" by v_1).

```
algorithm edge-contract;
begin
    C := {v₁};
    i := 1;
    while i < n do
    begin
        i := i + 1;
        if vᵢ ∈ V(G) and (v₁, vᵢ) ∈ E(G) then begin
            G' := the graph obtained by contracting (v₁, vᵢ) on v₁;
            if |E(G')|/|V(G')| ≥ m/n then
            begin
                C := C ∪ {vᵢ};
                i := 1;
                G := G';
            end;
        end;
    end;
end.
```

Once again we associate with this procedure a natural decision problem that is defined as follows.

Problem: EDGE CONTRACT.

Istance: A graph G with a numbering on the vertices $v_1, \ldots, v_n$ and a vertex v_i.

Question: Is v_i contained in the node set C computed by the edge-contract algorithm?

Theorem 3. *The* EDGE CONTRACT *problem is P-complete.*

4 Approximation Algorithms for Random Graphs

The aim of this section is to show that approximate solutions for the HIGH DEGREE SUBGRAPH problem can be obtained by parallel algorithms having polylogarithmic *expected* time and using a polynomial number of processors.

We consider the random graph model $\mathcal{G}(n, P(edge) = p(n))$ (for any positive integer n) which is the space of graphs with n vertices and where edges are chosen independently and with uniform probability equal to $p(n) = \frac{k}{n}$. Fundamental results on random graphs and their relevence in computer science can be found in [4]. Throughout the paper, for brevity sake, the fact that $\lim_{n \to \infty} [Prob(\Pi(\mathcal{G}(n, P(edge) = p(n)))$ is false)$] = 0$, where $\Pi(\mathcal{G}(n, P(edge) = p(n)))$ is an arbitrary boolean predicate, will be denoted as: *with high probability* $\Pi(\mathcal{G}(n, P(edge) = p(n)))$ *is true.*

4.1 The Trivial Cases

In this section, we focus on the case when either $p(n) < \frac{4}{n}$ or $p(n) \geq \frac{\log^2 n}{n}$. The regular behaviour of the graph families arising from such probability functions permit us to construct approximate solutions for the HIGH DEGREE SUBGRAPH problem by means of trivial techniques. In particular, when $p(n) = k(n)/n \geq \frac{\log^2 n}{n}$, from the well known Chernoff's bound to the tail of the binomial distribution, it is easy to prove that, with high probability, *all* vertices have degree which is arbitrarily "close" to the value $k(n)$ and thus the approximate solution for the HIGH DEGREE SUBGRAPH problem is the graph itself. In the other case, the following result shows that, with high probability, the optimum value is bounded by a constant and thus trivially approximable.

Lemma 4. *If $k \leq 4$ then, with high probability, we have that $d^*(G) \leq 12$.*

4.2 The Irregular Case

In this section we analyse the more interesting case of $p(n) = \frac{k}{n}$, i.e., when $4 \leq k < \log^2 n$. In this case, a family of *sparse* graphs is generated since the expected number of edges is $m = O(n \log^2 n)$. This choice of $p(n)$ yields a rather complex, irregular family of random graphs. For an exhaustive analysis of the combinatorial structure of such random graphs we refer to [4, 1]; however, concerning the HIGH DEGREE SUBGRAPH problem, it is sufficient to observe that, if we choose randomly a graph G in this family then, with probability not tending to 0 for $n \to \infty$, G has a fixed portion of its vertices whose degree is substantially "far" from the expected degree determined by $p(n)$.

Consequently, an efficient parallel algorithm for the HIGH DEGREE SUBGRAPH problem must consider this irregular behaviour.

- *Overall description of the algorithm*

We present an approximation parallel algorithm whose expected running time is $O(\log n \log \log n \log k)$ with nk processors: the algorithm returns, when k is constant, a subgraph H having minimum degree not smaller than $\frac{1}{6}d^* - o(1)$, where d^* denotes the measure of an optimum solution; furthermore, when $k = k(n) \to \infty$, then the subgraph H, generated by the algorithm, has minumum degree not smaller than $\frac{1}{2}d^* - o(1)$.

The algorithm is based on a different. more sophisticated version of the *node-deletion* procedure. Informally, after a preliminary parallel step which removes nodes having degree less than λ, where λ is a convenient monotone function depending on $p(n)$, the algorithm performs a sequence of consecutive *phases* whose individual execution can be easily performed in $O(\log k)$ expected parallel time. In each Phase t, a new subset Q_t of vertices is marked and removed in parallel. More precisely the set Q_t consists of all vertices whose degree induced by the subset $\bigcup_{i=0}^{t-1} Q_i$ is greater than γ, where γ is another function of $p(n)$. The choice of function γ determines the expected work performed in each phase. The initial set Q_0 is the vertex set of the graph G' generated by the preliminary step. The global procedure terminates at Phase T_f if and only if the set Q_{T_f+1} is empty and returns, if not empty, the subgraph induced by the vertex subset $V - \bigcup_{i=0}^{T_f} Q_i$; if, however, this subgraph is empty then a new execution of the global algorithm will be performed on graph G'. Notice that, in spite of the *node-deletion* procedure case, in our algorithm there is no *density* test and no explicit use of combinatorial lemmas. Clearly, the hardest task here is in showing that the algorithm has the desired performances. In particular, we prove that, after a constant number of executions of the global algorithm, a non empty subgraph will be generated. Furthermore, we will show that the expected number of phases is $O(\log n \log \log n)$ and that the obtained solution represents a "good" approximation of the optimum one. In order to prove the $O(\log n \log \log n)$ bound we provide an interesting proof technique: a) we define a property $\Pi(p(n))$ on the subgraph space of the input graph G with edge probability function $p(n)$; b) we then prove that, if $\frac{4}{n} \le p(n) < \frac{\log^2 n}{n}$, with high probability no subgraph with more than $\log n \log \log n$ vertices can exist having Property $\Pi(p(n))$; c) finally, we show that if the algorithm performs t consecutive phases then we can construct, from the execution of the algorithm, a subgraph with $\Theta(t)$ vertices having Property $\Pi(p(n))$. Points b) and c) thus prove the bound on the expected number of phases.

- *The algorithm and its expected performances*

algorithm $A1$;
begin
    $Z := 0$;
    $G_Z := G$;
    $W := \emptyset$;
    while $W = \emptyset$ **do**
    begin
        $Z := Z + 1$;

$V_Z := V_{Z-1} - \{v \in V_Z : d(v) \leq (\frac{1}{2} - \epsilon)k\}$:
if $V_Z = V_{Z-1}$ **then** $W := V_Z$
else
begin
    Consider the subgraph G_Z induced by V_Z;
    $S_0(Z) := \{v \in V_Z : d(v) \leq (\frac{1}{2} - \epsilon)k\}$:
    $i := 1$;
    $Q_i(Z) := \{v \in V_Z - S_{i-1}(Z) : d_{S_{i-1}(Z)}(v) \geq \delta k + \sqrt{k} + 3\}$;
    while $i \leq \log n \log \log n$ **and** $Q_i(Z) \neq \emptyset$ **do**
    begin
        $S_i(Z) := S_{i-1}(Z) \bigcup Q_i(Z)$;
        $i := i + 1$;
        $Q_i(Z) := \{v \in V_Z - S_{i-1}(Z) : d_{S_{i-1}(Z)}(v) \geq \delta k + \sqrt{k} + 3\}$;
    end;
    if $Q_i(Z) = \emptyset$ **then**
    begin
        $T_{fin} := i$;
        $W := V_z - S_{T_{fin}}$;
    end;
    end;
end;
return the subgraph G_W induced by W;
end.

We first prove that algorithm $A1$ is correct; informally, the following results show that, after a constant number of iterations of the first "while"-loop, algorithm $A1$ with high probability generates a set W which is not empty.

Lemma 5. *For any $k \geq 4$ and for any θ $(0 < \theta < 1)$ a positive constant $C(\theta)$ exists such that there exists, with high probability. a step $Z_f \leq C(\theta)$ for which:*
$|S_0(Z)| \leq \theta |V_Z|$.

Lemma 6. *For any $k \leq \log^2 n$ and for any θ $(0 < \theta < 1)$, with high probability, any subgraph with q vertices contains at most*

$$\frac{k}{n} \frac{q^2}{2} e^{1 + \frac{20}{\theta^3}} + \theta q \sqrt{k} + (1 + \theta)q$$

edges.

Theorem 7. *If $|S_0(Z)| \leq \theta |V_Z|$ and $\theta \leq \frac{\delta}{2} e^{-1 - \frac{20}{1/8}}$ then, with high probability, $|S_{T_{fin}}(Z)| \leq 2|S_0(Z)|$ and, consequently, the set W is not empty.*

Let us now determine the approximation ratio obtained by algorithm $A1$.

Lemma 8. *If $k \geq 4$ then, with high probability, any subgraph with q vertices contains at most $\frac{3qk}{2}$ edges. Moreover, if $k = k(n) \to \infty$ then, with high probability, any subgraph with q vertices contains at most $\frac{qk}{2}(1 + o(1))$ edges.*

Theorem 9. *Let $d^{A1}(G)$ be the minimum degree of the subgraph generated by algorithm $A1$ on input G. Then. the approximation ratio $\frac{d^{\bullet}(G)}{d^{A1}(G)}$ is with high probability bounded by*

$$\frac{3k}{max\{1, \left(\frac{1}{2} - \epsilon - \delta\right)k - \sqrt{k} - 3\}}.$$

Moreover, if $k = k(n) \to \infty$. then the ratio is with high probability bounded by $2 + \alpha(n)$, where $\alpha(n)$ tends to 0 as $n \to \infty$.

Finally, we analyse the complexity of algorithm $A1$. To this aim, we will say that a vertex subset H is a $Z - set$ if its induced subgraph contains a connected subgraph B such that any vertex of H either belongs to $S_0(Z)$ or it has degree restricted to B not smaller than three.

We will now determine the expected number of phases of algorithm $A1$ by studying the probability of the event "H is a Z-set' in random graphs with edge probability $\frac{4}{n} \leq p(n) \leq \frac{\log^2 n}{n}$.

Lemma 10. *With high probability, no Z_f-subset exists having size $\log n \leq q \leq \log^2 n$.*

Theorem 11. *If for some fixed Z, the corresponding number of phases of algorithm $A1$ is at least t, then a Z-subset of size $\Theta(t)$ exists.*

¿From the above results we have that, with high probability, a constant Z exists, for which the solution W is not empty and the number of phases is $O(\log n \log \log n)$. Moreover, each phase can be performed in $O(\log k)$ expected parallel time using $O(nk)$ processors, since every node must only update its current position with respect to the sets $S_i(Z)$ and $Q_i(Z)$.

References

1. N. Alon and J.H. Spencer (1992), *The Probabilistic Method*, Wiley-Interscience Publication.
2. R. Anderson, and E.W. Mayr (1984), "A P-complete problem and approximation to it", Technical Report STAN-CS-84-1014, Department of Computer Science, Stanford University.
3. R. Anderson, and E.W. Mayr (1987), "Parallelism and greedy algorithms", in *Advances in Computing Research: Parallel and Distributed Computing*, F.P. Preparata (ed.), JAI Press, 17–38.
4. B. Bollobas (1985), *Random Graphs*, Academic Press.
5. G. Bongiovanni, P. Crescenzi, S. De Agostino (1995), "MAX SAT and MIN SET COVER Approximation Algorithms are P-Complete", *Parallel Processing Letters*, to appear.
6. D.P. Bovet, A. Clementi. P. Crescenzi, R. Silvestri (1995), "Parallel approximation algorithms for optimization problems", Technical Report SI/RR-95/09. Department of Computer Science, Rome University.
7. A. Clementi, J. Rolim, L. Kucera (1994), "A Note on Parallel Randomized Algorithms for Searching Problems". *DIMACS Series in Discrete Mathematics and Theoretical Computer Sciences*, American Mathematical Society, to appear.

8. P. Erdos (1963), "On the structure of linear graphs", *Israel Journal of Mathematics* **1**, 156–160.
9. R. Greenlaw (1989), "Ordered vertex removal and subgraph problems", *J. Comput. System Sci.* **39**, 323–342.
10. R. Greenlaw (1992), "A model classifying algorithms as inherently sequential with applications to graph searching", *Information and Computation* **97**, 133–149.
11. R. Greenlaw. H.J. Hoover, and W.L. Ruzzo (1995), *Limits to parallel computation: P-completeness theory*, Oxford University Press, to appear.
12. R.M. Karp and V. Ramachandran (1990), "Parallel algorithms for shared memory machines". in *Handbook of Theoretical Computer Science*, J. van Leeuwen (ed.), MIT Press/Elsevier, Vol. A, 869–941.
13. D. Kavvadias. G.E. Pantziou, P.G. Spirakis, C. D. Zaroliagis (1994), "Hammock-on-Ears Decomposition: A Technique for the Efficient Parallel Solution of Shortest Paths and Other Problems", *Proc. of the 19th MFCS*, LNCS, 462-472.
14. L.M. Kirousis. M.J. Serna, P. Spirakis (1989), "The parallel complexity of the connected subgraph problem", *Proc. 30th IEEE-FOCS*, 446–456.
15. R.E. Ladner (1975), "The circuit value problem is log-space complete for P" *SIGACT News* **7**, 18–21.
16. S. Nikoletseas. K. Palem, P. Spirakis. M. Yung (1994). "Short Vertex Disjoint paths and Multiconnectivity in Random Graphs: Reliable Networks Computing", *Proc. of the 21st ICALP*, LNCS, 508-519.
17. K. Wagner (1970), *Graphentheorie*, Bibliographische Inst. AG.

Constant Ratio Approximations of the Weighted Feedback Vertex Set Problem for Undirected Graphs
(Extended Abstract)

Vineet Bafna[*]   Piotr Berman [†]   Toshihiro Fujito [‡]

Abstract

We consider the weighted feedback vertex set problem for undirected graphs. It is shown that a generalized local ratio strategy leads to an efficient approximation with the performance guarantee of twice the optimal, improving the previous results for both weighted and unweighted cases. We further elaborate our approach to treat the case when graphs are of bounded degree, and show that it achieves even better performance, $2 - \frac{2}{3\Delta - 2}$, where Δ is the maximum degree of graphs.

1   Introduction

In this paper, we consider the weighted feedback vertex set problem for undirected graphs. For a graph $G = (V, E)$, a set F of vertices is a *feedback vertex set* *(FVS)* if it contains at least one vertex from every cycle in G. Every vertex is assigned with some weight, and the weight of a vertex set is the sum of weights of vertices in it. Our goal is stated as finding a FVS of as small a weight as possible.

This problem is one of fundamental importance in combinatorial optimization, and was one of the original *NP*–complete problems given by Karp [9]. An important application appears in the context of combinatorial circuit design where cycles are potentially troublesome, e.g., they can cause "racing condition": some circuit element might receive new inputs before it stabilizes. One way to avoid such condition is placing a clocked register at each cycle in the circuit, and in that case, we would like to keep the number of clocked registers as low

---

[*]DIMACS Center, Piscataway, NJ 08854. Supported by Special Year National Science Foundation grant BIR-9412594. email: bafna@dimacs.rutgers.edu

[†]Dept. of Computer Science and Engineering, The Pennsylvania State University, University Park, PA 16802 USA. email:berman@cse.psu.edu

[‡]Dept. of Electrical Engineering, Hiroshima University, 1-4-1 Kagamiyama, Higashi-Hiroshima 739 JAPAN. email: fujito@huis.hiroshima-u.ac.jp

as possible, which is equivalent to computing a minimum FVS. See Bar–Yehuda et al. [3] for other applications in constraint satisfaction problems and the area of Bayesian inference.

Recently, Bar–Yehuda et al. [3] have shown that the smallest cardinality FVS can be approximated within constant ratio of 4, greatly improving upon the previous best of $\sqrt{\log n}$ obtained by Monien and Schulz in 1981 [13]. They also consider the weighted case for which they obtain a performance of $\min\{4\log n, 2\Delta^2\}$ where Δ is the maximum degree of a vertex. In this paper, we will show that the constant ratio approximation can be carried out also for the *weighted* FVS problem, and in fact, better than theirs for the *unweighted* case; namely, the performance guarantee of twice the optimal. Our algorithm can be implemented to run in time $O(|E| + |V|\log|V|)$, using the same data structures as Dijkstra's single–source shortest paths algorithm. It has been brought to our attention that independent of our work, Becker and Geiger [4] also show a performance of 2 for the weighted case, with a different analysis that does not easily extend to the bounded degree case.

It is also worth observing that achieving a performance ratio better than 2 might be computationally hard. The problem of FVS is a typical node-deletion problem for hereditary properties [11], and a general hardness result for this class of problems is known [12]: The node–deletion problem for every nontrivial hereditary property is *MAX SNP*–hard; thus, it does not have a polynomial time approximation scheme unless $P= NP$ [1]. Moreover, the Vertex Cover problem can be reduced to FVS in an approximation preserving manner, so that any performance ratio $r < 2$ for FVS would also imply a ratio r for VC. Better approximation of VC has been extensively researched over the years, yet the currently best ratio for VC is $2 - \frac{\log\log n}{\log n}$ [2, 14] (in fact it was conjectured by Hochbaum [8] that approximating VC in a constant factor smaller than 2 is itself *NP*–hard).

Our approach is based on the "local approximation" principle, which appeared already in the Gavril's maximal matching approximation for the unweighted VC problem (cited in [6, page 134]) and which was explicitly formalized as the local ratio theorem by Bar–Yehuda and Even [2]. The use of this principle, however, has been usually limited to extraction of subgraphs of "uniform" weights from either unweighted or weighted graphs, which was used to obtain a $O(\log n)$ performance for weighted FVS [3]. The important technical contribution of this paper is a generalization of this theorem. We show that significantly better approximations can be achieved by "choosing" weights of our own for local structures. It is our hope that this technique is general and can be applied to other weighted optimization problems.

We also study the weighted FVS problem restricted to graphs of bounded degree Δ. This version remains *NP*–hard as well as *MAX SNP*–hard when $\Delta \geq 4$. We elaborate our approach by separating some special cases and introducing general approximation preserving reductions for this problem. It will be then shown that a performance ratio improves to $2 - \frac{2}{3\Delta-2}$. Very recently, Berman and Lee [5] have significantly extended the ideas in this paper to improve the performance for the bounded degree case. In particular, they achieve a performance of 1.4 for $\Delta = 3$ and 1.58 for $\Delta = 4$.

Note that the weighted VC in bounded degree graphs can be approximated within ratio $2 - \frac{3}{2\Delta+4}$ [7]. It remains an open question whether one can approximate FVS with *exactly* the same ratio as VC, whether the degree is bounded or not. However, for bounded degree, the connection between the two problems is somewhat tenuous as known approximations do not preserve degree.

Definitions and Notation. The input weighted graph is $G = (V, E, w)$, where V is the set of vertices, E is the set of edges and w is a nonnegative weight function defined on V. In the algorithm we will consider subgraphs of G defined by weight functions. Let $\mathcal{R}_+$ be the set of nonnegative reals. Given $w : V \rightarrow \mathcal{R}_+$, let $V(w) \equiv \{u \in V | w(u) > 0\}$. A subgraph of G induced by U is denoted by $G(U)$, and moreover, $G(w)$ denotes $G(V(w))$; $d(u, U)$ denotes the number of neighbors of u in U, and $d(u, w) \equiv d(u, V(w))$.

For $G = (V, E, w)$ a collection of weight functions defined on V, $\{w_i, i = 1, \cdots, k\}$, is a *decomposition* of w if $\sum_i w_i(u) \leq w(u)$ for each $u \in V$. A weight function w is *degree–proportional* if for some $c > 0$ we have $w(u) = c(d(u, w) - 1)$ for every $u \in V(w)$. A cycle C is *semi-disjoint* in $G(w)$ if $C \subseteq V(w)$ and for every vertex $u \in C$, with at most one exception, $d(u, w) = 2$.

In section 2 of this extended abstract, we present our algorithm for approximating weighted FVS. We extend upon these ideas in section 3 to get an improved performance for the bounded–degree case.

2 Approximating within twice the optimal

Figure 1 presents our algorithm for the FVS problem. A call of the recursive function Feedback(w) computes a FVS for $G(w)$. At the beginning of the call vertices that have degree ≤ 1 are removed from $G(w)$ (implicitly, by setting their weight to 0), as they are surely not needed in a FVS. Next, w' is obtained by by calling Reduce(w, w') which extracts a degree–proportional weight function w_i from w; If $G(w)$ has a semi-disjoint-cycle C, then $V(w_i)$ is the set of vertices in C. Otherwise, $V(w_i) = V(w)$, and w_i is the largest degree–proportional function bounded by w. The vertices of $V(w) - V(w')$ become "candidates" for membership in the FVS being computed. The output FVS is computed as the result of Feedback(w'), to which the candidates are inserted only "when needed" (see the test in the for–loop).

For an input graph $G = (V, E, w)$ and an output F_{out} of the algorithm, the code of Feedback() easily implies the following, which will be used later to apply the local ratio theorem.

Lemma 1 (a) $w \geq \sum_i w_i$; (b) $w(F_{out}) = \sum_i w_i(F_{out})$.

Similarly, it is easy to prove the following by induction on the depth of the recursion and case analysis.

Input: a weighted graph $G = (V, E, \mathbf{w})$
Output: a FVS F_{out} for G

main()
$i \leftarrow 0$ (* i is used for analysis only *)
return Feedback($\mathbf{w}$)$\cup(V - V(\mathbf{w}))$

procedure Feedback(w)
begin
$i \leftarrow i + 1$
if $V(w) = \emptyset$ **then**
   return $\emptyset$
else begin
   while there exists $u \in V(w)$ s.t. $d(u, w) \leq 1$ **do**
      reset $w(u)$ to 0
   Reduce(w, w')
   $F \leftarrow$ Feedback(w')
   $Comp \leftarrow V(w') - F$
   for every u in $V(w) - V(w')$ **do**
      if $\{u\} \cup Comp$ contains a cycle **then**
         Add u to F
      else
         Add u to $Comp$
   end
   return F
end

procedure Reduce(w, w')
begin
   if there is a semi–disjoint cycle C in $G(w)$ **then**
      $V_i \leftarrow C$
   else
      $V_i \leftarrow V(w)$
   $\delta \leftarrow \min\{w(v)/(d(v, V_i) - 1) | v \in V_i\}$
   $w_i(v) \leftarrow \delta(d(v, V_i) - 1)$ if $v \in V_i$, and 0 otherwise
   $w' \leftarrow w - w_i$
end

Figure 1: A 2–Approximation Algorithm for Feedback Vertex Set Problem

Lemma 2 $F_{out} \cap V(w_i)$ *is a minimal cardinality FVS for $G(w_i)$.*

Observe that $G(w_i)$ can have two forms:

1. a simple cycle with all the vertices having equal weights, or

2. a graph of degree–proportional weight, with no vertices of degree ≤ 1 and no semi–disjoint cycles.

In the first case, the fact that $F_{out} \cap V(w_i)$ is a minimal FVS for $G(w_i)$ implies that it is also an optimal FVS for $G(w_i)$. The second case is more interesting. We show that for a graph with a degree–proportional weight function, a *minimal cardinality FVS has a weight that is within a factor of 2 from the optimum.*

Clearly, it suffices to prove it for the case when for every $u \in V, w(u) = d(u) - 1$, where $d(u)$ denotes the degree of u; we say that such a weight function is *degree–defined*. We will show it using a potential function $p : V \to \mathcal{R}$, s.t. $p(u) = \frac{d(u)}{2} - 1$ for every $u \in V$. Assume that w is the degree–defined weight function defined on V.

Lemma 3 *For an arbitrary FVS F of G, $w(F) \geq p(V)$ (Proof by induction on $|F|$).*

Lemma 4 *If G contains no semi–disjoint cycles and F is an FVS of minimal cardinality, then $w(F) \leq 2p(V)$.*

The proof sketch is as follows: observe that $G(V - F)$ is a forest and by minimality, each vertex $u \in F$ is *blocked* by a tree in the forest, i.e. each vertex in F has two edges incident on some tree $T \subseteq V - F$. Let e_T be the number of edges with one end in T and another in F. The next two statements are part of the proof of Lemma 4, and presented here independently because we will also apply them in the next section.

Fact 5 *Each tree T in $G(V - F)$ can block at most $\lfloor \frac{1}{2} e_T \rfloor$ vertices.*

Lemma 6 *Let T be an arbitrary tree in G. Then, $p(T) = \frac{1}{2} e_T - 1$.*

Every vertex $u \in F$ is a priori allocated a potential of $(d(u) - 2)/2$. If G contains no semi–disjoint cycles, $e_T \geq 3$ where T is a tree blocking u. Fact 5 and Lemma 6 then imply that each vertex in F can receive an extra potential of $(\frac{1}{2} e_T - 1)/\lfloor \frac{1}{2} e_T \rfloor \geq \frac{1}{2}$ from $V - F$. Therefore, for each contribution of $d(u) - 1$ to $w(F)$ we have at least $(d(u) - 1)/2$ in $p(V)$, and Lemma 4 follows.

Summarizing, for every i we have $w_i(F_{out}) \leq 2w_i(opt(G(w_i)))$ where $opt(G)$ denote an optimal FVS for G. This allows us to apply the following generalization of local ratio theorem of Bar–Yehuda and Even [2], and as a consequence, the desired performance of our algorithm can be proven.

Theorem 7 (Local Ratio Theorem) *For $G = (V, E, w)$ let $\{w_i, i = 1, \cdots, k\}$ be a decomposition of w and F be any FVS for G s.t. $w(F) = \sum_i w_i(F)$. Then,*

$$\frac{w(F)}{w(opt(G))} \leq \max_i \{\frac{w_i(F)}{w_i(opt(G(w_i)))}\}$$

Theorem 8 *The algorithm finds a FVS F_{out} for $G = (V, E, \mathbf{w})$ s.t. $\mathbf{w}(F_{out}) \leq 2\mathbf{w}(opt(G))$.*

3   Improved performance for bounded degree graphs

The bottleneck for the better performance of our algorithm is the ratio given in Lemma 4. To prove it, we observe that a vertex u of F with weight k has potential $\frac{1}{2}k - \frac{1}{2}$; additionally, it receives at least $\frac{1}{2}$ of the potential from a tree T in $G(V - F)$ that blocks it. We may modify the algorithm to assure that the extra potential received from T is at least $\frac{2}{3}$. This will assure the weight/potential ratio strictly below 2. Recall Fact 5 and Lemma 6. It is clear that as e_T increases, the extra potential that T can supply gets closer to $\frac{1}{2}$ per edge: it is $(e_T - 2)/(2e_T) = \frac{1}{2}(1 - 1/e_T)$. Hence, for $e_T \geq 6$ we have no problem; for $e_T = 5$, T can send $\frac{1}{3}$ on each edge to a vertex blocked by T and $\frac{1}{6}$ along each of the remaining edges (see Fig. 2). What we need to handle in a special manner is the case when $e_T \leq 4$.

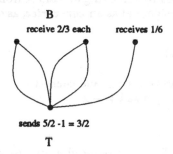

Figure 2: T with $e_T = 5$

For a tree $T \subseteq V - F$ with $e_T \leq 4$ we consider a subgraph induced by T together with the set B of vertices blocked by T. We can assume w.l.o.g. that no vertex of degree 2 is adjacent to another degree–2 vertex (later we will see why). Then, it is easy to see that $|T \cup B| \leq 9$ (the case of $|T \cup B| = 9$ is given in Fig. 3). The new algorithm will attempt to perform some reduction steps on each connected subgraph $G(A)$ with $1 < |A| \leq 9$. Below we refer to such A's as *small–subgraphs*. For each small subgraph A we need to know two FVS's, F_1^A and F_2^A. F_1^A is an optimal FVS for $(G(A), w)$, F_2^A is also an optimal FVS, but with restriction that no path going through $A - F_2^A$ may connect two edges that straddle A and $V - A$ (e.g., if two such edges are adjacent to the same vertex $u \in A$, then u forms a path, of length 0, connecting them, hence, u must belong to F_2^A).

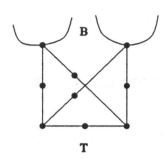

Figure 3: T with $|B \cup T| = 9$

3.1 Ear reduction

Define an *ear* as a small–subgraph A in which all edges straddling A and $V - A$ are either adjacent to the same vertex in A, or there are at most two such edges. A has properties similar to so–called an open or closed *ear* in respect to $V - A$ [10].

The algorithm proceeds by removing all ears A from the graph and replacing them by a single vertex designated as an *ear–vertex*, as described in the procedure *ReduceEar(A)*.

> Procedure *ReduceEar(A)*
> compute F_1^A, F_2^A
> Replace A by an ear-vertex u_A, and connect it
> to edges straddling A and $V - A$
> $w(u_A) = w(F_2^A) - w(F_1^A)$.

After computing an FVS F for the reduced graph, we execute *InsertEar(u_A)* for every ear–vertex u_A.

> Procedure *InsertEar(u_A)*
> if $u_A \in F$ then
> $F \leftarrow F - \{u_A\} \cup F_2^A$
> else
> $F \leftarrow F \cup F_1^A$

Lemma 9 below shows that the ear reduction preserves any performance ratio ≥ 1. Observe that the ear reduction strictly generalizes both *branchy reduction* [3] (the case of $|A| = 2$) and semi–disjoint cycle reduction. For this reason, degree-2 vertices form an independent set when the ear reduction is not applicable anymore.

Lemma 9 *Let (G', w') be a graph obtained from (G, w) by the reducing ear A, and that* InsertEar*(u_A) subsequently changes an FVS F' of G' into F. Let F^* and F_0^* be optimal FVS's for (G, w) and (G', w'), respectively. Then F is a FVS for G and $w(F) \leq rw(F^*)$, where $r = w'(F')/w'(F_0^*)$.*

3.2 Local lower bound reduction

The goal of our algorithm is to provide a FVS within ratio $r_\Delta = 2(1 - \frac{1}{3\Delta-2})$. Define a *good–small–subgraph* as a subgraph A satisfying $w(F_2^A) \le r_\Delta w(F_1^A)$.

If the algorithm cannot apply the ear reduction, it looks for a good–small–subgraph. For such A, the algorithm computes an FVS F of $(G(V - A), w)$ and returns $F \cup F_2^A$. The definition of F_2^A assures that the latter is a FVS for $G(V)$. By its very nature, the local lower bound reduction preserves performance ratio r_Δ.

3.3 Degree–proportional weight reduction

If none of the two previous reductions is applicable, the algorithm finds w_1, the largest degree–proportional weight s.t. $w_1 \le w$, and proceeds as in the second case of Reduce(). However, it later inspects the resulting FVS F to eliminate the situations described at the beginning of this section: a vertex of F receiving only $\frac{1}{2}$ of the extra potential unit from $V - F$.

Before we describe the way we modify F we have to explain exactly how the potential is transferred from $V - F$ to F. Let T be a component tree of $V - F$ and B be the set of vertices blocked by T. Consider an edge going from T to $V - T$. If $e_T \ge 5$, T sends $\frac{2}{3}$ to each vertex in B and $\frac{1}{6}$ along each of the edges going from T to $F - B$. We can do the same when $e_T = 4$ and $|B| \le 1$ or $e_T = 3$ and $|B| = 0$. The case of $e_T = 2$ does not happen at this stage as it allows for the ear reduction. When $e_T = 4$ and $|B| = 2$ or $e_T = 3$ and $|B| = 1$, T sends only $\frac{1}{2}$ to each element of B. Thus, we can have vertices of F with a deficit: such a vertex is blocked by T that gives it only $\frac{1}{2}$, and it is not adjacent to any edge carrying $\frac{1}{6}$. Note that deficit vertices do not occur in graphs of degree ≤ 3, as in these graphs the last two cases lead to the ear reduction (see Fig. 4).

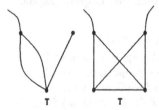

Figure 4: T with $e_T \le 4$ in degree–3 graphs

If $u \in B$ is a vertex with deficit but B contains another vertex that has a "surplus", i.e., receiving $\frac{1}{2} + \frac{2}{6}$ of the potential we shift $\frac{1}{6}$ of the potential to u from its "buddy" in B. If a certain vertex u has a deficit and no buddy with surplus, we change F and the allocation of the potential. Let F_1^A be an optimal FVS for $(G(A), w)$ where $A = T \cup B$; we replace B by F_1^A in F and send $\frac{1}{6}$ of the potential along each edge from A to $V - A$. We need to show two facts: (1) the new F is still a FVS for G, and (2) F_1^A has so small a weight that $p(A)$ can "justify" $w(F_1^A)$ and $\frac{1}{6}$'s sent along the edges to $V - A$. We prove these two facts in the next two lemmas.

Lemma 10 *After replacing B by F_1^A, F remains a FVS for $(G(V), w)$.*

The procedure reduces (G, w) to (G, w_0) where $w = w_1 + w_0$ and w_1 is a maximal degree–proportional weight s.t. $w_1 \leq w$. In our analysis, even if some replacement occurs we take $w_0(F)$ unchanged. Thus, if some replacement takes place changing $F \cap A$ from B to F_1^A we associate $w(F_1^A) - w_0(B)$ as the weight of $F \cap A$ in (G, w_1). The next lemma states that we can pay by $p(A)$ in (G, w_1) for such a weight as well as $\frac{1}{6}$ each along out–going edges if $w(F_1^A)$ is smaller than $w(B)$ by a factor of $\sqrt{3}$.

Lemma 11 *Assume that w_1 is the degree–defined weight function for G. Let m be the number of edges straddling A and $V - A$. If $G(A)$ is not a semi–disjoint cycle in G and if $w(F_1^A) \leq \frac{w(B)}{\sqrt{3}}$, then $p(A) \geq \frac{(w(F_1^A) - w_0(B))}{\sqrt{3}} + \frac{m}{6}$.*

Using these two lemmas we can prove that the degree–proportional weight reduction preserves performance ratio r_Δ.

Lemma 12 *Let w_1 be the largest degree–proportional weight function s.t. $w_1 \leq w$, and let $w_0 = w - w_1$. Also let F^* and F_0^* be optimal FVS's for (G, w) and (G, w_0), respectively. Suppose F_0 is a FVS for (G, w_0) s.t. $w_0(F_0) \leq r_\Delta w_0(F_0^*)$. Then, degree–proportional weight reduction constructs a FVS F from F_0 s.t. $w(F) \leq r_\Delta w(F^*)$.*

In the proof of Lemma 12 we use the fact that we do not have "deficit vertices" when $\Delta = 3$, and that $r_\Delta > \sqrt{3}$ for $\Delta \geq 4$; the latter allows us to apply Lemma 11.

Finally, we present the algorithm $\Delta\text{-}FVS()$ and prove that it provides an improved berformance for bounded degree graphs.

$\Delta\text{-}FVS(G(V, E))$
1. For all ears A, $G = ReduceEar(A)$
2. For all good–small–subgraphs A, $G = G(V - A)$
3. Find a maximal degree–proportional weight $w_1 \leq w$ and recursively compute a FVS F_0 for $(G, w - w_1)$.
4. Construct a minimal FVS F for (G, w) from F_0 (i.e., $F \supseteq F_0$) as in procedure Feedback() (section 2).
5. For all small–subgraph A with a deficit vertex, $F = (F - A) \cup F_1^A$
6. For all good–small–subgraphs A, $F = F \cup F_2^A$.
7. For all ear–vertices u_A, $F = InsertEar(u_A)$.
8. Return F

Theorem 13 *Algorithm $\Delta\text{-}FVS$ finds an FVS for G with performance ratio $2 - \frac{2}{3\Delta - 2}$, where Δ is the maximum degree of G.*

4 Running time

The running time of algorithm *Feedback()* is dominated by the time required to maintain the vertex with the maximal degree–proportional weight. Therefore its running time is $O(|E| + |V|\log|V|)$. For Δ-*FVS()*, observe that once we have applied *ReduceEar()*, each vertex can test if it is part of a small–subgraph in constant time. This implies a running time of $O(|V|^2)$.

References

[1] S. Arora, C. Lund, R. Motwani, and M. Sudan and M. Szegedy. Proof verification and intractability of approximation problems. In *33rd IEEE Symp. on Foundations of Computer Science*, 1992.

[2] Bar-Yehuda, R. and S. Even. A local-ratio theorem for approximating the weighted vertex cover problem. In *Annals of Discrete Mathematics 25*. North–Holland, 1985.

[3] R. Bar-Yehuda, D. Geiger, J. Naor and R. M. Roth. Approximation algorithms for the vertex feedback set problem with applications to constraint satisfaction and Bayesian inference. In *Proc. of the 5th Annual ACM-SIAM Symp. on Discrete Algorithms*, pages 344–354, 1994.

[4] A. Becker and D. Geiger. Approximation algorithms for the loop cutset problem. In *Uncertainity in Artificial Intelligence*, 9, 1994.

[5] P. Berman. [personal communication]

[6] M. R. Garey and D. S. Johnson. *COMPUTERS AND INTRACTABILITY: A Guide to the Theory of NP-Completeness*. W. H. Freeman and co., 1979.

[7] M.M. Halldórsson. Approximations via partitioning. Technical report, Japan Advanced Inst. of Sci. and Tech., March 1995.

[8] D.S. Hochbaum. Efficient bounds for the stable set, vertex cover and set packing problems. *Discrete Applied Mathematics*, 6:243–254, 1983.

[9] R.M. Karp. Reducibility among combinatorial problems. In R. E. Miller and J. W. Thatcher, editors, *Complexity of Computer Computations*, pages 85–103. Plenum Press, New York, 1972.

[10] L. Lovász and M.D. Plummer. *Matching Theory*. North-Holland, 1986.

[11] J.M. Lewis and M. Yannakakis. The Node–deletion problem for hereditary properties is NP–complete. *Journal of Computer and System Sciences*, 20:219–230, 1980.

[12] C. Lund and M. Yannakakis. The approximation of maximum subgraph problems. In *Proc. of 20th International Colloquium on Automata, Languages and Programming*, pages 40–51, 1993.

[13] B. Monien and R. Schulz. Four approximation algorithms for the feedback vertex set problem. In *Proc. of the 7th Conference on Graph Theoretic Concepts of Computer Science*, pages 315–326, 1981.

[14] B. Monien and E. Speckenmeyer. Ramsey numbers and an approximation algorithm for the vertex cover problem. *Acta Informatica*, 22:115–123, 1985.

[15] C. Papadimitriou and M. Yannakakis. Optimization, approximation and complexity classes. *Journal of Computer and System Sciences*, 43:425–440, 1991.

Greedy Approximations of Independent Sets in Low Degree Graphs

Magnús M. Halldórsson[1] * and Kiyohito Yoshihara[2]

[1] Science Institute, University of Iceland, IS-107 Reykjavik, Iceland.
[2] Department of Computer Science, Tokyo Institute of Technology.

Abstract. We investigate the power of a family of greedy algorithms for the independent set problem graphs of maximum degree three. These algorithm iteratively select vertices of minimum degree, but differ in the secondary rule for choosing among many candidates. We present two such algorithms that run in linear time, and show their performance ratios to be 3/2 and 9/7 ≈ 1.28, respectively. This also translates to good ratios for other classes of low-degree graphs.

We also show certain inherent limitations in the power of this family of algorithm: any algorithm that greedily selects vertices of minimum degree has a performance ratio at least 1.25 on degree-three graphs, even if given an oracle to choose among candidate vertices of minimum degree.

1 Introduction

An *independent set* of a graph G is a subset of vertices in which no two are adjacent. The MAX INDEPENDENT SET problem is that of finding an independent set of maximum cardinality. It is one of the core $\mathcal{NP}$-hard problems [4], and thus, polynomial time exact algorithms are unlikely to exist. It is therefore interesting to explore algorithms that produce solutions that are not always optimal but are close to optimal. The quality of an approximation algorithm is generally measured by the *performance ratio*, or the maximum ratio of the size of an optimal solution (the size of the maximum independent set) to the size of the solution found by the algorithm.

In this paper we focus on a central case of bounded-degree graphs, namely when the maximum degree is at most three. Since the independent set problem is polynomial solvable when maximum degree is two, this problem can be thought of as the initial frontier of $\mathcal{NP}$-hardness of the problem. Also, many of the results for higher degrees use reductions to lower degree cases, in which the degree-three plays the role of the basis case [3, 7, 5, 2], and improvements for that case translate to improvements for all odd degrees.

Let us review the known results about approximating independent sets in degree-three graphs. Hochbaum [9] presented an algorithm with a 1.5 ratio, that runs in time proportional to bipartite matching or $O(n^{1.5})$. Berman and Fürer

---

* Research partly performed at Japan Advanced Institute of Science and Technology, IBM Tokyo Research Lab, and Max Planck Institut fuer Informatik.

[3] presented a powerful local search approach that attains a performance ratio of 1.25. This has recently been brought down to 1.2 by Berman and Fujito [2] using additional tricks. The disadvantage of this approach is a phenomenally high time complexity: the analysis of [3, 2] yields a bound of at least $n^{2^{100}}$, while even with a tighter analysis [7] the complexity appears to be no less than n^{50}. The approach can be scaled down for a weaker ratio of $1.33 + \epsilon$ in time $O(\exp(1/\epsilon)n)$ [7, 5].

This paper focuses on the *minimum-degree greedy* paradigm:

Select a vertex of minimum degree, add the vertex to the solution, remove the vertex and its neighbors, and repeat until the graph is empty.

The approach is non-deterministic in the choice of a particular vertex of minimum degree. The basic algorithm, **Greedy**, which selects an arbitrary minimum degree vertex, was analyzed in detail by Halldórsson and Radhakrishnan [6]. In particular, the performance ratio on degree-three graphs was shown to be 5/3. We consider here greedy algorithms with more goal-directed selection rules.

In Section 3, we present a modified greedy algorithm that prefers eliminating more than the minimum number of edges in each reduction, and show its performance ratio to be 3/2. The analysis is straightforward and serves as a warmup for the following algorithm.

In Section 4, we present another greedy-like algorithm that seeks portions of the graph for which its rules perform optimally, and show its performance ratio to be $9/7 \approx 1.28$.

In Section 5, we show that any algorithm in our greedy paradigm will have a performance ratio of at least 1.25 on degree-three graphs. Thus, the whole family has limitations which are nearly matched by our second algorithm.

Finally, additional results that could not be discussed in detail are mentioned in Section 6. In the next section, we describe the notation used in the paper, along with the bound on the size of the optimal solution used in the analysis.

2 Preliminaries

The input graph $G = (V, E)$ is assumed to be of maximum degree three. Let n denote the number of vertices, m the number of edges, and α the independence number (i.e. size of the optimal independent set).

For an algorithm A for MAX INDEPENDENT SET, the size of the solution produced on G is denoted by $A(G)$; when the algorithm is non-deterministic or depends on the ordering among the vertices of the graph, we assume a worst-case. The *approximation ratio* $\rho_A(G)$ is defined as $\rho_A(G) \overset{\text{def}}{=} \alpha(G)/A(G)$. The *performance ratio* of A is defined as the maximum approximation ratio over all input graphs, or $\rho_A \overset{\text{def}}{=} \max_G \rho_A(G)$. We are primarily interested in the limit of this value as n goes to infinity.

We let I denote a fixed but arbitrary maximum independent set in G. Let *Out* denote the number of edges with both endpoints in $V - I$.

Upper bounding the optimal solution

In order to analyze the relative value of a heuristic solution compared with an optimal solution, a good upper bound of the optimal solution is essential. The number Out of edges outside some maximum cardinality solution I plays a crucial role.

Lemma 1. *For a degree-three graph* G, $\alpha(G) \leq n - m/3 - Out/3$.

Proof. Each edge has either one endpoint in I or both endpoints in $V - I$. The total number of endpoints in $V - I$ is at most $3(n - |I|)$. Thus,

$$m \leq 3(n - |I|) - Out,$$

which, when rearranged, yields the claim. ∎

3 A Modified Greedy Algorithm, MoreEdges

In this section, we propose a modified version of the greedy method, named **MoreEdges**, which considers the degrees of vertices adjacent to a vertex as a criteria for selecting the vertices. The criteria is:

> When minimum degree is two, select – whenever possible – a vertex with a neighbor of degree three.

Theorem 2. *The performance ratio of* **MoreEdges** *on degree-three graphs is* $3/2$.

The operation of the algorithm can be broken up into *reductions*, each of which consists of the addition of a single vertex to the current solution and the deletion of this and neighboring vertices along with the incident edges. If the recursive description of the algorithm is made iterative, a reduction corresponds to a single iteration. An (i, j)-*reduction* refers to one where $i - 1$ vertices and j edges are deleted.

The form of possible reductions are given in Figure 1, with selected vertices in black, their neighbors (which are also deleted) in grey, and other incident vertices in white. $(2, 4)$ and $(2, 5)$ reductions appear in two different guises.

We consider the following measures of each reduction r. Here, we fix some maximum cardinality independent set I for comparison.

$n(r)$	Number of vertices deleted
$e(r)$	Number of edges deleted
$\alpha(r)$	Number of the deleted vertices that belong to I
$Out(r)$	Number of deleted edges with both endpoints in $V - I$.

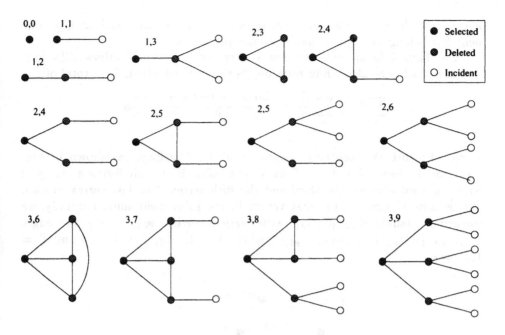

Fig. 1. The forms of the various reductions.

r	$n(r)$	$e(r)$	$Out(r)$	$\alpha(r)$	$f(r)$
$(0,0)$	1	0	0	1	4
$(1,x)$	2	1	0	1	6
$(2,3)$	3	3	1	1	6
$(2,4)$	3	4	0	1†	6
$(2,5+)$	3	5	0	2	6
$(3,6)$	4	6	3	1	4
$(3,7)$	4	7	1	2	6
$(3,8)$	4	8	1	2	5
$(3,9)$	4	9	0	3	6

Table 1. Bounds on measures of the reductions performed by MoreEdges.

Our primary *cost* measure of each reduction r is given by:

$$f(r) = 3n(r) - e(r) - Out(r) + \alpha(r).$$

Table 1 gives conservative bounds for these measures on each type of reduction. The values for $Out(r)$, $e(r)$ are lower bounds, while those for $\alpha(r)$ and $f(r)$ are upper bounds.

The values in Table 1 are easily verified from Figure 1, with the exception of the α value of $(2,4)$. When a $(2,4)$-reduction occurs, the graph necessarily consists of disjoint cycles. The algorithm will add the same number of vertices to the solution from a given cycle, no matter which vertex it starts with. Thus, it causes no harm if we assume that it chooses a vertex for which at most one neighbor belongs to a given maximum independent set.

The claim of the upper bound now follows easily from the f-values of Table 1, and Lemma 1. The algorithms finds one vertex per reduction, for a total of:

$$\sum_r 1 \geq \sum_r \frac{f_r}{6} = \frac{3n - e - Out + \alpha}{6} \geq \frac{4}{6}\alpha.$$

Lower bound We construct a hard graph for MoreEdges as shown in Figure 2, in the form of a chain of six vertex units. Each unit forms a six-cycle with one cord between the third and the fifth vertex. The last vertex in each unit is also adjacent to the first vertex in the subsequent unit. Formally, we construct a family of graphs G_q, with vertices vertices $v_{i,1}, \ldots, v_{i,6}$ and edges $(v_{i,j}, v_{i,j+1}), (v_{i,3}, v_{i,5}), (v_{i',6}, v_{i'+1,1})$, where $i = 1, \ldots q$, $j = 1, \ldots 6$ and $i' = 1, \ldots q - 1$.

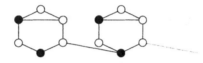

Fig. 2. Initial portion of a hard graph for MoreEdges.

MoreEdges may be assumed to select the first and the third vertex of each unit, while the optimal solution will contain the second, fourth and sixth. Only on the last unit will MoreEdges also find three vertices. Hence, the performance ratio of MoreEdges is no better than $3q/(2q+1) = 3/2 - \theta(1/n)$.

4 An Improved Greedy Algorithm, Simplicial

We consider in this section a still stronger member of the greedy paradigm. In particular, the algorithm performs the following two types of transformations whenever possible.

Branchy reduction When two vertices v and u of degree two are adjacent, some optimal solution will contain exactly one of these vertices. We can transform the graph into a graph G' that contains all vertices but v and u and has the other neighbors of v and u adjacent. The solution of the heuristic will contain the heuristic solution on G' along with one of v and u. To ensure that multi-edges do not appear, we insist that no third vertex be adjacent to both v and u. This is a case of a *delayed-commitment* reduction, whose effect is optimal.

Simplicial reduction A *simplicial* vertex is one whose neighborhood forms a clique. An optimal solution can contain at most one vertex from this open neighborhood, hence selecting a simplicial vertex is always optimal.

Simplicial reductions appear as $(1, 1 - 3)$, $(2, 3 - 5)$ and $(3, 6)$-reductions. Branchy reductions appear as $(1, 2)$-reductions. These tricks have earlier been used in [2].

```
Simplicial(G)
  repeat
    perform reductions in the following order of preference:
        1. branchy, simplicial, (2, 6)
        2. (3, 8)
        3. (3, 9)
    until done
  end
```

By keeping track of the shape of the neighborhood of each vertex in the current graph, the algorithm can be implemented in linear time.

This algorithm attains a ratio of $9/7 \approx 1.28$ on degree-three graphs. In comparison, the previous best ratio claimed for an algorithm with low-polynomial time complexity [6, 7, 5] was $4/3 + 1/\epsilon$ [5]. Further, by using this algorithm as the subroutine for degree three graphs in the schema of [7] (originating in [3]), we obtain similar improvements for the independent set problem in other classes of bounded-degree graphs.

Theorem 3. *The performance ratio of* Simplicial *on degree-three graphs is* $9/7$.

Using the same measures of the reductions as for MoreEdges, our *cost measure* for this algorithm is given by:

$$g(r) = 6n(r) - 2e(r) - 2Out(r) + \alpha(r).$$

Unfortunately, this measure is too large on $(1, 1)$ and $(2, 3)$ reductions. We alleviate this problem by considering short sequences of reductions, or *idioms* as we call them, and showing that the cost measure on these combinations behave as desired.

Let us view the execution of the algorithm as a string of reductions. We argue that that this string can be lexically partitioned into restricted types of idioms. We use the notation of regular expressions.

Claim 4. *Idioms involving* $(1, 1)$ *and* $(2, 3)$ *reductions must be of the form:*

1. $\{(2, 6) + (3, 8) + (3, 9)\} \{(1, 2) + (2, 4)\}^* (1, 1)$,
2. $\{(1, 3) + (2, 5) + (2, 6) + (3, 9)\} \{(1, 2) + (2, 4)\}^* (2, 3)$, *or*
3. $(3, 8)\{(1, 2) + (2, 4)\}^*(2, 3), \{(2, 5) + (2, 6)\}$.

The following observations have bearing on Claim 4.

1. The only reductions that can precede a $(2, 3)$ reduction are: $(1, 3)$, $(2, 5)$, $(2, 6)$, or $(3, 9)$. This ignores $(1, 2)$ and $(2, 4)$-reductions which may be interspersed in various ways.

2. Following the sequence $[(3,8),(2,3)]$, the remaining graph will be cubic except for a single vertex. Hence, only $(2,5)$ or $(2,6)$ reductions can immediately follow. Further, $[(3,8),(2,3),(2,5),(2,3)]$ is not possible.

Further observe that a $(3,7)$-reduction is no longer possible, since some two of the vertices in the reduction would give rise to a $(3,8)$-reduction.

We generalize the measures of reductions to measures of idioms, by taking the sum of the measure of each reduction in the idiom. Further, for an idiom π, let $t(\pi)$ denote the number of reductions within π.

π	$n(\pi)$	$e(\pi)$	$Out(\pi)$	$\alpha(\pi)$	$g(\pi)$	$g(\pi)/t(\pi)$
$(0,0)$	1	0	0	1	7	7
$(1,1)$	2	1	1	1	9	9
$(1,2)$	2	2	0	1	9	9
$(1,3)$	2	3	0	1	7	7
$(2,4)$	3	4	1	1	9	9
$(2,5)$	3	5	1	1	7	7
$(2,6)$	3	6	0	2	8	8
$(3,8)$	4	8	1	2	8	8
$(3,9)$	4	9	0	3	9	9
$\{(2,6),(2,3)\}$	6	9	2	2	16	8
$\{(3,9),(2,3)\}$	7	12	2	3	17	8.5
$\{(1,3),(2,3)\}$	5	6	1	2	18	9
$\{(2,5),(2,3)\}$	6	8	2	2	18	9
$\{(3,8),(2,3),(2,5)\}$	10	16	3	4	26	8.66
$\{(3,8),(2,3),(2,6)\}$	10	17	2	5	27	9

Table 2. Bounds on measures of the reductions performed by Simplicial.

Table 2 lists lower bounds for $Out(\pi)$ and upper bounds for $\alpha(\pi)$ and $g(\pi)$ for the fundamental idioms π in the alphabet. For the idioms that include $(2,3)$, we have omitted the interspersed $(1,2)$ and $(2,4)$ reductions, and counted them as individual idioms. The idioms involving $(1,1)$ have also been compacted.

The following values are different from Table 1 or are different from the sums of the values of the individual reductions in the idiom.

1. α and Out for $(2,4)$ and $\alpha(2,5)$: Notice that only the former form of these reductions in Figure 1 can now appear due to the preference for the delayed-commitment reduction.
2. Out for $(1,1)$: Whichever idiom a $(1,1)$-reduction appears in, an additional edge must be outside of I.
3. The Out values of $(2,5)(2,3)$ and $(3,9)(2,3)$ and α values of $(2,6)(2,3)$ and $(3,9)(2,3)$: The reasons are clear when we look at the subgraphs induced by these pairs of reductions. $(1,2)$ and $(2,4)$-reductions yield optimal results on the subgraph induced by the deleted vertices, thus, having interspersed

159

within an idiom does not affect any argument about the independence number or outside edges in the subgraph induced by the idiom.

The theorem now follows from Lemma 1 along with the fact that $g(\pi)/t(\pi)$ in Table 2 is always at most 9. The number of vertices found by the algorithm is:

$$\sum_{\pi} t(\pi) \geq \sum_{\pi} \frac{g(\pi)}{9} \geq \frac{6n - 2e - 2Out + \alpha}{9} \geq \frac{7}{9}\alpha.$$

Lower bound We construct a hard graph for Simplicial as in Figure 3. First build a unit with seven vertices, $v_1, \ldots v_7$ forming a cycle with chords between second and fourth, third and sixth, and fifth and seventh. All vertices are of degree three, except v_1. The hard graph is obtained by adding one vertex u connected to the v_1's of three units.

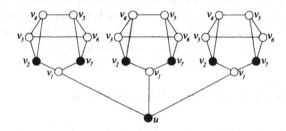

Fig. 3. A hard graph for Simplicial.

For the initial choice, the algorithm will prefer a $(3, 8)$-reduction over a $(3, 9)$-reduction, and may choose any vertex but u and v_1 vertices. One tie-breaking choice is to select v_2, followed by v_7, u, and two vertices from each of the two remaining units, for a total of seven. On the other hand, the optimal solution consists of the first, third and fifth vertex of each unit, for a total of nine.

5 Algorithm with Advice, Ultimate

The two algorithms we considered have a common basic strategy of removing a vertex of minimum degree in the current graph at each stage. We can easily see by their hard graphs that the weaknesses of the greedy algorithms appear when they have several ways of choosing a minimum-degree vertex. We are left to wonder if perfect advice for choosing among minimum-degree alternatives suffices to obtain an optimal solution.

In this section, we study the power of algorithms that are given the additional benefit of an oracle for selecting among alternatives. The only requirement is that the algorithm must choose one of the minimum degree vertices at any step. This algorithm, Ultimate, can use the existence or non-existence of edges arbitrarily far away for deciding whether to select a given minimum-degree vertex.

Theorem 5. *Any greedy algorithm that selects vertices of minimum degree must have performance ratios at least: 1.25, for degree-three graphs; 1.1$\overline{1}$, for cubic graphs; and 16/15 ≈ 1.06, for triangle-free cubic graphs.*

This reveals a limitation on the power of the family of greedy algorithms.

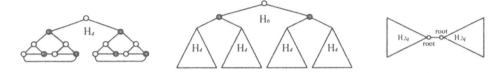

Fig. 4. Construction of H_4, H_6, and G_∞.

We first form a subgraph H_4 as on the left in Figure 4. The size of the maximum independent set of H_4 is six, whereas Ultimate finds only five vertices. One optimal solution consists of the shaded vertices. We construct a pseudo binary tree H_6 with four H_4's as leaves, as illustrated in the center of Figure 4. We form a pseudo binary tree H_{2q} with $2q(q \geq 2)$ levels by repeating the same operation $q - 2$ times.

The size of the solution found by the algorithm is:

$$HEU(H_{2q}) \;=\; 1+4{\cdot}HEU(H_{2(q-1)}) \;=\; 1+4+4^2+\cdots+4^{q-2}+4^{q-1}5 = \frac{4^{q-1}16 - 1}{3},$$

while the size of the optimal solution is:

$$OPT(H_{2q}) \;=\; 2+4{\cdot}OPT(H_{2(q-1)}) \;=\; 2(1+4+4^2+\cdots+4^{q-2})+4^{q-1}6 = \frac{4^{q-1}20 - 2}{3}.$$

As q grows, the ratio of HEU to OPT approaches 5/4.

We can also obtain a hardness result for regular (i.e. cubic) graphs. Join two H_{2q}s as on the right of Figure 4 in order to make the entire graph cubic, and call the resulting graph G_∞.

An optimal solution of G_∞ contains all vertices on the even levels of each H_{2q}. Suppose that Ultimate picks any vertex in a H_{2q} at the first step. It is easy to verify that Ultimate then proceeds optimally on that half of the graph. That leaves the other H_{2q} left, for which the algorithm will, by induction, be non-optimal.

Thus,

$$\rho(G_\infty) = \frac{2 \cdot OPT(H_{2q})}{OPT(H_{2q}) + HEU(H_{2q})} = \frac{20 + 20}{20 + 16} = \frac{10}{9} = 1.\overline{1}.$$

We can obtain similar hardness results for triangle-free graphs (or more generally graphs of high odd girth), by replacing the triangles at the bottom of H_5 by a five-cycle (or an appropriately large odd cycle) and connecting the pairs together as needed.

6 Cubic graphs

We have additionally considered cubic graphs (i.e. 3-regular graphs, where all vertices are of degree three) and further restrictions to graphs of high odd girth (i.e. containing no short odd cycle). Due to lack of space, we have omitted all details from this section.

We can give an approximation-preserving reduction to the degree-three case that shows that the problem remains $\mathcal{NP}$-hard and MAX SNP-hard (hard to approximate within some fixed constant greater than one) even under these strong restrictions.

On cubic graphs, Greedy attains a ratio of $3/2$ and MoreEdges attains a ratio of $17/12 \approx 1.42$. MoreEdges achieves further improved ratios of $29/21 \approx 1.38$ on triangle-free cubic graphs and more generally, $4/3 + 1/(9k + 3)$ on cubic graphs of odd girth $2k + 1$. These bounds are all tight.

Constant odd girth does not help Greedy to attain better asymptotic performance ratios, while it will find an optimal solution of bipartite regular graphs. Simplicial also performs no better no cubic graphs, as we saw in Figure 3, while the case of large odd girth is left open.

Acknowledgments

We are much indebted to Professor Osamu Watanabe and Professor Jaikumar Radhakrishnan for informative comments and discussions.

References

1. S. Arora, C. Lund, R. Motwani, M. Sudan, and M. Szegedy. Proof verification and hardness of approximation problems. *FOCS 1992*.
2. P. Berman and T. Fujito. On the approximation properties of independent set problem in degree 3 graphs. *WADS 1995*.
3. P. Berman and M. Fürer. Approximating maximum independent set in bounded degree graphs. *SODA 1994*.
4. M. R. Garey and D. S. Johnson. *Computers and Intractability: A Guide to the Theory of NP-completeness*. Freeman, 1979.
5. M. M. Halldórsson. Approximating discrete collections via local improvements. *SODA 1995*.
6. M. M. Halldórsson and J. Radhakrishnan. Greed is good: Approximating independent sets in sparse and bounded-degree graphs. *STOC 1994*. To appear in *Algorithmica*.
7. M. M. Halldórsson and J. Radhakrishnan. Improved approximations of independent sets in bounded-degree graphs. *SWAT 1994*.
8. M. M. Halldórsson and J. Radhakrishnan. Improved approximations of independent sets in bounded-degree via subgraph removal. *Nordic J. Computing*, 1(4):475–492, 1994.
9. D. S. Hochbaum. Efficient bounds for the stable set, vertex cover, and set packing problems. *Disc. Applied Math.*, 6:243–254, 1983.

Practical Logic
(Invited Presentation)

John N. Crossley

Department of Computer Science, Monash University,
Clayton, Australia 3168.
jnc@cs.monash.edu.au

Abstract

This talk is about getting correct programs by plagiarizing proofs that others have already produced. The method provides an alternative to verification as practised to the south of here in Brisbane. In ancient times in the East mathematicians worked out their mathematics to solve (practical) problems. In the West they concentrated on abstract proofs with calculations almost as an afterthought. At the beginning of this century the Intuitionist mathematicians refused to accept any mathematics which did not give a construction of the object which was being proved to exist. The present day industry of extracting programs from proofs started by using this logic. (Incidentally the logic was formulated in a way that was anathema to the Intuitionists themselves.) There is a general approach which most people seem to follow. Unfortunately we have no algorithm to implement this approach. However we shall show how to apply the general approach in several different contexts including Buss's logic which gives exactly programs for polynomial-time computable functions and Linear Logic which allows the measurement of resources consumed. All of this depends on actually having proofs—and worse than that—proofs in the formal system. We shall also present a framework for getting over this difficulty and using real proofs—the sort you find in books. This last part has not yet been implemented but we hope by the time of the talk that the others will have been.

Acknowledgement

This work was undertaken with the assistance of David Albrecht, Frank Baeuerle, John Jeavons and Colin Rhodes.

An Approximation Algorithm for MAX 3-SAT

Takao Ono[1] Tomio Hirata[1] Takao Asano[2]

[1] School of Engineering, Nagoya University
[2] Department of Information and System Engineering, Chuo University

Abstract. In this paper we present a 0.80-approximation algorithm for MAX 3-SAT. Previously 0.75- or 0.755-approximation algorithms were known for MAX SAT. Thus, we make slight improvement by limiting MAX SAT to MAX 3-SAT. Since approximating MAX 3-SAT within 112/113 is NP-complete, our result means that the best approximation ratio is between 0.80 and 112/113.

1 Introduction

We consider the Maximum 3-Satisfiability Problem(MAX 3-SAT), a restricted version of the Maximum Satisfiability Problem(MAX SAT). MAX SAT is known to be NP-complete, and 0.75- or 0.755-approximation algorithms for MAX SAT were proposed in [7, 3, 2]. But no good approximation algorithms for MAX 3-SAT are known. In this paper we present a 0.80-approximation algorithm for MAX 3-SAT. Thus we make slight improvement by limiting MAX SAT to MAX 3-SAT.

Since MAX 3-SAT is MAX SNP-complete[5], for some constant $\alpha < 1$ no α-approximation algorithms for MAX 3-SAT exist unless P = NP. In fact, approximating MAX 3-SAT within the approximation ratio 112/113 is known to be NP-complete[1]. Our result means that the best approximation ratio is between 0.80 and 112/113.

Our result is inspired by the recent work of Goemans and Williamson who used an algorithm for semidefinite optimization problems to obtain an approximation algorithm for MAX 2-SAT. Their algorithm is 0.878-approximation. In their algorithm, a maximization problem of a polynomial of degree 2 is treated. but in the case of MAX 3-SAT we have to treat a maximization problem of a polynomial of degree 4. Furthermore, the analysis of the approximation ratio is not trivial.

In Sect. 2, some definitions are given. We present a 0.770-approximation algorithm for MAX 3-SAT in Sect. 3. In Sect. 4 we analyze its approximation ratio. Combining this algorithm with Johnson's we show that the 0.80-approximation ratio can be achieved in Sect. 5. Conclusion with some remarks is in Sect. 6.

2 Definitions

Let $\{x_1, \ldots, x_n\}$ be a set of variables. A literal is either a variable x_i or its negation $\bar{x}_i$. A clause is a disjunction of literals. An instance I of MAX 3-SAT is a collection of clauses $C_1, \ldots, C_L$ with the corresponding weights $w_1, \ldots, w_L$, where each clause has at most three literals. We assume that no variables appear more than once in a clause, that is, we do not allow a clause like $\{x_1, \bar{x}_1, x_2\}$.

For an algorithm A, $W_A(I)$ denotes the sum of the weights of the clauses that are satisfied by the assignment produced by A for an input I. The approximation ratio of A is α if $W_A(I)$ is at least α times the weight $W_{OPT}(I)$ of the optimal assignment for any instance I. We call A an α-approximation algorithm.

3   0.770-Approximation Algorithm

In this section we present a 0.770-approximation algorithm for MAX 3-SAT. This algorithm solves MAX 3-SAT in the same way as Goemans and Williamson's algorithm for MAX 2-SAT. The algorithm consists of the following three steps.

1. Translate an instance of MAX 3-SAT into a multivariate polynomial, that is, we regard MAX 3-SAT as a maximization problem of a multivariate polynomial.
2. Reduce the maximization problem to a convex programming problem with relaxation and solve it.
3. Construct a truth assignment of the original MAX 3-SAT problem from a solution to the convex programming problem.

We explain each of the above steps in the following subsections.

3.1   Translation of an Instance into a Multivariate Polynomial

We translate an instance of MAX 3-SAT into a multivariate polynomial in the following way. First, we construct for each clause C_l a term c_l of the polynomial in two steps. (i) If C_l consists of one literal v_1, the term c_l is v_1'. If C_l consists of two literals v_1 and v_2, the term c_l is $1 - (1 - v_1')(1 - v_2')$. If C_l consists of three literals v_1, v_2 and v_3, the corresponding term c_l is $1 - (1 - v_1')(1 - v_2')(1 - v_3')$. (ii) When a literal v_i is a variable x_k, replace v_i' with $(1+y_k)/2$. When a literal v_i is a negation $\bar{x}_k$ of a variable, replace v_i' with $(1 - y_k)/2$.

Next, we define a multivariate polynomial $f(y_1, \ldots, y_n) = \sum_{l=1}^{L} w_l c_l$. Then, if we think that $y_i = +1$ when $x_i = $ true and $y_i = -1$ when $x_i = $ false, we can see that $c_l = 1$ if and only if $C_l = $ true and $c_l = 0$ if and only if $C_l = $ false. Thus we can regard MAX 3-SAT as a maximization problem of $f(y_1, \ldots, y_n)$ subject to $|y_i| = 1$ for all $i = 1, 2, \ldots, n$.

It is easily seen that c_l is a linear combination of the terms $1 \pm y_i$, $1 \pm y_i y_j$, and $1 \pm y_i y_j y_k$ with positive coefficients, and MAX 3-SAT is equivalent to the following optimization problem. From here, we simply write $a_i^{\pm}(1 \pm y_i)$ instead of $a_i^+(1 + y_i) + a_i^-(1 - y_i)$.

(P): Maximize $f(y_1, \ldots, y_n) = \sum_i a_i^{\pm}(1 \pm y_i) + \sum_{i<j} b_{ij}^{\pm}(1 \pm y_i y_j) + \sum_{i<j<k} c_{ijk}^{\pm}(1 \pm y_i y_j y_k)$ subject to $|y_i| = 1$ for all $i = 1, 2, \ldots, n$, where all $a_i^{\pm}$, $b_{ij}^{\pm}$ and $c_{ijk}^{\pm}$ are nonnegative.

We further rewrite the objective function as follows. First, we replace y_i with $y_0 y_i$, where $|y_0| = 1$, in order to eliminate all terms of degree 1.

By this replacement each term is rewritten as:

$$1 \pm y_i \Rightarrow 1 \pm y_0 y_i$$
$$1 \pm y_i y_j \Rightarrow 1 \pm y_i y_j \text{ (not changed)}$$
$$1 \pm y_i y_j y_k \Rightarrow 1 \pm y_0 y_i y_j y_k \ .$$

Thus the objective function is rewritten as follows, where all $a_{ij}^{\pm}$ and $b_{ijkl}^{\pm}$ are nonnegative:

$$f(y_0, \ldots, y_n) = \sum_{i<j} a_{ij}^{\pm}(1 \pm y_i y_j) + \sum_{i<j<k<l} b_{ijkl}^{\pm}(1 \pm y_i y_j y_k y_k) \ .$$

Next, we replace $1 \pm y_i y_j y_k y_l$ with an equivalent term $2 - (y_i y_j \mp y_j y_k)^2/2$. After this replacement (P) is equivalent to the following problem (P').

(P'): maximize $f(y_0, \ldots, y_n) = \sum_{i<j} a_{ij}^{\pm}(1 \pm y_i y_j) + \sum_{i<j<k<l} b_{ijkl}^{\pm}(2 - (y_i y_j \mp y_k y_l)^2/2)$ subject to $|y_i| = 1$ for all $i = 0, \ldots, n$, where all $a_{ij}^{\pm}$ and $b_{ijkl}^{\pm}$ are nonnegative.

The following lemma holds for the coefficients $b_{ijkl}^{\pm}$ of the objective function $f(y_0, \ldots, y_n)$ of (P').

Lemma 1. $\sum_{i<j<k<l}(b_{ijkl}^{+} + b_{ijkl}^{-}) = W_3/8$, where W_3 is the sum of the weights corresponding to all the clauses with 3 literals.

Proof. Terms $1 \pm y_i y_j y_k y_l$ are generated only from the clauses with 3 literals. If we assume that C_l has 3 literals v_1, v_2, and v_3, the corresponding term c_l has the form $1 - (1 - v_1')(1 - v_2')(1 - v_3')$ and the coefficient of $1 \pm y_i y_j y_k$ is $1/8$ from the way of replacing v_i''s. Using this fact, we can see that $\sum_{i<j<k<l}(b_{ijkl}^{+} + b_{ijkl}^{-})$ is $1/8$ of the sum of the weights of all 3-literal clauses, or $W_3/8$. □

3.2 Reducing to a Convex Programming Problem

Since (P') is too hard to solve, we introduce a relaxation to (P'). We use an $(n + 1)$-dimensional vector v_i ($\|v_i\| = 1$) instead of y_i and we replace a product $y_i y_j$ with an inner product $v_i \cdot v_j$. With this relaxation, (P') reduces to the following problem (Q).

(Q): maximize $\sum_{i<j} a_{ij} \pm (1 \pm v_i \cdot v_j) + \sum_{i<j<k<l} b_{ijkl}^{\pm}(2 - (v_i \cdot v_j \mp v_k \cdot v_l)^2/2)$ subject to $\|v_i\| = 1$ for all $i = 0, \ldots, n$.

We introduce new variables y_{ij} for $v_i \cdot v_j$. Then, matrix (y_{ij}) is symmetric and positive semidefinite. The following problem is equivalent to (Q).

(Q'): maximize $\sum_{i<j} a_{ij}^{\pm}(1 \pm y_{ij}) + \sum_{i<j<k<l} b_{ijkl}^{\pm}(2 - (y_{ij} \mp y_{kl})^2/2)$ subject to the condition that the matrix (y_{ij}) is symmetric and positive semidefinite, and $y_{ii} = 1$ for all $i = 0, \ldots, n$.

Since (Q') is a convex programming problem, we can find an approximate solution $\bar{y}_{ij}$ within a constant error ϵ in polynomial time[6]. We can find an approximate solution $\bar{v}_i$ to (Q) using Cholesky decomposition of the matrix $(\bar{y}_{ij})$. At this point we have an approximate solution to (Q) in polynomial time.

3.3 Finding an Approximate Solution to MAX 3-SAT

We use probabilistic method to find an approximate solution $\bar{y}_i$ to (P') from $\bar{v}_i$, an approximate solution to (Q). Taking a random vector r, let $y_i = +1$ if $\bar{v}_i \cdot r > 0$, or else $y_i = -1$. We show in Sect. 4 that the expectation of the objective function of (P') is greater than 0.770 times the optimal sum of the weights. We can derandomize this method following the same way as [2].

4 Analysis of the Approximation Ratio

We analyze the approximation ratio of the algorithm. We use following notations.

Let $W_k(k = 1, 2, 3)$ be the sum of the weights of the clauses that contain exactly k literals. For (P'), Z_P^* denotes the value of the objective function of the optimal solution and $\bar{Z}_P$ denotes the value of the approximate solution. $\bar{Z}_{P2}$ and $\bar{Z}_{P4}$ denote the first and second sums in $\bar{Z}_P$, respectively. Similarly, we use Z_Q^*, $\bar{Z}_Q$, $\bar{Z}_{Q2}$, and $\bar{Z}_{Q4}$ for (Q'). We denote by θ_{ij} the angle between two vectors $\bar{v}_i$ and $\bar{v}_j$, which are a part of an approximate solution to (Q). $E[X]$ denotes the expectation of a random variable X. Note that we can think that $\bar{Z}_Q = Z_Q^*$ since $Z_Q^* - \epsilon \le \bar{Z}_Q \le Z_Q^*$ and ϵ can be arbitrarily small.

The approximation ratio is $E[\bar{Z}_P]/Z_P^*$. Since (Q') is a relaxed version of (P), $\bar{Z}_Q = Z_Q^* \ge Z_P^*$. Therefore $E[\bar{Z}_P]/Z_P^* \ge E[\bar{Z}_P]/\bar{Z}_Q$, and we concentrate on $E[\bar{Z}_P]/\bar{Z}_Q$. If we have constants α, β, γ ($\alpha > \beta$) such that $E[\bar{Z}_{P2}] \ge \alpha\bar{Z}_{Q2}$, $E[\bar{Z}_{P4}] \ge \beta\bar{Z}_{Q4}$, and $\bar{Z}_{Q4} \le \gamma\bar{Z}_Q$, we can find a lower bound of $E[\bar{Z}_P]/Z_P^*$ since $E[\bar{Z}_P]/\bar{Z}_Q = (E[\bar{Z}_{P2}] + E[\bar{Z}_{P4}])/\bar{Z}_Q \ge (\alpha\bar{Z}_{Q2} + \beta\bar{Z}_{Q4})/\bar{Z}_Q \ge \alpha(1 - \gamma) + \beta\gamma$. Now we will find α, β, and γ.

Lemma 2–Lemma 5 are the same as in [2].

Lemma 2. *Suppose that $\|x\| = 1$ and $\|y\| = 1$. For random vector r with $\|r\| = 1$, $\Pr\{\text{sign}(r \cdot x) \ne \text{sign}(r \cdot y)\} = \arccos(x \cdot y)/\pi$.*

Note: Because $\bar{y}_i = (r \cdot \bar{v}_i)/|r \cdot \bar{v}_i|$ and $\bar{y}_j = (r \cdot \bar{v}_j)/|r \cdot \bar{v}_j|$, it is easy to see that $\Pr\{\bar{y}_i\bar{y}_j = -1\} = \Pr\{\bar{y}_i \ne \bar{y}_j\} = \theta_{ij}/\pi$ and $\Pr\{\bar{y}_i\bar{y}_j = 1\} = \Pr\{\bar{y}_i = \bar{y}_j\} = 1 - \theta_{ij}/\pi$ from this lemma.

We determine the value of α from Lemma 3 and Lemma 4.

Lemma 3. $E[(1 - \bar{y}_i\bar{y}_j)/2] = \theta_{ij}/\pi$, $E[(1 + \bar{y}_i\bar{y}_j)/2] = 1 - \theta_{ij}/\pi$.

Lemma 4. *Suppose that $\alpha = \min_{0 \le \theta \le \pi}(\theta/\pi)/((1 - \cos\theta)/2) = 0.878\ldots$,*

$$\theta/\pi \ge \alpha(1 - \cos\theta)/2 \ ,$$
$$1 - \theta/\pi \ge \alpha(1 + \cos\theta)/2 \ .$$

Lemma 5. $E[\bar{Z}_{P2}] \ge \alpha\bar{Z}_{Q2}$ *for α in Lemma 4.*

Next, we determine the value of β from Lemma 6 and Lemma 7.

Lemma 6. $E[1 - (\bar{y}_i\bar{y}_j + \bar{y}_k\bar{y}_l)^2/4] = (\theta_{ij} + \theta_{kl})/\pi - 2\theta_{ij}\theta_{kl}/\pi^2$, $E[1 - (\bar{y}_i\bar{y}_j - \bar{y}_k\bar{y}_l)^2/4] = 1 - (\theta_{ij} + \theta_{kl})/\pi + 2\theta_{ij}\theta_{kl}/\pi^2$.

Proof.

$$1 - \frac{(\bar{y}_i\bar{y}_j + \bar{y}_k\bar{y}_l)^2}{4} = \begin{cases} 0 & \text{if } \bar{y}_i\bar{y}_j = \bar{y}_k\bar{y}_l \\ 1 & \text{if } \bar{y}_i\bar{y}_j \neq \bar{y}_k\bar{y}_l \end{cases} .$$

From the note of Lemma 2,

$$\begin{aligned}
E[1 - (\bar{y}_i\bar{y}_j + \bar{y}_k\bar{y}_l)^2/4] &= \Pr\{\bar{y}_i\bar{y}_j \neq \bar{y}_k\bar{y}_l\} \\
&= (1 - \theta_{ij}/\pi)(\theta_{kl}/\pi) + (\theta_{ij}/\pi)(1 - \theta_{kl}/\pi) \\
&= (\theta_{ij} + \theta_{kl})/\pi - 2\theta_{ij}\theta_{kl}/\pi^2 .
\end{aligned}$$

Similarly,

$$\begin{aligned}
E[1 - (\bar{y}_i\bar{y}_j - \bar{y}_k\bar{y}_l)^2/4] &= \Pr\{\bar{y}_i\bar{y}_j = \bar{y}_k\bar{y}_l\} \\
&= 1 - \Pr\{\bar{y}_i\bar{y}_j \neq \bar{y}_k\bar{y}_l\} \\
&= 1 - (\theta_{ij} + \theta_{kl})/\pi + 2\theta_{ij}\theta_{kl}/\pi^2 . \qquad \square
\end{aligned}$$

Lemma 7.

$$\min_{\substack{0 \leq \theta \leq \pi \\ 0 \leq \phi \leq \pi}} \frac{(\theta + \phi)/\pi - 2\theta\phi/\pi^2}{1 - (\cos\theta + \cos\phi)^2/4} = \frac{1}{2} ,$$

$$\min_{\substack{0 \leq \theta \leq \pi \\ 0 \leq \phi \leq \pi}} \frac{1 - (\theta + \phi)/\pi + 2\theta\phi/\pi^2}{1 - (\cos\theta - \cos\phi)^2/4} = \frac{1}{2} .$$

Proof. We can prove both equalities in the same way, and we prove only the first equality here. $(\cos\theta + \cos\phi)^2/4 \geq (1 - 2\theta/\pi)(1 - 2\phi/\pi)$ implies $1 - (\cos\theta + \cos\phi)^2/4 \leq 2\big((\theta + \phi)/\pi - 2\theta\phi/\pi^2\big)$, which implies $\big((\theta + \phi)/\pi - 2\theta\phi/\pi^2\big)/\big(1 - (\cos\theta + \cos\phi)^2/4\big) \geq 1/2$. Hence we will show the inequality

$$(\cos\theta + \cos\phi)^2/4 \geq (1 - 2\theta/\pi)(1 - 2\phi/\pi) .$$

This is obvious when $(1 - 2\theta/\pi)(1 - 2\phi/\pi) \leq 0$, thus we consider the case $(1 - 2\theta/\pi)(1 - 2\phi/\pi) > 0$. Suppose $1 - 2\theta/\pi > 0$. It means $\theta < \pi/2$ and $\phi < \pi/2$. Because $\cos\theta \geq 1 - 2\theta/\pi > 0$ and $\cos\phi \geq 1 - 2\phi/\pi > 0$, we have $(1 - 2\theta/\pi)(1 - 2\phi/\pi) \leq \cos\theta\cos\pi \leq (\cos\theta + \cos\phi)^2/4$. If $1 - 2\theta/\pi < 0$, which means $\theta > \pi/2$ and $\phi > \pi/2$, then $\cos\theta \leq 1 - 2\theta/\pi < 0$ and $\cos\phi \leq 1 - 2\phi/\pi$. Thus $(1 - 2\theta/\pi)(1 - 2\phi/\pi) \leq \cos\theta\cos\phi \leq (\cos\theta + \cos\phi)^2/4$. Now the lemma holds. $\square$

Lemma 8. $E[\bar{Z}_{P4}] \geq (1/2)\bar{Z}_{Q4}$.

Proof. From Lemma 6 and Lemma 7, we have $E[1 - (\bar{y}_i\bar{y}_j + \bar{y}_k\bar{y}_l)^2/4] = (\theta_{ij} + \theta_{kl})/\pi - 2\theta_{ij}\theta_{kl}/\pi^2 \geq (1/2)(1 - (\cos\theta_{ij} + \cos\theta_{kl})^2/4) = (1/2)(1 - (\bar{y}_{ij} + \bar{y}_{kl})^2/4)$ and $E[1 - (\bar{y}_i\bar{y}_j - \bar{y}_k\bar{y}_l)^2/4] = 1 - (\theta_{ij} + \theta_{kl})/\pi + 2\theta_{ij}\theta_{kl}/\pi^2 \geq (1/2)(1 - (\cos\theta_{ij} - \cos\theta_{kl})^2/4) = (1/2)(1 - (\bar{y}_{ij} - \bar{y}_{kl})^2/4)$. Therefore $E[\bar{Z}_{P4}] = E[\sum_{i<j<k<l} b_{ijkl}^{\pm}(2 - (\bar{y}_i\bar{y}_j \mp \bar{y}_k\bar{y}_l)^2/2] \geq (1/2)\sum_{i<j<k<l} b_{ijkl}^{\pm}(2 - (\bar{y}_{ij} \mp \bar{y}_{kl})^2/2) = (1/2)\bar{Z}_{Q4}$. $\square$

From Lemma 8, we can select $1/2$ for β.

Finally, we determine the value of γ.

Lemma 9. $Z_P^* \geq W_1/2 + 3W_2/4 + 7W_3/8$.

Proof. We extend the proof of [4] for MAX SAT, taking account of the weights. The objective function of (P) is $f(y_1, \ldots, y_n) = \sum_{l=1}^{L} w_l c_l$. Since this function is linear with respect to each variable, for any $y_i (i = 1, \ldots, n)$ $f(y_1, \ldots, y_{i-1}, -1, y_{i+1}, \ldots, y_n) + f(y_1, \ldots, y_{i-1}, +1, y_{i+1}, \ldots, y_n) = 2f(y_1, \ldots, y_{i-1}, 0, y_{i+1}, \ldots, y_n)$. Therefore

$$Z_P^* = \max_{y_1, \ldots, y_n \in \{-1, +1\}} f(y_1, \ldots, y_n)$$
$$\geq \max_{y_2, \ldots, y_n \in \{-1, +1\}} f(0, y_2, \ldots, y_n)$$
$$\cdots$$
$$\geq \max_{y_n \in \{-1, +1\}} f(0, \ldots, 0, y_n)$$
$$\geq f(0, 0, \ldots, 0) .$$

When all variables are 0, the value of the term c_l is $1 - 1/2^k$, where the clause C_l has k literals. And we have $Z_P^* \geq f(0, 0, \ldots, 0) = W_1/2 + 3W_2/4 + 7W_3/8$. $\quad\square$

Lemma 10. $\bar{Z}_{Q4}/\bar{Z}_Q \leq 2/7$.

Proof. From Lemma 1,

$$\bar{Z}_{Q4} = \sum_{i<j<k<l} b_{ijkl}^{\pm} (2 - (\bar{y}_{ij} \mp \bar{y}_{kl})^2/2)$$
$$\leq \sum_{i<j<k<l} 2b_{ijkl}^{\pm}$$
$$= 2 \sum_{i<j<k<l} (b_{ijkl}^{+} + b_{ijkl}^{-}) = W_3/4 .$$

From Lemma 9, $\bar{Z}_Q \geq Z_P^* \geq W_1/2 + 3W_2/4 + 7W_3/8 \geq 7W_3/8$. Thus, $\bar{Z}_{Q4}/\bar{Z}_Q \leq (W_3/4)/(7W_3/8) = 2/7$. $\quad\square$

Now we can find the approximation ratio of the above algorithm.

Theorem 11. *The approximation ratio of the above algorithm is $\alpha - (\alpha - 1/2)\bar{Z}_{Q4}/\bar{Z}_Q$.*

Proof. From Lemma 5 and Lemma 8, $E[\bar{Z}_{Q2}] \geq \alpha \bar{Z}_{Q2}$ and $E[\bar{Z}_{Q4}] \geq (1/2)\bar{Z}_{Q4}$. Thus,

$$\frac{E[\bar{Z}_P]}{Z_P^*} \geq \frac{E[\bar{Z}_P]}{\bar{Z}_Q}$$
$$= \frac{E[\bar{Z}_{P2}] + E[\bar{Z}_{P4}]}{\bar{Z}_Q}$$

$$\geq \frac{\alpha \bar{Z}_{Q2} + (1/2)\bar{Z}_{Q4}}{\bar{Z}_Q}$$

$$= \frac{\alpha(\bar{Z}_Q - \bar{Z}_{Q4}) + (1/2)\bar{Z}_{Q4}}{\bar{Z}_Q}$$

$$= \alpha - \left(\alpha - \frac{1}{2}\right)\frac{\bar{Z}_{Q4}}{\bar{Z}_Q} . \qquad \Box$$

Corollary 12. *The above algorithm is .770-approximation.*

Proof. Since $\bar{Z}_{Q4}/\bar{Z}_Q \leq 2/7$ from Lemma 10, $E[\bar{Z}_P]/Z_P^* \geq \alpha - (\alpha - 1/2)(2/7) = 0.770\ldots$. Thus that algorithm is .770-approximation. $\qquad \Box$

5 A 0.80-Approximation Algorithm

In this section we show that a combination of the above algorithm and Johnson's algorithm can achieve 0.80-approximation. The following algorithm A is 0.80-approximation: Algorithm A simultaneously executes the above algorithm and Johnson's algorithm. It outputs the better solution.

We assume that A is a-approximation and determine the value of a. Let $W_J(I)$ be the weight of the solution which Johnson's algorithm produces to the instance I. When $W_J(I)/W_{OPT}(I) \geq a$, $W_A(I)/W_{OPT}(I) \geq a$. Thus we concentrate on the case when $W_J(I)/W_{OPT}(I) < a$.

It is easily seen that $W_J(I) \geq W_1/2 + 3W_2/4 + 7W_3/8$. Clearly

$$a(W_1 + W_2 + W_3) \geq aW_{OPT}(I) > W_J(I) \geq \frac{W_1}{2} + \frac{3W_2}{4} + \frac{7W_3}{8} .$$

Thus

$$\left(a - \frac{1}{2}\right)W_1 + \left(a - \frac{3}{4}\right)W_2 > \left(\frac{7}{8} - a\right)W_3 .$$

This implies

$$W_1 + \frac{3}{2}W_2 > \frac{7/8 - a}{a - 1/2}W_3 .$$

Therefore

$$aW_{OPT}(I) > W_J(I)$$

$$\geq \frac{W_1}{2} + \frac{3W_2}{4} + \frac{7W_3}{8}$$

$$> \frac{7/8 - a}{2a - 1}W_3 + \frac{7}{8}W_3$$

$$= \left(\frac{7 - 8a}{8(2a - 1)} + \frac{7}{8}\right)W_3$$

$$= \frac{3a}{4(2a - 1)}W_3 .$$

Now

$$\frac{\bar{Z}_{Q4}}{\bar{Z}_Q} \leq \frac{W_3}{4\bar{Z}_Q}$$

$$\leq \frac{W_3}{4W_{OPT}(I)}$$

$$\leq \frac{2a - 1}{3} .$$

This means that the 0.770-approximation algorithm produces the solution whose weight is greater than

$$\left[\alpha - \frac{(\alpha - 1/2)(2a - 1)}{3} \right] W_{OPT}(I) .$$

Since $W_J(I) < aW_{OPT}(I)$, this weight must be greater than $aW_{OPT}(I)$. Thus $\alpha - (\alpha - 1/2)(2a - 1)/3 \geq a$. Solving this inequality we have $a \leq 0.802\ldots$. Therefore A is a 0.80-approximation algorithm.

6   Concluding Remarks

We have presented a 0.80-approximation algorithm for MAX 3-SAT. Previous algorithms were 0.75-approximation or 0.755-approximation for general MAX SAT, so we have made slight improvement by limiting MAX SAT to MAX 3-SAT.

For MAX 3-SAT, approximating within the ratio 112/113 is known to be NP-complete. The approximation ratio of our algorithm is much smaller than 112/113. A challenging problem is to narrow this gap.

References

1. M. Bellare, S. Goldwasser, C. Lund, and A. Russell. Efficient probabilistically checkable proofs and applications to approximation. In *Proc. 25th STOC*, pages 294–304, 1993.
2. Michel X. Goemans and David P. Williamson. .878-approximation algorithms for MAX CUT and MAX 2SAT. In *Proc. 26th STOC*, pages 422–431, 1994.
3. Michel X. Goemans and David P. Williamson. New 3/4-approximation algorithms for the maximum satisfiability problem. *SIAM Journal of Disc. Math.*, 7(4):656–666, November 1994.
4. David S. Johnson. Approximation Algorithms for Combinatorial Problems. *Journal of Comput. and Sys. Sci.*, 9:256–278, 1974.
5. Christos H. Papadimitriou and Mihalis Yannakakis. Optimization, approximation, and complexity classes. In *Proc. 20th STOC*, pages 229–234, 1988.
6. Pravin M. Vaidya. A new algorithm for minimizing convex function over convex sets. In *Proc. 30th FOCS*, pages 338–343, 1989.
7. Mihalis Yannakakis. On the approximation of maximum satisfiability. In *Proc. 3rd SODA*, pages 1–9, 1992.

Learning of Restricted RNLC Graph Languages

Sei'ichi Tani

Dept. of Mathematics
Tokai University
1117 Kitakaname Hiratsuka-Shi
Kanagawa-Ken 259-12, Japan
sei-ichi@sm.u-tokai.ac.jp

Koichi Yamazaki

Dept. of Computer Sci. and Inform. Math.
University of Electro-Communications
1-5-1 Chofugaoka, Chofu-Shi
Tokyo 182, Japan
yamazaki@cs.uec.ac.jp

Abstract

We consider learning of graph languages (sets of graphs) which are generated by restricted regular node label controlled (for short RNLC) graph grammars. We show that for any restricted RNLC graph languages L, given the Parikh image of L, one can construct a restricted RNLC graph grammar G such that $L(G) = L$ in polynomial time. We present an algorithm to construct a restricted RNLC graph grammar which generates an unknown restricted RNLC graph language using restricted superset queries and restricted subset queries. We also showed that for a fixed nonnegative integer t, this algorithm halts in polynomial time when the Parikh image of the unknown restricted RNLC graph languages has at most t-periods.

1 Introduction

In this paper, we consider the learning problem of graph languages (sets of graphs). In computational learning theory, many researchers have investigated learnability or non-learnability for many objects such as sets of strings, semilinear sets, and formulas. There exists few reports of learning theory for sets of graphs [7]. There are several ways to define sets of graphs by finite devices. The main ones are graph grammars [9, 10, 13, 14], graph rewriting systems[5, 12], graph automata[6], obstruction sets[8], and so on. Using graph grammars as devices to define sets of graphs, there are merits in applications for practical problems. In this paper, we use a type of graph grammars as a representation language in learning of graph languages.

One of the well-known graph grammar models is the node-label-controlled (NLC) graph grammars. The class of regular NLC (RNLC) graph languages is a natural subclass of NLC graph languages. In this paper, we consider the learning problem of the graph languages generated by restricted RNLC graph grammars. A graph grammar G is restricted RNLC graph grammar, if G has the following properties:
(1) G is a symmetric RNLC graph grammar, namely $(a, b) \in conn$ iff $(b, a) \in conn$, where $conn$ is the connection relation,

(2) for all nonterminal label A and all terminal label a, $(a, A) \in conn$,
(3) the start graph (axiom) consists of a node.
For example, the set of complete k-partite graphs is generated by a restricted RNLC graph grammar.

In computational learning theory, learning problems are investigated in several learning models. D. Angluin introduced the notion of exact-learning via queries[2, 3]. We consider exact-learning of restricted RNLC graph languages via queries. In this paper, we present an algorithm to construct a restricted RNLC graph grammar which generates a given unknown restricted RNLC graph language, using restricted superset and restricted subset queries. Furthermore, we show that for a fixed nonnegative integer t, this algorithm halts in polynomial time when the Parikh image of the given unknown restricted RNLC graph languages has at most t-periods.

In section 2, we investigate the relationship between the restricted RNLC graph languages and those Parikh image in order to show the correctness of our learning algorithm. Let L be a restricted RNLC graph language and $\Psi(L)$ be Parikh image of L. We show that given $\Psi(L)$, one can construct a restricted RNLC graph grammar G such that $L(G) = L$ in polynomial time. In section 3, we present the learning algorithm for restricted RNLC graph languages.

2 Restricted RNLC graph grammars and Parikh mapping

2.1 Restricted RNLC graph grammars

We consider *finite undirected node labeled graphs* without *loops* and without *multiple* edges. The set of all graphs over Σ is denoted by $\mathcal{G}_\Sigma$.

Definition 2.1. For a set of *labels* Σ, a *graph* X (*over* Σ) is specified by V_X, E_X, and φ_X, where V_X is a finite nonempty set of *nodes*, E_X is a subset of $\{\{x, y\} \mid x, y \in V_X, x \neq y\}$, called set of *edges*, and φ_X is a function from V_X into Σ, called the *labeling function*.

Definition 2.2 A *node-label-controlled* (NLC) *graph grammar* is a system $G = (\Sigma, \Delta, P, conn, Z_{ax})$, where Σ is a finite nonempty set of *labels*, Δ is a nonempty subset of Σ (the set of *terminals*), P is a finite set of pairs (d, Y) where $d \in \Sigma - \Delta$ and $Y \in \mathcal{G}_\Sigma$ (the set of *productions*), *conn* is a relation between Σ and Σ (the *connection relation*), and $Z_{ax} \in \mathcal{G}_\Sigma$ (the *axiom*).

The set $\Sigma - \Delta$ is referred to as the set of *nonterminals*. A node x is a *terminal* (*nonterminal* respectively) *node*, if x is labeled by elements of Δ ($\Sigma - \Delta$ respectively).

Now, we explain the method of derivation of a graph. Let $G = (\Sigma, \Delta, P, conn, Z_{ax})$ be an NLC graph grammar, $p = (d, W)$ be a production in P, X and Y be graphs over Σ such that $V_X \cap V_Y = \emptyset$ and Y be isomorphic to W, and x

be a node labeled by d in X. Then, a graph Z is derived from the graph X by the production (d, W) in the following way:

Step 1: Delete the node x (and the edges which are incident with x) from X, (Notice that x is labeled by d.)

Step 2: Add to Y instead of x, (Notice that Y is a copy of W.)

Step 3: Connect every node y in Y to the neighbor x' of x by an edge iff $(\varphi_Y(y), \varphi_X(x')) \in conn$ holds.

Consequently, graph Z is obtained, where

$$V_Z = V_{X-x} \cup V_Y,$$
$$E_Z = E_{X-x} \cup E_Y \cup \{\{x', y\} \mid x' \text{ is a neighborhood of } x \text{ in } X, y \in V_Y,$$
$$(\varphi_Y(y), \varphi_X(x')) \in conn\},$$
$$\varphi_Z(u) = \begin{cases} \varphi_{X-x}(u) & \text{if } x \in V_{X-x}, \text{ and} \\ \varphi_Y(u) & \text{if } x \in V_Y. \end{cases}$$

Then we say that "X concretely derives Z (in G, by the production p)", and denoted by $X \Rightarrow_{G\ p} Z$ or simply by $X \Rightarrow_p Z$. A sequence of successive concrete derivation steps in G

$$D : X_0 \Rightarrow_{p_{i_1}} X_1 \Rightarrow_{p_{i_2}} \cdots \Rightarrow_{p_{i_n}} X_n$$

where $n \geq 1$ and the sets V_{X_0} and V_{Y_i} $(1 \leq i \leq n)$ are pairwise disjoint, is referred to as *a concrete derivation in G (from X_0 to X_n)*, and the sequence of applied productions $(p_{i_1}, p_{i_2}, \cdots, p_{i_n})$ is termed *applied productions (in the derivation)*.

A graph X *directly derives* a graph Z *(in G)*, denoted by $X \leadsto_G Z$, if there is a graph Z' such that Z' is isomorphic to Z and X concretely derives Z' in G. The symbol $\leadsto_G^*$ denotes the reflexive and transitive closure of $\leadsto_G$. If $X \leadsto_G^* Z$, then we say that X *derives* Z *(in G)*. The *graph language generated by G*, denoted by $L(G)$, is the set $\{X \in \mathcal{G}_\Delta \mid Z_{ax} \leadsto_G^* X\}$. A set L of graphs is *an NLC graph language*, if there is an NLC graph grammar G such that $L = L(G)$.

Definition 2.3 A *regular* NLC (RNLC) graph grammar is an NLC graph grammar $G = (\Sigma, \Delta, P, conn, Z_{ax})$ such that every production of G is either of the form (A, H) or the form (A, a), where $A \in \Sigma - \Delta$, $a \in \Delta$, $H = (V, E)$ is a graph such that $V = \{x, y\}$, $\varphi_H(x) = B \in \Sigma - \Delta$, $\varphi_H(y) = a$, $E = \{\{x, y\}\}$.

We denote the production which has form of (A, H) by $(A, \{B, a\})$. An RNLC graph grammar G is "symmetric" if $(a, b) \in conn$ iff $(b, a) \in conn$. For any RNLC graph grammar G, $|G|$ denotes the sum of the number of labels and the number of productions.

Definition 2.4 A graph grammar G is said to be *restricted RNLC graph grammar* if G has the following properties:
(1) G is a symmetric RNLC graph grammar,
(2) for all nonterminal label A and all terminal label a, $(a, A) \in conn$,

(3) the start graph (axiom) consists of a node.

Example 2.1 Let $G_{ex} = (\Sigma, \Delta, P, conn, Z_{ax})$ be a restricted RNLC graph grammar, where $\Sigma = \{ S, A, B, C, D, E, F, G, a_1, a_2, a_3 \}$, $\Delta = \{a_1, a_2, a_3\}$, $conn = \{(a_1, a_2), (a_2, a_1), (a_2, a_3), (a_3, a_2)\} \cup \{(X, Y) | X$ is a nonterminal label, Y is a terminal label $\} (\cup \{(Y, X) | X$ is a nonterminal label, Y is a terminal label $\})$, Z_{ax} consists of a node with label S, and P is depicted in Fig. 1.

$$(S, \overset{B}{\circ}\!\!-\!\!\overset{a_3}{\circ} | \overset{E}{\circ}\!\!-\!\!\overset{a_3}{\circ} | \overset{a_2}{\circ} | \overset{G}{\circ}\!\!-\!\!\overset{a_2}{\circ})$$

$$(B, \overset{C}{\circ}\!\!-\!\!\overset{a_3}{\circ}) \quad (C, \overset{A}{\circ}\!\!-\!\!\overset{a_2}{\circ}) \quad (A, \overset{B}{\circ}\!\!-\!\!\overset{a_3}{\circ} | \overset{a_2}{\circ})$$

$$(E, \overset{F}{\circ}\!\!-\!\!\overset{a_2}{\circ}) \quad (F, \overset{D}{\circ}\!\!-\!\!\overset{a_1}{\circ}) \quad (D, \overset{E}{\circ}\!\!-\!\!\overset{a_3}{\circ} | \overset{G}{\circ}\!\!-\!\!\overset{a_2}{\circ})$$

$$(G, \overset{a_1}{\circ})$$

Figure 1:

Then graphs depicted in Fig. 2 can be generated by G_{ex}.

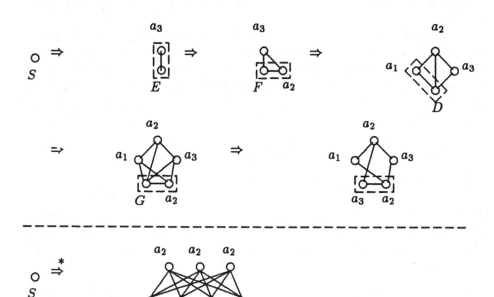

Figure 2:

2.2 Parikh mapping of restricted RNLC graph languages

In this section, we give the definition of Parikh mapping of restricted RNLC graph languages and some related properties which will be used in this paper.

Definition 2.5 Let us fix the set of terminal labels $\Delta = \{a_1, \ldots, a_n\}$ and define a mapping, called *Parikh mapping*, from $\mathcal{G}_\Delta$ into $\mathbf{N}^n$ given by

$$\Psi(H) = (\sharp_{a_1}(H), \sharp_{a_2}(H), \ldots, \sharp_{a_n}(H)),$$

where $\sharp_{a_i}(H)$ denotes the number of occurrences of a_i in the graph H. For any $L \subseteq \mathcal{G}_\Delta$, define $\Psi(L) = \{\Psi(H) \mid H \in L\}$. $\Psi(L)$ is called *Parikh image* of L.

Definition 2.6 A set of the form

$$M = \{\alpha_0 + n_1\alpha_1 + \cdots + n_m\alpha_m : n_j \geq 0 \text{ for } 1 \leq j \leq m\},$$

where $\alpha_0, \ldots, \alpha_m$ are elements of $\mathbf{N}^n$, is said to be a *linear* subset of $\mathbf{N}^n$. Each α_i is called *period* of M. A *semilinear* is a finite union of linear sets.

Proposition 2.1 *For any RNLC graph language L, $\Psi(L)$ is a semilinear set.*

Proof. Two languages L_1 and L_2 are called *letter equivalent* if $\Psi(L_1) = \Psi(L_2)$. It is known that for any language L $\Psi(L)$ is a semilinear set if and only if L is a letter equivalent to a regular set. It is clear that any graph language generated by an RNLC graph grammar is letter equivalent to a regular set by Definition 2.3. $\square$

For any terminal label $a \in \Sigma$, a production $p \in P$ is called *a-production* if a appears in right- hand side of p.

Proposition 2.2 *Let G_1, G_2 be restricted RNLC graph grammars with a common connection relation, $D_1 = (p_{i_1}, p_{i_2}, \ldots, p_{i_n})$ be an applied productions in a derivation in G_1 and $D_2 = (q_{j_1}, q_{j_2}, \cdots, q_{j_n})$ be an applied productions in a derivation in G_2. If for any terminal label a the number of appearances of a-production in D_1 is equal to one of a-production in D_2, then the graph by derived by D_1 is isomorphism to the graph derived by D_2.*

Proof. Let G be a restricted RNLC graph grammar with a connection relation *conn* and H be a graph generated by G. And let a and b be node labels in H, x be a node with label a in H and y be a node with label b in H.

By the conditions (1) and (3) of Definition 2.4, without loss of generality we can assume that x is yield before y is yield. Let z be the nonterminal node such that y is yield by rewriting of z.

First we will show the following Fact 1 :

Fact 1 : x is adjacent to y iff $(a, b) \in conn$.

For such nodes x, y, and z, the following property hold :

x is adjacent to y iff

(1) x is adjacent to z, and

(2) $(b, a) \in conn$.

The condition (2) of Definition 2.4 guarantees that x is adjacent to z. Hence Fact 1 holds.

Let H_1 be the graph generated by the derivation D_1 and H_2 be the graph generated by D_2. Since for any terminal label a the number of appearances of a-production in D_1 is equal to one of a-production in D_2, we can obtain the following Fact 2 :

Fact 2 : $\Psi(H_1) = \Psi(H_2)$.

Therefore, from fact 1 and 2, H_1 is isomorphic to H_2. $\qquad\qquad\qquad\Box$

The above proposition means that a graph in a restricted RNLC graph language is characterized by the connection relation and the number of applications of a-production for each terminal label a. Hence a restricted RNLC graph language is also characterized by the number of nodes with label a for each terminal label a when the connection relation is fixed. Therefore, we can obtain the following lemma.

Lemma 2.3 *Let G_1, G_2 be restricted RNLC graph grammars with a common connection relation. If $\Psi(L(G_1)) = \Psi(L(G_2))$ then $L(G_1) = L(G_2)$.*

Proof. It is sufficient to show $L(G_1) \subseteq L(G_2)$. Let H be a graph in $L(G_1)$. From $\Psi(L(G_1)) = \Psi(L(G_2))$, there exist a graph H' in $L(G_2)$ such that $\Psi(H') = \Psi(H)$. Hence, from Proposition 2.2 and the condition that G_1, G_2 have a common connection relation, H' is isomorphism to H. $\qquad\qquad\qquad\Box$

A regular string grammar G is *CNF left linear* if every production of G is either of the form $A \to Ba$ or the form $A \to a$ for some nonterminal A, B and terminal a.

Lemma 2.4 *Let L be a restricted RNLC graph language with a set of terminal labels Δ. When a semilinear set S which is the Parikh image of L and conn are given, one can construct a restricted RNLC graph grammar $G = (NT \cup \Delta, \Delta, P, conn, Z)$ such that $L = L(G)$.*

Proof. We present a polynomial time algorithm MG to construct a restricted RNLC graph grammar $G = (NT \cup \Delta, \Delta, P, conn, Z)$ such that $L = L(G)$ when a semilinear set S such that $S = \Phi(L)$ are given as input. Assume that $S = M_1 \cup \cdots \cup M_m$. Let $M_i = \{\alpha_{i_0} + n_{i_1}\alpha_{i_1} + \cdots + n_{i_{r_i}}\alpha_{i_{r_i}} \mid n_{i_j} \geq 0$ for $1 \leq j \leq r_i\}$ for each $1 \leq i \leq m$. Let $y_{i_0}, y_{i_1}, \cdots, y_{i_{r_i}}$ be strings in Σ^* whose images under Ψ are, respectively, $\alpha_{i_0}, \alpha_{i_1}, \cdots, \alpha_{i_{r_i}}$. First MG constructs left linear string grammar G'' with the productions:

$\qquad S \to A_1 \mid \cdots \mid A_m,$

$\qquad A_i \to A_i\, y_{i_j} \mid y_{i_0}$ for all $1 \leq i \leq m$, $1 \leq j \leq r_i$.

Next MG constructs a CNF left linear string grammar G' such that $L(G') = L(G'')$ from G''. Finally, MG constructs graph productions of P form each string production in G' by following way ; Transform a string production form of $A \to Ba_i$ ($A \to a_i$) into the graph production form of $(A, \{B, a_i\})$ ((A, a_i) respectively). Let NT be the set of nonterminal labels appearing in P. MG outputs $G = (NT \cup \Delta, \Delta, P, conn, Z)$. It is clear that MG halts in polynomial time. From Lemma 2.3, we can obtain that the output G of MG generates L. $\square$

The above G_{ex} in the example has the following semilinear set $M_1 \cup M_2$:

$$M_1 = \begin{pmatrix} 0 \\ 1 \\ 0 \end{pmatrix} + n_{1_1} \begin{pmatrix} 0 \\ 1 \\ 2 \end{pmatrix} , \quad M_2 = \begin{pmatrix} 1 \\ 1 \\ 0 \end{pmatrix} + n_{2_1} \begin{pmatrix} 1 \\ 1 \\ 1 \end{pmatrix} .$$

From above lemma, if we know appropriate *conn*, we can obtain a restricted RNLC graph grammar such that $L = L(G)$.

3 Learning of restricted RNLC graph languages

3.1 The problem

Angluin introduced the notion exact-learning via queries[2, 3]. In this paper, we consider the problem of learning restricted RNLC graph languages via queries. Suppose L_u be a given unknown target graph language. We assume that the set of terminal labels Δ are known.

The following two types of queries are used as access of an unknown target graph languages in learning procedure:

1. *Restricted subset queries:* The input is a conjectured restricted RNLC graph grammar G_c and the output is *yes* if $L(G_c) \subseteq L_u$ and *no* otherwise.

2. *Restricted superset queries:* The input is a conjectured restricted RNLC graph grammar G_c and the output is *yes* if $L_u \subseteq L(G_c)$ and *no* otherwise.

The answer for unrestricted version of each type of queries is not only *yes* or *no* but also with a *counterexample* in case of *no*, where the counterexample is an element in $L_u - L(G_c)$ $(L(G_c) - L_u)$ if the type of the query is subset query (superset query respectively).

In the rest of this section, we consider the problem to construct a restricted RNLC graph grammar G which generates the unknown graph languages using restricted subset queries and restricted superset queries. When we consider the efficiency of learning procedure, the size of the learning problem have to be decide. Usually, the size of minimum representation of target language is used as the size of the learning problem. (For example, each of regular grammar, regular expression, and deterministic finite automata is a representation set for regular sets.) Occasionally, the efficiency of learning procedure depends on what representation set are used (for example, see [3]). For instance, the learning problem of regular sets from membership and equivalence queries[1] using deterministic finite automata is efficient[2], but if $\mathbf{RP} \neq \mathbf{NP}$, one using nondeterministic finite

---

[1] Let L_u be an unknown target language. *Membership query*: The input is an instance x and the output is *yes* if $x \in L_u$ and *no* otherwise. *Equivalence query*: The input is a conjectured representation L_c and the output is *yes* if $L_u = L_c$ and *no* otherwise. When the answer is *no*, a *counterexample* is also supplied, that is, an instance $x \in (L_u - L_c) \cup (L_c - L_u)$. The choice of counterexamples is assumed to be arbitrary.

automata is not efficient[4]. We use the restricted RNLC graph grammars as the representation of graph languages L and $\min\{|G|\,|\,L(G) = L\}$ as the size of L.

3.2 The learning algorithm

Here, we demonstrate our learning algorithm for restricted RNLC graph languages using restricted superset and restricted subset queries, which is called LA-rRNLC. Takada showed the learning algorithm LA-SLS for the family of semilinear sets from restricted subset and restricted superset queries[11]. Our learning procedure LA-rRNLC uses Takada's LA-SLS as a subprocedure. Let L be an unknown target restricted RNLC graph language. Since the set of terminal labels Δ is fixed, the number of connection relations k is $2^{|\Delta|C_2}$. Let $conn_1, conn_2, \cdots, conn_k$ be distinct connection relations each other. First, for any $i \in \{1, \ldots, k\}$, we define a learning algorithm LA-rRNLC_$conn_i$ to output a restricted RNLC graph grammar G which generates an unknown target restricted RNLC graph language, when $conn_i$ is the real connection rule. Each LA-rRNLC_$conn_i$ simulates LA-SLS. When LA-SLS makes a restricted subset (superset) query with a semilinear set S_c, LA-rRNLC_$conn_i$ constructs a restricted RNLC graph grammar $G_c = (NT \cup \Delta, \Delta, P, conn_i, Z)$ from S using the procedure MG described in the proof of Lemma 2.4, makes a restricted subset (superset) query with G_c, and return the answer of the query to LA-SLS (See Fig. 3). If $conn_i$ is equivalent to the connection relation of the unknown tar-

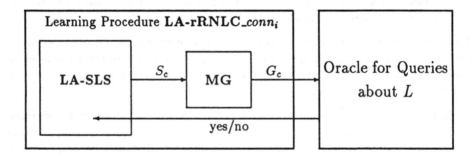

Figure 3:

get, LA-rRNLC_$conn_i$ outputs a restricted RNLC graph grammar G such that $L = L(G)$ and consumes polynomial time of the running time of LA-SLS from Lemma 2.4.

Our learning algorithm LA-rRNLC is as follows:

Procedure LA-rRNLC

$\qquad j := 1$
$\qquad$ repeat

j-th step of LA-rRNLC_$conn_1$
j-th step of LA-rRNLC_$conn_2$
$$\vdots$$
j-th step of LA-rRNLC_$conn_i$
$$\vdots$$
j-th step of LA-rRNLC_$conn_k$
$j := j + 1$
    until some LA-rRNLC_$conn_i$ halts and outputs a grammar G
    output G
end.

LA-rRNLC halts and outputs a correct grammar, because $conn_i$ is equivalent to the connection relation of the unknown target. Hence, we can obtain the following lemma.

Lemma 3.1 *Using restricted superset queries and restricted subset queries for an unknown restricted RNLC graph language L_u, the learning algorithm LA-rRNLC eventually terminates and outputs a restricted RNLC graph grammar G such that $L(G) = L_u$.*

Let t be a nonnegative integer. We say that a semilinear set S is *t-periods semilinear set* if S is represented by a finite union of linear sets which have at most t periods. Takada showed that the learning algorithm LA-SLS identifies any t-periods semilinear sets S of $\mathbf{N}^k$ from restricted subset and restricted super set queries in polynomial time where k is a fixed dimension[11]. From this fact and Lemma 3.1, we obtain the following theorem.

Theorem 3.2 *Let t be a fixed nonnegative integer. For any restricted RNLC language L_u whose Parikh image has at most t-periods, the learning algorithm LA-rRNLC halts and outputs a restricted RNLC graph grammar G such that $L(G) = L_u$ using restricted superset queries and restricted subset queries in polynomial time.*

References

[1] IJ.J. Aalbersherg, A. Ehrenfeucht, and G. Rozenber, On the membership problem for regular DNLC grammars, *Discrete Applied Mathematics* 13:79-85 (1986).

[2] D. Angluin, Learning regular sets from queries and counter examples, *Information and Computation*, 75 (1987), pp.87-106.

[3] D. Angluin, Queries and concept learning, *Machine Learning*, 2 (1987), pp.319-342.

[4] Angluin, D. and M. Kharitonov: When Won't Membership Queries Help? in *Proc. of 23nd STOC*, ACM, 1991, pp.444-454.

[5] M. Bauderon, and B. Courcelle, Graph expressions and graph rewritings, *Math. Systems Theory.* 20:83-127 (1987).

[6] F.J. Brandenburg and K. Skodinis, Edge-marking Graph Automata and Linear Graph Languages, unpublished manuscript (1994).

[7] C. Domingo and J. Shawe-Taylor, Positive and Negative Results for Learning Minor Closed Graph Classes, unpublished manuscript (1995).

[8] J.V. Leeuwen, Graph algorithms, in *Handbook of theoretical computer science vol. A*, (ed. J.V. Leeuwen), (1990), pp.527-631.

[9] D. Janssens and G. Rozenberg, On the structure of node label controlled graph languages, *Inform. Sci.* 20:191-216 (1980).

[10] G. Rozenberg and E. Welzl, Boundary NLC graph grammars - basic definitions, normal forms, and complexity, *Inform. Control* 69:136-167 (1986).

[11] Y. Takada, Learning semilinear sets from examples and via queries, *Theoretical Computer Science* 104 (1992), pp.207-233.

[12] T. Uchida, T. Shoudai, and S. Miyano, Parallel Algorithms for Refutation Tree Problem on Formal Graph Systems, *IEICE TRANS. INF. & SYST.*, vol.E78-D, no.2 (1995), pp.99-112.

[13] K.Yamazaki, A normal form problem for the unlabeled boundary NLC graph languages, *Inform. Comput.* 120:1-11 (1995).

[14] K.Yamazaki and T.Yaku, A Pumping Lemma and Structure of Derivations in the Boundary NLC Graph Languages, *Inform. Sci.* 75:81-97(1993).

OPTIMAL INFORMATION DELIVERY

Christos H. Papadimitriou[1], Srinivas Ramanathan, and P. Venkat Rangan

{christos, venkat, sramanat}@cs.ucsd.edu

University of California at San Diego

ABSTRACT: *Minimizing transmission and storage costs for delivering large amounts of data at specific locations and times is an important aspect of video-on-demand applications. We formulate a broad class of optimization problems of this sort as Steiner tree problems in a time-space product graph. We show that the general problem is NP-complete, we develop a polynomial-time algorithm for an important special case, and a fast heuristic that achieves an approximation ratio of two. Experimental results suggest that the heuristic is usually within a few percentage points of the optimum.*

1. INTRODUCTION

Emerging video-on-demand applications [Lo] call for the delivery, at specified times, of the same information at many sites. Since the amounts of data involved are huge, storage and transmission costs can be considerable. By cleverly "caching" information at cheap storage sites for future delivery to other sites can be very beneficial: It has been observed [PRR] that such caching can reduce the delivery costs very considerably. The resulting optimization problem differs from traditional manufacturing and inventory control problems (where transportation and storage costs for delivery of goods are also optimized) because in the present case the "goods" are data, which can be duplicated at zero cost. As a result, the optimization problem in question is closer to the *Steiner tree problem* than to the minimum-cost (or multicommodity) flow problems that typically arise in inventory control.

In a typical situation, we have a network —usually, a tree representing a storage hierarchy of less and less central information storage sites, with the customers as leaves. Leaves in this network require the delivery of the same information but at different times —say, leaf i requires delivery at time t_i. Since transmission delays are a rather controllable part of the problem, we assume that transmission is instantaneous. The transmission costs of the links of the network, as well as the storage costs at the internal nodes, are known and given —and possibly time-dependent.

It turns out that determining the optimal delivery strategy is an extremely complex problem. Consider for example the situation in Figure 1. There are three users who have requested the same data at three different viewing times (1pm, 2pm, and 11pm, respectively), and three servers with different storage

---

[1] Research supported by an NSF grant and by the Greek Ministry of Research program ΠΕΝΕΔ-91Δ648.

costs. The network costs between servers are also indicated —all costs are time-independent in this example. In the Figure we show two of the possible delivery strategies in this case. Strategy (b) caches the data at the middle server for the second user, and at the lower server for the third. Strategy (c), the optimal strategy, caches the data at the lower server for the first user, and the root for the third. Conceivably, a strategy could even send data upstream for cheap storage between viewing times. If network and storage costs change with time, even more complex strategies may emerge as optimal.

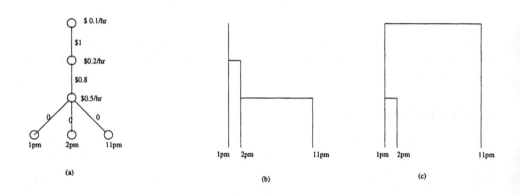

Figure 1

There are certain important *epochs* in this problem: The desired delivery times, and the times at which network and storage costs change. One basic observation (Lemma 1) is the following: *An optimal strategy may be assumed to transmit data only at these epochs* (otherwise, it can be transformed into a strategy with no worse cost that conforms to this normal form). As a result, we can formulate the problem in terms a *time-space product graph*, the Cartesian product of the network with the set of epochs. The graph must be directed, to reflect the irreversibility of time. We must find a *minimum Steiner tree* on this graph, that is, the shortest subgraph connecting the root (the central server where the data resides in the beginning) with all delivery nodes (pairs consisting of a user with the corresponding delivery time).

The Steiner tree problem for graphs is NP-complete [Ka], and can be approximated within a ratio of $\frac{11}{6}$ [Ze]. The present problem however is different: On the negative side, the underlying graph is *directed;* on the positive side, the graph is very restricted (it is the product of a tree by an independent set), and thus hopefully specialized techniques may apply.

In Section 3 we give a rather complete characterization of the complexity of this problem. We show that the general problem is NP-complete (Theorem

3), and in fact that it remains NP-complete even if the network is a star. The case in which the network is a tree and the costs are time-invariant is left open here —we conjecture that it is NP-complete. An important special case is when the network is in fact a straight line —a hierarchy of storage servers— plus many leaves emanating from the last server. We give a polynomial dynamic programming algorithm for this case, even when the network costs and storage costs are time-dependent. The complexity of the algorithm is $O(n^3m^2)$, where n is the number of customers (plus the number of epochs at which cost changes occur), and m is the number of servers (Theorem 4).

Finally, in Section 4 we develop an approximation algorithm for this problem. The algorithm is a simple *greedy* algorithm *"service the next request by the smallest additional cost possible, without changing the way in which the previous requests have been serviced."* We prove (Theorem 5) that the greedy heuristic yields a solution which is never more than twice the optimum. Interestingly, and despite this somewhat disappointing performance guarantee, experimentation with this algorithm shows that it is rarely more than 1-2% above the optimum. We fully expect it to be a most useful technique in practice —especially since it achieves such a good performance *on-line*, that is, with no advance knowledge of user requests.

For a more complete description of the motivating application of this problem to multimedia systems (as well as for a preliminary announcement of some of these results without proofs) see [PRR] and [PRRS].

2. DEFINITIONS

We are given a tree $T = (V, E)$. The leaves of the tree are called *users*, and the internal nodes are called *servers*. There is a distinguished server r called the *root* (intuitively, where the data resides at time zero). Each user i has been assigned a nonnegative integer *time* $t_i \geq 0$. We assume that there is a user with $t_1 = 0$. Each edge $[i, j]$ of the tree has been assigned a *network cost* N_{ij}, and each server i a *storage cost* S_i. Since we shall allow these costs to be *time-dependent*, we define N_{ij} to be a set of pairs of the form $\{(T_\ell, c_\ell) : \ell = 1, \ldots, k\}$, where $0 = T_1 < T_2 < \cdots < T_k$ are nonnegative integers, and $c_\ell \geq 0$ is the cost between T_ℓ and $T_{\ell+1}$ (after T_ℓ if $\ell = k$). The network cost at time $t \geq 0$, $N_{ij}(t)$, is thus defined to be the unique c such that $(c, T_\ell) \in N_{ij}$ and either $\ell = k$ and $t \geq T_\ell$, or $T_\ell \leq t \leq T_{\ell+1}$. Similarly for storage costs.

A *delivery schedule* for a tree T is a set of *storage intervals* of the form (s, t, t'), where s is a server and $0 \leq t < t'$ (meaning, intuitively, that the data is stored in s between times t and t'), and a set of *transmissions* of the form (i, j, t) (meaning that the data is transmitted from i to j at time t) where $[i, j] \in E$ and $t \geq 0$, such that the following hold:

(1) For each user i there is a transmission (s, i, t_i) for some server s (a transmission delivering the data at the right time).

(2) For each transmission (i, j, t) either $i = r$, or there is another transmission (k, i, t) into i at the same time t, or there is a storage interval (i, t', t'') with $t' \leq t \leq t''$ (intuitively, the data must be at the origin of the transmission at

the time of the transmission; recall that we are assuming that transmissions are instantaneous).

The *cost* of a transmission (i, j, t) in a delivery schedule is $N_{ij}(t)$, and that of storage interval (s, t, t') is $\int_t^{t'} S_s(\tau)d\tau$ —easily computed by adding over the piecewise constant function S_s. The cost of a delivery schedule is the total cost of all its storage intervals and all of its transmissions. We seek the delivery schedule with minimum cost.

Let Z be the set of all *epochs* of the problem, that is, the set of all viewing times t_i of the users, and the set of all cost-change times T_j. The following is a most helpful normalization:

Lemma 1: *There is an optimal delivery schedule in which for all storage intervals (s, t, t') $t, t' \in Z$, and for all transmissions (i, j, t) $t \in Z$.*

Proof: By progressively transforming any delivery schedule into a standard one at no cost increase. We do this by first moving all transmissions whose time t is not an epoch to *either* the epoch that is immediately before t, *or* the epoch that is immediately after t. While increasing t, we also increase all start and the finish times of storage intervals that coincide with t. Moving t by ϵ incurs an extra cost of $c \cdot \epsilon$, where c is a real number. This means that one of the two directions would incur a nonpositive cost. We move t to the next epoch in this direction. $\square$

Lemma 1 allows us to concentrate on a finite directed graph G which captures the problem completely. The set of vertices of G is $V \times Z$ —all vertex-epoch pairs. The directed edges of G are these of two kinds: the edge $((s, t), (s, t'))$, with cost $S_s(t) \cdot (t' - t)$ for any server s and any two *consecutive* epochs $t < t'$ (that is, there is no other epoch $t'' \in Z$ with $t < t'' < t'$); and the edge $((i, t), (j, t))$ for any edge $[i, j]$ of the tree and epoch t, with cost $N_{ij}(t)$ (notice that both edges $((i, t), (j, t))$ and $((j, t), (i, t))$ are present). A *Steiner tree* of G is a subgraph of G which is a tree rooted at $(r, 0)$ with leaves all nodes of the form (i, t_i), where i is a leaf. The cost of the Steiner tree is the total cost of its edges.

Lemma 2: *The cost of the optimal delivery schedule of T is precisely the cost of the optimal Steiner tree of G.*

Proof: It is easy to see that any delivery schedule contains a Steiner tree of the graph G. Also, any Steiner tree of G is a delivery schedule (edge $((i, t), (j, t))$ in the Steiner tree corresponds to a possible transmission (i, j, t), while an edge $((s, t), (s, t'))$ corresponds to a possible storage interval (s, t, t'); and vice-versa). Since costs are nonnegative, we may restrict our optimization to Steiner trees. $\square$

In the next two sections we shall consider the computational problem of computing the optimum Steiner tree of the graph G.

3. METROPOLITAN NETWORKS

We first show the following negative result:

Theorem 3: *The optimum delivery schedule problem is NP-complete when the costs are time-varying, even if the tree is a star. In the time-invariant case, the problem becomes NP-complete when the network is allowed to be an arbitrary graph.*

Proof: The proof is by a reduction from, predictably, the *Steiner tree problem* for weighted undirected graphs. Given a graph $G = (V, E)$ with a set $M \subseteq V$ of *mandatory* nodes and a weight $w[e]$ for each edge $e \in E$, we are asked to find the smallest subgraph $S \subseteq E$ of G that connects all nodes in M. We transform this into an instance of the optimum delivery schedule problem as follows: We form a tree which is a star with a root r and with set of leaves V. All storage and network costs prohibitive *except for very short "windows of opportunity," which simulate the edges of G* (and their weights). In particular, and let T be an one-to-one mapping from the set $E \times \{1, 2, \ldots, |V|^2\}$ to the positive even integers. If node $u \neq r$ has k edges leaving it, say $e_1, \ldots, e_k$, then the network cost of the link $[v, r]$ is of the form $\{[0, C], [T[e_1, 1], w[e_1]], [T[e_2, 1] + 1, C], [T[e_2, 1], w[e_2]], [T[e_2, 1] + 1, C], \ldots, [T[e_k, 1], w[e_k]], [T[e_k, 1]+1, C], [T[e_1, 2], w[e_1]], \ldots, [T[e_k, |V|^2]+1, C]\}$, where C is a prohibitively large cost. The storage costs are 0 at the leaves and C at the root. Thus, any transmission through the root must correspond to a "synchronized" transmission between two leaves $[u, v]$, where $[u, v] = e \in E$, at a time of the form $T[e, j]$ (or $T[e, j] + 1$) for some $j \leq d$. There is also a transmission cost 0 at time 0 between the root r and a distinguished node $m \in M$.

It is clear that any delivery schedule in this setting must correspond, through the synchronized transmissions from leaf to leaf, to a subgraph of G. Since the nodes in M correspond to users (with zero storage cost), the subgraph must contain all of M. Since we have up to d "windows of opportunity" for each edge, the optimum schedule will correspond to a tree (any cycle can be removed, or, equivalently, any subgraph can be simulated), and must start with the free transmission from r. The equivalence of the two problems follows.

By allowing arbitrary networks, we can make the transmission costs time-invariant; the root must be a node in M. $\square$

It is open whether the optimum delivery schedule problem is NP-complete in the time-invariant case of trees.

Suppose that the tree is a *path* of m servers, with n users adjacent to the last server. This represents an important and most interesting special case, which are called in [PRR] *metropolitan networks*. Metropolitan networks, with small m, are a plausible architecture for information delivery networks. Also, they arise as an important subproblem, when one tries to decouple and optimize independently the various paths of a large information delivery tree.

We next develop a dynamic programming algorithm for finding the optimum delivery schedule.

The dynamic programming recurrence is quite tricky to formulate. For clarity we shall develop the recurrence in the time-invariant case (the general case is not much harder to solve in this way). Order the users in nondecreasing times, $0 = t_1 \leq t_2 \leq \cdots$. Consider a user i and a server S_j, and suppose that the transmission to i at time t_i will come from S_j (that is, in the optimal delivery schedule there is a storage interval (S_j, t, t') with $t \leq t_j \leq t'$, and transmissions $(S_j, S_{j+1}, t_i), (S_{j+1}, S_{j+2}, t_i), \ldots (S_m, i, t_i)$; equivalently, in the optimal Steiner tree the path from $(r, 0)$ to (i, t_i) goes through an edge going into (S_j, t_i)). Without loss of generality, there is exactly one such j. Define now $C[i, k, j]$, where $k \geq i$ is another user, as follows: $C[i, k, j]$ is the cost of the optimum Steiner tree of G with root (S_j, t_i) and leaves $(i, t_i), (i+1, t_{i+1}), \ldots, (k, t_k)$ (see Figure 2). *This quantity is the fundamental subproblem solved by the dynamic programming recurrence.*

There are two cases: Either all paths from (S_j, t_i) to all these users pass through edge $((S_j, t_i), (S_{j+1}, t_i))$, or some of them (say, to users $\ell, \ell+1, \ldots, k$, for some $i < \ell \leq k$) go through $((S_j, t_i), (S_{j-1}, t_i))$ (see Figure 2). We call the cost in the first case $C_d[i, k, j]$, and the second $C_u[i, k, j]$:

$$C_d[i, k, j] = \min_{j \leq r \leq m} [NC(S_j, i) + SC(S_r)(t_{i+1} - t_i) + C[i+1, k, r]].$$

We have used here $NC(S_j, i)$ to denote the sum of all storage costs (at time t_i) between S_j and i —recall that we are assuming they are time-invariant. In the second case,

$$C_u[i, k, j] = \min_{1 \leq r \leq j, i < \ell \leq k} [NC(S_r, i) + SC(S_r)(t_\ell - t_i) + C_d[i, \ell-1, r] + C[\ell, k, r]].$$

Finally,

$$C[i, k, j] = \min\{C_d[i, k, j], C_u[i, k, j]\}.$$

Naturally, the optimal total cost is precisely $C[1, n, 0] = C_d[1, n, 0]$.

Theorem 4: *The optimum delivery schedule for metropolitan networks with m servers and n users can be found in $O(m^2 n^3)$ time.*

Proof: By carrying out these recurrences, in increasing $k - i$, after a precomputation of the shortest paths in G. $\square$

4. THE GREEDY HEURISTIC

Let us restrict now our attention to the case in which the network and storage costs are *time-invariant*. The *greedy heuristic* considers the users in increasing viewing times t_i. At the first step (recall that $t_i = 0$) we have no choice but to deliver the data to i at time 0. At the ith step, $i > 0$, we determine the least expensive manner in which user i can be serviced, given the schedule for servicing users $1, \ldots, i-1$. That is, we determine *the least incremental storage and transmission costs* that would be incurred if some server $S_j \in [S_1, ..., S_m]$ were to transmit the data at time t_i to user i. This is a simple shortest path calculation from any vertex (S_i, t) of G, such that the data is stored at S_i at time t, to (i, t_i), and can be carried out in $O(m)$ time because of the special structure of the weighted graph G.

Theorem 5. *The greedy heuristic produces a delivery schedule with cost at most twice the optimum.*

Sketch: We analyze the following *variant* of the greedy heuristic: User i gets the data via shortest path from user $i - 1$ (as opposed from any other server, as in the greedy heuristic). It is easy to see that the greedy heuristic produces a schedule whose cost is no larger than that of this variant. We shall show that this variant has cost no more than twice the optimum.

In proof, notice first that the variant of the greedy heuristic produces a *shortest branching* [PS] of the weighted directed graph G induced by the nodes (i, t_i) and the source $(S_1, 0)$ —this is due to the fact that G contains no cycle involving two such nodes, and hence the induced graph is a DAG.

Consider next any Steiner tree of G, and *double* its edges. It is not hard to see that the resulting graph contains such a branching (more precisely, results in a branching by omitting certain directed edges). The bound follows. $\square$

Experimentation with the greedy algorithm (reported in [PRRS]) reveals that its empirical performance is excellent. In several thousands of reasonably large examples of metropolitan networks (with G with up to 2,500 vertices), with random viewing times, network and storage costs, first solved exactly by dynamic programming, the greedy heuristic produced solutions with cost far closer to the optimum than is suggested by Theorem 5. The relative error was typically between 1% and 2%, and the maximum relative error observed (for a relatively small instance) was 10%.

REFERENCES

[Ka] R. M. Karp "Reducibility among Computational Problems," in *Complexity of Computer Computations*, R. E. Miller and J. W. Thatcher (eds.), Plenum 1972.

[Lo] S. Loeb "Architecting Personalized Delivery of Multimedia Information," *Comm. ACM*, *35*, 12, pp. 39–48, 1992.

[PRR] C. H. Papadimitriou, S. Ramanathan, and P. V. Rangan "Information Caching for Delivery of Personalized Video Programs on Home Entertainment Channels", *Proceedings of the IEEE International Conference on Multimedia Computing and Systems*, May 15-19, 1994, Boston.

[PRRS] C. H. Papadimitriou, S. Ramanathan, P. V. Rangan, and S. Sampathkumar "Multimedia Information Caching for Personalized Video-on-Demand", *Computer Communications*, (to appear).

[PS] C. H. Papadimitriou, K. Steiglitz *Combinatorial Optimization: Algorithms and Complexity*, Prentice-Hall, 1982.

[Ze] A. Z. Zelikovski "An $\frac{11}{6}$-approximation Algorithm for Steiner Trees", *Algorithmica*, *9*, 5, pp. 463–470, 1993.

On the Complexity of Testing for Catastrophic Faults*

N. Santoro, J. Ren and A. Nayak

School of Computer Science

Carleton University

Ottawa, Canada, K1S 5B6

Abstract

In this paper, the problem of determining a fault pattern to be catastrophic for unidirectional linear array with $k-$link redundancy is studied. First, we establish a necessary and sufficient condition for a fault pattern to be catastrophic. Based on this necessary and sufficient condition, we derive an efficient testing algorithm whose complexity is $O(mk)$, where k is the number of bypass links, and m is the number of faults. This testing scheme, which improves the existing $O(mk \log k)$ bound, is based on a novel "geometric" approach.

1   Introduction

In a linear array of N processing elements, one faulty element in any location is sufficient to stop the flow of information from one side to the other. A common approach for achieving fault tolerance in such systems is through the incorporation of redundant links (called *bypass links*) in a regular fashion. These links can be activated in a reconfiguration phase to bypass faulty elements [4-5].

There are some inherent limits in this approach. Making the realistic assumption that the length of the longest link is small with respect of the number of processing elements, there are sets of faults occurring in strategic locations which affect the chip in an unrepairable way, regardless of the amount of redundancy, and cannot be overcome by any clever reconfiguration process, see [9]. These sets of faults are called *Catastrophic Fault Patterns* (CFP), and have been extensively studied in the literature [3,6-12].

It is not surprising that the most important question with regards to CFP is the *Testing Problem*; that is, the problem of determining whether a fault pattern is catastrophic. Any solution to this problem is called a *testing scheme*. The efficiency of a testing scheme depends on many parameters: the number m of faults in the pattern, the number k of bypass links at each element, the lengths

*This work was supported in part by Natural Sciences and Engineering Research Council of Canada under Operating Grant A2415.

of the bypass links. It is known that, to be catastrophic, the number m of faults in the pattern must be greater than or equal to the length g of the longest bypass link [9].

The Testing problem has been sudied for the specific case $m = g$ (*minimal fault patterns*) [6,8,9,11] for both unidirectional and bidirectional links. The general case of arbitrary link redundancy (i.e., $k \geq 1$) and arbitrary number of faults (i.e., $m \geq g$) has been considered in [2]. Based on a graph-theoretic interpretation, different bounds have been established depending on whether the links are uni- or bidirectional. Namely, in the case of bidirectional links they show that the problem has a simple $O(mk)$ solution; on the other hand, if the links of the array are unidirectional, the proposed testing scheme is rather complex and requires time $O(mk \log k)$.

This difference in bounds is not significant from a practical viewpoint; however, it raises the interesting theoretical question of whether the Testing Problem is computationally more difficult in the case of unidirectional links.

In this paper, we present an answer to this question by proving that testing can be done in time $O(mk)$ also for arrays with unidirectional links.

The testing scheme achieving the bound is based on a novel "geometric" approach. As in [7], fault patterns are mapped to a Boolean matrix; unlike previous work, the faults are treated as geometric objects (segments) which are functionally *dead*. A necessary and sufficient condition, based on dead segments, is established for a fault pattern to be catastrophic.

Based on this geometric characterization, we develop a testing algorithm which can decide whether or not a given fault pattern is catastrophic in time $O(mk)$, improving the existing $O(mk \log k)$ bound.

Our solution actually solves the classical problem of finding a (not necessarily the shortest) obstacle-avoiding rectilinear path (e.g., [1]) in a two-dimensional directed grid where each node has $k - 1$ additional outgoing links (for details see [11]).

2   Preliminaries

Let $A = \{x_1, x_2, \ldots, x_N\}$ denote a one-dimensional array, where each $x \in A$ represents a processing element (PE) and there exists a regular link between x_i and x_{i+1}, $1 \leq i < N$. Any link connecting x_i and x_j where $j > i + 1$ is said to be a bypass link. The length of a bypass link, connecting x_i and x_j, is the distance in the array between x_i and x_j; i.e., $|j - i|$. Given an integer $g \in (1, N]$ and an array A of size N, A is said to have link redundancy g, if for every $x_i \in A$ with $i \leq N - g$ there exists a link between x_i and x_{i+g}. The array has link redundancy $G = \{g_1, g_2, \ldots, g_k\}$ where $g_j < g_{j+1}$ and $g_j \in (1, N]$, if A has link redundancy $g_1, g_2, \ldots, g_k$. At two ends of A, two special processors, called I (for Input) and O (for Output), are responsible for the I/O functions of the system. We assume that I is connected to $x_1, \ldots, x_k$, while $x_{N-g_k+1}, \ldots, x_N$ are connected to O. Given a linear array A of size N, a fault pattern for A is a set of integers $F = \{f_1, f_2, \ldots, f_m\}$ where $m \leq N$, $f_j < f_{j+1}$ and $f_j \in [1, N]$. An assignment of a fault pattern F to A means that for every $f \in F$, x_f is faulty. A window W_F of a fault pattern F is a subset of F that starts f_1 and finishes with f_m. The width ω_F of a fault pattern F is the number of PEs between and including the first and last fault in F. That is, if $F = \{f_1, \ldots, f_m\}$ then $\|W_F\| = \omega_F = f_m - f_1 + 1$.

Definition 1 *A fault pattern F is catastrophic for an array A with link redundancy G if I and O are not connected in the presence of such an assignment of faults.*

Example 1. Example of a catastrophic fault pattern of 5 faults with link redundancy $\{1, 4\}$ is given in Figure 1.

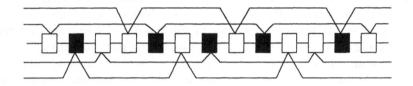

Fig. 1. A fault pattern of size 5 with link redundancy $\{1, 4\}$

We now describe the Boolean matrix representation for fault patterns introduced in [10]. This representation will be instrumental in establishing the geometric characterization of catastrophic fault patterns and deriving an efficient testing scheme.

Consider a linear array A with a link redundancy $G = \{g_1, g_2, \ldots, g_k\}$ and a fault pattern $F = \{f_1, f_2, \ldots, f_m\}$. Without loss of generality, let $f_1 = 1$. We represent F by a Boolean matrix W of size $(\omega_F^+ \times g_k)$ where $\omega_F^+ = \lceil \omega_F / g_k \rceil$.

In the matrix representation, each $f_l \in F$ is mapped into $W[x_l, y_l]$ where $x_l = \lfloor f_l - 1/g_k \rfloor$ and $y_l = f_l - 1 \bmod g_k$. From now on we will use the coordinate pair (x_l, y_l) to denote $W[x_l, y_l]$. Notice that $W[1, 1] = 1$ which indicates the location of the first fault.

Example 2. The Boolean matrix representation of the fault pattern $F = \{f_1, f_2, f_3, f_4, f_5\} = \{(1, 1), (1, 4), (2, 2), (2, 4), (3, 3)\}$ is given in Figure 2.

$$
\begin{bmatrix}
1 & 0 & 0 & 1 \\
0 & 1 & 0 & 1 \\
0 & 0 & 1 & 0 \\
0 & 0 & 0 & 0
\end{bmatrix}
$$

Fig. 2. A Boolean matrix of a catastrophic fault pattern of 5 faults with link redundancy $\{1, 4\}$

3 Segments, Gaps and Shadows

3.1 Dead Segments

In this section, we introduce the notion of segments which will be used in the development of an efficient testing scheme. Before that we will define "dead elements" and then use them to define "dead segments".

Definition 2 *A dead element is*

(1) a faulty element

(2) a non-faulty element (x, y), if $\exists i$, such that $y - g_i \geq 1, (x, y - g_1), \ldots, (x, y - g_i), (x - 1, y - g_{i+1} + g_k), (x - 1, y - g_{i+2} + g_k), \ldots, (x - 1, y)$ are all dead elements; or $(x - 1, y - g_1 + g_k), (x - 1, y - g_2 + g_k), \ldots, (x - 1, y)$ are all dead elements.

A live element is an element which is not a dead element.

Definition 3 *A dead segment is*

(1) a set of dead elements $S = \{(x, y), (x, y + 1), \ldots, (x, y + d)\}$, where $0 \leq d \leq g_k - 1$; or

(2) if $S_1 = \{(x, y), \ldots, (x, y + d)\}$ and $S_2 = \{(x, y'), \ldots, (x, y' + h)\}$ are dead segments and if $y' = y + d + 1$, then $S = \{(x, y), \ldots, (x, y + h')\}$ is a dead segment.; or

(3) if $S_1 = \{(x, y), \ldots, (x, g_k)\}$ and $S_2 = \{(x+1, 1), \ldots, (x+1, d)\}$ are dead segments where $1 \leq d \leq g_k$, then $S = \{(x, y), \ldots, , (x, g_k), (x + 1, 1), \ldots, (x + 1, d)\}$ is a dead segment.

3.2 A Necessary and Sufficient Condition

We now establish a necessary and sufficient condition for a fault pattern, with arbitrary number of faults for an arbitrary link redundancy, to be catastrophic. This condition is based on the notion of segments given in the previous section.

A live path is a path on which every element is live. Given live elements x and y, y is reachable from x if there exists a live path from x to y. The set of entry points of row (i) is: $E_i = \{$live elements in row (i) reachable from some element in $E_{i-1}\}$ if $i > 1$, and $E_1 = \{$live elements in row $(1)\}$ We claim that for every row the set of entry points coincides with the set of live elements.

Property 1 *Let L_i be the set of live elements of row (i), then $E_i = L_i$.*

Theorem 1 *A fault pattern is catastrophic if and only if it has at least one dead segment of g_k elements.*

Proof. (if part) By contradiction, let the fault pattern be catastrophic, and $\forall i \ L_i \neq \emptyset$. Consider the element sequence $l_1, l_2, \ldots, l_{Last}, l_i \in L_i, l_i = (x_i, y_i)$, where l_{Last} is an arbitrary element in $L_{Last} \neq \emptyset$; and $l_i, i < Last$, is recursively constructed from l_{i+1} as follows: L_{i+1} (which by Property 1 is E_{i+1}) is the set of live elements in row($i + 1$) reachable from some live element in row(i). Thus for every $a \in L_{i+1}$, there exists at least an element $b \in L_i$ such that a is reachable from b. Choose l_i be the element in L_i from which l_{i+1} is reachable. Therefore, the sequence of $l_1, l_2, \ldots, l_{Last}$, defines a path of live elements, l_{Last} is reachable from l_1; thus contradicts the fact that the fault pattern is catastrophic.

(only if part) By contradiction, let $L_j = \emptyset$ and the fault pattern be not catastrophic. Since the fault pattern is not catastrophic, then there exists $l_1, l_2, \ldots, l_{Last}$ such that $l_i \in L_i$ and $l_i \in$ row(i), contradicting $L_j = \emptyset$. $\square$

3.3 Gaps and Shadows

It is possible for the Boolean matrix to have several contiguous rows (called "gap") which do not contain any faulty elements. Should this be the case, the given fault pattern can be partitioned into smaller fault patterns such that, the Boolean matrix of each of these fault patterns does not have any gap. The original fault pattern is catastrophic if and only if at least one of these smaller fault patterns is catastrophic; this is due to the following theorem.

Theorem 2 *[2] Given the Boolean matrix representation of a fault pattern $F = \{(x_1, y_1), (x_2, y_2), \ldots, (x_n, y_n)\}$ there exists i such that $x_{i+1} - x_i > 2$, let $F_1 = \{(x_1, y_1), (x_2, y_2), \ldots, (x_i, y_i)\}$ and $F_2 = \{(x_{i+1}, y_{i+1}), (x_{i+2}, y_{i+2}), \ldots, (x_n, y_n)\}$; then F is catastrophic if and only if F_1 or F_2 is catastrophic.*

We now define the "shadow" of a segment.

Definition 4 *Let S be a segment, and S' be a segment at distance d from S. Then S is called the shadow of S' at distance d if*
    1) $|S| = |S'|$, and
    2) $\forall(x, y) \in S$ either $(x, y - d) \in S'$ or $(x - 1, y - d + g_k) \in S'$.

Given a segment S, let $x(S)$ be the x coordinate of the first element of S, $y_F(S)$ be the y coordinate of the first element of S, $y_L(S)$ be the y coordinate of the last element of S.

Definition 5 *Let Ω_i, Ω_j be two segments, $\Omega_i < \Omega_j$ if $y_L(\Omega_i) < y_F(\Omega_j)$ and $x(\Omega_i) = x(\Omega_j)$, or $x(\Omega_i) < x(\Omega_j)$.*

Certain segments can be concatenated to form larger segments. The definition for concatenation of segments now follows.

Definition 6 *Let Ω_i, Ω_j be two segments with $y_L(\Omega_i) + 1 = y_F(\Omega_j)$ and $x(\Omega_i) = x(\Omega_j)$, or $y_L(\Omega_i) = g_k$, $y_F(\Omega_j) = 1$, and $x(\Omega_i) + 1 = x(\Omega_j)$. The concatenation of Ω_i and Ω_j, denoted by $\Omega_i @ \Omega_j$, is the segment Ω where $y_F(\Omega) = y_F(\Omega_i)$, $y_L(\Omega) = y_L(\Omega_j)$, and $x(\Omega) = x(\Omega_i)$.*

Lemma 1 *Let S be a segment and S_{g_i} be the segment at distance g_i from S, $(1 \leq i \leq k)$. The dead segments in S are those in the set $\{faulty\ elements\ in\ S\} \cup \{\prod_{i=1}^{g_k} shadows\ of\ dead\ segments\ of\ S_{g_i}\ at\ distance\ g_i\}$*

Definition 7 *Let $DS[i]$ and $DS[i+1]$ be the set of the dead segments of row(i) and row(i+1), respectively. The concatenation of $DS[i]$ and $DS[i+1]$, denoted by $DS[i] \bullet DS[i+1]$, is the set of dead elements defined as follows: let $a = Max\{y_L(\Omega) : \Omega \in DS[i]\}$, $b = Min\{y_L(\Omega) : \Omega \in DS[i+1]\}$; and let Ω_a and Ω_b be the corresponding segments. Then*

$$DS[i] \bullet DS[i+1] = \begin{cases} DS[i] \cup DS[i+1] \cup \{\Omega_a @ \Omega_b\} - \{\Omega_a, \Omega_b\} & \text{if } a = g_k \text{ and } b = 1 \\ DS[i] \cup DS[i+1] & \text{otherwise} \end{cases}$$

4   An Efficient Testing Scheme

4.1   The Algorithm

The algorithm proceeds as follows. The given fault pattern F is decomposed into patterns which do not contain any gap. By Theorem 2, F is catastrophic if and only if at least one of the pattern is catastrophic. By Theorem 1, any such pattern F' is catastrophic if and only if there exists a dead segment of size g_k. Thus, the algorithm constructs all the dead segments; in particular, it constructs $DS[i]$ (the set of dead segments of row(i)), given $DS[i-1]$ (the set of dead segments of row(i-1)). By Definition 3, a segment S is dead if all k segments at distance $g_1, g_2, \ldots, g_k$ from S are dead. An important aspect of the algorithm which is crucial to its efficiency is that the dead segments of row(i) can only be found among the shadows of the dead segments of row(i-1). Therefore, we can disregard all elements of row(i) which are not in the shadows of the distance equal to the longest bypass link.

Let S be a shadow at distance g_k from some segment in $DS[i-1]$. The algorithm determines which parts of the segments in $DS[i-1] \bullet DS[i]$ are at distance $g_1, g_2, \ldots, g_{k-1}$ from S and compute the intersection of all these parts. Any element in S is dead if and only if it belongs to this intersection. Another aspect of the algorithm crucial for its efficiency is that once the above described process has been completed for S, no backtracking is done when considering the next segment in $DS[i-1]$. If, at any time, a dead segment of size g_k is encountered, the fault pattern is catastrophic. If all dead segments have been constructed and no dead segment of size g_k has been found, the pattern is not catastrophic.

More in detail, given a Boolean matrix representation of a fault pattern F. The algorithm first checks if F is empty. If it is empty, the algorithm stops and reports that the fault pattern is not catastrophic; otherwise, the algorithm scans linearly the fault pattern to detect the existence of gaps (i.e., if $x_{i+1} - x_i > 2$). When the first gap is found, it decomposes F into two fault patterns F_1 and F_2 such that F_1 does not contain any gap. Then the algorithm calls a procedure (*DeadSegment*) to check if F_1 is catastrophic. If F_1 is catastrophic, the algorithm stops; otherwise, it recursively call itself with F_2 as its input. Note that if F does not contain any gap, then $F_1 = F$ and $F_2 = \emptyset$. The procedure *DeadSegment* is as follows:

DeadSegment(F)

<u>Input</u>: A Boolean representation of F which does not contain any gaps.

<u>Output</u>: TRUE if F is catastrophic; FALSE if F is not catastrophic.

<u>Data Structure Used</u>: $DS[i]$ is an ordered (by the (x, y) coordinates of the starting element) set of dead segments of row(i), and it is implemented as a linked list. For simplicity, let $DS[i](k)$ denote the k-th dead segment in $DS[i]$. $DS[i] \bullet DS[j]$ is the concatenation of $DS[i]$ and $DS[j]$, and it is also implemented as a linked list. For simplicity, let $(DS[i] \bullet DS[j])(k)$ denote the k−th dead segment in the concatenation of $DS[i]$ and $DS[j]$.

Step 1: Initially, *DeadSegment*() checks if $|F| \geq g_k$. If so, it continues; otherwise, it returns FALSE (by the characterization of catastrophic fault patterns given in [9], the pattern is not catastrophic).

Step 2: Then, *DeadSegment*() scans F to form, for each row, the set of dead segments lying on that row. If, at any time, *DeadSegment*() encounters a dead segment encompassing the entire row, it returns TRUE. (by Theorem 1, the pattern is catastrophic). If all dead segments in F have been built and no dead segment of size g_k is found, *DeadSegment*() returns FALSE (by Theorem 1, the pattern is not catastrophic). The set of dead segments for row(1), $DS[1]$, is composed of the segments formed by the faulty elements in row(1). Given the set $DS[i-1]$ of dead segments for row$(i-1)$, the set $DS[i]$ of dead segments for row(i) is constructed as follows:

```
L0: form an initial set DS[i] by merging faulty elements in row(i);
     p = 1;
L1: if DS[i − 1] = ∅ then DS[i] is done;
     α = |DS[i − 1]|;
     for (1 ≤ d ≤ k) l_d = 2;
     for (1 ≤ t ≤ α) {
          S_k ← DS[i − 1](t);
          Ω ← shadow of S_k at distance g_k;
          for (k − 1 ≥ j ≥ 1) {
              l_j = max{l_j, l_{j+1}};
L2:           S_j ← (DS[i − 1] • DS[i])(l_j);
              Ω_j ← shadow of S_j at distance g_j;
              if Ω_j < Ω then
                      if l_j = |DS[i − 1] • DS[i]| then goto L3;
                          else {l_j = l_j + 1; goto L2;}
              if Ω < Ω_j then goto L3;
              Ω ← Ω ∩ Ω_j;
          }
          MergeSegment(Ω, DS[i], p);
L3:       continue;
     }
```
and where the *MergeSegment*() is as follows:

MergeSegment($\Omega, DS[i], p$)

Input: $\Omega, DS[i], p$;

Output: $DS[i], p$;

Outline: this procedure first scans $DS[i]$ starting from p to find the location for Ω; then inserts Ω into $DS[i]$. The procedure then returns the modified $DS[i]$ and the location of the next segment p.

> while $(y_F(\Omega) > y_L(DS[i](p)))$ $p = p + 1$;
> $y'_F = \min(y_F(\Omega), y_F(DS[i](p)))$;
> $q = p$;
> while $(y_L(\Omega) > y_F(DS[i](q)))$ $q = q + 1$;
> $y'_L = \max(y_L(\Omega), y_L(DS[i](q - 1)))$;
> replace all segments from p to $q - 1$ with the new segment
> $\{(i, y'_F), (i, y'_L)\}$ in $DS[i]$;
> $p =$ pointer to this new segment;
> return $DS[i]$ and p.

Note that $p + 1$ is the operation of next(p), and $p - 1$ is the operation of previous(p).

4.2 Analysis

Property 2 *Algorithm correctly determines whether or not a fault pattern F is catastrophic.*

Proof. F is decomposed into patterns which do not contain any gap. By Theorem 2, F is catastrophic if and only if at least one of these patterns is catastrophic. Thus, to prove the correctness, it suffices to prove that the algorithm correctly determines whether any such pattern F' is catastrophic. By Theorem 1, F' is catastrophic if and only if there exists a dead segment of size g_k. Thus, it suffices to show that the algorithm correctly constructs all the dead segments; in particular, it suffices to show that $DS[i]$ will be constructed correctly given $DS[i - 1]$. By Definition 3, a segment S is dead if all k segments at distance $g_1, g_2, \ldots, g_k$ from S are dead. First of all, observe that it suffices to verify the above condition only for the segments which are shadows at distance g_k from the segments in $DS[i - 1]$. In fact, any element of row(i), which is a shadow at distance g_k of a live element, is live; in other words, the only possible candidates for dead elements (and, thus, dead segments) are those which are the shadow at distance g_k of a dead segment in $DS[i - 1]$. Let S be a shadow at distance g_k from some segment in $DS[i - 1]$. To determine whether an element of S is dead, we must determine whether its shadows at distance $g_1, g_2, \ldots, g_{k-1}$ are dead; that is, we must determine whether its shadows at distance $g_1, g_2, \ldots, g_{k-1}$ are in $DS[i - 1] \bullet DS[i]$. The algorithm first determines which parts of the segments in $DS[i - 1] \bullet DS[i]$ are at distance $g_1, g_2, \ldots, g_{k-1}$ from S and compute the intersection of all these parts. Any element in S is dead if and only if it belongs to this intersection. Since this is done for every row, the claim follows. $\square$

Property 3 *For any row(i) in the Boolean representation of a fault pattern, let n_i be the number of faults in that row, and $|DS(i)|$ be the number of dead segments in that row, then $|DS(i)| \leq n_i + 1$.*

Proof. By contradiction, let $|DS(i)| > n_i + 1$. Then there are at least two dead segments, S_1 and S_2 which do not contain any faults. Without loss of generality, let $S_1 < S_2$; thus $y_F(S_2) > 1$. In other words, $w = (i, y_F(S_2) - 1)$ exists and is live. Since S_2 does not contain any faulty element, all elements in S_2 are reachable by p. This contradicts the fact that S_2 is a dead segment. Thus, the statement holds. □

Property 4 *Algorithm requires $O(kn)$ time, where k is the number of bypass links, and n is the number of faults.*

Proof. The given fault pattern F is decomposed into patterns which do not contain any gap. This is done in time $O(n)$. In the worst case, the procedure *DeadSegment* will be applied to each such pattern. For any input F' of *DeadSegment*:

1) the number of rows in F' is at most $2n'$ where n' is the number of faults in F'. This is because F' does not contain any gap. By Property 3, $|DS(F')|$, the number of dead segments in F' is at most $\sum_i^{2n'} (n_i' + 1) = n' + 2n' = 3n'$. Thus, $|DS(F')|$ is $O(n')$.

2) DeadSegment construct the overall set DS row by row. It uses $k-1$ pointers (the l_i's). At each step of the execution, each pointer can move forwards ("advance") or not ("stay"); it cannot ever move backwards. The number of "advances" for each pointer is at most $|DS(F')|$, so is the number of "stays". Thus, each pointer requires atmost $2|DS(F')| = O(n)$ operations, for a total of $O(kn)$ time for $k - 1$ pointers.

3) As for the cost of the merging operation, the initial merge (executed in step L0 of the algorithm) requires $O(n_i')$ operations. The total cost of merging new dead segments with $DS[i]$ by calling *MergeSegment* is $O(n_i')$ since the entire $DS[i]$ is only scanned once. The total cost for F', is $O(n')$.

Therefore, the total cost for F' is $O(kn')$. Thus, the total cost for each pattern F' into which F has been decomposed is $O(kn')$. Since these patterns are disjoint, the total cost of Algorithm is $O(kn)$, where k is the number of bypass links and n is the number of faults. □

Theorem 3 *Algorithm correctly determines whether or not a fault pattern is catastrophic in $O(kn)$ time, where k is the number of bypass links at each element, n is the number of faults.*

Proof. The theorem follows from Properties 2 and 4. □

5 References

1. Apostolico, A., Atallah, M. J., Larmore, L. and McFaddin, H. S., "Efficient Parallel Algorithms for String Editing and Related Problems," *SIAM J. Comput.*, Vol. 19, 1990, pp. 968-998.

2. De Prisco, R. and Monti, A., "On Reconfigurability of VLSI Linear Arrays," in *Proc. Workshop on Algorithms and Data Structures*, Lecture Notes in Computer Science No. 709, Springer-Verlag, 1993, pp. 553-564.

3. De Prisco, R. and De Santis, A., "Catastrophic Fault Patterns for a VLSI Linear Array with Unidirectional Bypass Links," Internal Report, Dipartimento di Informatica, Università di Salerno, Italy, 1992.

4. Koren, I. and Pradhan, D. K., "Introducing Redundancy into VLSI Designs for Yield and Performance Enhancement," in *Proc. 15th Int'l Conf. on Fault-Tolerant Computers*, 1985, pp. 330-335.

5. Li, H. F., Jayakumar, R. and Lam, C., "Restructuring for Fault-Tolerant Systolic Arrays," *IEEE Trans. on Computers*, Vol. C-38, No. 2, 1989, pp. 307-311.

6. Nayak, A., Pagli, L. and Santoro, N., "Recognition of Catastrophic Faults," in *Proc. Int'l Workshop on Defect and Fault Tolerance in VLSI Systems*, Dallas, 1992, pp. 70-79.

7. Nayak, A., Pagli, L. and Santoro, N., "Efficient Construction of Catastrophic Patterns for VLSI Reconfigurable Arrays", *Integration - the VLSI journal*, Vol. 15, No. 2, Oct. 1993, pp. 133-150.

8. Nayak, A., Pagli, L. and Santoro, N., "Combinatorial and Graph Problems arising in the Analysis of Catastrophic Fault Patterns," in *Proc. 23rd Southeastern Int'l Conf. on Combinatorics, Graph Theory, and Computing; Congressus Numerantium 88* (Utilitas Mathematica), 1992, pp. 7-20.

9. Nayak, A., Santoro, N. and Tan, R., "Fault-Intolerance of Reconfigurable Systolic Arrays," in *Proc. 20th Int'l Symp. on Fault-Tolerant Computers*, Newcastle upon Tyne, 1990, pp. 202-209.

10. Pagli, L. and Pucci, G., "Reliability Analysis of Redundant VLSI Arrays," *Information Processing Letters*, 50(1994), pp. 337-342.

11. Ren, J., "Geometric Characterization of Fault Patterns in Linear Systolic Arrays," M.C.S. Thesis, School of Computer Science, Carleton University, Canada, 1994.

12. Sipala, P., "Faults in Linear Arrays with Multiple Bypass Links," Research Report No. 18, Dipartimento di Informatica, Università Degli Studi di Trieste, Italy, 1993.

Scheduling Parallel Tasks with Individual Deadlines*

Oh-Heum Kwon and Kyung-Yong Chwa

Dept. Computer Science, KAIST
Kusong-dong 373-1, Yusong-gu, Taejon 305-701, Republic of Korea
E-mail:ohkwn@jupiter.kaist.ac.kr, kychwa@gayakum.kaist.ac.kr

Abstract. In this paper, we consider the problem of scheduling independent parallel tasks with individual deadlines so as to maximize the sum of the works of the tasks which complete their executions before deadlines. We propose two polynomial time heuristic algorithms for non-malleable parallel tasks and malleable tasks with linear speedups, respectively. The approximation factors of two algorithms are $5 + \epsilon$ and 4.5, respectively, where ϵ is an arbitrary positive constant.

1 Introduction

A *parallel task* is one that can be executed by multiple processors. All the processors allotted to a task are required to execute that task in unison and simultaneously. In contrast, a *sequential task* is one that can be executed only by a single processor. A huge amount of work has been devoted to the problem of scheduling sequential tasks. Recently, with the development of parallel systems, the problems of scheduling parallel tasks are being noticed and studied by several authors [3, 5, 7, 8]. This paper is also concerned with the parallel task scheduling problem.

In general, two categories are used to classify parallel tasks. A *non-malleable parallel task* is one that requires a specific number of processors for a specific units of time, while a *malleable parallel task* is one that can be executed on any number of processors, with its execution time being a function of the number of processors allotted to it.

Malleable tasks can be further classified by the nature of the processing time function. In this paper, we restrict our attention into a special case of malleable tasks, namely, *malleable tasks with linear speedups*, where the task execution times describe *linear* speedups up to some specified maximum degree of parallelism, and are constant thereafter. Formally, a task is specified by a tuple (t_i, p_i), where p_i is the maximum degree of parallelism and t_i is the corresponding processing time. Then, the processing time function τ_i is described as follows:

$$\tau_i(x) = \begin{cases} \frac{t_i p_i}{x} & x \le p_i \\ t_i & x > p_i. \end{cases}$$

---

* This research was partially supported by Electronics and Telecommunications Research Institute under Contract No.NN15260

This implies that a task designed to be executed on some specific number (say, p_i) of processors can be simulated on a smaller number of processors while preserving the work of that task. This assumption was introduced by [3] in the context of scheduling parallel tasks in PRAMs, hypercubes, and meshes.

Various measures are used to estimate the efficiencies of the schedules generated by any algorithm. The problem of minimizing *makespan*, one of the most important and standard measures, has been studied extensively. Notable works for this problem include [3, 5, 7]. Recently, a few authors [8] studied the problem of minimizing the *average completion time* which is another standard measure in the classical scheduling theory.

In the classical scheduling theory, some measures involving the notion of *deadline* and the corresponding optimization problems have been studied. One notable problem among them investigated by many authors [1, 2, 4, 6, 9] can be written as follows: *Suppose that each task is associated with a deadline d_i and a value v_i. In a schedule, if task T_i completes its execution before its deadline d_i, the scheduler obtains the corresponding value v_i. The goal of the scheduler is to obtain as much value as possible.*

In the followings, this problem will be referred as "general problem". This general problem can be easily shown to be NP-hard even for uniprocessor environments. In [6], Sahni introduced a fully polynomial approximation scheme for this general problem on a uniprocessor system.

An interesting subproblem of this general problem comes from restricting the value distributions of the tasks. Suppose that the tasks are either non-malleable or malleable with linear speedups. When the values are all equal to the works of the tasks, we say that the tasks have *uniform value densities*. This reflects a simple but reasonable assumption that a task is important than another one with the smaller work. Moreover, this is also related to the problem of maximizing the *utilization* of the target system. The sequential task scheduling problems adopting this assumption were investigated in [1, 9].

This restricted problem remains NP-hard for uniprocessor environments, since this problem embeds in itself a subset sum problem which is a well-known NP-hard problem. In this paper, we consider this restricted problem for parallel tasks and investigate approximation algorithms with provably reasonable performance bounds for both of non-malleable tasks and malleable tasks with linear speedups.

For any task set T and any scheduling algorithm A, let $A(T)$ denote the sum of the works of the tasks which are completed before their deadlines in the schedule generated by algorithm A, and let $O(T)$ denote that of the optimal algorithm. An algorithm A is said to have *an approximation factor of ρ* iff $O(T) \leq \rho A(T)$ for any task set T.

The results of this paper is as follows: For non-malleable tasks, we present a heuristic algorithm with an approximation factor of $5 + \epsilon$ for any constant ϵ. Especially, when each task has a bounded degree of parallelism, that is, $p_i \leq \alpha P, 0 < \alpha < 1$, for each task T_i, a part of the proposed algorithm can be used to produce a schedule of which the approximation factor is no more than $\frac{2}{1-\alpha}$. If this algorithm is adopted to the sequential task scheduling problem on multiple processors, then the approximation factor is to be 2. For malleable tasks with linear speedups, we present another heuristic algorithm with an approximation factor of 4.5.

2 Preliminaries

We are given P identical processors and a set of non-malleable tasks $T = \{T_i = (t_i, p_i, d_i) \mid 1 \leq i \leq n\}$, where t_i is the processing time, p_i is the number of required processors, and d_i is the deadline of the task. We assume that $d_1 \leq d_2 \leq \ldots \leq d_n$. For any subset T' of T, we introduce a notation $\|T'\|$ indicating the sum of the works of all tasks in T', that is, $\|T'\| = \sum_{T_i \in T'} t_i p_i$.

A schedule S for T is composed of tasks in any subset T' of T. Each task $T_i \in T'$ is associated with a starting time s_i with $s_i + t_i \leq d_i$ and a subset of processors σ_i with $|\sigma_i| = p_i$. We will say that task $T_i \in T'$ is *included* in S, while task $T_j \notin T'$ is *rejected* in S. We will use the same notation $\|S\|$ to indicate the sum of the works of the tasks that are included in S, that is, $\|S\| = \sum_{T_i \in T'} t_i p_i$.

A simple upper bound on $O(T)$ and its variants will be used throughout this paper. Let S_o be an optimal schedule for task set T. For any time instance $t \geq 0$, we can divide the tasks included in S_o into two groups; the first group G_1 consists of the tasks which are completed after t in S_o, and the second G_2 consists of the remaining tasks. Each task in G_1 has a deadline larger than t. Hence, $\|G_1\| \leq \sum_{\forall i \text{ with } d_i > t} t_i p_i$. Each task in G_2 terminates its execution before t. Hence, $\|G_2\| \leq Pt$. Consequently, we have, for any $t \geq 0$,

$$O(T) \leq \sum_{\forall i \text{ with } d_i > t} t_i p_i + Pt \tag{1}$$

This bound also holds for malleable tasks with linear speedups where p_i is taken to be the maximum degree of parallelism and t_i is the corresponding processing time of the task.

Before proceeding, we will introduce a simple heuristic algorithm, called EDF(Earliest Deadline First), for scheduling non-malleable tasks. This algorithm will be used as a subroutine in the algorithm presented in the subsequent sections.

To describe EDF algorithm, we will introduce two notions; *idle* and *free* processors. For any schedule S, a processor p is said to be *idle* at time t if p is executing no task at time t in S. Moreover, a processor is said to be *free* at time t if, for any time $t' \geq t$, p is idle at t' in S.

EDF algorithm is as follows: Tasks are considered by the order of non-decreasing deadlines. When task T_i is considered, the scheduler tries to find the earliest time instance t with $t \leq d_i - t_i$ at which p_i or more processors are free. If succeeded, T_i is scheduled at that time to those processors; otherwise, T_i is simply rejected.

This simple heuristic EDF guarantees no performance bound in general. However, we will show that, in Lemma 1, imposing the following two restrictions C1 and C2 on the tasks results in a constant approximation factor of EDF heuristic. C1. (Bounded Degree of Parallelism Constraints) $\forall i, p_i \leq \alpha P$; C2. (Large Slack Time Constraints) $\forall i, t_i \leq \beta d_i$.

Lemma 1. *For task system T satisfying two constraints C1 and C2, $EDF(T) \geq (1-\alpha)(1-\beta)O(T)$.*

Proof. Suppose that $T_k, 1 \leq k \leq n$, is the task with the largest index which is rejected by EDF algorithm. Using (1), we have $O(T) \leq \sum_{k+1 \leq i \leq n} t_i p_i + d_k P$. Let S be the partial schedule constructed by EDF so far just before rejecting T_k. Suppose for contradiction that, for any time instance t with $t \leq (1-\beta)d_k$, αP or more processors are idle in S. Note that those processors must be also free at t in S since no task would be started after t in S. Consequently, the task T_k could be scheduled at time t, which is a contradiction.

Hence, we can conclude that the total work included in S is no less than $(1-\alpha)(1-\beta)d_k P$. Moreover, tasks $T_{k+1}, T_{k+2}, \ldots, T_n$ are added to S in the final schedule generated by EDF. Now, we have $EDF(T) \geq \sum_{k+1 \leq i \leq n} t_i p_i + (1-\alpha)(1-\beta)d_k P$, which completes the proof. $\square$

3 Scheduling Non-malleable Tasks

In this section, we consider non-malleable tasks, and present a heuristic scheduling algorithm with an approximation factor of $5 + \epsilon$ for any constant ϵ.

In our algorithm, tasks are classified into two groups T_W and T_N, and considered separately, where T_W consists of tasks T_i with $p_i > \frac{1}{2}P$, and $T_N = T - T_W$. The subscripts W and N represents the terms "wide" and "narrow", respectively.

Suppose we are given two heuristics H_W and H_N with $H_W(T_W) \geq \frac{1}{c_1}O(T_W)$ and $H_N(T_N) \geq \frac{1}{c_2}O(T_N)$. Let T_O be the set of tasks which are included in the optimal schedule. Then, it is obvious that $\|T_O \cap T_W\| \leq O(T_W)$, and $\|T_O \cap T_N\| \leq O(T_N)$. Hence, $O(T) = \|T_O \cap T_W\| + \|T_O \cap T_N\| \leq O(T_W) + O(T_N)$. First, assume that $\frac{1}{c_1}O(T_W) \geq \frac{1}{c_2}O(T_N)$. Then, we have $O(T) \leq (1 + \frac{c_2}{c_1})O(T_W) \leq (c_1 + c_2)H_W(T_W)$. For the second case where $\frac{1}{c_1}O(T_W) < \frac{1}{c_2}O(T_N)$, we have $O(T) \leq (c_1 + c_2)H_N(T_N)$ by the symmetric argument.

Hence, if we are given two algorithms H_W and H_N with $H_W(T_W) \geq \frac{1}{1+\epsilon}O(T_W)$ and $H_N(T_N) \geq \frac{1}{4}O(T_N)$, the following simple algorithm achieves what we want: Firstly, partition the tasks into T_W and T_N; secondly, apply two heuristic H_W and H_N into each of these two task sets, separately; finally, choose the one including the larger work.

3.1 Scheduling Tasks in T_W

Consider any schedule S for task set T_W. Since each task in T_W requires more than $\frac{P}{2}$ processors, no two or more tasks can be executed simultaneously in S. Hence, the problem becomes in fact a single processor scheduling problem.

A simple transformation of the problem into the *general problem* on a single processor resolves this case. Each task $T_i = (t_i, p_i, d_i) \in T_W$ is transformed into a single processor task T_i' associated with the value equal to the work of the original task, that is, $T_i' = (t_i, d_i, v_i)$, where $v_i = t_i p_i$. We can solve this general problem with an approximation factor of $1 + \epsilon$ for any positive constant ϵ using FPAS introduced by Sahni [6],

3.2 Scheduling Tasks in T_N

We will apply a modified EDF heuristic into tasks in T_N. T_N is again partitioned into two groups: T_{NL} and T_{NS}, where T_{NL} consists of tasks T_i with $t_i \geq \frac{1}{2} d_i$ and T_{NS} consists of the remaining tasks. That is, T_{NS} consists of the tasks with relatively large slack times. Let $T_{NL} = \{T_i = (t_i, p_i, d_i) \mid 1 \leq i \leq n_{NL}\}$, and let $T_{NS} = \{T_j' = (t_j', p_j', d_j') \mid 1 \leq j \leq n_{NS}\}$. Suppose that $d_1 \geq d_2 \geq \cdots \geq d_{n_{NL}}$, and that $d_1' \leq d_2' \leq \cdots \leq d_{n_{NS}}'$. To avoid some trivial arguments, we assume a dummy task $T_{n_{NL}+1}$ with $d_{n_{NL}+1} = 0$, $t_{n_{NL}+1} = 0$ and $p_{n_{NL}+1} = P$.

Let $T_k, 1 \leq k \leq n_{NL}+1$, be the task with $\sum_{1 \leq i \leq k-1} p_i \leq P$ and $\sum_{1 \leq i \leq k} p_i > P$. We will introduce a function *"owner"* which associates each processor p with a task $T_i, 1 \leq i \leq k$, so that $owner(p) = i$ iff $\sum_{1 \leq j \leq i-1} p_j < p \leq \sum_{1 \leq j \leq i} p_j$. When $owner(p) = i$, we will say that task T_i *owns* processor p, and also, reversely, processor p *is owned* by T_i.

Algorithm H_N is as follows: Firstly, tasks $T_i, 1 \leq i \leq k - 1$, are scheduled into the processors they own at time 0, that is, T_i, is assigned to the processors numbered from $\sum_{1 \leq j \leq i-1} p_j + 1$ to $\sum_{1 \leq j \leq i} p_j$. Secondly, the tasks in T_{NS} are scheduled by EDF heuristic. Note that EDF heuristic is well-defined in this situation where some tasks in T_{NL} are already placed.

In Lemma 3 and 4 , we will prove that the algorithm presented guarantees an approximation factor of 4. To prove that, we need some additional definitions and notations. We introduce another function f_N which associates each processor p with the deadline of task $T_{owner(p)}$, that is, $f_N(p) = d_{owner(p)}$, where the subscript "N" represents the term "non-malleable". Function f_N makes a staircase-shaped upper boundary as shown in Figure 1. Let A denote the region under this boundary. Fact 2 asserts that at least a fourth of the area of region A is filled by tasks $T_1, T_2, \ldots, T_{k-1}$.

Fact 2. $\sum_{1 \leq i \leq k-1} t_i p_i \geq \frac{1}{4} \sum_{\forall p} f_N(p)$.

Proof. Omitted. □

Suppose that $T_l', 1 \leq l \leq n_{NS}$, is the task with the largest index which is rejected by EDF heuristic in the second step of the proposed algorithm. Another function g_N is defined as follows: $g_N(p) = \max\{f_N(p), d_l'\}$. Let B be the region bounded by this function g_N. Note that the area of B is $\sum_{\forall p} g_N(p)$.

Lemma 3. $O(T_N) \leq \sum_{l+1 \leq i \leq n_{NS}} t_i' p_i' + \sum_{\forall p} g_N(p)$

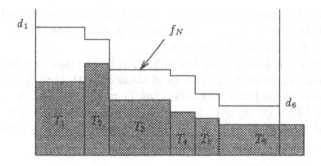

Fig. 1. Definitions of f_N

Proof. Let S_o be an optimal schedule for task set T_N. For each processor p, let $h(p)$ be the time instance at which the last task assigned to p which is in T_{NL} is terminated, if any; let $h(p) = 0$, otherwise. Since all the processors are identical, we may assume, without loss of generality, that $h(1) \geq h(2) \geq \cdots \geq h(P)$. At this point, observe that $h(p) \leq f_N(p) \leq g_N(p)$ for all p. That is, no task in T_{NL} is executed outside region B in the optimal schedule. Hence, the task which lies outside B must be a task in T_{NS}, and, moreover, has a deadline larger than d'_l. Hence, $O(T_N)$ is no more than $\sum_{l+1 \leq i \leq n_{NS}} t'_i p'_i + \sum_{\forall p} g_N(p)$, where the first term represents the sum of the work of the tasks in T_{NS} with the deadlines larger than d'_l, and the second term represents the area of region B. $\square$

Lemma 4. $H_N(T_N) \geq \sum_{l+1 \leq i \leq n_{NS}} t'_i p'_i + \frac{1}{4} \sum_{\forall p} g_N(p)$

Proof. Let S be a partial schedule constructed by H_N so far just before rejecting task T'_l. It is suffice for our purpose to show that $||S|| \geq \frac{1}{4} \sum_{\forall p} g_N(p)$. Consider a processor p with $g_N(p) > d'_l$. Task $T_{owner(p)}$ is assigned to p, and, thereby, p is busy during the time interval $[0, t_{owner(p)}]$ where $t_{owner(p)} \geq \frac{1}{2} d_{owner(p)} = \frac{1}{2} g_N(p)$. Moreover, with the exactly same arguments with Lemma 1, no $\frac{1}{2} P$ or more processors are idle at time t with $t \leq \frac{1}{2} d'_l$. From the combination of these two observations, the proof is completed. $\square$

Theorem 5. *For non-malleable tasks, there is a polynomial time scheduling algorithm with an approximation factor of $5 + \epsilon$ for any constant ϵ.*

4 Scheduling Malleable Tasks

In this section, we consider malleable tasks with linear speedups, and propose a heuristic scheduling algorithm with an approximation factor of 4.5.

Assume that the tasks are given in the form of $T = \{T_i = (t_i, p_i, d_i) \mid 1 \leq i \leq n\}$, where p_i is the maximum degree of parallelism and t_i is the corresponding processing time, that is, the processing time when the task is run on p_i processors. Task T_i can be run on any $p'_i \leq p_i$ processors while preserving the work of the task whenever the corresponding processing time does not exceed deadline d_i. In any schedule S of T, if p_i' processors are actually assigned into T_i, we say that the *execution parallelism* of T_i in S is p_i'.

Our algorithm is similar with H_N in Section 3. The tasks are divided into two groups, T_S and T_L, where T_S consists of tasks T_i with $t_i \leq \frac{d_i}{2}$ and $t_i p_i \leq \frac{d_i P}{4}$, and T_L consists of the remaining tasks. Let $T_L = \{T_i = (t_i, p_i, d_i) \mid 1 \leq i \leq n_L\}$, and let $T_S = \{T_j' = (t_j', p_j', d_j') \mid 1 \leq j \leq n_S\}$. Assume that $d_1 \geq d_2 \geq \cdots \geq d_{n_L}$, and also that $d_1' \leq d_2' \leq \cdots \leq d_{n_S}'$. From the conditions that a task belongs to T_L or T_S, Fact 6 and 7 are obvious.

Fact 6. *For any task $T_j' \in T_S$, we can always choose an appropriate execution parallelism p_j'' with $p_j'' \leq \frac{1}{2}P$ so that the corresponding processing time $t_j'' \leq \frac{d_j'}{2}$.*

Fact 7. *For any task $T_i \in T_L$, $t_i > \frac{d_i}{4}$. Moreover, if $p_i \leq \frac{P}{2}$, then $t_i > \frac{d_i}{2}$.*

Again, we assume a dummy task T_{n_L+1} with $t_{n_L+1} = 0$, $d_{n_L+1} = 0$ and $p_{n_L+1} = P$, and assume that $\sum_{1 \leq i \leq k-1} p_i \leq P$ and $\sum_{1 \leq i \leq k} p_i > P$. A function *owner* which specifies for each processor an index of a task in T_L is also defined as follows:

$$owner(p) = q \qquad \text{if } \sum_{1 \leq i \leq q-1} p_i < p \leq \sum_{1 \leq i \leq q} p_i.$$

Moreover, define function f_M which associates each processor with the deadline of the "owner" task, that is, $f_M(p) = d_{owner(p)}$. Let A be the region defined by f_M, and let $\Delta = P - \sum_{1 \leq i \leq k-1} p_i$. Note that, in this case, no bound on Δ holds.

The first step of our algorithm is to schedule the tasks in T_L so that at least a fourth of the area of A is filled by the tasks. However, in this step of our presentation, we leave this phase of the algorithm as a black box for a while, and continue to present the next part of the algorithm which deals the tasks in T_S.

The second step is to schedule the tasks in T_S. The algorithm is basically the same with EDF heuristic. However, before scheduling them, the execution parallelism of each task must be determined. From Fact 6, for each task $T_i' \in T_S$, we can always find an appropriate p_i'' with $p_i'' \leq \frac{P}{2}$ and $t_i'' \leq \frac{d_i'}{2}$ while preserving the work of the task, that is, $t_i'' p_i'' = t_i' p_i'$. If more than one choices are possible, take an arbitrary one. Now, schedule the tasks using simply EDF heuristic.

Let T_l' be the task with the largest index among tasks which are rejected by EDF algorithm, and let S be the schedule constructed so far just before rejecting T_l'. Let g_M be a function defined by $g_M(p) = \max\{f_M(p), d_l'\}$. And also, let B be the region bounded by function g_M. Lemma 8 is just a repetition of Lemma 3, and the proof will be omitted.

Lemma 8. $O(T) \leq \sum_{l+1 \leq i \leq n_S} t_i' p_i' + \sum_{\forall p} g_M(p)$

In the followings, we will present the skipped part of our algorithm which schedules the tasks in T_L, and, simultaneously, will prove the following inequality for the partial schedule S.

$$\|S\| \geq \frac{1}{4.5} \sum_{\forall p} g_M(p) \tag{2}$$

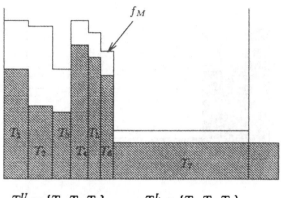

$$T_L^U = \{T_1, T_2, T_3\} \qquad T_L^L = \{T_4, T_5, T_6\}$$

Fig. 2. T_L^U, T_L^L and function f_M

CASE I. $\exists$ a task $T_i, 1 \leq i \leq k - 1$, with $p_i \geq \frac{P}{2}$

Let T_i be such a task. We simply reject all tasks T_j with $j > i$, and assign tasks $\{T_1, T_2, \ldots, T_i\}$ to the processors they own at time 0. That is, $T_j, 1 \leq j \leq i$, is assigned to the processors numbered from $\sum_{1 \leq h \leq j-1} p_h + 1$ to $\sum_{1 \leq h \leq j} p_h$. The proof of inequality (2) in this case will be ommitted in this version of the paper.

Before going on, let focus on task T_k. Let $\hat{p}_k$ be the minimum degree of parallelism of T_k under the condition that the corresponding processing time $\hat{t}_k$ does not exceed deadline d_k. That is, $(\hat{t}_k, \hat{p}_k)$ is a pair with $\hat{t}_k \hat{p}_k = t_k p_k$, $\hat{t}_k \leq d_k$ and $\frac{t_k p_k}{\hat{p}_k - 1} > d_k$. Moreover, we assume that there is no task $T_i, 1 \leq i \leq k - 1$, with $p_i \geq \frac{P}{2}$. Remind that this assumption implies that $t_i \geq \frac{d_i}{2}, 1 \leq i \leq k - 1$.

CASE II. $\sum_{1 \leq i \leq k-1} p_i + \hat{p}_k \leq P$

In this case, all tasks in $\{T_1, T_2, \ldots, T_k\}$ are allotted to the processors they own at the time 0. Task $T_i, 1 \leq i \leq k - 1$, is assigned into p_i processors it owns, while task T_k is assigned into the remaining $P - \sum_{1 \leq i \leq k-1} p_i$ processors. Note that T_k can be run on those processors without violating the deadline. The proof is obvious from the fact that (a) $t_i \geq \frac{d_i}{2}, 1 \leq i \leq k - 1$, (b) $t_k \geq \frac{d_k}{4}$, and also (c) no $\frac{P}{2}$ or more processors are idle at time t with $t \leq \frac{d_i}{2}$.

CASE III. $\sum_{1 \leq i \leq k-1} t_i p_i \geq \frac{1}{4} \sum_{\forall p} f_M(p)$

In this case, we assign all tasks in $\{T_1, T_2, \ldots, T_{k-1}\}$ into the processors they own at time 0. Note that $t_i \geq \frac{d_i}{2}, 1 \leq i \leq k - 1$. Hence, the same argument with Lemma 3 holds for this case.

CASE IV. $\sum_{1 \leq i \leq k-1} t_i p_i < \frac{1}{4} \sum_{\forall p} f_M(p)$ and $\sum_{1 \leq i \leq k-1} p_i + \hat{p}_k > P$

This case is the most difficult part of our proof. First of all, concentrate on Fact 9 given below.

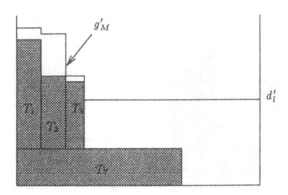

Fig. 3. An example of CASE IV

Fact 9. *The followings are hold;* (a) $\sum_{1 \leq i \leq k-1} p_i < \frac{P}{2}$; (b) $p_k > \frac{P}{2}$; (c) $t_k p_k > \Delta d_k \geq \frac{1}{2} \sum_{\forall p} f_M(p)$.

Proof. Omitted. $\square$

Tasks $\{T_1, T_2, \ldots, T_{k-1}\}$ is classified into two groups: T_L^U consists of tasks T_i with $d_i - t_i \geq d_k$ and T_L^L consists of tasks T_j with $d_j - t_j < d_k$. At this time, we give new indices to the tasks so that $T_L^U = \{T_1, T_2, \ldots, T_h\}$ and $T_L^L = \{T_{h+1}, T_{h+2}, \ldots, T_{k-1}\}$, where the non-increasing order of the deadlines of the tasks are maintained within each of two groups.

Of course, this new index scheme break the condition that $d_1 \geq d_2 \geq \ldots \geq d_{k-1}$ maintained so far. However, it gives no impact on the argument given below. Functions $owner, f_M$ and g_M are also adjusted according to new index scheme. Note that f_M and g_M are no longer non-increasing functions. Figure 2 shows the adjusted functions f_M and g_M. However, the sum $\sum_{\forall p} f_M(p)$ and $\sum_{\forall p} g_M(p)$ remain constant.

The algorithm is as follows: We simply reject all tasks in T_L^L, and assign the tasks in T_L^U and T_k as shown in Figure 3. That is, task T_k is assigned into the processors numbered from 1 to p_k at time 0, and the tasks in T_L^U are assigned to the processors into which T_k is already assigned at the time t_k. This is always possible since $\sum_{1 \leq i \leq h} p_i < p_k$ (from Fact 9, (a) and (b)) and also $d_i - t_i \geq d_k \geq t_k, 1 \leq i \leq h$.

First, consider the special case where $d_k \geq d_l'$. In this case, $\sum_{\forall p} g_M(p) = \sum_{\forall p} f_M(p) \leq 2t_k p_k \leq 2\|S\|$, which completes the proof of inequality (2). From now on, assume that $d_k < d_l'$.

We introduce a new function g_M' which is defined as follows:

$$g_M'(p) = \begin{cases} d_l' & \text{if } \sum_{1 \leq i \leq h} p_i < p \leq \sum_{1 \leq i \leq k-1} p_i \\ g_M(p) & \text{otherwise} \end{cases}$$

We claim that (a) $\|S\| \geq \frac{1}{4} \sum_{\forall p} g_M'(p)$, and (b) $\sum_{\forall p} g_M(p) - \sum_{\forall p} g_M'(p) \leq \frac{1}{2}\|S\|$. Note that the combination of these two facts implies inequality (2).

First, consider (a). For any processor p with $g'_M(p) > d'_l$, a task in T_L^U is assigned into p, and, from the nature of EDF heuristic, no $\frac{P}{2}$ or more processors are idle at time t with $t < \frac{d'_l}{2}$ in S. Hence, claim (a) obviously holds.

For claim (b), we have $\sum_{\forall p} g_M(p) - \sum_{\forall p} g'_M(p) = \sum_{h+1 \leq i \leq k-1}(d_i - d'_l)p_i \leq \sum_{h+1 \leq i \leq k-1}(d_i - d_k)p_i \leq \sum_{h+1 \leq i \leq k-1} t_i p_i \leq \frac{1}{4}\sum_{\forall p} f_M(p) \leq \frac{1}{2}t_k p_k \leq \frac{1}{2}||S||$. Consequently, we have Lemma 10 for our heuristic algorithm H described so far.

Lemma 10. $H(T) \geq \sum_{l+1 \leq i \leq n_S} t'_i p'_i + \frac{1}{4.5}\sum_{p \in P} g_M(p)$

Theorem 11. *For malleable tasks with linear speedups, there is a polynomial time scheduling algorithm with an approximation factor of 4.5.*

5 Concluding Remarks

Several authors considered the on-line version of this problem for sequential tasks [1, 4, 9]. In on-line scheduling, each task is associated with an unknown ready time r_i. For parallel task scheduling problem, it is easy to show that no on-line algorithm can achieve any constant approximation factor. We consider a special case where tasks have a common deadline D, and, moreover, each task satisfies two conditions C1 and C2; that is, $p_i \leq \alpha P$ and $t_i \leq \beta(D - r_i), \forall i$. We can prove that a simple algorithm based on FCFS (First Come First Serviced) strategy achieves an approximation factor of $1 + \frac{1}{(1-\alpha)(1-\beta)}$. The details will be omitted in this version of the paper.

References

1. S. Baruah, G. Koren, D. Mao, B. Mishra, A. Raghunathan, L. Rosier, D. Shasha, and F. Wang. On the competitiveness of on-line real-time task scheduling. *The Journal of Real-Time Systems*, 4:125–144, 1992.
2. R. Conway, W. Maxwell, and L. Miller, *Theory of Scheduling*, Addison-Wesley, 1967.
3. A. Feldmann, M. Kao, J. Sgall, and S. Teng. Optimal online scheduling of parallel jobs with dependencies. *STOC*, 642–651, 1993.
4. G. Koren and D. Shasha. Moca: a multiprocessor on-line competitive algorithm for real-time system scheduling. *Theoretical Computer Science*, 128:75–97, 1994.
5. W. Ludwig and P. Tiwari. Scheduling malleable and nonmalleable parallel tasks. *SODA*, 167–176, 1994.
6. S. K. Sahni. Algorithms for scheduling independent tasks. *JACM*, 116–127, 1976.
7. J. Turek, J. Wolf and P. Yu, Approximate algorithms for scheduling parallelizable tasks. *SPAA*, 323–332, 1992.
8. J. Turek, W. Ludwig, J. L. Wolf, L. Fleischer, P. Tiwari, J. Glasgow, U. Schwiegelshohn and P. S. Yu, "Scheduling parallelizable tasks to minimize average response time," *SPAA*, 200–209, 1994.
9. G. Woeginger. On-line scheduling of jobs with fixed start and end times. *Theoretical Computer Science*, 130:5–16, 1994.

Orders of Gauss Periods in Finite Fields

Joachim von zur Gathen and Igor Shparlinski

FB Mathematik-Informatik, Universität-GH Paderborn, Germany,
gathen@uni-paderborn.de
and
School of MPCE, Macquarie University, Australia,
igor@mpce.mq.edu.au

Abstract. We show that Gauss periods of special type give an explicit polynomial-time construction of elements of exponentially large multiplicative order in some finite fields. It can be considered as a step towards solving the celebrated problem of finding primitive roots in finite fields in polynomial time.

1. One of the most important unsolved problems in the theory of finite fields is designing a fast algorithm to construct primitive roots in a finite field $\mathbb{F}_q$ of q elements. All known algorithms for this problem work in two stages:

1. Find a "small" set $\mathcal{M} \subseteq \mathbb{F}_q$ guaranteed to contain a primitive root of $\mathbb{F}_q$.
2. Test all elements of $\mathcal{M}$ for primitivity.

In many cases, we have quite good algorithms for the first stage, especially if one employs the Extended Riemann Hypothesis (see Chapter 3 of [12]). Unfortunately, at the current state of the art, the second stage requires the integer factorization of $q - 1$ and is not known to be doable in polynomial time. It is demonstrated in [13] that we can find a primitive root of $\mathbb{F}_q$ in time $O(q^{1/4+\varepsilon})$ for any $\varepsilon > 0$. Even the Extended Riemann Hypothesis (ERH) cannot help too much here.

On the other hand, for many applications, instead of a primitive root just an element of high multiplicative order is sufficient. Such applications include but are not limited to cryptography, coding theory, pseudo random number generation and combinatorial designs. As a specific example we point out the sparse polynomials interpolation algorithms [1, 15] where instead of a primitive root just an element of large order can be used (after some simple adjustments of the other parameters). Also, it is often enough to solve such a problem for some sufficiently "dense" sequence of fields, rather than for all fields.

In this paper we design, under these two relaxations of the original problem, fast deterministic algorithms.

The work was motivated by and can be considered as a continuation of [2] whose key tool are special Gauss periods over finite fields which are shown to generate *normal bases*. The experimental results in that such periods often produce primitive roots and thus generate *primitive normal bases*, or at least have high multiplicative order.

Here we concentrate on Gauss periods of type $(n, 2)$ which are defined as follows. Put $r = 2n + 1$. For a prime power q, let β be a primitive rth root of unity over $\mathbb{F}_q$. Then

$$\alpha = \beta + \beta^{-1} \in \mathbb{F}_{q^{\varphi(r)}},$$

where $\varphi(r)$ is the Euler function, is called the Gauss period of type $(n, 2)$ over $\mathbb{F}_q$.

We note that for a given n, to find the characteristic polynomial of α over $\mathbb{F}_q$, first of all one should find the characteristic polynomial of β (which, essentially is factorizing $X^r - 1$ over $\mathbb{F}_q$), the rest is can be done with the help of linear algebra only (thus in polynomial time). While a deterministic polynomial time algorithm to factorize polynomials over finite fields is not known yet, still there are many very efficient (probabilistic and deterministic, under the ERH and unconditional etc.) algorithms. We just mention that $X^n - 1$ can be completely factorized over $\mathbb{F}_q$ with the following number of arithmetic operations over $\mathbb{F}_q$:

- $(n \log q)^{O(1)}$ probabilistically,
- $n^{O(1)} q^{1/2} \log q$ deterministically,
- $(n \log q)^{O(1)}$ deterministically under the ERH.

More precise forms of these assertions and further details can be found in [12], Section 1.1. It is also useful to recall that if q is a primitive root modulo a prime r then the characteristic polynomial of β is known, it is $X^{r-1} + \ldots + X + 1$. The last remark is related to the construction of Theorem 1 below.

In this paper we do not estimate the complexity of constructing α (which as we mentioned above can be easily done) but rather concentrate on the complexity of finding n for which the corresponding α is of large period.

Gauss periods of type (n, k) for $k > 2$ can be defined similarly and are of great interest as well, but unfortunately at the moment we cannot give any lower bounds on their multiplicative order. It remains an interesting open problem.

2. Let A be the set of all integers $a \in \mathbb{N}$ for which Artin's conjecture holds in the following form:

$$\pi_a(x) \geq x/C(a) \log^2 x, \qquad x \geq x_0(a) \tag{1}$$

where $\pi_a(x)$ is the number of primes up to x for which a is a primitive root. It is known that A contains the odd powers of all but at most two prime numbers [3], and of all prime numbers under the ERH [4]; many other relevant results can be found in [8]. For any $a \in \mathbb{N}$, clearly $a^2 \notin A$.

Theorem 1. *For any prime power* $q = p^k \in A$ *and any sufficiently large* N *there is an integer* n *with* $N \leq n \leq M$ *where* $M = 3C(q)N \log N$ *such that the Gauss period* $\alpha \in \mathbb{F}_{q^n}$ *of type* $(n, 2)$ *over* $\mathbb{F}_q$ *generates a normal basis of* $\mathbb{F}_{q^n}$ *over* $\mathbb{F}_q$ *and has multiplicative order at least*

$$2^{(2n)^{1/2} - 2}.$$

For any $\varepsilon > 0$, *such* n *can found with the following number of bit operations:*

- $O\big(C(q)\exp\big[(1+\varepsilon)\log^{1/2}M\log^{1/2}\log M\big]\big)$ *probabilistically,*
- $O(M^{5/4+\varepsilon})$ *deterministically,*
- $O(M^{6/5+\varepsilon})$ *deterministically under the ERH.*

Proof. First of all we estimate the number of primes r in the interval $2N+1 \le r \le 2M+1$ for which q is a primitive root.

More precisely, we show that for N large enough

$$\pi_a(2M+1) > \pi(2N) + \frac{M}{C(q)\log^2(2M)} \qquad (2)$$

Indeed, assume that N is such that

$$M \ge x_0(q)$$

and

$$3\big(\log(2N) - 3/2\big)\log N > 2\log^2\big(6C(q)N\log N\big).$$

Then,

$$0.5\pi_a(2M+1) \ge \frac{M}{C(q)\log^2(2M)} \ge \frac{3N\log N}{\log^2\big(6C(q)N\log N\big)} > \frac{2N}{\log(2N) - 3/2}.$$

We have,

$$\frac{2N}{\log(2N) - 3/2} \ge \pi(2N)$$

by [9], (3.4). Therefore $\pi_a(2M+1) - \pi(2N) \ge 0.5\pi_a(2M+1)$ and (2) follows.

It follows from (2) that $\pi_a(2M+1) > \pi(2N)$, hence there exists a prime number r with $2N+1 \le r \le 2M+1$ and such that q is a primitive root modulo r.

In order to find such r we use the following procedure.

For a given r, $2N+1 \le r \le 2M+1$ we first of all test its primality. Then we factorize $r-1$ and check if q is a primitive root modulo r by testing if

$$q^{(r-1)/l} \not\equiv 1 \pmod{r}$$

for each prime divisor l of $r-1$. The latter can be done in polynomial time. Primality testing and finding the integer factorization of $r-1$ can be both done with the following number of bit operations:

- $O\big(\exp\big[(1+\varepsilon)\log^{1/2}M\log^{1/2}\log M\big]\big)$ *probabilistically* [7],
- $O(M^{1/4+\varepsilon})$ *deterministically* [11],
- $O(M^{1/5+\varepsilon})$ *deterministically under the ERH* [10].

Also, for the probabilistic algorithm we can select r at random in the interval $[2N+1, 2M]+1$ with the probability of success at least

$$\frac{\pi_a(2M+1) - \pi_a(2N)}{2M - 2N} \ge \frac{\pi_a(2M+1) - \pi(2N)}{2M} \ge \frac{1}{2C(q)\log^2 M},$$

by (2). For the deterministic algorithm we test all numbers $r \in [2N+1, 2M+1]$.

Now, let r be a prime number with $2N + 1 \leq r \leq 2M + 1$ and such that q is a primitive root modulo r, and $n = (r-1)/2$. Then $\varphi(r) = 2n$.

Let $\beta \in \mathbb{F}_{q^{2n}}$ be a primitive rth root of unity, and $\alpha = \beta + \beta^{-1} \in \mathbb{F}_{q^n}$ the corresponding Gauss period of type $(n, 2)$. It generates a normal basis for $\mathbb{F}_{q^n}$ over $\mathbb{F}_q$ [2].

Let

$$h = \lfloor \sqrt{r} \rfloor - 1, \quad H = \{1, ..., h\} \subseteq \mathbb{F}_r, \quad S = \{\operatorname{ind} a : a \in H\} \subseteq \mathbb{Z}_{r-1},$$

where we identify $\mathbb{F}_r$ with $\{0, \ldots, r-1\}$, and $\operatorname{ind} a$ is the index (or discrete logarithm) of $a \in \mathbb{F}_r$ in base q. Now let $U, U' \subseteq S$ be two different subsets of S, and

$$u = \sum_{s \in U} q^s, \quad u' = \sum_{s \in U'} q^s.$$

We show that $\alpha^u \neq \alpha^{u'}$. This implies that we have at least $2^{|S|} = 2^h$ distinct powers of α, and thus the order of α is at least

$$2^h \geq 2^{(2n)^{1/2} - 2}.$$

We may suppose that $U \cap U' = \emptyset$. Assume that $\alpha^u = \alpha^{u'}$. Then

$$0 = \alpha^u - \alpha^{u'} = \prod_{s \in U} (\beta + \beta^{-1})^{q^s} - \prod_{s \in U'} (\beta + \beta^{-1})^{q^s}$$

$$= \beta^{-u} \prod_{s \in U} (\beta^{2q^s} + 1) - \beta^{-u'} \prod_{s \in U'} (\beta^{2q^s} + 1).$$

Since β is an rth root of unity, we may reduce the exponents modulo r, and with

$$E = \{q^s \bmod r : s \in U\}, \quad E' = \{q^s \bmod r : s \in U'\} \subseteq H,$$

$$e = \sum_{t \in E} t, \quad e' = \sum_{t \in E'} t \in \mathbb{N},$$

we have $E \cap E' = \emptyset$ and

$$0 = \beta^{-e} \prod_{t \in E} (\beta^{2t} + 1) - \beta^{-e'} \prod_{t \in E'} (\beta^{2t} + 1).$$

We may assume that $e' \geq e$, and let

$$f = x^{e'-e} \prod_{t \in E} (x^{2t} + 1) - \prod_{t \in E'} (x^{2t} + 1) \in \mathbb{F}_q[x].$$

Then $f(\beta) = 0$, and

$$\deg f \leq 2e' \leq 2 \sum_{i \in H} i = h(h+1) \leq r - \sqrt{r} < r - 1.$$

Since q is primitive modulo r, β has degree $r - 1$ over $\mathbb{F}_q$, and therefore the polynomial f is zero. If $e' > e$, then $f(0) = -1$. Thus $e' = e$. But then the monomial x^{2t} occurs in f with nonzero coefficient, where $t = \min(E \cup E')$. This contradiction proves that $\alpha^u \neq \alpha^{u'}$. $\qquad \square$

We note that under the ERH, from the asymptotic formula for $\pi_a(x)$ of [4], one can get a slightly better estimate for n, namely apparently one can take $M = cN \log \log \log q$, provided that $N > q^C$ with some absolute constants c and C. Unfortunately C does not seems to be effectively computable and it is not clear how to use this better bound in order to design a faster algorithm. On the other hand, all constants occurred in [3] are effective and thus lead to an effective form of (1) for odd powers of all but at most two prime numbers.

3. In the next theorem we eliminate the condition that q belongs to A and consider a denser sequence of n. The price we pay is losing the property of α being normal and a weaker lower bound.

Theorem 2. *There is an absolute constants $C > 0$ such that for any prime power $q = p^f$, $f \in \mathbb{N}$, and any sufficiently large $N \geq 2$ there are integers m and n with*

$$m = \varphi(2n + 1), \qquad N \leq m \leq N + O(N/\log^C N),$$

and such that the Gauss period $\alpha \in \mathbb{F}_{q^m}$ of type $(n, 2)$ over $\mathbb{F}_q$ has multiplicative order at least

$$2^{c_q m^{1/2} - d_q}$$

where

$$(c_q, d_q) = \begin{cases} (\sqrt{7.5}\, q^{-3}, 9), & \text{if } p = 2 \text{ or } p \geq 7, \\ (\sqrt{17.5} q^{-5}, 25), & \text{if } p = 3, \\ (\sqrt{10.5} q^{-4}, 13), & \text{if } p = 5. \end{cases} \tag{3}$$

For any $\varepsilon > 0$, such n and m can be found within $O(\log^{2+\varepsilon} N)$ bit operations.

Proof. We define

$$(l_1, l_2) = \begin{cases} (3, 5), & \text{if } p = 2 \text{ or } p \geq 7, \\ (5, 7), & \text{if } p = 3, \\ (3, 7), & \text{if } p = 5. \end{cases}$$

and put

$$\psi_q = \frac{l_1 l_2}{\varphi(l_1 l_2)}.$$

We note that if k_1, k_2 are positive integers then

$$l_1^{k_1} l_2^{k_2} = \psi_q \varphi(l_1^{k_1} l_2^{k_2}).$$

Let r_0 be the smallest integer greater than $\psi_q N/l_1 l_2$ of the form $r_0 = l_1^{m_1} l_2^{m_2}$, where m_1, m_2 are nonnegative integers.

Tijdeman's result [14] on the distribution of numbers containing only a fixed set of primes in their factorization implies that

$$\frac{\psi_q N}{l_1 l_2} \leq r_0 \leq \frac{\psi_q N}{l_1 l_2} + O(N/\log^C N)$$

with some absolute constant $C > 0$. Thus if we define $r = r_0 l_1 l_2$, $n = (r-1)/2$ and $m = \varphi(r) = \psi_q^{-1} r$, then $N \leq m \leq N + O(N/\log^C N)$.

To estimate the complexity we put $R = \psi_q N / l_1 l_2$ and note that $r_0 = l_1^{m_1} l_2^{m_2} \leq l_1 R$. Therefore, $0 \leq m_i \leq K_i$ where

$$K_i = \left\lfloor \frac{\log(l_1 R)}{\log l_i} \right\rfloor, \qquad i = 1, 2$$

To find r_0, for each

$$k_1 = 0, \ldots, K_1,$$

by using binary search on the interval $0 \leq k_2 \leq K_2$ we compute k_2 with

$$l_1^{k_1} l_2^{k_2-1} \leq R \leq l_1^{k_1} l_2^{k_2}.$$

For each k_1 it can be done in $O(\log^{1+\varepsilon} N)$ bit operation. From obtained $K_1 = O(\log N)$ numbers we select the smallest one which is r_0.

Let t be the order of q modulo r,

$$h = \lceil r(2t)^{-1/2} - t_0 r/t \rceil, \quad S = \{s: \; 0 \leq s < t \text{ and } 1 \leq (q^s \bmod r) \leq h\}.$$

Let t_0, t_1 and t_2 be the order q modulo $l_1 l_2$, l_1 and l_2, respectively. We define γ_i as the largest power of l_i which divides $q^{t_i} - 1$, $i = 1, 2$. Korobov [5, 6] shows that

$$t \geq t_0 l_1^{-\gamma_1} l_2^{-\gamma_2} r.$$

We have $l_i^{\gamma_i} < q^{t_i}$, $i = 1, 2$, and

$$t_0 \leq t_1 t_2 \leq (l_1 - 1)(l_2 - 2) \tag{4}$$

Therefore,

$$t \geq t_1 t_2 q^{-t_1 - t_2} r.$$

Taking into account that the function $x q^{-x}$ is monotonically decreasing for any $q \geq 3$ and that $t_i \leq l_i - 1$, $i = 1, 2$, we obtain,

$$t \geq (l_1 - 1)(l_2 - 1) q^{-l_1 - l_2 + 2} r \tag{5}$$

for $q \geq 3$. For $q = 2$ one can verify that

$$\min_{\substack{1 \leq t_1 \leq 2, \\ 1 \leq t_2 \leq 4}} t_1 t_2 2^{-t_1 - t_2} = 2^{-3} = (3-1)(5-1)2^{-3-5+2},$$

thus (5) holds for any q. Therefore, $t \geq C_q r$ where

$$C_q = \begin{cases} 8q^{-6}, & \text{if } p = 2 \text{ or } p \geq 7, \\ 24q^{-10}, & \text{if } p = 3, \\ 12q^{-8}, & \text{if } p = 5. \end{cases}$$

Hence $t \geq C_q \psi_q m = 2c_q^2 m$, where $c_q = (C_q \psi_q / 2)^{1/2}$, as in (3).

Also, Korobov [5, 6] demonstrates that

$$\big||S| - th/r\big| \leq t_0.$$

Thus we have, $|S| \leq th/r + t_0 \leq (t/2)^{1/2}$, therefore

$$2h|S| \leq 2\big(r(2t)^{-1/2} - t_0 r/t\big)(t/2)^{-1/2} < r.$$

Let $U, U' \subseteq S$ be two different subsets and

$$u = \sum_{s \in U} q^s, \qquad u' = \sum_{s \in U'} q^s.$$

As before, we have a primitive rth root of unity β over $\mathbb{F}_q$, $\alpha = \beta + \beta^{-1} \in \mathbb{F}_{q^m}$, and as in the proof of Theorem 1 it follows that $\alpha^u \neq \alpha^{u'}$. Therefore, we have at least $2^{|S|}$ distinct powers of α. From

$$|S| \geq th/r - t_0 \geq (t/2)^{1/2} - 2t_0 - t/r \geq (t/2)^{1/2} - 2t_0 - 1$$

and from the inequality (4) we obtain the result. $\qquad\qquad\qquad\qquad\square$

More generally, it can be shown that if $r = 2n + 1$ is relatively prime to q and q generates a group of bounded index in the group of units modulo r, the same construction produces α of order at least

$$\exp(c(q)n^{1/2}/\log n).$$

This is based on another estimate of Korobov

$$|S| = th/r + O(r^{1/2}\log r);$$

thus h can be chosen of order $r^{1/2}\log r$.

Certainly, the constants in Theorem 2 can be improved. We do not do it because we believe that our subexponential lower bound can be essentially improved, perhaps up to $\exp\big(C(q)m\big)$ with some constant $C(q) > 0$ depending on q.

Acknowledgment. Part of this paper was written during a visit by the second author in October 1994 to the University of Paderborn, whose hospitality is gratefully acknowledged. The authors would like to thank Francesco Pappalardi for a useful discussion of the effective computability issues related to Artin's conjecture.

References

1. M. Clausen, A. Dress, J. Grabmeier and M. Karpinski, "On zero testing and interpolation of k-sparse multivariate polynomials over finite field", *Theor. Comp. Sci.*. **84** (1991), 151–164.

2. S. Gao, J. von zur Gathen and D. Panario, "Gauss periods and fast exponentiation in finite fields", *Proceedings LATIN '95, Springer Lecture Notes in Comp. Sci.*, **911** (1995), 311–322.

3. D. R. Heath-Brown, "Artin's conjecture for primitive roots", *Quart. J. Math.*, **37** (1986), 27–38.

4. C. Hooley, "On Artin's conjecture", *J. Reine Angew. Math.*, **225** (1967), 209–220.

5. N. M. Korobov, "Exponential sums with exponential functioms and the distribution of digits in periodic fractions", *Matem. Zametki*, **8** (1970), 641–652 (in Russian).

6. N. M. Korobov, "On the distribution of digits in periodic fractions", *Matem. Sbornik*, **89** (1972), 654–670 (in Russian).

7. H. W. Lenstra and C. Pomerance, "A rigorous time bound for factoring integers", *J. Amer. Math. Soc.*, **5** (1992), 483–516.

8. W. Narkiewicz, *Classical problems in number theory*, Polish Sci. Publ., Warszawa, 1986.

9. J.B. Rosser and L. Schoenfeld, "Approximate functions for some functions of prime numbers ", *Illinois J. Math.* **6** (1962) 64–94.

10. R. J. Schoof, "Quadratic fields and factorization", *Computational Methods in Number Theory*, Amsterdam, 1984, 235–279.

11. D. Shanks, "Class number, a theory of factorization and genera", *Proc. Symp. in Pure Math.*, Amer. Math. Soc., Providence, 1971, 415–420.

12. I. Shparlinski *Computational and algorithmic problems in finite fields*, Kluwer Acad. Publ., Dordrecht, 1992.

13. I. Shparlinski, "On finding primitive roots in finite fields", *Theor. Comp. Sci.* (to appear).

14. R. Tijdeman, "On the maximal distance between integers composed of small primes", *Compos. Math.*, **28** (1974), 159–162.

15. K. Werther, "The complexity of sparse polynomials interpolation over finite fields", *Appl. Algebra in Engin., Commun. and Comp.*, **5** (1994), 91–103.

A Hard Problem That is Almost Always Easy

George Havas and B.S. Majewski

Key Centre for Software Technology, Department of Computer Science, University of Queensland, Queensland 4072, Australia

Abstract. NP-completeness is, in a well-defined sense, a worst case notion. Thus, 3−colorability of a graph, for a randomly generated graph, can be determined in constant expected time even though the general problem is NP-complete. The reason for this is that some hard problems exhibit a structure where only a small (perhaps exponentially small) fraction of all possible instances is intractable, while the remaining large fraction has a polynomial time solution algorithm. We add a new problem to the list of NP-complete problems that are solvable in average polynomial time.

1 Introduction

Some provably intractable tasks, including NP-complete problems, may be solvable in acceptable time in practice. Different possibilities exist. Firstly, the instances that actually occur may always be small, hence even an exponential time solution runs sufficiently fast. This seems to be the case with input-output decomposition, a decomposition technique for acyclic graphs, proved to be NP-complete by Tarjan (1984). Pichai, Sezer & Šiljak (1984) have argued, however, that this NP-completeness result is not relevant to their situation since large instances are not considered in practice.

A stronger result arises when a given difficult problem has a large class of instances that are solvable in polynomial time. For example, Hamiltonian circuit has a polynomial time algorithm due to Karp (1975) and subsequently improved by Angluin & Valiant (1977) that succeeds in $O(n^2 \log n)$ time with probability $1 - O(n^{-c})$ for a constant c. This may yield a method that runs fast in practice, but still does not result in average polynomial time complexity. The reason is as follows. A dynamic programming routine that guarantees finding a Hamiltonian circuit, if one exists, has complexity $O(n^2 2^n)$. If we want to have an algorithm that always answers correctly the question about the existence of a Hamiltonian cycle for a given graph we may start with the $O(n^2 \log n)$ algorithm and, when it fails to provide an answer, call the exponential time dynamic programming routine. As the latter is executed with probability as high as $\Omega(n^{-c})$, this yields a contribution as high as $2^n n^{2-c}$ to the expected running time, still leaving this approach exponential.

Finally, there exists a class of problems which, although NP-complete, have average polynomial time complexity. Wilf (1984) has shown that for graphs where each edge is present with however small but fixed nonzero probability, 3-colorability can be solved in expected constant time. The reason is that such graphs are almost certain to contain a large number of 4-cliques, and the standard backtrack search algorithm can be expected to detect one in constant time.

We consider the extended gcd problem which, for a given vector of positive numbers $a = [a_i]_{i=1}^n$, asks for an integer vector $x = [x_i]_{i=1}^n$, such that $\sum_{i=1}^n x_i a_i = \gcd(a_1, a_2, \ldots, a_n)$. (Implementations, applications and the importance of the problem are discussed by Havas & Majewski (1995).) Solving this problem for any vector x is easy; the challenge arises when we want to minimize some measure of x. In this paper we concentrate on just one measure, the L_0 metric, which is equal to the number of nonzero elements in x. By finding an x optimal with respect to this metric we obtain a sparsest possible solution to the extended gcd problem.

Recent results of Majewski & Havas (1994, Theorem 1) show that finding a solution to the extended gcd problem which is optimal with respect to the L_0 metric is NP-complete. The proof relies on a polynomial time transformation from MINIMUM COVER (see Garey & Johnson (1979, problem SP5)). This contradicts practical experience in that the gcd's of large sets of randomish numbers can be found with the use of 2 or 3 numbers. Indeed, as indicated by a result of E. Cesàro (see Theorem 3), we have a substantial chance that just two of these numbers will have gcd equal to 1. Thus on the one hand we expect to be able to find a short x by simply inspecting a few pairs of numbers, while on the other hand we have a proof that sometimes it may take exponential time to find a short solution x. These seemingly contradictory statements can be easily explained. The majority of extended gcd problems can be solved in polynomial time. However, the polynomial transformation from MINIMUM COVER targets only a small subset of difficult instances of the problem, as illustrated in Fig. 1.

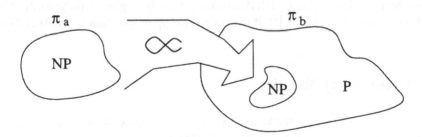

Fig. 1. NP-completeness of π_b by polynomial time transformation from a known NP-complete problem π_a.

We prove that, under the assumption of a uniform and random distribution for the input numbers, the optimization version of the extended gcd problem for the L_0 metric is polynomial in n and $\log(\max_i\{a_i\})$.

Before we do so we clarify some definitions, following Johnson (1984). For a given decision problem D we denote the set of instances of size n by I and the probability that a random such instance is x by $\mu(x)$, and we let $T(x)$ be the running time of the algorithm under consideration for input x. In one view, an algorithm is considered to have average polynomial running time if

$$E(T) = \sum_{x \in I} \mu(x) T(x) = O\left(n^{O(1)}\right).$$

As pointed out by Johnson, such a definition has a serious drawback: unlike the standard notion of worst case polynomial time, it is not machine independent. Instead the following approach should be used. An algorithm has average polynomial time complexity if for some $c > 0$

$$E^*(T) = \sum_{x \in I} \mu(x) \sqrt[c]{T(x)} = O(n).$$

A randomized decision problem D together with a probability distribution μ on problem instances is considered easy if it has an algorithm with average polynomial time complexity for distribution μ. Then AP (Average P) is the class of all easy randomized decision problems.

It is worth mentioning that Levin (1986) considered the question whether all decision problems in NP are easy, i.e., if they belong to AP. The answer seems to be no. For a number of problems it has been shown that whole classes of methods almost always run in exponential time: INDEPENDENT SET (Chvátal, 1977); GRAPH k–COLORABILITY (McDiarmid, 1979); and KNAPSACK (Chvátal, 1980), to mention just a few. Thus, these problems seem to be hard on average. To capture this notion Levin introduced a class called "random NP-complete", subsequently renamed DNP (Distributional NP), and proved that RANDOM TILING belongs to this class. This means that RANDOM TILING, and any other problem belonging to DNP, can be solved in expected polynomial time if and only if all problems in DNP can. A major open question is whether DNP $\subseteq$ AP.

2 Complexity Results

In this section we use the RAM model of computation with uniform cost measure as defined by Aho, Hopcroft & Ullman (1974). In particular we assume that basic arithmetic operations (such as addition, multiplication, modulo, and accessing an array) take constant time. Carrying out the analysis for other computational models, such as Turing Machines or Markov algorithms, although more laborious, leads to the same results, with different constants.

Consider the time complexity of a simple exhaustive search through all possible k–tuples of numbers, $2 \leq k \leq n$. We assume that $a_i \neq a_j$ for $i \neq j$. (If this is not the case, all duplicates can be easily eliminated in linear time, once the a_i's are sorted.) Furthermore, we stipulate that $\gcd(a_1, a_2, \ldots, a_n) = 1$. (Again, if this is not the case we can guarantee this condition in $O(n + \log(\max_i\{a_i\}))$ time by first computing the gcd then dividing through by it.)

The algorithm, presented in Fig. 3, calls the procedure *kSearch* (Fig. 2) for increasing k's, starting with $k = 2$. As soon as a subset with the gcd 1 is discovered, the search is abandoned and the subset is returned. This allows the main procedure to provide the user with a proof that the found subset indeed gives a solution. In the implementation presented in Fig. 3 the subset is discarded and its cardinality, representing the size of the optimum with respect to the L_0 metric solution, is returned. In the following paragraphs we show that the expected complexity of this superficially naïve approach is polynomial in n and $\log(\max_i\{a_i\})$, even though the worst case complexity is clearly exponential in those parameters.

```
procedure kSearch(k)
    p := k;
    for i := 1, 2 .. k do
        s_i := i;
    end for;
    loop
        if gcd(a_{s_1}, a_{s_2}, ..., a_{s_k}) = 1 then
            return {a_{s_1}, a_{s_2}, ..., a_{s_k}};
        end if;
        if s_k = n
            then p := p - 1;
            else  p := k;
        end if;
        exit if p < 1;
        for i := k, k - 1 .. p do
            s_i := s_p + i - p + 1;
        end for;
    end loop;
    return ∅;
end kSearch;
```

Fig. 2. Lexicographic search through all k-subsets

Lemma 1. *The time complexity of a single execution of the procedure kSearch is* $O\big(\binom{n+1}{k}(k + \log a_n)\big)$.

Proof. Consider initially the number of times elements of the index array s are modified. Initially, the algorithm executes k assignments to s_i, $1 \le i \le k$. In the inner **for** loop s_k is modified for each of $\binom{n}{k}$ combinations, s_{k-1} is modified once for each combination with $s_k = n$, ie, $\binom{n-1}{k-1}$ times, s_{k-2} is modified once for each combination with $s_{k-1} = n - 1$ and $s_k = n$, ie, $\binom{n-2}{k-2}$ times, and so on.

Thus the total number of times we access the array s is

$$k + \sum_{j=0}^{k} \binom{n-j}{k-j} = \binom{n+1}{k} + k.$$

For each of these combinations the gcd of the elements $a_{s_1}, a_{s_2}, \ldots, a_{s_k}$ has to be computed. By Theorem 1 of Bradley (1970) it takes at most $O(k+\log(\min_i\{a_i\}))$ constant time steps. If we assume that the sequence is sorted, as the k-subsets are generated in lexicographic order, the last expression takes the form $O(k + \log(a_{n-k+1})) = O(k + \log(a_n))$. In fact we can make the algorithm somewhat more clever, computing the gcd's at a minimum possible cost by maintaining another array, g. However, as efficiency is not the main issue and in order to simplify the analysis, we split the computations into two separate operations. Thus the time complexity can be estimated as $O(((\binom{n+1}{k}) + k) + \binom{n}{k}(k + \log a_n)) = O((\binom{n+1}{k})(k + \log a_n))$. $\qquad\square$

Observe that this estimate is suitable for small k. For k approaching n the estimate is unnecessarily high by a factor of n. For $k > n/2$ we can use a similar algorithm to generate complements of $(n-k)$-subsets, thus maintaining average constant time per subset.


```
        Sort the sequence a: a₁ ≤ a₂ ≤ ⋯ ≤ aₙ;
        if a₁ = 1 then
            return 1;
        end if;
Step 1:
        if kSearch(2) ≠ ∅ then
            return 2;
        end if;
Step 2:
        for k := 3 .. n − 1 do
            if kSearch(k) ≠ ∅ then
                return k;
            end if;
        end for;
        return n;
```

Fig. 3. Optimum zero metric solver

Lemma 2. *Let $a = \langle a_1, a_2, \ldots, a_n \rangle$ be a sequence of random integers. The probability that there exist a pair a_i, a_j with $\gcd(a_i, a_j) = 1$ is $1 - (1 - 6/\pi^2)^{\binom{n}{2}}$.*

Proof. To prove the lemma we use the following theorem, due to E. Cesàro, 1881 (Knuth, 1973, p. 301):

Theorem 3. *For two integers chosen at random the probability that their gcd is 1 is $6/\pi^2 = 1/\zeta(2)$.*

From Theorem 3, for each pair (a_i, a_j) the probability of their gcd being 1 is $1/\zeta(2)$. There are $\binom{n}{2}$ pairs among n numbers, thus the probability that one or more pairs have gcd 1 is equal to the complement of the probability that none of the pairs has gcd 1. The latter probability is $(1 - 1/\zeta(2))^{\binom{n}{2}}$, which yields the desired result. $\qquad\square$

Theorem 4. *Under the assumption of a uniform and random distribution for the input numbers, the algorithm shown in Fig. 3 belongs to AP.*

Proof. The algorithm starts by sorting the sequence and testing if the smallest number in it is equal to 1, at the cost of $O(n \log n)$ constant time operations. Step 1 of the algorithm is executed with probability $1 - \Pr(a_1 = 1)$ which may be assumed to be 1 if numbers are selected from a sufficiently large universe. Step 2, however, is activated only if Step 1 fails. By Lemma 2 this occurs with probability $(1 - 1/\zeta(2))^{\binom{n}{2}}$, which is less than $\left(\frac{2}{5}\right)^{n^2/3}$ for $n \geq 3$. Notice that we are interested only in cases with $n \geq 3$, thus for all practical purposes we may assume that the inequality always holds. The complexity of Step 1 is $O\left(\binom{n+1}{2}(2 + \log(a_n))\right) = O(n^2 \log(a_n))$. The complexity of Step 2 is

$$\sum_{k=3}^{n} \binom{n+1}{k} (k + \log(a_n)) = (n+1)(2^n - 2(n+1))$$

$$+ \left(2^{n+1} - \frac{n(n-3)}{2} - 3\right) \log(a_n)$$

$$< 2^{n+1} (n + \log(a_n)).$$

From this it follows that the expected time complexity of the algorithm of Fig. 3, for $n \geq 3$, is bounded from above by

$$E(T) \leq n \log n + \left(1 - \left(1 - \frac{1}{\zeta(2)}\right)^{\binom{n}{2}}\right) O(n^2 \log a_n)$$

$$+ \left(1 - \frac{1}{\zeta(2)}\right)^{\binom{n}{2}} O(2^{n+1} (n + \log(a_n)))$$

$$< O(n^2 \log a_n) + 2(n + \log(a_n)) \left(\frac{2}{5}\right)^{n^2/3} 8^{n/3}.$$

The last expression, again for $n \geq 3$, is bounded from above by

$$O\left(n^2 \log(a_n)\right) + 2(n + \log(a_n)) \left(\frac{4}{5}\right)^n.$$

The above construction proves that $E(T) = O((n \log(a_n))^{O(1)})$ which, as noted in the previous section, is a computational model dependent proof of average time complexity. For the assumed model of computation, taking $c = 2$ yields the desired result of $E^*(T) = O(n \log(a_n))$. As noted at the beginning of this section, for other models of computation, we obtain the same result by increasing c by a constant. □

The above analysis does not take into account the fact that the algorithm, after failing to detect a pair with the gcd 1, may stop in polynomial time for a k−tuple, with $k = O(1)$. In fact, the probability that the algorithm will require scanning $(k+1)$−tuples is exponentially smaller that the probability that it will fail for j−tuples, with $1 \leq j \leq k-1$ and succeed for k−tuples.

It is also interesting to note that the expected number of pairs that need to be inspected before we find the first pair with gcd 1 is $\zeta(2) < 2$. The algorithm terminates with probability $1 - \frac{1}{n}$ after inspecting $\ln(n)/\left(2\ln(\pi) - \ln(\pi^2 - 6)\right) \approx 1.068\ln(n)$ pairs of numbers. Thus the expected running time of the algorithm is $O(n + \log(\max\{a_i\}))$, and it terminates almost always in $O(n + \ln(n)\log(\max\{a_i\}))$ time. (These results apply to a version that does not sort the input numbers. The term n comes from a verification process that looks for $a_i = 1$.)

3 Conclusion

Under the assumption of a uniform and random distribution for the input numbers, we have shown that the extended gcd optimization problem with respect to the L_0 metric belongs to AP. Moreover, the algorithm is expected to terminate very rapidly. It would be interesting to consider the behavior of the method under different distributions, more or less commonly encountered in practice.

Another question worth investigating would be to ask if a similar result holds for any other norm. For example, as shown by Havas & Majewski (1995), applying the Fincke-Pohst algorithm to the solution returned by the LLL gcd method is certain to obtain an optimal solution with respect to the Euclidean norm, and in practice it seems to terminate reasonably quickly.

References

Aho, A.V. and Hopcroft, J.E. and Ullman, J.D. (1974), *The Design and Analysis of Computer Algorithms*. Addison-Wesley Pub. Co., Reading, Mass.

Angluin, D. and Valiant, L.G. (1977), "Fast Probabilistic Algorithms for Hamiltonian Circuits and Matchings", *Proceedings 9th Annual ACM Symposium on Theory of Computing*, pp. 30–41.

Bradley, G.H. (1970), "Algorithm and Bound for the Greatest Common Divisor of n Integers", *Communications of the ACM*, **13**, 433–436.

Chvátal, V. (1977), "Determining the Stability Number of a Graph", *SIAM J. on Computing*, **6**, 643–662.

Chvátal, V. (1980), "Hard Knapsack Problems", *Operations Research*, **28**, 1402–1411.

Garey, M.R. and Johnson, D.S. (1979), *Computers and Intractability: A Guide to the Theory of NP-completeness*. W.H. Freeman, San Francisco.

Havas, G. and Majewski, B.S (1995), "Extended Gcd Calculation", Research Report **TR0325**, The University of Queensland, Brisbane.

Johnson, D.S. (1984), "The NP-Completeness Column: An Ongoing Guide", *Journal of Algorithms*, **5**, 284–299.

Karp, R.M. (1975), "On the Complexity of Combinatorial Problems", *Networks*, **5**, 45–68.

Knuth, D.E. (1973), *The Art of Computer Programming, Vol. 2: Seminumerical Algorithms* (2nd edition). Addison-Wesley, Reading, Mass.

Levin, L. (1986), "Average Case Complete Problems", *SIAM J. on Computing*, **15**, 285–286.

Majewski, B.S. and Havas, G. (1994), "The Complexity of Greatest Common Divisor Computations": *Algorithmic Number Theory* pp. 184–193. Springer-Verlag.

McDiarmid, C.J.H. (1979), "Determining the Chromatic Number of a Graph", *SIAM J. on Computing*, **8**, 1–14.

Pichai, V. and Sezer, M.E. and Šiljak, D.D. (1984), "Reply to "Input-output Decomposition of Dynamic Systems is NP-complete"", *IEEE Transactions on Automatic Control*, **29**(9), 864.

Tarjan, R.E. (1984), "Input-output Decomposition of Dynamic Systems is NP-complete", *IEEE Transactions on Automatic Control*, **29**(9), 863–864.

Wilf, H.S. (1984), "An $O(1)$ Expected Time Graph Coloring Algorithm", *Information Processing Letters*, **18**, 119–122.

Computing the Tutte Polynomial of a Graph of Moderate Size

Kyoko Sekine[1], Hiroshi Imai[1] and Seiichiro Tani[2]

[1] Department of Information Science, University of Tokyo
[2] Nippon Telegraph and Telephone Corporation

Abstract. The problem of computing the Tutte polynomial of a graph is #P-hard in general, and any known algorithm takes exponential time at least. This paper presents a new algorithm by exploiting a fact that many 2-isomorphic minors appear in the process of computation. The complexity of the algorithm is analyzed in terms of Bell numbers and Catalan numbers. This algorithm enables us to compute practically the Tutte polynomial of any graph with at most 14 vertices and 91 edges, and that of a planar graph such as 12×12 lattice graph with 144 vertices and 264 edges.

1 Introduction

The problem of computing the Tutte polynomial of a graph has interested many researchers in recent years. This is because many counting problems in graph theory and related areas are reduced to this computation. For example, the Tutte polynomial of a graph at some specific points gives the number of spanning trees, the number of forests, the number of acyclic orientations, etc.[9] Moreover the following invariants are known to be some special cases of the Tutte polynomial of a graph and a matroid [9].

- the chromatic polynomial and flow polynomial of a graph
- the partition function of a Q-state Potts model
- the Jones polynomial of an alternating link
- the weight enumerator of a linear code over $GF(q)$

However, in general the computation problem of the Tutte polynomial is #P-hard except in some special cases [9]. Recently, a polynomial-time randomized approximation scheme was proposed for dense graphs [2] and densely connected graphs [4]. However, as summarized in section 2, any known algorithm which can exactly compute the Tutte polynomial requires at least time proportional to the number of spanning trees of a given graph. The number of spanning trees is exponential with respect to the numbers of vertices and edges in most cases, and such algorithms cannot solve the problem for moderate-size graphs.

In this paper, we present a new algorithm by exploiting a fact that many 2-isomorphic minors appear in computing the Tutte polynomial of a graph by the so-called edge deletion/contraction formula. The Tutte polynomials of 2-isomorphic graphs are identical, so that it is redundant to compute the Tutte polynomials of them each. Testing the 2-isomorphism between two graphs is hard in general. However focusing on the correspondence between edges which have the same order induced by the edge ordering of the original graph G, we can find 2-isomorphic minors efficiently. Furthermore, this 2-isomorphism test with some additional restriction reduces to the problem of testing the equivalence of partitions on the vertex subset. This reduction also enables us to analyze the time and space complexity of this method by using the Bell number and the Catalan number which are much less than the number of spanning trees.

By this algorithm, we can compute the Tutte polynomial of any graph with at most 14 vertices and $\binom{14}{2} = 91$ edges, and that of a planar graph such as 12×12 lattice graph with 144 vertices and 264 edges. In practice, it enables us to compute the BDD of spanning trees for these graphs in about an hour by a standard workstation. Furthermore, the Jones polynomial of an alternating link of moderate size can be computed by our algorithm.

2 Tutte Polynomial of a Graph

In this section we first describe necessary definitions and discuss the time complexity of computing the Tutte polynomial by existing methods.

Let $G = (V, E)$ be a simple connected undirected graph with a vertex set V and an edge set E. For an edge e in E, we denote by $G \backslash e$ the graph obtained by deleting e from G, and by G/e the graph obtained by contracting e from G. For $E' = E - (E^c \cup E^d)$ with $E^c \cap E^d = \emptyset$, the graph obtained from G by contracting edges in E^c and deleting edges in E^d is called a *minor* of G with an edge set E'. We denote by $V(E')$ the set of vertices of edges in E' and the graph $G' = (V(E'), E')$ is called a subgraph of G on E'.

The Tutte polynomial of a graph G is a two-variable polynomial defined by

$$T(G; x, y) = \sum_{E' \subseteq E} (x - 1)^{\rho(E) - \rho(E')} (y - 1)^{|E'| - \rho(E')}.$$

where $\rho: 2^E \to \mathbf{Z}$ is the rank function of a graph G. That is, $\rho(E')$ is the rank of a subgraph G': the number of vertices ,$|V(E')|$, minus the number of connected components of the subgraph.

For an edge e in E, the following recursive formula holds.

$$T(G; x, y) = \begin{cases} xT(G/e; x, y) & e : \text{coloop} \\ yT(G \backslash e; x, y) & e : \text{loop} \\ T(G \backslash e; x, y) + T(G/e; x, y) & \text{otherwise} \end{cases}$$

Here, a loop is an edge connecting the same vertex, and a coloop is an edge whose removal decreases the rank of the graph by 1. By definition, the Tutte polynomial of a loop is y, that of a coloop is x and that of an empty graph is 1.

Two graphs G_1 and G_2 are *2-isomorphic* if there is a bijection $\psi : E_1 \to E_2$ such that the edge set $E_1' \subseteq E_1$ is a spanning tree in G_1 if and only if $\psi(E_1')$ is a spanning tree in G_2. Therefore if G_1 and G_2 are isomorphic they are also 2-isomorphic. Below, we will focus on consider the 2-isomorphism between minors of a graph. This is because 2-isomorphic graphs have the same Tutte polynomial [8].

The Tutte polynomial can be computed by the above definition. However, the computation merely based on the definition requires enumerating all the subsets of E. On the other hand, the original definition by Tutte [8] is in the following form,

$$T(G; x, y) = \sum_{\text{spanning trees } T} x^{r(T)} y^{s(T)}$$

where $r(T)$ and $s(T)$ are the number of edges satisfying some properties for the spanning tree T and, given a spanning tree, these two numbers can be computed in $O(|V||E|)$ time. In most graphs, the number of spanning trees is exponential in $|V|$ and $|E|$ which dominates the time complexity. Given a graph, its spanning trees can be enumerated very efficiently ([5], etc.), and we have the following.

Theorem 1. *Based on the original definition by Tutte, the Tutte polynomial of a graph $G = (V, E)$ can be computed in $O(|V||E|T(G; 1, 1))$ time.*

By definition, $T(G; 1, 1)$ counts the number of spanning trees of G. For a complete graph K_n of n vertices, $T(K_n; 1, 1)$ is equal to n^{n-2}. In addition, $|V||E|$ may be reduced to $|E|$.

By using the recursive formula, we can also compute the Tutte polynomial. We may order edges arbitrarily and this process forms an expansion tree (Fig.1). The root corresponds to the graph G, and each parent has at most two children. For each path from a root to a leaf in the expansion tree, a set of contracted edges corresponds to a spanning tree of G one-to-one, i.e., the number of leaves equals the number of spanning trees. The depth of the expansion tree is $|E|$. Using this expansion tree we obtain the following bound which is better than any other known bound [10].

Theorem 2. *Using the recursive formula, the Tutte polynomial of a graph $G = (V, E)$ can be computed in $O(|E|T(G; 1, 1))$ time.*

3   Sharing Computations for 2-Isomorphic Minors

In this section, we will show that there exists an algorithm which does not take time proportional to the number of spanning trees.

Suppose we apply the recursive formula in the order of $e_1, e_2, \ldots, e_m$ ($m = |E|$) in a top-down fashion as in the expansion tree described in the previous section. Nodes in the i-th level in the expansion tree correspond to minors of G on $\{e_{i+1}, e_{i+2}, \ldots, e_m\}$ (the 0-th level is a root). Since the Tutte polynomial is the same for 2-isomorphic graphs, we may represent 2-isomorphic minors among them by one of these members. However, testing the 2-isomorphism of arbitrary graphs is hard in general, and finding all 2-isomorphic minors may be difficult.

The 2-isomorphism between two graphs whose edges have an identity map can be determined in linear time. For this reason, we may restrict ourselves just

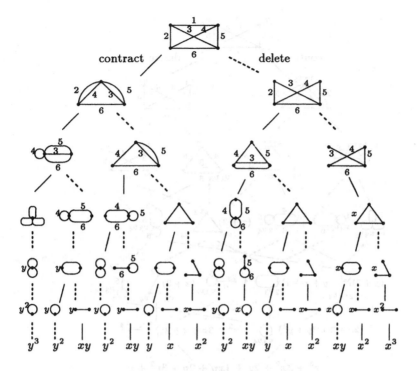

Fig. 1. Expansion tree for K_4

to find 2-isomorphic minors whose corresponding edges have the same order in the original graph G. By this modification, the expansion tree becomes an acyclic graph (edges are directed from a parent to a child). See an example of the BDD for K_4 in Fig.2. This acyclic graph has a single source (the original graph G) and the m-th level may be regarded as a single sink. We call this acyclic graph the BDD of all spanning trees of a graph G with respect to the edge ordering $e_1, e_2, \ldots, e_m$. This is because this acyclic graph has strong connection with the BDD (Binary Decision Diagram) representing all the spanning trees of a graph. BDD has been demonstrated as a tool for efficient Boolean function manipulation which is required in VLSI design, etc. [1, 3]. The size of this BDD is defined to be the number of its nodes. The width of this BDD is defined to be the maximum among the numbers of nodes at each level.

Rigorously, the BDD of all spanning trees of G is defined via the following construction algorithm, where S_i is the set of nodes in the i-th level and in this algorithm we identify nodes of S_i with their corresponding minors.

$S_0 := \{G\}$;

for $i := 1$ **to** m **do**

  begin

    $S_i := \emptyset$;

    for each minor $\widetilde{G}$ in S_{i-1} **do**

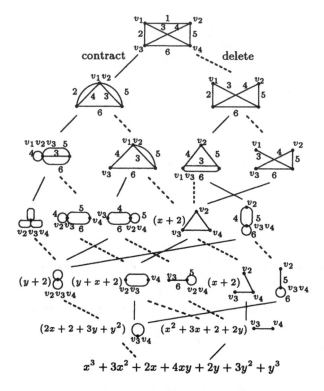

Fig. 2. BDD of spanning trees for K_4

begin
    if e_i is a loop in $\widetilde{G}$ **then** $M(\widetilde{G}) := \{\widetilde{G}\backslash e_i\}$
    else if e_i is a coloop in $\widetilde{G}$ **then** $M(\widetilde{G}) := \{\widetilde{G}/e_i\}$
    else (comment: e_i is neither a loop nor a coloop) $M(\widetilde{G}) := \{\widetilde{G}\backslash e_i, \widetilde{G}/e_i\}$;
    for each minor $\widetilde{G}_{e_i}$ in $M(\widetilde{G})$ **do**
    begin
        check if there is a 2-isomorphic graph with the same edge ordering
        in S_i;
        if there is such a 2-isomorphic graph $\hat{G}$ in S_i then construct an
        edge from the node representing $\widetilde{G}$ to the node representing $\hat{G}$;
        otherwise, add $\widetilde{G}_{e_i}$ to S_i and construct an edge from the node
        representing $\widetilde{G}$ to the node of $\widetilde{G}_{e_i}$;
    end
    end
end;

From the BDD of all spanning trees of G, the Tutte polynomial can be
computed as follows, where a two variable polynomial $t(v; x, y)$ is associated
with each node v in the BDD.

```
t(source; x, y) := 1;
for i := 1 to m do
  begin
    for all nodes u in S_i do t(u; x, y) := 0;
    for each node v in S_{i-1} do
      begin
        if v has two children u, w then
          begin
            t(u; x, y) := t(u; x, y) + t(v; x, y);
            t(w; x, y) := t(w; x, y) + t(v; x, y)
          end
        else (comment: v has only one child u)
          if e_i is a loop then t(u; x, y) := t(u; x, y) + yt(v; x, y)
          else (comment: e_i is a coloop) t(u; x, y) := t(u; x, y) + xt(v; x, y);
      end
  end;
t(sink; x, y) is T(G; x, y).
```

We have described a new algorithm computing the Tutte polynomial of G. Next, we analyze the size of the BDD. In so doing, by imposing some condition on the edge ordering, we can evaluate the number of nodes in the BDD utilizing that condition and also can test the 2-isomorphism more easily.

Let $e_1, e_2, \ldots, e_m$ be an ordering of edges in E. Let $E_i = \{e_1, e_2, \ldots, e_i\}$, and $\overline{E}_i = \{e_{i+1}, e_{i+2}, \ldots, e_m\}$ $(i = 1, \ldots, m)$. We say this edge ordering has a *connected property* if for each $i = 1, \ldots, m$, the subgraph of G on $\overline{E}_i$ is connected.

The i-th *elimination front* $\widetilde{V}_i$ is a vertex subset consisting of vertices v such that v is incident to some edges in E_i and some others in $\overline{E}_i$. Suppose that for $E^c \subseteq E_i$, an equivalence relation on $\widetilde{V}_i$ such that two vertices are in the same equivalence class if and only if when the edges in E^c are contracted they are unified into one vertex. Then consider a partition of $\widetilde{V}_i$ into the equivalence classes by this relation. We call this partition the i-th *elimination partition* induced by E^c.

Theorem 3. *Let G_1 and G_2 be two minors of G on the same edge set $\overline{E}_i$ such that for $E_\alpha^c \cup E_\alpha^d = E_i$, $E_\alpha^c \cap E_\alpha^d = \emptyset$, G_α is obtained by contracting edges in E_α^c and deleting edges in E_α^d $(\alpha = 1, 2)$.*

(i) If the i-th elimination partition induced by E_1^c and that induced by E_2^c are identical, G_1 and G_2 are 2-isomorphic with the same edge ordering induced by the edge ordering of the original graph G.

(ii) Suppose that the edge ordering has a connected property. Then if G_1 and G_2 are 2-isomorphic with the same edge ordering induced by the edge ordering of the original graph G, the i-th elimination partition induced by E_1^c and that induced by E_2^c are identical.

Proof. (i) Since two partitions are identical, there is a bijection $\phi : V_1 \to V_2$ such that u and v in V are unified into one vertex uv in V_1 iff u and v are unified into $\phi(uv)$ in V_2, and for the rest vertex w in V, w corresponds to w' in V_1 iff w corresponds to $\phi(w')$ in V_2. In addition, since G_1 and G_2 are minors on the same edge set $\overline{E}_i$, G_1 and G_2 are isomorphic. Hence they are 2-isomorphic.

(ii) Suppose two partitions are not identical. Then there is a pair of vertices u and v in the i-th elimination front such that for $\alpha = 1$, u and v are in the same set at the partition, while for $\alpha = 2$, they are in different sets at the partition (interchange 1 and 2 if necessary). This means that u and v are unified into one vertex in G_1 while they are not unified in G_2. Since the edge ordering has a connected property, in $\overline{E}_i$ there exists a path connecting u and v. Then, there exists a spanning tree of G_2 containing the path, while there is no spanning tree in G_1 containing the path. Therefore, G_1 and G_2 are not 2-isomorphic.

This theorem can be used not to check the 2-isomorphism more easily but also to analyze the size of the BDD. In order to see t\.e latter issue, we consider a complete graph K_n of n vertices, since the bound on K_n is an upper bound for the other simple connected graphs.

For the complete graph K_n of n vertices, order the vertices from 1 to n. Then, represent each edge by a tuple (u, v) where u and v are numbers attached to their endpoints and $u < v$, and order edges in the increasing lexicographic order of (u, v). This ordering is called a canonical edge ordering of a complete graph. This has a connected property.

Let $L(G, i)$ be the number of nodes in the i-th level of the BDD of all spanning trees of G for the canonical edge ordering. The Bell number B_n is the number of partitions of a set of n elements. Using this number, the size of each level can be bounded as follows.

Theorem 4. (i) $L(K_n, i) \leq 2^i$. Especially, $L(K_n, i) = 2^i$ for $0 \leq i \leq n - 2$ and $L(K_n, n - 1) = 2^{n-1} - n + 1$.

(ii) For $n \geq 4$, $L(K_n, i) \leq \frac{1}{2}(3^{n-1} + 1) - (n-1)2^{n-2} + \binom{n-1}{2}$ for $n \leq i < 2n - 3$ and $L(K_n, 2n - 3) = \frac{1}{2}(3^{n-2} + 1) - (n - 2)2^{n-3} + \binom{n-2}{2}$.

(iii) $L(K_n, i) \leq B_j$ for $n(n-1)/2 - j(j-1)/2 \leq i < n(n-1)/2 - (j-1)(j-2)/2$, and $L(K_n, i) = B_j$ for $n(n-1)/2 - (n - \lfloor(n+1)/3\rfloor)(n - \lfloor(n+1)/3\rfloor - 1)/2 \leq i$.

Proof. (i) $L(K_n, i) \leq 2^i$ is trivial. It is also trivial that equality holds for $0 \leq i \leq n - 2$. When $i = n - 1$, the vertices which are adjacent to v_1 by contracted edges are unified into one vertex in a minor, i.e., at the partition of $n - 1$ vertices every set has exactly one element, or one set has more than one element and the others has exactly one element. Then $L(K_n, n - 1)$ becomes $1 + \sum_{r=2}^{n-1} \binom{n-1}{r}$.

(ii) Similar to the $i = n - 1$ when $i = 2n - 3$, since the vertices which are adjacent to v_1 or v_2 by contracted edges are unified into one vertex in a minor, i.e., at the partition at most two sets has more than one element and the others has exactly one element. Then $L(K_n, 2n - 3)$ is equal to $1 + \sum_{r=2}^{n-2} \binom{n-2}{r} + \sum_{r=4}^{n-2} \binom{n-2}{r}(2^{r-1} - 1 - r)$.

(iii) For such i and j, the i-th elimination front consists of j vertices, and then from Theorem 3 the number of nodes in the i-th level is bounded by the Bell number B_j. Moreover when all edges which are incident to $\lfloor(n + 1)/3\rfloor$ vertices are contracted or deleted, every case occurs and equality holds.

Table 1. The size of BDD of spanning trees of K_n

n	width of BDD	Bell number B_{n-2}	size of BDD	number of spanning trees
2	1	–	2	1
3	2	(1)	6	3
4	5	(2)	20	16
5	14	(5)	67	125
6	42	(15)	225	1296
7	130	(52)	774	16807
8	406	(203)	2765	262144
9	1266	(877)	10292	4782969
10	3926	[4140]	39891	10^8
11	15106	[21147]	160837	$\approx 2.36 \times 10^9$
12	65232	115975	673988	$\approx 6.20 \times 10^{10}$
13	279982	678570	2932313	$\approx 1.79 \times 10^{12}$
14	1191236	4213597	13227701	$\approx 5.67 \times 10^{13}$

Corollary 5. *For $n \geq 12$, the width of the BDD of spanning trees of K_n for the canonical edge ordering is bounded by B_{n-2}*

Proof. $B_{n-2} > \frac{1}{2}(3^{n-1} + 1) - (n-1)2^{n-2} + \binom{n-1}{2}$ for $n \geq 12$.

This bound is not so tight. We have computed the size of BDD of spanning trees of K_n up to $n = 14$ by the algorithm proposed here (Table 1).

Lemma 6. *For any simple connected graph G, there exists an edge ordering with a connected property such that the size of BDD of spanning trees of G with respect to the edge ordering is less than or equal to the size of BDD of spanning trees of a complete graph of the same number of vertices with respect to the canonical edge ordering.*

Proof. Choose a leaf of a spanning tree and regard this as its root. Then, order vertices from 1 to $|V|$ in postorder in this rooted spanning tree. As in the case of complete graphs, order edges in the increasing lexicographic order. Then this is a canonical edge ordering of G.

Corollary 7. *For any simple connected graph G with n vertices ($n \geq 10$), there exists an edge ordering for which the width of the BDD of spanning trees of G is bounded by B_{n-2}.*

Note that, in computing the Tutte polynomial of a graph with n vertices via the BDD of its all spanning trees, the space complexity is bounded by the width of the BDD and hence by B_{n-2}. This is another advantage of this algorithm.

4 Computing the Tutte Polynomial of a Planar Graph

In this section, we will show that the algorithm proposed in the previous section solves the problem of computing the Tutte polynomial of a planar graph, which itself is still #P-hard, very efficiently.

Table 2. BDD of spanning trees of $k \times k$ lattice graph $L_{k,k}$ of $n = k^2$ vertices

k	n	width of BDD $(= C_{k+1})$	size of BDD	number of spanning trees
2	4	2	8	4
3	9	5	47	192
4	16	14	252	100352
5	25	42	1260	557568000
6	36	132	6002	$\approx 3.26 \times 10^{13}$
7	49	429	27646	$\approx 1.99 \times 10^{19}$
8	64	1430	124330	$\approx 1.26 \times 10^{26}$
9	81	4862	549382	$\approx 8.32 \times 10^{33}$
10	100	16796	2395385	$\approx 5.69 \times 10^{42}$
11	121	58786	10336173	$\approx 4.03 \times 10^{52}$
12	144	208012	44232654	$\approx 2.95 \times 10^{63}$

For a $k \times k$ lattice graph $L_{k,k}$ with $n = k^2$ vertices, order the vertices in a row-major order i.e., from the top row to the bottom row, and for each row from left to right. Then, a canonical edge ordering of a lattice graph is defined by the same way for a complete graph. This has a connected property. In addition, the Catalan number C_{k+1} is defined to be $\frac{1}{k+1}\binom{2k}{k}$.

Theorem 8. $L(L_{k,k}, i) \leq C_{k+1}$. *Furthermore, equality holds for* $\lfloor \frac{k}{2} \rfloor (2k-1) \leq i \leq 2(k^2 - k) - k$ *and* $L\left(L_{k,k}, 2(k^2 - k) - j\right) = C_{j+2}$ *for* $0 \leq j < k$.

Proof. Consider k vertices on a line in the plane and suppose that some pairs among k vertices are connected by an edge without crossing in the upper half plane. Then the k vertices are partitioned into the connected components. Catalan number C_{k+1} is the total number of such partitions of k vertices. For the canonical edge ordering of the lattice graph, since for each column at most one vertex belongs to elimination front, the size of the elimination front is bounded by k. The vertices which are connected by a path consisting of contracted edges are unified to one vertex in a minor, i.e., such vertices belongs to the same set in a partition. Moreover by planarity these paths do not cross each other. Clearly the number of partitions obtained in this way is at most C_{k+1}. Equality holds when all edges incident to $\lfloor \frac{k}{2} \rfloor \times k$ vertices are contracted or deleted. For $i = 2(k^2 - k) - j$, elimination front has $j + 1$ vertices.

Again, we have computed the size of BDD of spanning trees of $L_{k,k}$ up to $k = 12$, i.e., up to 144 vertices, by the algorithm proposed here (Table 2).

For a planar graph, by using the planar separator theorem, a good edge ordering exists and can be computed efficiently as follows [6].

Theorem 9. *For a simple connected planar graph G of n vertices, there exists an edge ordering such that any elimination front consists of $O(\sqrt{n})$ vertices, and such an edge ordering can be found in $O(n \log n)$ time.*

The edge ordering obtained from the planar separator theorem does not necessarily satisfy a connected property. Without satisfying a connected property Theorem 3(ii) may not hold. However, Theorem 3(i) holds and the 2-isomorphism whose edges have an identity map can be tested by decomposing the graph into the 3-connected components. Hence the arguments concerning the bound, etc., still carry over to this case.

Theorem 10. *The size of the BDD of all spanning trees of a planar graph G with n vertices is bounded by $O(2^{O(\sqrt{n})})$.*

Proof. For $k \geq 4$, Catalan number $C_{k+1} \leq 4^{k-2}$. By planarity and Theorem 9 replacing k with $O(\sqrt{n})$, the width of the BDD is $2^{O(\sqrt{n})}$.

It should be noted that the Jones polynomial of an alternating link, whose computation is #P-hard can be computed by substituting variables in the Tutte polynomial of a planar graph [7, 9].

Acknowledgment

The authors would like to thank Kazuyoshi Hayase for giving comments on the earlier draft of this paper. Part of this work of the second author was supported by the Grant-in-Aid of the Ministry of Education, Science and Culture of Japan.

References

1. S. B. Akers: Binary Decision Diagrams. *IEEE Trans. on Computers*, Vol.C-27(1978), pp.509–516.
2. N. Alon, A. Frieze and D. J. A. Welsh: Polynomial Time Randomised Approximation Schemes for the Tutte Polynomial of Dense Graphs. *Proceedings of the IEEE Annual Symposium on Foundations of Computer Science*, 1994, pp.24–35.
3. R. E. Bryant: Graph Based Algorithms for Boolean Function Manipulation. *IEEE Trans. on Computers*, Vol.C-35(1986), pp.677–691.
4. D. R. Karger. A Randomized Fully Polynomial Time Approximation Scheme for the All Terminal Network Reliability Problem. *Proceedings of the 27th Annual ACM Symposium on Theory of Computing*, 1995, pp.11–17.
5. A. Shioura, A. Tamura and T. Uno: An Optimal Algorithm for Scanning All Spanning Trees of Undirected Graphs. *SIAM Journal on Computing*, to appear.
6. S. Tani: *An Extended Framework of Ordered Binary Decision Diagrams for Combinatorial Graph Problems*. Master's Thesis, Department of Information Science, University of Tokyo, 1995.
7. M. B. Thistlethwaite: A Spanning Tree Expansion of the Jones Polynomial. *Topology*, Vol.26 (1987), pp.297–309.
8. W. T. Tutte: A Contribution to the Theory of Chromatic Polynomials. *Canadian Journal of Mathematics*, Vol.6 (1954), pp.80–91.
9. D. J. A. Welsh: *Complexity: Knots, Colourings and Counting*. London Mathematical Society Lecture Note Series, Vol.186, Cambridge University Press, 1993.
10. H. S. Wilf: *Algorithms and Complexity*. Prentice-Hall, 1986.

More Efficient Parallel Flow Algorithms

Jürgen Dedorath *   Jordan Gergov **   Torben Hagerup **

Abstract. We develop an arsenal of tools for improving the efficiency of parallel algorithms for network-flow problems and apply it to a maximum-flow algorithm of Goldberg, a blocking-flow algorithm of Shiloach and Vishkin, and a maximum-flow algorithm of Ahuja and Orlin. Depending on the exact model of computation and the time available for the computation, we achieve a polylogarithmic reduction in the time-processor product. In particular, this leads to the first parallel implementations with optimal speedup of the corresponding sequential algorithms.

1 Introduction

A *network* is a tuple (G, cap, s, t), where $G = (V, E)$ is a directed graph, $cap : E \to \mathbb{R}$ is a function that assigns to each edge of G a nonnegative value called its *capacity*, and s and t are distinguished vertices of G called the *source* and the *sink*, respectively. A network models a system of interconnected pipes, in a concrete or abstract sense, designed to carry a commodity from one point, represented by s, to another point, represented by t. The capacity of an edge represents the maximum carrying capacity of the corresponding pipe.

Without loss of generality we will assume that the network is symmetric, i.e., $(v, w) \in E \Leftrightarrow (w, v) \in E$, for all $v, w \in V$. Given an assignment f of a nonnegative value $f(e)$ to each edge e of G, representing an actual carrying load, such that $f(v, w) = -f(w, v)$, for all $(v, w) \in E$ (antisymmetry constraint), the *excess* $e(v)$ of a vertex $v \in V$ with respect to f is defined as $\sum_{(u,v) \in E} f(u, v)$, i.e., as the net flow into v. f is called a *preflow* if $f(e) \leq cap(e)$ for all $e \in E$ (capacity constraint) and $e(v) \geq 0$ for all $v \in V \setminus \{s\}$. If, additionally, $e(v) = 0$ for all $v \in V \setminus \{s, t\}$ (flow-conservation constraint), f is called a *flow*, and $e(t)$ is called the *value* of f. A flow represents a situation in which the commodity to be transported enters the network only at s and leaves the network only at t. A *maximum flow* is a flow whose value is maximum, among the values of all flows.

The *residual capacity* $r(e)$ of an edge $e = (v, w)$ with respect to a preflow f is defined as $cap(e) - f(e)$, i.e., as the amount of additional flow that the edge can carry or that can be returned over (w, v). The edge is said to be *saturated*

---

* Bayerische Hypotheken- und Wechsel-Bank AG, O/EDV-AE3-W1, D–81925 München, Germany. E-mail: dedorath@Hypo.DE. Part of the research was carried out while this author was with the Max-Planck-Institut für Informatik.

** Max-Planck-Institut für Informatik, D–66123 Saarbrücken, Germany. E-mail: gergov@mpi-sb.mpg.de, torben@mpi-sb.mpg.de. Supported by the ESPRIT Basic Research Actions Program of the EU under contract No. 7141 (project ALCOM II).

if its residual capacity is zero, and a flow is *blocking* if every path in G from s to t contains at least one saturated edge.

One of the best sequential algorithms for computing maximum flows is the dynamic-trees algorithm of Goldberg and Tarjan [12], which runs on networks with n vertices and m edges in $O(nm \log(n^2/m))$ time, while the *maximum-distance discharge* (MDD) algorithm of the same authors was shown by Cheriyan and Maheshwari [8] to have a worst-case running time of $\Theta(n^2\sqrt{m})$. If the edge capacities are integral and bounded by $U \geq 2$, the excess scaling maximum-flow algorithm of Ahuja and Orlin [1] achieves a running time of $O(nm + n^2 \log U)$ (see also [2, 3]). A parallel implementation of the MDD algorithm that runs in $O(n^2 \log(2m/n + p)(\sqrt{m}/p))$ time on a CREW PRAM with $p \leq \sqrt{m}$ processors was given by Goldberg [11]. Observing that Goldberg's parallel algorithm is slower than the sequential algorithm of [12] unless $2m/n + p = \Omega(n^{1/4})$, we may as well replace $\log(2m/n + p)$ by $\log n$ in his time bound. For $m/n \geq 2$, a parallel version of the algorithm of Ahuja and Orlin can be executed in $O(n^2 \log p \log U)$ time on an EREW PRAM with $p = \lceil m/n \rceil$ processors [1]. For our purposes, it is convenient to express the resource bounds of a parallel algorithm by giving its (minimum) running time and its time-processor product, i.e., the number of operations executed. For Goldberg's algorithm, these parameters are $O(n^2 \log n)$ time and $O(n^2\sqrt{m} \log n)$ operations, while for the algorithm of Ahuja and Orlin they are $O(n^2 \log(m/n) \log U)$ time and $O(nm \log(m/n) \log U)$ operations.

A problem closely related to the computation of a maximum flow is that of computing a blocking flow. As demonstrated by Dinic [10], a blocking flow in an n-vertex acyclic network can be computed in $O(n^2)$ time. A parallel algorithm for this problem that runs in $O(n \log n)$ time using $O(n^2 \log n)$ operations on an EREW PRAM was described by Shiloach and Vishkin [18, 19]. This algorithm and both parallel maximum-flow algorithms mentioned above can be seen to fall short, by a factor of $\Theta(\log n)$ (Goldberg and Shiloach/Vishkin) and between $\Theta(\log(m/n))$ and $\Theta(\log(m/n) \log U)$ (Ahuja/Orlin), of being implementations with optimal speedup of the corresponding sequential algorithms, i.e., of executing the same number of operations, up to a constant factor. Our goal in this paper is to reduce this logarithmic factor or to eliminate it altogether. While this factor may appear insignificant compared to the overall operation bound, it should be noted that the factor by which a parallel algorithm deviates from optimal speedup is precisely the minimum number of processors needed for a meaningful execution of the algorithm (if fewer processors are available, the sequential algorithm will be faster). The viability of parallel network-flow algorithms in a practical setting may therefore depend critically on the techniques developed here. Our methods are fairly general and may find applications in other parallel network-flow algorithms. We will describe them here in the context of the MDD algorithm. After suitable modifications, our techniques also apply to the blocking-flow algorithm of Shiloach and Vishkin and the parallel excess scaling algorithm of Ahuja and Orlin.

In attempting to reduce the time-processor product of the MDD algorithm, we face two bottlenecks. The first of these is caused by $O(n^2\sqrt{m})$ searches in binary trees with $n^{O(1)}$ leaves. Following [18], Goldberg [11] pays a cost of $O(\log n)$ per search, which leads to his operation bound of $O(n^2\sqrt{m}\log n)$. We here note that the $O(n^2\sqrt{m})$ searches are not arbitrary and show how, due to their special structure, they can be carried out with a total of $O(n^2\sqrt{m})$ operations, i.e., with constant amortized cost per search.

A basic operation in network-flow algorithms is a *push* over an edge e of value a, which simply increases $f(e)$ by a. The second bottleneck of the MDD algorithm is due to the execution of $O(n^2\sqrt{m})$ pushes. After pushes have been executed simultaneously over many edges, each vertex v must determine its new excess by summing the flow "quanta" contributed by the pushes over its incoming edges. A standard optimal summation algorithm, applied to the adjacency list of v, could be employed for this purpose, but this would be much too expensive, since often pushes are executed only over a tiny fraction of the incoming edges. The solution of [11], also employed in [18], is to let the flow contributions destined for v ascend a binary tree, adding two contributions when they meet. The cost of this is $O(\log n)$ per push, or $O(n^2\sqrt{m}\log n)$ altogether. In order to counter this bottleneck, we observe that the execution of pushes reduces to sorting and, moreover, to integer sorting, namely to sorting the flow quanta by their destination vertices, after which the new excesses are readily computed.

The cost of integer sorting on a PRAM depends on the exact PRAM variant employed and on the time allowed for the computation and lies between $\Theta(1)$ and $\Theta(\log n)$ per element to be sorted, which translates into a total operation bound for our implementation of the MDD algorithm of between $O(n^2\sqrt{m})$ and $O(n^2\sqrt{m}\log n)$. Due to the great and perhaps confusing variety of best known integer-sorting results for the PRAM (shown in Table 1), we obtain a corresponding variety of results for computing maximum flows.

The maximum-flow algorithm of Goldberg works in $O(n+m)$ space. However, this space bound is achieved through the use of the parallel 2-3-*tree data structure* of [17], which is incompatible with our reduction in the time-processor product (this can be viewed as a third bottleneck). We describe a simple parallel data structure that achieves the same space bound without using 2-3 trees and without detriment to the time-processor product.

2 Parallel Implementation of the MDD Algorithm

An important family of algorithms for the maximum-flow problem is the class of *preflow-push algorithms* introduced by Goldberg and Tarjan [12]. The MDD algorithm is a specific instance of the generic preflow-push algorithm. A preflow-push algorithm constructs a maximum flow by successive modifications of a preflow in the input network.

We call an edge e *residual* if its residual capacity $r(e)$ is positive. For a given preflow, a *valid distance labeling* is a function d from the set of vertices to the nonnegative integers such that $d(t) = 0$, $d(s) = n$, and $d(v) \leq d(w) + 1$ for all residual edges (v, w). Further, we say that a vertex $v \notin \{s, t\}$ is *active* if its excess $e(v)$ is positive. A residual edge $e = (v, w)$ is called *admissible* if $d(v) = d(w)+1$.

A preflow-push algorithm begins by constructing an initial preflow in G and a distance labeling valid for it by setting $f(v, w)$ to $cap(v, w)$ if $v = s$, to $-cap(w, v)$ if $w = s$, and to zero otherwise, and by setting $d(s)$ to n and $d(v)$ to zero for all $v \neq s$. It then repeatedly applies so-called *push* and *relabel* operations to active vertices in G. Eventually no operation applies, at which point the algorithm has transformed the initial preflow into a maximum flow in G. A *push* operation on v is applicable only over an admissible edge (v, w); it increases $f(v, w)$ and $e(w)$ by $\min\{e(v), r(v, w)\}$ and decreases $f(w, v)$ and $e(v)$ by the same amount. A *relabel* operation on v sets its label $d(v)$ to $\min_{(v,w) \in E}\{d(w) + 1 \mid r(v, w) > 0\}$; it is applicable to an active vertex v if there are no admissible edges out of v. It can be shown that no vertex is relabeled more than $2n - 1$ times, and that $d(v)$ remains bounded by $2n - 1$ for all $v \in V$.

We consider an implementation of MDD based on a combination of *push* and *relabel* operations called *discharge*. A *discharge* operation applied to an active vertex v is a possibly empty sequence σ of *push* operations on v such that after the application of σ, either the whole excess of v is exhausted, i.e., $e(v) = 0$, or a *relabel* operation is applicable to v. In the latter case, the *discharge* operation is finished by a relabeling of v. The MDD algorithm constructs a maximum flow by repeated application of *discharge* operations to active vertices v whose distance $d(v)$ is maximum [12]. Its parallel version computes a maximum flow by applying *discharge* operations in parallel to all active vertices with maximum distance [11]. In the context of the parallel MDD algorithm, we can think of a *discharge* operation on v as the computation of a set Q_v of "pushes" (an edge (v, w) of G and the amount of flow to be pushed over (v, w)) and their subsequent execution. An outline of the parallel version of MDD is provided in Fig. 1. During the computation we maintain a data structure for the sets $D_i = \{v \in V \mid d(v) = i \text{ and } e(v) > 0\}$, for $i = 0, \ldots, 2n - 1$. Further, *max* denotes the maximum of $d(v)$ over all active vertices.

1   *Initialize*;
2   **while** the set of active vertices is nonempty **do**
3   **for all** $v \in D_{max}$ **do in parallel**
4   compute the set of all pushes Q_v out of v;
5   execute all pushes in $\bigcup_{v \in D_{max}} Q_v$;
6   relabel all vertices in D_{max} whose excess is not exhausted;
7   update $D_0, \ldots, D_{2n-1}$ and the value of *max*;

Fig. 1. Outline of the parallel MDD algorithm.

Theorem 1. *Let $s(n)$, $t(n)$ and $q(n)$ be resource bounds such that n integers in the range $1..n$ can be sorted in $t(n)$ time using $nq(n)$ operations and $s(n)$ space on some PRAM model, and assume that $\log n \cdot \max\{\log^* n - \log^* q(n), 1\} = O(t(n))$. Then a maximum flow in a network with n vertices and m edges can be computed in $O(n^2 t(n))$ time using $O(n^2 \sqrt{m} q(n))$ operations and $O(n + m + s(n))$ space on the same PRAM model. On the CRCW PRAM the condition $\log n \cdot \max\{\log^* n - \log^* q(n), 1\} = O(t(n))$ can be replaced by $\log n / \log \log n = O(t(n))$.*

The complicated condition $\log n \cdot \max\{\log^* n - \log^* q(n), 1\} = O(t(n))$ is satisfied for all EREW and CREW PRAM sorting algorithms known at the time of writing, which have either $t(n) = \Omega(\log n \log^* n)$ or $t(n) = \Omega(\log n)$ and $q(n) = \Omega(\log \log n)$. Similarly, a CRCW PRAM algorithm that works in $o(\log n/\log \log n)$ time must use a superpolynomial number of processors [6] and therefore is of little interest.

Combining different integer-sorting results (Table 1) with Theorem 1 yields a variety of results for computing maximum flows. Note, in particular, that algorithms with optimal speedup can be obtained from the first and last lines of the table. As another example, one of the best integer-sorting results for the EREW PRAM (line 2) is that n integers of size polynomial in n can be sorted in $O((\log n)^{3/2}/(\log \log n)^{1/2})$ time using $O(n(\log n)^{1/2}(\log \log n)^{1/2})$ operations and $O(n)$ space. Correspondingly, we obtain an EREW PRAM maximum-flow algorithm that runs in $O(n^2(\log n)^{3/2}/(\log \log n)^{1/2})$ time using $O(n^2\sqrt{m}(\log n)^{1/2}$ $(\log \log n)^{1/2})$ operations and $O(n + m)$ space. This algorithm is slower than Goldberg's CREW PRAM algorithm [11] by a factor of about $\Theta(\sqrt{\log n})$, but more efficient by a similar factor.

Model	Time	Operations	Source
EREW	$O(t),\ t \geq 2\log n$	$O(n + n\log n/\log(t/\log n))$	[9, 20, 16]
EREW	$O(t),\ t \geq \log n \log \log n$	$O(n(\log n)^{1/2}(\log \log n)^{1/2} + n(\log n)^2/t)$	[4]
CREW	$O(t),\ t \geq \log n$	$O(n(\log n)^{1/2} + n\log n/2^{t/\log n})$	[4]
CRCW	$O(\log n/\log \log n)$	$O(n\log \log n)$	[7]
CRCW	$O(\log n/\log \log n)$	$O(n)$	[5]

Table 1. The best known bounds for sorting n integers in the range $1 .. n$ on a PRAM. The algorithm cited in the last line is randomized and has a failure probability of $2^{-n^{\Omega(1)}}$; the remaining algorithms are deterministic.

At a high level, the proof of Theorem 1 proceeds as follows. A *stage* of the MDD algorithm is the parallel application of *discharge* operations to all active vertices with maximal distance, i.e., to the vertices in D_{max} (Fig. 1, lines 3–7). As a simple potential-function argument shows, a maximum flow is constructed in $O(n^2)$ stages [12]. We will show that the (parallel) time of a single stage of our implementation of the MDD algorithm is dominated by the time of integer sorting. Recall that our main goal is to design an implementation of the MDD algorithm with a small time-processor product. We divide a single stage into three *phases*. The implementation and analysis of the first phase (lines 3–4), which we call *saturation search*, is explained in more detail in Section 2.1. We show how to do saturation search in $O(t(n))$ time per stage and using a total of $O(n^2\sqrt{m}q(n))$ operations. The second phase consists of lines 5–6. Once the pushes involved have been determined in the first phase, we proceed as follows to actually send flow between the vertices. First a prefix summation allows representations of all flow quanta to be sent anywhere to be stored compactly in an array. Sending the flow quanta now is a matter of sorting them by their destination, an integer in the range $1 .. n$, and subsequently processing them in simple ways derived from prefix summation. Using standard arguments, it can be shown that the cost of relabeling is dominated by that of integer sorting, i.e., the second phase takes $O(t(n))$ time per stage and a total of $O(n^2\sqrt{m}q(n))$ operations. Finally, in the third phase (line 7), the data structure representing $D_0, \ldots, D_{2n-1}$ is updated. The corresponding problem of maintaining dynamically changing sets of total size bounded by n within the resource bounds of Theorem 1 is addressed in Section 2.2. Finally, maintaining max is easy, since it can decrease only as a result of a push, and then only by 1.

2.1 Saturation Search

Our implementation of *discharge* makes use of the notion of a *current edge* [12]. Initially, the current edge of a vertex v is the first edge on its adjacency list $e_1, \ldots, e_\ell$. A *push* operation on v is always applied over its current edge e_c. If e_c is not admissible, it is replaced as the current edge of v by the next edge on the adjacency list. If e_c is the last edge on the list, v is relabeled and the first edge becomes current.

A saturation-search query at v essentially asks for a maximal set of edges that can be saturated with the available excess $e(v)$. For ease of discussion, we will assume below that the search comes to an end without "hitting" the end of the adjacency list of v. Define h^* as the smallest nonnegative integer such that the total residual capacity of the admissible edges in $\{e_c, \ldots, e_{c+h^*}\}$ is at least $e(v)$. In the following we consider a particular vertex v and concentrate on the implementation of a query asking for h^*. Saturation search (at v) easily reduces to the computation of h^*.

The basic observation underlying our improvement is that the value returned by a query directly reflects a certain amount of progress made by the MDD algorithm. If this value is large, the algorithm makes substantial progress, since in one go many edges will be saturated. We conclude from this that queries associated with little progress should be cheap, whereas queries associated with substantial progress can be allowed to be quite expensive. More precisely, if a saturation-search query at v takes $O(h^* + q(n))$ operations, then the total cost of all queries at all vertices will be $O(nm + n^2\sqrt{m}q(n)) = O(n^2\sqrt{m}q(n))$. This is because the work at nonadmissible edges is bounded by $O(nm)$ (a vertex is relabeled at most $2n - 1$ times, and therefore each of its outgoing edges is "skipped" at most $2n$ times), and the number of pushes performed is $O(n^2\sqrt{m})$. Similarly, we can allow time $t(n)$ per phase. In the sequel we present three solutions to this problem. The worst case of our saturation-search procedures is not better than the cost of the solution in [11], but their amortized cost is smaller.

Log-Star Saturation Search. We first test whether $h^* \leq q'$, where $q' = \min\{q(n), t(n)\}$, by inspecting the edges $e_c, \ldots, e_{c+q'}$, which takes $O(q') = O(t(n))$ time. If the answer is affirmative, we are done. If not, the search proceeds in a different manner. In addition to the simple sequence of residual capacities used in the linear search, we use a data structure consisting of a complete binary tree T_v with $e_1, \ldots, e_\ell$ at its leaves, and with each node storing the total residual capacity of the admissible edges at its leaf descendants. We can determine whether $h^* < h$, for any given $h \geq 2$, by adding the values stored at left siblings of the nodes on the leaf-to-root path in T_v starting in e_{c+h} and comparing the result with the target value of the query. This can certainly be done in $O(\log n)$ time; if h^* is small, however, it can be done faster, as we now demonstrate.

Let x be the lowest common ancestor of e_c and e_{c+h} in T_v. The question of whether $h^* < h$ can be answered by adding the residual capacity of e_c, the values stored at the right siblings of nodes $\neq x$ on the path from e_c to x, and the values stored at the left siblings of nodes $\neq x$ on the path from e_{c+h} to x. Moreover, all such siblings occur at heights $\leq \log h$, since otherwise the descendants of such a sibling would separate e_c and e_{c+h} by a distance of more than h. This implies that the question of whether $h^* < h$ can be answered with $O(\log h)$ operations.

We use the above to test whether $h^* < h_1^2$, $h^* < h_2^2$, etc., until the answer to some test comes out affirmative. Here $h_0 = \lfloor \sqrt{q'} \rfloor$, and $h_{j+1} = 2^{h_j}$, for $j \geq 0$. Each test is executed in parallel for all vertices that are still participating; before the test, these vertices are compacted, so that they can be accessed efficiently by the available processors. Assume that the conclusion reached in the case of v is that $h^* \geq h_{j_0}^2$, but $h^* < h_{j_0+1}^2$. Then the total cost of the tests is $O(\sum_{j=1}^{j_0+1} \log h_j^2) = O(\log h_{j_0+1}^2) = O(h_{j_0}) = O(\sqrt{h^*})$. Furthermore, h^* can subsequently be determined exactly by binary search at a cost of $O((\log h_{j_0+1})^2) = O(h^*)$ operations, each of $O(\log h_{j_0+1})$ stages of binary search being executed as the test above. The cost of compacting vertices can be seen to be dominated by that of the tests proper.

Although the cost in terms of operations of the search above is linear in the progress made, as desired, it may be too slow, since we would like to execute each stage in $O(\log n)$ time, while for large values of h^* the search, as just described, may require $\Theta((\log n)^2)$ time. As soon as the search has established that $h^* \geq \log n$, however, we can switch from many partial leaf-to-root traversals to a single root-to-leaf traversal, using the values stored at the nodes of T_v to navigate to the correct leaf e_{c+h^*}, at a cost in operations and time of $O(\log n)$. Simply by testing whether $h^* \geq \log n$ before the binary search (if this is not already excluded), we can therefore reduce the worst-case time bound for the search itself to $O(\log n)$. The sequence $h_1, h_2, \ldots$ has $O(\max\{\log^* n - \log^* q(n), 1\})$ elements, so that the time needed for repeated compactions is $O(\log n \cdot \max\{\log^* n - \log^* q(n), 1\})$. For the EREW and CREW PRAMs, this is $O(t(n))$ by assumption. For the CRCW PRAM, we offer the following solution working in $O(\log n/\log \log n) = O(t)$ time.

Let b be a positive integer with $b = \Theta(\sqrt{\log n/\log \log n})$. We run the EREW PRAM algorithm, but replacing the first element in the sequence $h_1, h_2, \ldots$ that is greater than b by b, and the element following it by 2^b; if and when the algorithm discovers that $h^* \geq 2^{2b}$ by carrying out the corresponding test, it is aborted. If $h^* < 2^{2b}$, the EREW PRAM algorithm is not aborted, and it computes h^* in $O(b^2) = O(\log n/\log \log n)$ time using $O(h^* + q(n))$ operations. Otherwise we switch from a sequential binary search with each test carried out sequentially to a parallel multiway search with each test carried out in parallel. More precisely, we replace the binary search by k-ary search, where $k = 2^{\Theta((\log\log n)^2)}$, and instead of doing a single test in $\Theta(\log n)$ (worst-case) time with one processor, we do it in $O(\log\log n)$ time with $O(\log n)$ processors, one processor per node on the relevant leaf-to-root path in T_v, which is easy if T_v is represented in a suitably regular way. The time needed to complete the k-ary search is $O(\log n/\log k \cdot \log\log n) = O(\log n/\log\log n)$, and the number of operations is $O(\log n/\log k \cdot k \cdot \log n) = o(2^b)$, i.e., less than what can be afforded once h^* is known to be at least 2^{2b}. The compaction of participating vertices can be replaced by the approximate compaction of [14], which uses a total of $O((\log\log n)^3 \log^* n) = O(t(n))$ time.

An update of T_v following a push out of v can be carried out in $O(\log n)$ time. By deferring all such updates to the next relabeling of v, we can reduce this to $O(1)$ per push. Provided that we maintain the flow over the current edge of v and over all edges out of v preceding it in the adjacency list of v separately, the search procedures described above are easily adapted to this complication.

Simple Saturation-Search Procedures. The next two solutions are much simpler than log-star saturation search; however, their time or number of operations is slightly larger. *Naive saturation search*, so called because it uses no data structure beyond the sequence of residual capacities, works as follows: Let h^* be as in the previous subsection, i.e., after the query under consideration h^* edges will be saturated. Then, for $j = 1, 2, \ldots$, we test whether $h^* \leq 2^j$, stopping when the answer first comes out affirmative, and compacting the vertices still participating between each pair of successive test rounds. The test is carried out via a parallel computation of the 2^j relevant prefix sums, which uses $O(2^j)$ operations and $O(j)$ time, and which also computes h^* exactly if $h^* \leq 2^j$. Assume that the test concludes that $h^* > 2^{j_0}$, but $h^* \leq 2^{j_0+1}$. Then the total number of operations expended over all tests is $O(\sum_{j=1}^{j_0+1} 2^j) = O(2^{j_0}) = O(h^*)$, and the total time spent is $O(\sum_{j=1}^{j_0+1} j) = O(j_0^2) = O((\log n)^2)$.

A solution tailored to the MDD algorithm is based on the observation that the number of saturation-search queries such that the current edge is nonadmissible after the query is $O(nm)$. This is because for a fixed v the number of queries of this type is $O(n \deg(v))$. If we first check whether the current edge is admissible and its residual capacity is larger than the excess at v and perform a root-to-leaf search in T_v only if the answer is negative, we can bound the total work spent in saturation-search queries by $O(n^2\sqrt{m} + nm\log n)$. Further, updates of a current edge e_c in T_v need only be executed when e_c becomes saturated.

2.2 Space Requirements

Except for the maintenance of the sets $D_0, \ldots, D_{2n-1}$, it is easy to implement the MDD algorithm to run in $O(n + m + s(n))$ space. The size of D_i remains bounded by n, for $i = 0, \ldots, 2n - 1$, so a representation of $D_0, \ldots, D_{2n-1}$ in $O(n^2)$ space is straightforward. Since the sets $D_0, \ldots, D_{2n-1}$ form a partition of V, however, they can in fact be represented in a total of $O(n)$ space through a parallel adaptation of a scheme of [15]. We provide a sketch.

The elements of each set D_i are stored in the first cells of a *segment*, possibly with unused cells at the end of the segment. When D_i expands through the insertion of new elements, these are simply placed in the unused part of the segment, as long as there is room for them. When the segment is full, a new segment of size twice the new cardinality of D_i is allocated, the new elements of D_i are stored there, and the remaining elements of D_i are gradually transferred from the old segment to the new one, after which the old segment is abandoned. Provided that the effort expended on the migration from the old segment to the new one is made proportional to the number of (future) insertions into D_i, with a suitable factor of proportionality, it can be arranged that a new migration need never start before the previous one has been completed. Thus D_i is spread over at most two segments, which ensures efficient access to its elements. Concurrently with the allocation of new segments from a large *base segment*, we run a garbage collector that reclaims abandoned segments and moves the segments in use closer to the beginning of the base segment. A slight adaptation of the argument of [15] shows that provided that the effort devoted to garbage collection is made proportional to the total number of insertions, with a suitable factor of proportionality, a base segment of size $O(n)$ suffices. Both garbage collection and the simultaneous allocation of new segments reduce to prefix summation, and thus parallelize effortlessly.

3 Other Applications

The techniques presented here are applicable also to other parallel network-flow algorithms. The parallel excess scaling algorithm of Ahuja and Orlin [1] computes a maximum flow in $O(n^2 \log(m/n) \log U)$ time using $O(nm \log(m/n) \log U)$ operations. Using log-star saturation search we can prove the following theorem.

Theorem 2. *A maximum flow in a network with n vertices, $m \geq 2n$ edges and capacities bounded by $U \geq 2$ can be computed in $O(n^2 \log(m/n) \log U)$ time using $O(nm + n^2 \log U)$ operations and $O(n + m)$ space on an EREW PRAM.*

Applying our techniques to the blocking-flow algorithm of Shiloach and Vishkin [18, 19], we obtain the next theorem.

Theorem 3. *Let $s(n)$, $t(n)$ and $q(n)$ be resource bounds such that n integers in the range $1..n$ can be sorted in $t(n)$ time using $nq(n)$ operations and $s(n)$ space on some PRAM model. Then a blocking flow in an n-vertex acyclic network can be computed in $O(n(t(n) + \log n \cdot \max\{\log^* n - \log^* q(n), 1\}))$ time using $O(n^2 q(n))$ operations and $O(n^2 + s(n))$ space on the same PRAM model.*

The second line of Table 1 combined with Theorem 3 results in an EREW PRAM algorithm that computes a blocking flow in $O(n(\log n)^{3/2}/(\log \log n)^{1/2})$ time using $O(n^2 (\log n)^{1/2} (\log \log n)^{1/2})$ operations and $O(n^2)$ space, which is about $\Theta(\sqrt{\log n})$ slower but about $\Theta(\sqrt{\log n})$ more efficient than the solution given in [18, 19]. An interesting consequence of Theorem 3 and a reduction of Goldberg and Tarjan [13] is a more efficient parallel algorithm for computing minimum-cost circulations in networks.

References

1. R. K. Ahuja and J. B. Orlin, A fast and simple algorithm for the maximum flow problem, *Oper. Res.* **37** (1989), pp. 748–759.
2. R. K. Ahuja, J. B. Orlin, and R. E. Tarjan, Improved time bounds for the maximum flow problem, *SIAM J. Comput.* **18** (1989), pp. 939–954.
3. R. K. Ahuja, J. B. Orlin, C. Stein, and R. E. Tarjan, Improved algorithms for bipartite network flow, *SIAM J. Comput.* **23** (1994), pp. 906–933.
4. S. Albers and T. Hagerup, Improved parallel integer sorting without concurrent writing, Tech. Rep. No. MPI-I-94-137, Max-Planck-Institut für Informatik, Saarbrücken, 1994. A preliminary version appeared in Proc. 3rd Annual ACM-SIAM Symposium on Discrete Algorithms (SODA 1992), pp. 463–472.
5. H. Bast and T. Hagerup, Fast parallel space allocation, estimation and integer sorting, Tech. Rep. No. MPI-I-93-123, Max-Planck-Institut für Informatik, Saarbrücken, 1993. Also *Inform. and Comput.*, to appear.
6. P. Beame and J. Hastad, Optimal bounds for decision problems on the CRCW PRAM, *J. Assoc. Comput. Mach.* **36** (1989), pp. 643–670.
7. P. C. P. Bhatt, K. Diks, T. Hagerup, V. C. Prasad, T. Radzik, and S. Saxena, Improved deterministic parallel integer sorting, *Inform. and Comput.* **94** (1991), pp. 29–47.
8. J. Cheriyan and S. N. Maheshwari, Analysis of preflow push algorithms for maximum network flow, *SIAM J. Comput.* **18** (1989), pp. 1057–1086.
9. R. Cole and U. Vishkin, Deterministic coin tossing with applications to optimal parallel list ranking, *Inform. and Control* **70** (1986), pp. 32–53.
10. E. A. Dinic, Algorithm for solution of a problem of maximum flow in a network with power estimation, *Soviet Math. Dokl.* **11** (1970), pp. 1277–1280.
11. A. V. Goldberg, Processor-efficient implementation of a maximum flow algorithm, *Inform. Process. Lett.* **38** (1991), pp. 179–185.
12. A. V. Goldberg and R. E. Tarjan, A new approach to the maximum-flow problem, *J. Assoc. Comput. Mach.* **35** (1988), pp. 921–940.
13. A. V. Goldberg and R. E. Tarjan, Finding minimum-cost circulations by successive approximation, *Math. of Oper. Res.* **15** (1990), pp. 430–466.
14. T. Hagerup, Fast deterministic processor allocation, *J. Algorithms* **18** (1995), pp. 629–649.
15. T. Hagerup, K. Mehlhorn, and J. I. Munro, Maintaining discrete probability distributions optimally, in Proc. 20th International Colloquium on Automata, Languages and Programming (ICALP 1993), Springer Lecture Notes in Computer Science, Vol. 700, pp. 253–264.
16. C. P. Kruskal, L. Rudolph, and M. Snir, Efficient parallel algorithms for graph problems, *Algorithmica* **5** (1990), pp. 43–64.
17. W. Paul, U. Vishkin, and H. Wagener, Parallel computation on 2-3 trees, *RAIRO Theoret. Inform.* **17** (1983), pp. 397–404.
18. Y. Shiloach and U. Vishkin, An $O(n^2 \log n)$ parallel MAX-FLOW algorithm, *J. Algorithms* **3** (1982), pp. 128–146.
19. U. Vishkin, A parallel blocking flow algorithm for acyclic networks, *J. Algorithms* **13** (1992), pp. 489–501.
20. R. A. Wagner and Y. Han, Parallel algorithms for bucket sorting and the data dependent prefix problem, in Proc. International Conference on Parallel Processing (1986), pp. 924–930.

Linear-time In-place Selection in Less than $3n$ Comparisons

Svante Carlsson[*] and Mikael Sundström[*]

Abstract. By developing and exploiting new in-place techniques, we show that finding the element with the median value out of n elements stored in an array can be performed in-place in $(2.95 + \epsilon)n$ (for any $\epsilon > 0$) comparisons and in linear time. This is arbitrarily close to the upper bound for the same problem without space-restrictions. To make the algorithm competitive we also try to minimize the number of element moves performed by the algorithm since this is the other critical operation. This has resulted in a trade-off between the number of comparisons and the number of moves. By minimizing the sum of the critical operations we achieve an algorithm that uses at most $3.75n$ comparisons and $9n$ moves for finding the median in-place. This is, in principle, twice as good as earlier attempts on implicit selection for both of the operations.

1 Introduction

The problem of selecting the element with a given rank in a set of n elements was shown to be solvable in $\Theta(n)$ time by Blum et al. in 1973 [2]. This was a surprise to the research community since it was believed that the general selection problem was as difficult as sorting. They showed that the number of comparisons needed for the problem of median finding is at most $5.43n + o(n)$. This result was improved in 1976 by Schönhage, Paterson, and Pippenger [7] to $3n + o(n)$ comparisons. It was first in 1994 anyone could improve the $3n$ upper bound despite many attempts from researchers all over the world. By modification, of the algorithm by Schönhage et al., Dor and Zwick achieved an upper bound of $2.95n$ comparisons for the problem [5]. There is still, however, a large gap between the upper bound and the lower bound on $2n$ comparisons that is due to Bent and John [1], also an old result from 1985.

In this context, we would like to study how well this can be implemented in-place. This is interesting both from a theoretical and from a practical point of view. A theoretically interesting question is how much of the information gathered by the algorithm can be implicitly stored in the in-place ordering of the elements and how much information needs to be recomputed or stored externally. From a practical point of view we would also like to minimize the extra storage used. One reason is that by limiting the extra storage we will be able to keep a larger part of the data set in faster memory, which will result in a faster

---

[*] Division of Computer Science, Luleå University of Technology, S-971 87 LULEÅ, Sweden. E-mail: {Svante.Carlsson, Mikael.Sundstrom}@sm.luth.se

algorithm. Another reason is that in an implicit data structure we are able to take advantage of localities of memory accesses. When we are using pointer representations we can never be sure where elements that are going to be accessed close in time are going to be stored, which will lead to several page faults that will slow the algorithm down.

By implicit (or in-place) selection we mean that we are only allowed to use a constant amount of additional space, apart from the array storing the elements. In this paper, we show that the heavy space restriction only have a marginal effect on the number of comparisons for finding the median and still only using linear time for the algorithm. We can get arbitrarily close to Dor and Zwick's upper bound on $2.95n$ comparisons. The best algorithm, in the sum of the number of comparisons and moves, that we have found takes $3.75n$ comparisons and $9n$ moves, which is about twice as good for both operations as for the previously best result of slightly less than $7n$ comparisons and $19n$ data moves by Lai and Wood [6].

2 Overview of the Algorithm

Our in-place selection algorithm accepts as input an n elements array $A = a_0, a_2, \ldots, a_{n-1}$ of elements drawn from a totally ordered universe. We will assume, in the rest of the paper, that the elements in A are distinct. However, by using three way comparisons, our algorithm, with minor modifications, also handles repetitions among the elements in A, at no extra cost.

Based on similar ideas as the selection algorithm by Schönhage, Paterson and Pippenger [7], our algorithm relies heavily upon a *spider factory* used in the mass production of k-spiders. A k-spider is a partial order consisting of a special element c called the center, k elements that are smaller than c, and k elements that are larger than c. We will locate the spider factory in the leftmost part of the input array. The rest of the array will be organized in as many *spider sites* of size $2k + 1$ as possible. A spider site can be either *prepared, in use* or *eliminated*. A prepared spider site contains raw material to the spider factory, a spider site in use contains the elements of a k-spider while an eliminated spider site contains elements that no longer are considered as median candidates.

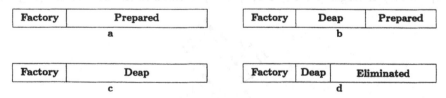

Fig. 1. The state of A : a. after preparation, b. during propagation, c. at the end of propagation, and d. during elimination.

The first phase of our algorithm is the *preparation phase*, in which the spider factory is built and all spider sites are prepared (Fig. 1 a). When a spider site is prepared, the elements stored in it are refined into the kind of raw material (e.g. pairs) accepted by the factory.

The next phase of the algorithm is the *propagation phase*, in which k-spiders are produced until all the prepared spider sites are used. A k-spider is produced by *extracting* elements from the factory and swapping them with the elements from the leftmost prepared spider site, followed by a *reconstruction* of the factory. By using the center as key, the produced k-spider is then inserted into a double ended priority queue of previously produced k-spiders. Since we, logically, have an *array* of k-spiders, we will use an implicit priority deque called deap [4]. In Fig. 1 b-c, we see how the prepared spider sites are consumed while the deap grows.

In the final phase of our algorithm, the *elimination phase*, we will eliminate elements until only a few are still in consideration as median candidates. Elimination is accomplished by moving the smallest and largest k-spiders to the spider sites at the right end of the deap as they are deleted from the deap. By swapping elements between the spider sites, the right spider site becomes eliminated. The left spider site is then prepared and used in the production of a new k-spider, which is then inserted into the deap. As one k-spider is inserted for every two deleted, the deap shrinks during this phase Fig. 1 d. The $o(n/\lg n)$ elements remaining when elimination can not proceed are sorted with heapsort to find the median.

To make the selection algorithm implicit there are mainly two problems that need to be solved, namely, how to construct an efficient spider factory in-place and how to implement the priority deque of k-spiders while limiting the number of moves.

3 The Implicit Spider Factory

The key to our main result is the implementation of the spider factory. It consists of a finite partial order, called *hyperpair*, with a distinguished element, the *center*, defined recursively by: H_0 is a single element, and H_{i+1} is obtained from two disjoint copies of H_i by comparing their centers and taking the smaller ($i+1 = 1, 4, 6, 8, \ldots$) or larger ($i+1 = 2, 3, 5, 7, \ldots$) of these as the new center (Fig. 2).

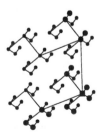

Fig. 2. A H_6 hyperpair with the center shown as a square and the related elements shown as larger dots.

The hyperpair was also used by Schönhage et al. [7], and Dor and Zwick [5]. We use it in a slightly different fashion by relying on two basic operations, *pair* and *unpair*, to build a H_{2h}, $2^{h-1} < k \leq 2^h - 1$, and then produce k-spiders by extracting them from the H_{2h} while simultaneously reconstructing the hyperpair.

The basic idea behind our representation of hyperpairs is to store the elements of a H_i in 2^i consecutive locations of A, with one H_{i-1} sub-hyperpair stored, recursively, in the 2^{i-1} first locations, and the second in the last 2^{i-1} locations. By using a *bit array* $B = b_1, b_2, \ldots$, we record the relation between the centers of the sub-hyperpairs in the bit with the same index as the first element of the second sub-hyperpair.

The problem with the above *basic representation* is that the center of a H_i can not be found in constant time. By examining one bit we can decide in which half of the hyperpair the center resides, and by repeating this i times we will find the center. However, if we, after comparing the centers of the H_{i-1}s, store the one that is taken as the center of the H_i being under construction in the first location (by swapping the centers of the H_{i-1}s, if necessary), we will be able to find the center of the H_i without examining any bits. We are also able to restore the sub-hyperpairs since we have recorded the outcome of the comparison, and, hence, also if we have performed a swap or not. This is called the *swap representation*.

In the swap representation, a pair (i, j) is sufficient to identify a H_i hyperpair stored at position j (the position of the center is j). Also, it is necessary that the two H_is used to build a H_{i+1} hyperpair are adjacent at positions j_1 and j_2 satisfying $|j_1 - j_2| = 2^i$ and $\min(j_1, j_2) \equiv 0 \pmod{2^{i+1}}$. We define the basic operations as follows:

- *pair* $(i + 1, j)$ builds $(i + 1, j)$ by comparing the centers a_j and a_{j+2^i} of (i, j) and $(i, j + 2^i)$ respectively. If a_j is taken as the new center, we flip b_{j+2^i} to record this. Otherwise, a_j and a_{j+2^i} are swapped.
- *unpair* $(i + 1, j)$ reverses the effect of a previous call *pair* $(i + 1, j)$. If $b_{j+2^i} = 0$, it is flipped and $\langle j, j + 2^i \rangle$ is returned. Otherwise, a_j and a_{j+2^i} are swapped and $\langle j + 2^i, j \rangle$ is returned.

Since our factory is a hyperpair with 2^{2h} elements we will need a bit array of size $2^{2h} - 1$ for B. We will use a pair of adjacent elements in A to emulate a bit, by placing them in ascending or descending order. This standard technique for encoding bits in-place results in one comparison for each bit examined, and one swap for each bit flip. We extend our *factory space* to $3 \cdot 2^{2h}$ locations in A to contain also the bit array representation.

Lemma 1. *Performing pairing in the swap representation requires one element comparison and two moves, while unpairing requires one bit comparison and two moves. The factory is built in $2\left(2^{2h} - 1\right)$ comparisons and $4\left(2^{2h} - 1\right)$ moves.*

4 k-spider Production

In this section we describe how a k-spider is produced and also present an analysis, which amounts in the following lemma.

Lemma 2. *For any non negative integer, x, there exists a spider factory of size $\Theta\left(k^2\right)$, in A, that, given pairs as raw material, produces a k-spider in $\left(4 + 6 \cdot 2^{-x}\right)k + o(k)$ comparisons and $\left(2^{x+1} + 2 + 24 \cdot 2^{-x}\right)k + o(k)$ moves.*

We begin by describing how a k-spider is produced using a simple factory that accepts singletons as raw material.

Production is performed by first *decomposing* the H_{2h} constituting the factory into its center c and a disjoint set of hyperpairs $H_0, H_1, \ldots, H_{2h-1}$, followed by swapping c with the element at the first location of the spider site. The factory is decomposed in $\Theta(h)$ time by repeatedly calling *unpair*.

To obtain k elements that are smaller than c in the k-spider, we will *downward process* the hyperpairs $H_1, H_2, H_4, \ldots, H_{2(h-1)}$, which all have centers smaller than c. Downward processing a H_i at position j is accomplished by calling $d\_process(i, j)$,

$$
\begin{aligned}
d\_process(0, j) &= \text{swap } a_j \text{ with an element from the spider site} & (1.1)\\
d\_process(i, j) &= \langle j', j'' \rangle \leftarrow unpair(i, j) & (2.1)\\
&\quad d\_process(i - 1, j') & (2.2)\\
&\quad \text{if } i = 2, 3, 5, 7, \ldots \text{ then } d\_process(i - 1, j'') & (2.3)\\
&\quad pair(i, j) & (2.4)
\end{aligned}
$$

which recursively swaps its center and the elements (guaranteed to be) smaller than its center with raw material from the spider site. As the recursion rolls up, a new H_i is built at the same position. To complete the k-spider we extract k elements larger than c by *upward processing*. This is accomplished by applying a procedure similar to $d\_process$ to the hyperpairs $H_0, H_3, H_5, \ldots, H_{2h-1}$, which have centers larger than c.

Lemma 3. *By downward processing, 2^i elements that are smaller than c can be extracted from the $H_{2i}, i > 0$, hyperpair, and by upward processing, 2^i elements that are larger than c can be extracted from the $H_{2i+1}, i > 0$, hyperpair.*

Proof. Let $f_e(i)$ be the number of elements extracted by downward processing a H_i hyperpair. In (1.1), the singleton which is smaller than c is simply swapped with an element (raw material) at an appropriate location in the spider site. A hyperpair (i, j), $i > 0$, with center $c' < c$, is split in (2.1). The restored sub-hyperpair $(i - 1, j')$, which, by the definition of *unpair*, has c' as center, is then recursively processed in (2.2) obtaining $f_e(i - 1)$ elements. If the condition in (2.3) is satisfied, the center of $(i - 1, j'')$ is smaller than c', by the definition of *pair*, and, hence, $(i - 1, j'')$ can also be recursively processed to extract another $f_e(i)$ elements. A new (i, j) hyperpair is built in (2.4). Solving the recurrence $f_e(2) = 2$, $f_e(2i) = 2f_e(2(i - 1))$, yields the lemma for downward processing. The proof for upward processing is similar. □

From Lemma 3, it follows that k smaller and k larger elements can be extracted, provided $k \leq 2^h - 1$. We show in [3] how our way of representing hyperpairs makes it possible to implement the processing procedures iteratively with constant extra storage.

The k-spider now completed but the hyperpairs $H_0, H_0, H_1, \ldots, H_{2h-1}$ must be assembled into a H_{2h} before the next k-spider can be produced. This is achieved in $\Theta(h)$ time, by repeatedly calling *pair*.

In the remainder of this section we will investigate the processing cost while extracting elements as pairs instead of singletons and by using pairs as raw material. The raw material in a prepared spider site will be organized in k pairs of adjacent elements stored in sorted order. That implicit representation of H_1s can be used also in the spider factory to avoid bit comparisons at the lowest level in the recursion. We can, as shown in the following lemma, store even larger hyperpairs in an implicit representation of a hyperpair of that size to reduce the number of bit comparisons arbitrarily close to 0.

Lemma 4. *If we represent H_{2x}s implicit in the spider factory we get a processing cost of $2 + 3 \cdot 2^{-x}$ comparisons and $2^x + 1 + 12 \cdot 2^{-x}$ moves per extracted element.*

Proof. The number of element comparisons performed in the downward processing of a H_{2i} and the upward processing of a H_{2i+1} are given by the recurrence $f_c(i) = 2 + 2f_c(i-1)$, since *pair* is called once at each level in the recursion and the recursion branches every second level. We now consider the downward and upward processing of a H_2, $\alpha < \beta < \gamma$, $\alpha < \delta$. In the former case, two singletons γ and δ are left after the pair $\alpha < \beta$ has been extracted, while, in the latter case, the pair $\beta < \gamma$ is extracted, leaving the pair $\alpha < \delta$. It follows that the cost for building a new H_2 from the elements left and a new pair from the spider site is 2 element comparisons for downward processing and 1 element comparison for upward processing. Also accounting for one element comparison used to pair two H_2s, during the upward processing of a H_3, yields the base case $f_c(1) = 2$ for both kinds of processing.

We get a slightly higher cost for upward processing a H_{2x+1} than for downward processing a H_{2x} (2^x elements are obtained in both cases) since the H_{2x+1} must be unpaired before the H_{2x} can be processed. It follows that the cost for upward processing is slightly higher than the cost for downward processing.

Let $f_b(i)$ be the number of bit comparisons and $f_m(i)$ be the number of moves required to upward process a H_{2i+1}. For comparisons, the base case is $f_b(x) = 1$, since one call to *unpair* is made when a H_{2x+1} is processed. For moves, we first consider the processing of a H_{2x}. Exactly 2^x elements are extracted. These are swapped with 2^x elements from the spider site. Since the H_{2x} is stored in an implicit representation, the reconstruction of the H_{2x}, from the elements left and the new elements from the spider site, requires, in the worst case, a complete reorganization of the $2^{2x} - 2^x$ elements of the H_{2x} that are left in the factory. It follows that the number of moves to process a H_{2x} is bounded by $2^{2x} + 2^x$. Also accounting for the calls to *unpair* and *pair*, performed in the processing of the H_{2x+1}, yields the base case $f_m(x) = 4 + 2^{2x} + 2^x$.

At each level in the recursion *unpair* and *pair* are called once, and, since the recursion branches ever second level, we get the recurrences $f_b(i) = 2 + 2f_b(i-1)$ and $f_m(i) = 8 + 2f_m(i-1)$. It is easily shown that $f_c(i) = 2 \cdot 2^i$, $f_b(i) \leq 2^i \cdot 3 \cdot 2^{-x}$ and that $f_m(i) \leq 2^i \left(2^x + 1 + 12 \cdot 2^{-x}\right)$. Since 2^i elements are extracted we get a cost of $2 + 3 \cdot 2^i$ comparisons and $2^x + 1 + 12 \cdot 2^{-x}$ moves per element. $\square$

We may choose x as an arbitrarily large value, and still only have a linear number of moves for extraction of k elements. Since $2k$ elements are extracted, and the decomposition and composition of the factory takes $\Theta(h)$ time, Lemma 2 follows from Lemma 4.

5 The Priority Deque and the Spider Sites

As described in Section 2 the centers of the k-spiders are inserted into a priority deque after the k-spider have been produced. Any implicit priority deque that supports insertion, extract-min and extract-max in $O(\lg t)$, where t is the number of elements in the deque, can be used. We have chosen the deap as our priority deque [4].

A question that arises, and must be answered, at this point is exactly what to insert in the deap. We can not afford to insert the whole k-spider since it would impose a severe overhead in the number of moves performed. On the other hand, if the center alone is inserted it would be difficult to remember the position of the remaining $2k$ elements without external pointers. Our approach is something in between these two extremes.

We store a k-spider in two parts called the *head* and the *tail*, where the head consist of the center and $\lceil \lg n \rceil$ pairs of the k-spider while the remaining $k - \lceil \lg n \rceil$ pairs constitutes the tail. By using the same technique for encoding bits, as used in the spider factory, we will encode the binary representation of the first location of the tail among the $\lceil \lg n \rceil$ pairs of the head. In this way, the head is provided with enough intelligence to find its own tail.

Now, consider the preparation of spider sites in the *preparation phase*. As mentioned in Section 2, we prepare all spider sites. A spider site is prepared by building k pairs, at a cost of k comparisons and $2k$ moves. Since the number of spider sites prepared is bounded by $n/(2k+1)$, and the spider factory is built in $\Theta(2^{2h})$ time, it follows that $0.5n + \Theta(2^{2h})$ comparisons and $n + \Theta(2^h)$ moves are performed during the preparation phase.

Now we turn our attention to the *propagation phase*. When a new k-spider is to be produced, the leftmost prepared head and prepared tail, are designated as spider site. When the k-spider is completed, a pointer to the tail is encoded in the head, followed by an insertion of the head in the deap.

Finally, we give a more detailed description of the behaviour of our algorithm during the *elimination phase*. In this phase, k-spiders are alternately produced and destroyed. In fact, one k-spider is produced for each pair of k-spiders destroyed.

It can be shown that the smallest center and the largest center, in the deap, are too small and too large, respectively, to be the median if the number of elements left in the algorithm is larger than $7k^3$. It follows that the k elements smaller than the smallest center and the k elements that are larger than the largest center are also too small or too large to be the median. These $2k + 1$ elements can, therefore, be eliminated. Elimination is accomplished by swapping the pairs from the smallest k-spider that are smaller than the smallest center, with the pairs from the largest k-spider that are smaller than the largest center. In this way, the site of the larger k-spider becomes eliminated, while the site of the smaller k-spider becomes prepared. The prepared spider site is then used in the production of a new k-spider. The cost for insertion, deletion and elimination is $\Theta(\lg n)$ comparisons and $k + \Theta(\lg^2 n)$ moves per k-spider.

Choosing $k = 2^h - 2$, where $h = \lceil 0.25 \lg n \rceil$, yields a negligible cost for sorting the $\Theta(k^3)$ elements left (when elimination can not proceed). The costs for inserting, deleting and eliminating a k-spider also becomes negligible except for k moves during elimination.

6 Improving the Factory with Grafting

Schönhage et al. obtained a major improvement in their algorithm by a process called *grafting*. In this section we will incorporate this process in our algorithm.

Grafting is performed prior to processing, and the idea is to repeatedly compare elements with the center until either k elements smaller than the center or k elements larger than the center are found. The grafted elements are obtained at a low cost compared to the, at most, k elements that have to be extracted from the factory to complete the k-spider.

As Schönhage et al., we will use pairs for grafting, and, in order to reduce the number of moves, performed by the grafting process, we will use the pairs that already resides in the prepared spider site currently in use. Grafting the pair $v < w$, to the center c, yields one of three outcomes. We either have (i) $v < w < c$, (ii) $c < v < w$, or (iii) $v < c < w$. Three counters, p_0, p_1, and s, are used to record the number of occurrences of outcomes (i), (ii), and (iii), respectively. Prior to the grafting, the k *pair locations*, in the spider site, are occupied by k pairs (raw material). We dedicate $\frac{k}{2}$ of these pairs and pair locations as *smaller pair candidates* and *smaller pair locations*, respectively. The remaining pairs and locations are referred to as *larger pair candidates* and *larger pair locations*, respectively. The grafting proceeds according to the following description until $s + 2\max(p_0, p_1) \geq k - 1$. If $p_0 > p_1$, let $v < w$ be the $p_0 + 1$st smaller pair candidate and compare w to c, and if $w > c$, compare v, also, to c. Otherwise, let $v < w$ be the $p_1 + 1$st larger pair candidate and compare v to c, and then compare w and c, also, if v turns out to be smaller than c. If outcome (i) occurs when $p_0 \leq p_1$ or outcome (ii) occurs when $p_0 < p_1$, the $p_0 + 1$st smaller pair candidate and the $p_1 + 1$st larger pair candidate are swapped. When outcome (iii) occurs, the grafted pair is swapped with the $k - \frac{s}{2}$th smaller pair candidate if s is even, or swapped with the $k - \lfloor \frac{s}{2} \rfloor$th larger pair candidate if s is odd.

When the grafting terminates, the pairs from outcome (i) and (ii) occupies the first p_0th smaller pair locations and the first p_1th larger pair locations, respectively, whereas the singletons resulting from outcome (iii) occupies the last $\lceil \frac{s}{2} \rceil$ smaller pair locations and the $\lfloor \frac{s}{2} \rfloor$ last larger pair locations. The pairs that are not related to the center constitutes the raw material that is used in the completion of the k-spider by processing hyperpairs in the factory. To enable recycling of pairs and singletons, the binary representation of s is encoded in the k-spider itself, by using $\lceil \lg k \rceil$ pairs of elements. For an extended description please refer to [3].

Lemma 5. *Selection in-place can be performed in* $(3 + 3 \cdot 2^{-x}) \cdot n + o(n)$ *comparisons.*

Proof. Every comparison performed by the grafting process yields at least one element related to the center of the k-spider. The cost for increasing $\max(p_0, p_1)$ by one during the grafting process is $\max(p_0, p_1) - \min(p_0, p_1) + O(1)$ comparisons. Hence, the cost for obtaining $2s + 2p_0 + 2p_1$ elements by grafting is $2\min(p_0, p_1) + p_0 + p_1 + 2s + O(1)$ comparisons. For every *two* occurrences of outcome (iii) we must charge one extra comparison for building a pair when these singletons are recycled, yielding an additional $0.5s$ comparisons. By Lemma 4, the $2k - 2s - 2p_0 - 2p_1$ elements, needed to complete the k-spider, are extracted from the factory at a unit cost of $2 + 3 \cdot 2^{-x}$ comparisons. Since $2 + 3 \cdot 2^{-x} \geq 2$, $s + 2\max(p_0, p_1) \geq k - 1$, and the number of k-spiders produced is bounded by $\frac{n}{k}$, we get a total production cost of $(2.5 + 3 \cdot 2^{-x})n + o(n)$ comparisons for the algorithm. Adding the $\frac{n}{2} + o(n)$ comparisons performed during the preparation phase yields the lemma. □

Obtaining a bound of $(2^x + 3 + 12 \cdot 2^{-x}) \cdot n + o(n)$ moves for the described algorithm is straight forward. The recycling cost is $1.5s$ moves and the proof is essentially the same as for comparisons. However, to obtain our best upper bound of $(2^x + 2 + 12 \cdot 2^{-x}) \cdot n + o(n)$ moves, while keeping the same bound for comparisons, we have used much more sophisticated techniques of moving elements. The description and analysis of these techniques takes about five pages and are, therefore, left out due to space restrictions [3].

By inserting $x = 2$ in the expressions above we get the best sum of comparisons and moves.

Theorem 6. *In-place selection can be performed in* $3.75n + o(n)$ *comparisons and* $9n + o(n)$ *moves.*

This result improves the previously best bounds for implicit selection of $6.7756n + o(n)$ comparisons and $18.6873n + o(n)$ moves by Lai and Wood [6].

7 Getting Below $3n$ Comparisons

Recently Dor and Zwick [5] improved the upper bound for selection to $2.95n + o(n)$ comparisons with an algorithm that is based on Schönhage et al. but with some new features. We will indicate how these new features can be incorporated in our in-place selection algorithm.

The algorithm uses sub-factories that accepts different kinds of partial orders as raw material. Partial orders of the same kind can be stored in sub-arrays called *stockpiles* in the in-place version. Each stockpile can be moved in time proportional to the distance by moving elements from one end of the stockpile to the other end. This can be achieved since the order between the partial orders in a stockpile is irrelevant, and since we can allow one partial order to be stored with one part in each end of the stockpile. If space, in some direction, must be made available for a stockpile (to move or to grow) the next stockpile in that direction is moved. This may, of course, result in that all stockpiles are moved, but can still be performed in constant time since both the number of stockpiles and the size of stored partial orders are bounded by a constant.

Another feature is that the spiders produced does not necessarily have k elements on each side of the center. This results from the use of a more powerful grafting, which yields many outcomes, and that, for each outcome, an optimal number of elements are extracted from the factory. The different kinds of outcomes from the grafting are managed by using stockpiles also in the spider sites. When a k-spider has been completed, the size of each stockpile is recorded in the k-spider using the same technique as described in Section 5.

Since the new features of Dor and Zwick's algorithm can be incorporated with our algorithm and since we can reduce the number of bit comparisons to an arbitrarily small value, by Lemma 4, our main theorem follows.

Theorem 7. *There exists an linear time in-place selection algorithm that finds the median in $(2.95 + \epsilon) \cdot n + o(n)$ comparisons and $\Theta(n)$ moves.*

8 Acknowledgement

We would like to thank Dr. Jingsen Chen for valuable discussions on the problem and for constructive comments on the presentation of these results.

References

1. S. W. Bent, J. W. John, Finding the median requires $2n$ comparisons, In *Proceedings of the 17th Annual Symposium on Theory of Computing*, pp. 213–216, 1985.
2. M. Blum, R. W. Floyd, V. Pratt, R. L. Rivest, and R. E. Tarjan, Time bounds for selection, *Journal of Computer and System Sciences*, 7:448–461, 1973.
3. S. Carlsson and M. Sundström, Linear-time In-place Selection in Less than 3n Comparisons, extended version available at <URL:http://www.sm.luth.se/~msm/reports/in-place.ps>
4. S. Carlsson, The Deap – A double-ended heap to implement double-ended priority queues, *Information processing Letters* 26 (1):33–36, 1987.
5. D. Dor and U. Zwick, Selecting the median, In *Proceedings of the 6th ACM-SIAM Symposium on Discrete Algorithms (SODA'95)*, San Francisco, California, January 1995.
6. T. W. Lai and D. Wood, Implicit Selection, In *SWAT 88*, pp. 14–23, 1988.
7. A. Schönhage, M. Paterson, and N. Pippenger, Finding the median, *Journal of Computer and System Sciences*, 13:184–199, 1976.

Heap Construction:
Optimal in Both Worst and Average Cases?

Svante Carlsson   Jingsen Chen

Department of Computer Science, Luleå University, S–971 87 Luleå, Sweden

Abstract. We investigate the complexity of constructing heaps. The heap construction problem has been extensively studied. However, there was no algorithm for building heaps that is optimal in both the worst and average cases *simultaneously*. In particular, the worst-case fastest algorithm, proposed by Gonnet and Munro, takes $1.625n$ comparisons to build an n-element heap (with an average cost of $1.5803n$ comparisons). The best known average-case upper bound of $1.5212n$ comparisons was derived by McDiarmid and Reed, which has a worst-case performance of $2n$ comparisons. Both algorithms require extra space and were conjectured to be optimal respectively in the worst and the average case. In this paper, we design a heap construction algorithm that takes at most $1.625n$ and $1.500n$ comparisons in the worst and average cases, respectively. Our algorithm not only improves over the previous best known average-case result by McDiarmid and Reed, but also achieves the best known worst-case upper bound due to Gonnet and Munro. Moreover, we also show that a heap on n elements can be constructed *in-situ* using at most $1.528n$ comparisons on average and $2n$ comparisons in the worst case. This is only $0.007n$ comparisons more than that of McDiarmid and Reed's algorithm, while the latter needs n bits of extra space.

1   Introduction

The study and construction of time and space efficient algorithms is one of the fundamental areas of modern computing. It would be desirable to have a single algorithm efficient in both worst- and average-case performances. However, this is considerably hard and it is solved only for a couple of problems, such as the problem of finding the minimum out of a set of elements, and the problem of finding the minimum and maximum simultaneously [6,12]. For some problems, such "universal" optimal algorithms even do not exist. For example, it has been shown that there is no algorithm for finding the second smallest of six elements that can be both worst- and average-case optimal [12]. In this paper, we investigate this interesting algorithmic aspect for the heap construction problem from the complexity point of view.

One of the fundamental data types in Computer Science is the priority queue. It has been useful in many applications [1,12]. A *priority queue* is a set of elements on which two basic operations are defined: inserting a new element into the set; and retrieving and deleting the minimum element of the set. Several data structures have been proposed for implementing priority queues. Probably the most elegant one is the heap [16]. The problem of constructing heaps has received much attention in the literature [7,8,10,12,13,15]. Considerable effort has been made to reduce the gaps between the lower and upper bounds, for both the worst and average cases, on the number of comparisons required to solve the construction problem.

The first algorithm for constructing heaps was proposed by Williams [16]. His method creates a heap by repeatedly inserting elements into an initially empty heap. It has been shown [2,16] that Williams' method for constructing an n-element heap requires $n \log n + \mathcal{O}(n)$ and $2.76n$ comparisons in the worst and average cases, respectively. By repeatedly merging smaller heaps into bigger ones, Floyd [8] presented a heap construction algorithm that takes $2n + o(n)$ comparisons in the worst case and $1.88n$ comparisons on the average [7,12]. Combining the ideas of the previous methods, Carlsson [3] reduced the expected number of comparisons for creating an n-element heap to $1.65n$ using no extra space; and the $1.521n$-algorithm was due to McDiarmid and Reed [13], which needs n bits of extra space. Both algorithms have a worst-case performance of $2n + o(n)$ comparisons. The worst-case fastest algorithm, developed by Gonnet and Munro [10], takes $1.625n$ comparisons to build an n-element heap. This algorithm can be modified slightly to ensure an average cost of $1.5803n$ comparisons with no affect on its worst-case complexity [10,13]. The best known lower bound for the heap construction is $1.5n$ comparisons [5] in the worst case and $1.3644n$ on the average [10].

In this paper, we shall present a heap construction algorithm that takes at most $1.625n$ and $1.500n$ comparisons in the worst and average cases, respectively. Our result not only improves the previous best known average-case upper bound of $1.521n$ comparisons which was conjectured to be optimal [13], but also achieves the best known worst-case upper bound [10]. Recall that one nice property about heaps is that the relative ordering information on the structure can be stored implicitly (i.e., using no extra space). One would certainly like to see that efficient heap construction algorithms demand as little space as possible. We shall show that a heap on n elements can be constructed *in-situ* using at most $1.528n$ comparisons on average. This is only $0.007n$ comparisons more than the previous best known upper bound due to McDiarmid and Reed [13], while the latter needs n bits of extra space.

2 Preliminaries

A (min-)heap is a binary tree with *heap-ordering*: (i) It has the *heap shape*; i.e., all leaves lie on at most two adjacent levels and all leaves on the last level occupy the leftmost positions and all other levels are complete; (ii) It is *min-ordered*: the key value associated with each node is not smaller than that of its parent. The minimum element is then at the root. A (max-)heap is defined similarly. A heap of size $2^{h+1} - 1$ ($h \geq 0$) is called a *full* heap.

In developing our algorithms for the heap construction, we will employ two data structures: the binomial tree [14] and the pseudo-binomial tree. A *binomial tree* of height k ($k \geq 1$) is a min-ordered tree comprising two binomial trees of height $k-1$ which are connected at their roots (i.e., making the root of one with the larger key value stored in its root being a child of the root of the other). A binomial tree of height zero is a tree with only one element. Notice that a binomial tree of height k has exact 2^k elements, and the class of binomial trees can also be represented by arrays. A *pseudo-binomial tree* is a binomial tree with one of its leaves missing. In other words, a pseudo-binomial tree of height k is of $2^k - 1$ elements and can be regarded as a binomial tree of size 2^k with one of its leaves being empty (or having value $+\infty$).

To search for efficient heap construction algorithms, one is often concerned with finding an elegant method which has a good performance on the "hardest" instance of the problems. It has been shown that if the cost to construct a full heap of size m is $c(m)$, then building a heap on any number n of elements takes at most $c(n)+\mathcal{O}(\log^2 n)$ comparisons [10]. Therefore, we shall limit our attention to the problem of constructing full heaps in this paper. Assume that all $n!$ linear orders on input data of size n are equally likely. If all the building blocks are random, then the bottom-up construction scheme, of repeatedly merging small structures to form bigger ones, preserves randomness [7,12]; namely, each of the all possible heaps on n elements is equally likely after the construction process. When performing insert operations, we assume that the new element is taken from a set whose elements are distinct mutually and distinct with those elements that have already been stored in the structure.

3 Fast in Both Worst and Average Cases

The heap construction algorithm [10], with the best known worst-case number of comparisons (namely $1.625n$), works by first building a binomial tree and then converts it recursively into a heap. The recursion step can be modified slightly to ensure an average cost of $1.5803n$ comparisons with no affect on its worst case complexity [10,13]. However, we can do even better by building a forest of of binomial trees instead of just one binomial tree of size n. Before presenting our main result of this paper, we shall demonstrate some properties of the pseudo-binomial tree and its relation to the heap. Recall that a binomial tree on n elements can be built in $n-1$ comparisons. By first creating a collection of binomial trees of sizes 2^i ($i = 0, 1, \cdots, k-1$) and then connecting the roots of the trees, we have that

Lemma 1. *Building a pseudo-binomial tree of size 2^k-1 costs 2^k-2 comparisons in the worst case.*

To transform a pseudo-binomial tree into a heap on the elements, we first examine pseudo-binomial trees of smaller sizes.

Lemma 2. *Given four binomial trees of sizes $1, 2, 4$, and 8, respectively, a 15-element heap can be built in 9 comparisons in the worst case and 7.7685 comparisons on average.*

Proof. Suppose that the roots of the binomial trees are r_1, r_2, r_4, and r_8, respectively; see Figure 1.

FIGURE 1: Build a heap of size 15.

In order to build a 15-element heap out of these binomial trees by using 7.7685 comparisons on the average, we first compare r_1 with r_2, next compare the smaller one with r_4, and finally the winner is compared with r_8. These three comparisons will lead to different pseudo-binomial trees each of size 15. With a careful examination, the total number of comparisons required by the transformation can be counted as follows case by case. Denote the binomial tree rooted at r_i by $\mathcal{B}_i$ for $i = 1, 2, 4$, and 8.

1. **r_1 is the smallest.**
   The probability of arriving at this point ($\mathcal{P}_1$ in Figure 2) is $\frac{1}{3} \times \frac{3}{7} \times \frac{7}{15} = \frac{1}{15}$.
   In this case, one can show that at most $3\frac{15}{49}$ comparisons are needed to build the left subheap of r_1 from binomial trees $\mathcal{B}_2$ and $\mathcal{B}_4$ plus the element x. Moreover, one more comparison must be done in order to build the right subheap of r_1.

2. **r_2 is the smallest.**
   The probability that this case ($\mathcal{P}_2$ in Figure 2) occurs equals $\frac{2}{3} \times \frac{3}{7} \times \frac{7}{15} = \frac{2}{15}$.
   In this case, we first make a comparison between x and y and then compare the winner to r_1. Hence, the average-case cost to transform $\mathcal{P}_2$ into a 15-element heap is at most $5\frac{1}{6}$ comparisons.

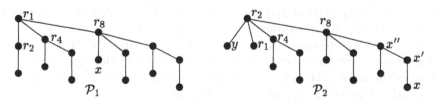

FIGURE 2: Either r_1 or r_2 is the smallest.

3. **r_8 is the smallest.**
   The probability that this event occurs is equal to $\frac{8}{15}$. The average number of comparisons for building the heap is $4\frac{4}{7}$.

4. **r_4 is the smallest.**
   The probability that either $\mathcal{Q}_1$ or $\mathcal{Q}_2$ in Figure 3 arises is $\frac{4}{7} \times \frac{7}{15} = \frac{4}{15}$ and one additional comparison is needed to establish the right subheap of r_4.

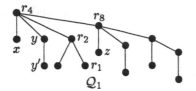 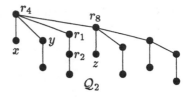

FIGURE 3: Cases arisen when r_4 is the smallest.

For constructing the left subheap, either $4\frac{40}{147}$ (starting from $\mathcal{Q}_1$) or $3\frac{19882}{28665}$ (starting from $\mathcal{Q}_2$) comparisons are sufficient on average. Consequently, the average number of comparisons for the case when r_4 is the smallest is at most

$$Pr(r_2 \prec r_1) \times 5\frac{40}{147} + Pr(r_1 \prec r_2) \times 4\frac{19882}{28665} = \frac{2}{3} \times 5\frac{40}{147} + \frac{1}{3} \times 4\frac{19882}{28665} = 5\frac{6817}{85995}$$

In summary, building a 15-element heap, starting from four binomial trees of sizes respectively $1, 2, 4,$ and 8, costs at most:

$$3 + Pr(r_1 \text{ is MIN}) \times 4\frac{15}{49} + Pr(r_2 \text{ is MIN}) \times 5\frac{1}{6} + Pr(r_8 \text{ is MIN}) \times 4\frac{4}{7}$$
$$+ Pr(r_4 \text{ is MIN}) \times 5\frac{6817}{85995}$$
$$= 3 + \frac{1}{15} \times 4\frac{15}{49} + \frac{2}{15} \times 5\frac{1}{6} + \frac{8}{15} \times 4\frac{4}{7} + \frac{4}{15} \times 5\frac{6817}{85995}$$
$$= 7\frac{991348}{1289925}$$

It should be clear that the worst-case number of comparisons done by the algorithm above is at most 9, which is implicit in the above proof.

With the above lemma, we can refine the idea behind Gonnet and Munro's algorithm [10] to slightly reduce the average number of comparisons performed by the algorithm. We sketch the refined algorithm below. First we build a binomial tree and then recursively convert the binomial trees of sizes 2^i into heaps of sizes $2^i - 1$, same as the algorithm in [10]. Notice that the modified algorithm will jump out of the recursion on the trees of size 32 and employ the method provided by Lemma 2. It is clear that this new algorithm attains the same worst-case complexity as its original, since the maximum number of comparisons used by the new algorithm to transform a binomial tree of size 32 into a 31-element heap is $9+5$ which is exactly the same as that of Gonnet and Munro's algorithm. The second addendum results from the cost to build the right subheap of the root of the binomial tree on 32 elements, which is at most $7\frac{2}{7}$ comparisons. Now, for such a tree of size 32, we can save at least

$$\left(9 - 7\frac{991348}{1289925}\right) + \left(8 - 7\frac{2}{7}\right) = \frac{2509877}{1289925}$$

comparisons on average over its worst-case complexity. The algorithm has thus an average-case cost of at most

$$1.625n - \frac{2509877}{1289925} \times \frac{n}{32} \doteq 1.5642n$$

comparisons for constructing full heaps. Thus, we obtain a slightly better average upper bound for the heap construction.

Theorem 3. *There is an algorithm that always correctly constructs an n-element heap by using $1.625n + \mathcal{O}(\log^2 n)$ comparisons in the worst case and $1.5642n + \mathcal{O}(\log^2 n)$ on the average.*

We have made some improvement on the average-case behavior of the worst-case fastest algorithm for the heap construction. However, we can go even further.

Algorithm 4. *Suppose the number of input elements is $n = 2^k - 1$.*

1. *Build binomial trees $\mathcal{B}_1, \mathcal{B}_2, \mathcal{B}_4, \cdots,$ and $\mathcal{B}_{2^{k-1}}$, where the size of $\mathcal{B}_i$ is i and its root is denoted by r_i, for $i = 1, 2, 4, \cdots, 2^{k-1}$.*
2. *Let $j := 1$.*
3. *Form a tree $\mathcal{B}_{2j+1}$ by taking the smaller of $\{r_j, r_{j+1}\}$ as the root (denoted by r_{2j+1}) and make the union of tree $\mathcal{B}_j$ and $\mathcal{B}_{j+1}$ plus the order relations between r_j and r_{j+1};*
   $j := 2j + 1$;
   Repeat this step until a pseudo-binomial tree $\mathcal{B}'$ on n elements is created.
4. *Let the subtrees of the root (i.e., the minimum element) of $\mathcal{B}'$ be $\mathcal{B}'_1, \mathcal{B}'_2, \mathcal{B}'_4,$ $\cdots, \mathcal{B}'_{2^{k-1}}$, where $\mathcal{B}'_i$ is a binomial tree of size i for all $i \neq m$ and $\mathcal{B}'_m$ is a pseudo-binomial tree of size $m - 1$ for some m.*

   (a) *If $\mathcal{B}'_{2^{k-1}}$ is a pseudo-binomial tree of size $2^{k-1} - 1$, then*

       i. *Build a heap of size $2^{k-1} - 1$ from $\mathcal{B}'_1, \mathcal{B}'_2, \mathcal{B}'_4, \cdots, \mathcal{B}'_{2^{k-2}}$ by applying Algorithm 4 recursively (starting from Step 2).*

       ii. *Transform $\mathcal{B}'_{2^{k-1}}$ into a heap of size $2^{k-1} - 1$ by applying Step 4 recursively.*

   (b) *If $\mathcal{B}'_{2^{k-1}}$ is a binomial tree of size 2^{k-1}, then*

       i. *Transform $\mathcal{B}'_{2^{k-1}}$ into a heap of size $2^{k-1} - 1$ plus one extra element x by building a heap of size $2^i - 1$ ($i = 2, 3, \cdots, k-2$) from binomial trees of sizes $2^0, 2^1, \cdots, 2^{i-1}$, respectively.*

       ii. *If $m > 1$, then*
           − *Build a binomial tree $\mathcal{B}''_m$ of size m from $x, \mathcal{B}'_1, \mathcal{B}'_2, \mathcal{B}'_4, \cdots, \mathcal{B}'_{\frac{m}{2}}$.*
           − *Transform $\mathcal{B}''_m, \mathcal{B}'_m, \mathcal{B}'_{2m}, \cdots, \mathcal{B}'_{2^{k-2}}$ into a heap of size $2^{k-1} - 1$ by applying Algorithm 4 recursively. (More precisely, we do not need to perform Step 1 and the first $\lfloor \log(m-1) \rfloor$ comparisons in Step 3).*

       iii. *If $m = 1$ (i.e., $\mathcal{B}'_1$ is an empty tree), then transform $x, \mathcal{B}'_2, \mathcal{B}'_4,$ $\cdots$, and $\mathcal{B}'_{2^{k-2}}$ into a heap of size $2^{k-1} - 1$ by applying Algorithm 4 recursively.*

Clearly, the above algorithm constructs a full heap of size n correctly. Let $C(n)$ be the worst-case number of comparisons done by Algorithm 4 and $T(n)$ the cost to transform a pseudo-binomial tree of size n into an n-element heap. Therefore, $C(n) = T(n) + n - 1$. Notice that a 7-element heap can be built in 3 comparisons form a binomial tree of size 4 and a pseudo-binomial tree of size 3 in the worst case. Hence,

$$
\begin{cases}
T(7) & = 2 \quad \text{and} \\
T(2^k - 1) & = 2 \cdot T(2^{k-1} - 1) + \lfloor \log(2^{k-1} - 1) \rfloor
\end{cases}
$$

if we perform Step 4(a) or

$$\begin{cases} T(7) & = 2 \qquad \text{and} \\[2mm] T(2^k - 1) & = (k-1) + T(2^{k-1} - 1) + \sum_{i=3}^{k-2} \left((i-1) + T(2^i - 1)\right) + 1 \end{cases}$$

if we run Step 4(b) for the heap construction. The latter recurrence is exactly the same as the previous one, which gives $T(2^k - 1) = \frac{5}{8} \cdot 2^k - k$. Hence, $C(n) = T(n) + n - 1 \le 1.625n$.

For the average-case complexity, $\bar{C}(n)$, of Algorithm 4, we can regard the element x specified by Step 4(a).i as a random element. This is because building a 7-element heap starting from three binomial trees of sizes 1, 2, and 4, respectively, takes at most $3\frac{2}{7}$ comparisons in the average case. Moreover, if the element of the binomial tree of size 1 is known to be larger than some other elements, then the cost to build the heap will be lower. Notice that $Pr(\|\mathcal{B}'_{2^{k-1}}\| = 2^{k-1}) = \frac{2^{k-1} - 1}{n} = \frac{n-1}{2n}$ for $n = 2^k - 1$, where $\|\mathcal{B}\|$ denotes the size of the tree $\mathcal{B}$. Therefore, the average-case complexity of Algorithm 4 can be computed by $\bar{C}(n) = \bar{T}(n) + n - 1$, where $\bar{T}(n)$ is the expected number of comparisons used by our algorithm to create an n-element heap starting from a pseudo-binomial tree of size n.

$$\bar{T}(7) \le 1\tfrac{2}{7}$$

$$\bar{T}(2^k - 1) = k - 1 + \frac{n-1}{2n}\left((k-1) + \bar{T}(2^{k-1} - 1) + \sum_{i=3}^{k-2}\left((i-1) + \bar{T}(2^i - 1)\right) + 1\right)$$

$$\qquad + \frac{n+1}{2n}\left((k-2) + \bar{T}(2^{k-1} - 1) + \bar{T}(2^{k-1} - 1)\right)$$

$$\quad = \frac{n-1}{2n}\left(\sum_{i=3}^{k-1}\bar{T}(2^i - 1) + \tfrac{1}{2}k(k-1) - (k-2)\right)$$

$$\qquad + \frac{n+1}{2n}\left(k - 2 + 2 \cdot \bar{T}(2^{k-1} - 1)\right)$$

$$\quad = \bar{T}(2^{k-1} - 1) + \tfrac{1}{2}\left(\sum_{i=3}^{k-1}\bar{T}(2^i - 1) + \tfrac{1}{2}k(k-1)\right) + s_k,$$

where

$$s_k = \frac{1}{2n}\left(2(k-2) + \bar{T}(2^{k-1} - 1) - \left(\sum_{i=3}^{k-2}\bar{T}(2^i - 1) + \tfrac{1}{2}k(k-1)\right)\right)$$

By computing $\bar{T}(2^k - 1) - \bar{T}(2^{k-1} - 1)$, we have that

$$\bar{T}(2^k - 1) = \tfrac{5}{2}\bar{T}(2^{k-1} - 1) - \bar{T}(2^{k-2} - 1) + \tfrac{1}{2}(k-1) + s_k - s_{k-1}$$

Assume that $\bar{T}(2^i - 1) \le \frac{1}{2} \cdot 2^i - i + \mathcal{O}(1)$ for $i < k$. Then we have with a simple computation that $\bar{T}(2^k - 1) \le \frac{1}{2} \cdot 2^k - k + \mathcal{O}(\log n)$. This proves the following result by combining the fact that full heaps are the "hardest" instances for the construction problem.

Theorem 5. *There is an algorithm that builds an n-element heap in $\frac{3}{2}n +$ $\mathcal{O}(\log^2 n)$ comparisons on average and $\frac{13}{8}n + \mathcal{O}(\log^2 n)$ comparisons in the worst case.*

Remark that Algorithm 4 makes two progresses. First, it improves the previous best average-case upper bound of $1.521n$ comparisons, the latter bound was conjectured to be optimal by McDiarmid and Reed [13]. Second, our algorithm achieves the best known worst-case upper bound for the heap construction as well.

4 Efficient in Both Time and Space Complexities

In this section we mainly concentrate on both the time and space complexities for efficiently constructing heaps. We shall present a simple algorithm that reduces the amount of space consumed for constructing heaps. Our algorithm consists of two steps. In the first step, we apply McDiarmid and Reed's algorithm [13] to build full heaps of smaller sizes. The second step employs Carlsson's algorithm [3] to handle the remaining elements. More precisely, we have

Theorem 6. *A heap on n elements can be constructed in*

$$1.521n + \left(\frac{8}{3(m+1)} + o\left(\frac{\log(m+1)}{(m+1)^2} \right) \right) n$$

comparisons on average by using m extra bits for any $m = 2^{k+1} - 1$, where $k \geq 1$ and $m \leq n$.

Proof. To achieve the desired time and space bounds, we first apply the algorithm in [13] (called Algorithm MR) to build heaps each of size m. Next, the algorithm in [3] (referred to as Algorithm S) will be employed. Consider the problem of constructing full heaps. Recall that Algorithm MR costs

$$2^{\lfloor \log n \rfloor} \sum_{i=1}^{\lfloor \log n \rfloor} \frac{1 - \frac{i}{2^i - 1}}{2^i}$$

comparisons, which is less than that of Algorithm S when a full heap of size n is built [3,13]. Let $n = 2^{h+1} - 1$ and $m = 2^{k+1} - 1$. Hence, the average number of comparisons done is at most

$$1.649271n - 2^{h-k} \cdot \left(2^k \cdot \sum_{i=1}^{k} \frac{1 - \frac{i}{2^i - 1}}{2^i} \right)$$

$$= 1.649271n + 2^h \cdot \left(-0.2559662 + \frac{8}{3 \cdot 2^k} + o\left(\frac{k}{2^{2k}} \right) \right)$$

$$= 1.521288n + \left(\frac{8}{3(m+1)} + o\left(\frac{\log(m+1)}{(m+1)^2} \right) \right) n$$

Notice that Algorithm S is an in place method while Algorithm MR requires m extra bits. The result follows.

The worst number of comparisons performed by the above method is at most $2n$. Theorem 6 gives a good average performance for heap constructions by investing some extra space. One specific case of the theorem occurs when $m = 123$. Such a situation demands only 123 bits of extra space, and the expected cost to construct an n-element full heap will be dropped down to

$$1.649271n - \frac{n}{2} \cdot \sum_{i=1}^{6} \left(\frac{i+1}{2^i} - \frac{i}{2^i - 1} \right) \doteq 1.5288009n$$

Corollary 7. *A heap on n elements can be constructed in at most $1.528801n + \mathcal{O}(\log^2 n)$ comparisons on average and $\mathcal{O}(1)$ extra space.*

Although our algorithm takes $0.007n$ comparisons more than the previous best known average-case upper bound [13], the latter algorithm uses n extra bits while our algorithm requires $\mathcal{O}(1)$ extra space.

5 Concluding Remarks

We have developed heap construction algorithms that are simple, fast in both the worst and average cases, and efficient in both the time and space complexities. We believe that both the worst upper bound of $1.625n$ and average upper bound of $1.5n$ comparisons in Theorem 5 are tight, although it is not clear how to prove it.

References

1. A.V. Aho, J.E. Hopcroft, and J.D. Ullman. *The Design and Analysis of Computer Algorithms.* Addison-Wesley, Reading, Massachusetts, 1974.
2. B. Bollobás and I. Simon. Repeated random insertion into a priority queue. *Journal of Algorithms*, 6:466–477, 1985.
3. S. Carlsson. Average-case results on heapsort. *BIT*, 27:2–17, 1987.
4. S. Carlsson. A variant of heapsort with almost optimal number of comparisons. *Information Processing Letters*, 24:247–250, 1987.
5. S. Carlsson and J. Chen. The complexity of heaps. In Proceedings of the Third Annual ACM-SIAM Symposium on Discrete Algorithms, pages 393–402. Orlando, Florida, USA, January 1992.
6. W. Cunto and J.I. Munro. Average case selection. *Journal of the ACM*, 36:270–279, 1989.
7. E.-E. Doberkat. An average case analysis of Floyd's algorithm to construct heaps. *Information and Control*, 61:114–131, 1984.
8. R.W. Floyd. Algorithm 245 - Treesort 3. *Communications of the ACM*, 7:701, 1964.
9. A.M. Frieze. On the random construction of heaps. *Information Processing Letters*, 27(2):103–109, 1988.
10. G.H. Gonnet and J.I. Munro. Heaps on heaps. *SIAM Journal on Computing*, 15:964–971, 1986.

263

11. R. Hayward and C.J.H. McDiarmid. Average case analysis of heap building by repeated insertion. *Journal of Algorithms*, 12(1):126–153, 1991.
12. D.E. Knuth. *The Art of Computer Programming, Vol. 3: Sorting and Searching.* Addison-Wesley, Reading, Massachusetts, 1973.
13. C.J.H. McDiarmid and B.A. Reed. Building heaps fast. *Journal of Algorithms*, 10:352–365, 1989.
14. J. Vuillemin. A data structure for manipulating priority queues. *Communications of the ACM*, 21:309–314, 1978.
15. I. Wegener. BOTTOM-UP-HEAPSORT, a new variant of HEAPSORT beating, on an average, QUICKSORT (if n is not very small). *Theoretical Computer Science*, 118:81–98, 1993.
16. J.W.J. Williams. Algorithm 232: Heapsort. *Communications of the ACM*, 7:347–348, 1964.

Fast Approximate Dictionary Matching

Fei Shi

Department of Computer Science, ETH Zentrum
CH-8092 Zurich, Switzerland

Abstract. In the approximate dictionary matching problem, a dictionary that contains a set of pattern strings is given. The user presents a text string and a tolerance k (k is a positive integer) and asks for all occurrences of all dictionary patterns that appear in the text with at most k differences to the original patterns. We present two algorithms for the problem. The first algorithm assumes that all patterns in the dictionary are of the same length. The second algorithm removes this assumption at the expense of a bit more complicated preprocess of the dictionary and slower query time. The basic idea behind our algorithms is to represent each dictionary pattern with one or two points in a $|\Sigma|^q$ - dimensional real space under the L_1-metric where Σ is the underlying alphabet and q a fixed integer and then organize these points with some spatial data structure to make subsequent searches with different texts of different lengths and different tolerance values fast. Although the approximate dictionary matching would be of enormous importance in molecular biological applications, no previous results for the problem are known.

1 Introduction

A dictionary is a set of pattern strings. The approximate dictionary matching problem with tolerance k (k is a positive integer) is to find all occurrences of any dictionary patterns in a given text with at most k errors (differences). More precisely, let $D = \{P_1, P_2, \cdots, P_z\}$ be the dictionary and $T = t_1 t_2 \cdots t_n$ the text, our goal is to find all patterns $P_i \in D$ and positions x and j within T such that the edit distance $D_{ed}(P_i, T_j)$ between P_i and some substring $T_j = t_x t_{x+1} \cdots t_j$ of T ($1 \le x \le j \le n$) is no more than k. The edit distance between two strings is defined to be the minimal number of deletions, insertions, and substitutions of symbols needed to transform one string into the other.

Aho and Corasick [AC75] introduced and solved the exact (static) dictionary matching problem. The exact *dynamic* version of the problem, where the dictionary can change over time by insertion and deletion of individual patterns, was first studied by Amir and Farach [AF91]. Aho and Corasick's algorithm is not suited for the dynamic paradigm, since adding or removing a dictionary string requires re-preprocessing the whole dictionary. Other related work on the exact dynamic dictionary matching problem can be found in [IS92], [AFM92] and [AFIPS93].

The approximate dictionary matching is of interest in many areas. For example, in molecular biology one often needs to compare a newly sequenced protein against a (huge) dictionary to see if it contains, allowing certain number of errors in a match, any patterns stored in the dictionary. Another example is searching for an interesting combination of bases in a DNA segment. Hereby, exact matching paradigm is inappropriate since on some occasions molecules of the same protein differ slightly among organisms of the same species. There are many other cases where one should tolerate certain number of errors. If the same message is transmitted repeatedly on the same communication channel it will be received differently on different occasions due to noise on the channel. The same word or computer instruction typed on different occasions may differ occasionally, due to human error.

To the best of our knowledge, the present paper is the first attempt to address the approximate dictionary matching problem. We assume a very large and relatively stable dictionary with a (subsequently given) small and variable text. Thus, the time for preprocessing and updating the dictionary is counted separately from that of scanning the text. The basic idea behind our approach is to represent each dictionary pattern by one or two points in a $|\Sigma|^q$ - dimensional real space under the L_1-metric where Σ is the underlying alphabet and q a positive integer to be stated later. The text is transformed into a set of points in the same space in a similar manner. The original problem then becomes that of the fixed radius near neighbor searching with the points corresponding to the text being the centers and the maximal distance allowed in a match the radius.

Sections 2 and 3 provide some preliminaries for our approach to the problem. We define the q-gram distance (for some fixed integer q) between two strings and show how to compute the q-gram distance between two strings in linear time and space. Since computing the edit distance between two strings requires quadratic time, we use the q-gram distance to estimate the edit distance. In Section 3 we recall a data structure, called the monotonous bisector tree, which we will use to organize the points corresponding to the dictionary patterns. In Section 4 we present our first algorithm for the problem under the assumption that all patterns in the dictionary are of the same length. We remove this assumption presenting another algorithm in Section 5. We conclude the paper by discussing some open problems suggested by this research in Section 6.

2 q-gram distance

Let Σ be a finite alphabet. Let Σ^* denote the set of all strings over Σ and Σ^q denote the set of all strings of length q over Σ in lexicographic order, for some integer q. A q-gram is any string $w = x_1 \cdots x_q$ in Σ^q.

Definition 1 (q-gram profile/distance) *The q-gram profile $G(S)$ of a string $S \in \Sigma^*$ is a vector of dimension $|\Sigma|^q$ whose i-th element equals the number of occurrences of the i-th q-gram in S. The q-gram distance $D_q(S_1, S_2)$ between two strings S_1 and S_2 in Σ^* is defined to be the L_1-distance of their q-gram profiles.*

The concept of q-grams dates back to Shannon [Sha48] and has been applied successfully in many variations in different string processing methods (see, e.g., [KR78], [OM88] and [Ukk91]).

Constructing a compact q-gram profile

Let $\Sigma = \{a_0, a_1, \cdots, a_{e-1}\}$ with $e = |\Sigma|$. Any string $X = x_1 x_2 \cdots x_q$ can be interpreted as an integer $f(X)$ in base - $|\Sigma|$ - notation:

$$f(X) = \sum_{i=1}^{q} \bar{x}_i e^{q-i} \tag{1}$$

where $\bar{x}_i = j$ if $x_i = a_j$, $i = 1, 2, \cdots, q$. Note that $f(X)$ is a bijection, i.e., $f(X) = f(X')$ if and only if $X = X'$. We call $f(X)$ the finger-print of X. When "scanning" a string $T = t_1 t_2 \cdots t_n \in \Sigma^*$, i.e, when considering q-grams in adjacent positions, the finger-prints can be computed quickly. More precisely, for $W_i = t_i t_{i+1} \cdots t_{i+q-1}$, $1 \le i \le n - q + 1$, we have

$$f(W_{i+1}) = (f(W_i) - \bar{t}_i e^{q-1})e + \bar{t}_{i+q} \tag{2}$$

By first computing $f(W_1) = \sum_{i=1}^{q} \bar{t}_i e^{q-i}$ and then applying Eq.(2) for $i = 1, 2, \cdots, n - q$, the finger-prints of all q-grams in T can be computed in time $O(n)$, assuming that multiplications, additions and subtractions of integers take constant time each, and that the values of e^1, e^2, $\cdots, e^{q-1}$ can be stored in a table. Each time when we computing the finger-print of a q-gram W_i, we record its occurrence in T in an array $G(T)[0 : e^q - 1]$ by incrementing $G(T)[f(W_i)]$ by 1.

So, the q-gram profile $G(T)$ of T of length n can be constructed in time $O(n)$ and space $O(|\Sigma|^q)$.

Since at most $n - q + 1$ elements of the q-gram profile $G(T)[0 : e^q - 1]$ of T are active (non-zero), the space requirement can be reduced to $O(n)$ with some standard hashing technique on the non-zero elements of $G(T)[0 : e^q - 1]$, resulting in a compact version of the q-gram profile $CG(T)$ of T. Thus, we have

Lemma 1 *The compact q-gram profile $CG(T)$ of a string T of length n can be computed in time and space $O(n)$.*

Computing q-gram distance

From Lemma 1, it is easy to see

Lemma 2 *Given the compact q-gram profiles of two strings T and P of length n and m respectively, the q-gram distance $D_q(T, P)$ between T and P can be computed in time $O(n + m)$.*

Computing the compact q-gram profiles of all substrings of length m of a string of length n

Let $T = t_1 t_2 \cdots t_n$. There are $n - m + 1$ substrings of length $m < n$ in T:

$$W_i = t_i t_{i+1} \cdots t_{i+m-1}, \ i = 1, 2, \cdots, n - m + 1;$$

there are $n - q + 1$ q-grams in T ($q < m$):

$$Q_j = t_j t_{j+1} \cdots t_{j+q-1}, \ j = 1, 2, \cdots, n - q + 1.$$

We know that the compact q-gram profile of a substring of length m can be computed in time $O(m)$. A straightforward way to compute the compact q-gram profiles of all substrings of length m of T of length n would take time $O(mn)$.

The problem with this brute force algorithm is that the computation of each q-gram would be repeated $m-q+1$ times, since each q-gram Q_j occurs in $m-q+1$ substrings T_i of length m.

We now describe our linear algorithm for computing the compact q-gram profiles of the substrings. Recall that $f(X)$ denotes the finger-print of the string X as determined by Eq.(1) and $G(X)$ the q-gram profile of X.

Observation $G(T_i)$ and $G(T_{i+1})$ differ at most in their $f(Q_i)$-th and $f(Q_{i+m-q+1})$-th elements. That is, if $f(Q_i) \neq f(Q_{i+m-q+1})$ then

$$G(T_{i+1})[f(Q_i)] = G(T_i)[f(Q_i)] - 1,$$
$$G(T_{i+1})[f(Q_{i+m-q+1})] = G(T_i)[f(Q_{i+m-q+1})] + 1, \text{ and}$$

the rest elements of $G(T_{i+1})$ are the same as those of $G(T_i)$; otherwise, $G(T_{i+1}) = G(T_i)$.

Thus, $G(T_{i+1})$ can be easily computed from $G(T_i)$ by performing

- $G(T_{i+1})[f(Q_i)] := G(T_i)[f(Q_i)] - 1;$
- $G(T_{i+1})[f(Q_{i+m-q+1})] := G(T_i)[f(Q_{i+m-q+1})] + 1;$
- the rest of $G(T_{i+1})$ is the same as that of $G(T_i)$.

Thus, if proper hashing technique (see, e.g., [FKS84]) on the non-zero elements of $G(T_i)[0 : |\Sigma|^q - 1]$ is used to produce the compact q-gram profile $CG(T_i)$, we can have

Lemma 3 *The compact q-gram profiles of all substrings of length m of a string of length $n \geq m$ can be computed in time $O(n)$.*

q-gram distance as an estimate of edit distance

It is known that any algorithm to compute the edit distance $D_{ed}(X, Y)$ between two strings X of length n and Y of length m needs time $\Omega(nm)$ (see, e.g., [WC76]). Our motivation to estimate the edit distance by q-gram distance is that the q-gram distance can be computed in linear time as has already been seen.

The following fact suggests that q-gram distance can be used as a lower bound for the edit distance and was successfully used by Ukkonen to speed up the algorithm for the k differences pattern matching problem [Ukk91].

Lemma 4 *Let $P = p_1 p_2 \cdots p_m$ and $T = t_1 t_2 \cdots t_n$ ($n \geq m$) be any two strings and q any positive integer, $q < m$. Then, for $1 \leq i \leq n$,*

$$D_q(P, t_{i-m+1} \cdots t_i)/2q \leq \min_{1 \leq j \leq i} D_{ed}(P, t_j \cdots t_i).$$

Our experiments with real DNA databases have shown that q-gram distance is a satisfactory lower bound for the edit distances of DNA sequences [BRSWW93].

3 Monotonous bisector trees

The data structure we choose to store the set of the compact q-gram profiles of the dictionary patterns is the so-called monotonous bisector tree (MBT) (cf. [NVZ92]).

Definition 2 (monotonous bisector tree) *Let S be a finite set of points in an arbitrary space E with distance function d, and let $x_1 \in S$. A monotonous bisector tree $MBT(S, x_1)$ is a binary tree recursively defined as follows:*

$1 \leq |S| \leq 2$: *$MBT(S, x_1)$ consists of a single node containing all elements of S;*
$|S| > 2$:   *The root w of $MBT(S, x_1)$ contains two points: x_1 and another point $x_2 \in S - \{x_1\}$. The children of W are $MBT(S_1, x_1)$ and $MBT(S_2, x_2)$ such that S_i contains all points in S that are closer to x_i than to x_j $(i, j = 1, 2; i \neq j)$ with respect to the distance function d. The points in S with equal distance to x_1 and x_2 are split into two halves, one for S_1 and one for S_2.*

Note that by the definition $S_1 \cap S_2 = \emptyset$ and $S_1 \cup S_2 = S$. The two subtrees $MBT(S_1, x_1)$ and $MBT(S_2, x_2)$ can be regarded as two clusters of S. For our purpose, we additionally store in each node of the tree the radius of the cluster corresponding to that node: $radius(S_i, x_i) := \max\{d(x, x_i)|x \in S_i\}$.

Lemma 5 ([NVZ92]) *1. In a MBT, the radii of the clusters corresponding to the nodes on a path from the root down to a leaf generate a monotonously decreasing sequence. 2. Let $S \subset IR^d$ be a finite set of n points. For any L_p-metric $(1 \leq p \leq \infty)$ an MBT with logarithmic height can be constructed in time $O(n \log n)$ and storage $O(n)$, assuming that the distance between any two points can be computed in constant time.*

In our case, the underlying space is the $|\Sigma|^q$ - dimensional real space of the q-gram profiles under the L_1-metric. However, the fact that this space is of high dimension does not adversely affect the efficiency of our algorithm, because the construction of the MBT is mainly based on distance relations between two points and is carried out with the compact q-gram profiles.

Fixed-radius near neighbor search

Monotonous bisector trees support efficiently a large variety of proximity queries. What we are interested in is the fixed-radius near neighbor search: we search for all the points in S with distance at most r to a given point $p \in E$. p is called the center and r the radius of the search ball. Starting from the root of the tree, suppose we are now at a node corresponding to the subtree $MBT(V, x)$. We can prune the subtree $MBT(V, x)$ from our search if $D_q(p, x) - radius(V, x) > r$. We accept the whole subtree if $D_q(p, x) + radius(V, x) < r$.

Furthermore, we check if the search ball intersects the bisector of x_1 and x_2 or not. If the search ball lies in one side of the bisector we can restrict our search to the subtree corresponding to this side. More precisely, the search can be restricted to the subtree $MBT(S_1, x_1)$ if $D_q(p, x_1) - D_q(p, x_2) < -2r$, and correspondingly, to the subtree $MBT(S_2, x_2)$ if $D_q(p, x_2) - D_q(p, x_1) < -2r$.

These considerations together with the guaranteed logarithmic height of the MBT enable an efficient fixed-radius near neighbor search on the average. Extensive experimental results show that the average query time for the fixed-radius near neighbor search is $\log n + s$ where s is the size of the output (cf. [NVZ92], [Zir90], [Zir92], [Ver92]).

4 Algorithm A

We are now ready to describe our algorithms for the approximate dictionary matching problem. The basic idea of our algorithms is to preprocess the dictionary into a monotonous bisector tree, to make subsequent searches with different texts of different lengths fast. Algorithm A assumes that all patterns in the dictionary $D = \{P_1, P_2, \cdots, P_z\}$ are of the same length m (This assumption will be removed in Algorithm B to be presented in next section).

Algorithm A

1. Preprocessing the dictionary
   (a) compute the set Γ of the compact q-gram profiles of all patterns in D.
   (b) construct the monotonous bisector tree MBT for the set Γ under the q-gram distance.
2. Searching
   For each substring $T_i = t_i \cdots t_{i+m-1}$ of T of length m, $i = 1, 2, \cdots, m-q+1$, Do
   (a) compute the compact q-gram profile τ_i of T_i.
   (b) select candidate patterns from D with respect to T_i by performing fixed-radius near neighbor search on the MBT with τ_i being the center and $2qk$ the radius.
   (c) check candidate patterns obtained in Step 2(b) one by one against $T_{i-k} = t_{i-k} \cdots t_{i+m-1}$ using some approximate string matching algorithm (e.g., Landau and Vishkin's algorithm [LV86]).
3. Updating the dictionary.

Note that Step 2(b) is based on Lemma 4.

Inserting a pattern P of length m into the dictionary is easy: we compute the compact q-gram profile of P and insert it into the MBT of the dictionary. Deleting a pattern from the dictionary is done similarly.

The complexity. Let M denote the sum of the lengths of all patterns in D and n the length of the text. By Lemma 1, Step 1(a) takes time and space $O(M)$. Since each comparison of two compact q-gram profiles needs time $O(m)$, by Lemma 5 Step 1(b) uses time $(M \log z)$ and storage $O(M)$ where z denote the number of patterns in D. Hence, preprocessing the dictionary into a monotonous bisector tree needs time $O(M \log z)$ and storage $O(M)$.

The entire loop of Step 2(a) takes time $O(n)$ by Lemma 3. As the expected query time for the fixed-radius near neighbor search on an MBT of N points is $t \log N + s$ where s is the size of the output and t the time needed to compute the distance between any two points (cf. [NVZ92], [Zir90], [Zir92], [Ver92]), the loop of Step 2(b) needs expected time $O(m((n-m)\log z + \sum_{j=1}^{n-m+1} s_i))$ where s_i denotes the number of q-gram profiles in the MBT contained by the L_1-ball with the q-gram profile τ_i being the center and $2qk$ the radius. If Landau and Vishkin's algorithm [LV86] is used, Step 2(c) takes $O(k(m+k)s)$ time where $s = \sum_{j=1}^{n-m+1} s_i$. Hence, the expected query time of Algorithm A is $O(m((n-m)\log z + k s))$.

Since each comparison of two compact q-gram profiles of patterns of length m needs time $O(m)$ and the MBT has height $O(log z)$, inserting /deleting a pattern of length m into/from the MBT of the dictionary takes time $O(m \log z)$.

5 Algorithm B

In this section we present another algorithm (Algorithm B) for the approximate dictionary matching problem, removing the assumption that all patterns in the dictionary are of the same length.

Lemma 6 *Any occurrence of a string $P = p_1 p_2 \cdots p_m$ in another string $T = t_1 t_2 \cdots t_n$ $(n \geq m)$ ending at position j with at most k differences implies that both of the following statements are true: 1. $D_q(P_1, t_{j-|P|+1} \cdots t_{j-|P|+|P_1|}) \leq 2qk$ where P_1 is any prefix of P and 2. $D_q(P_2, t_{j-|P_2|+1} \cdots t_j) \leq 2qk$ where P_2 is any suffix of P.*

Proof. 1. Define $G(X, j) = \min_{1 \leq i \leq j} D_{ed}(X, t_i \cdots t_j)$. Imagine we have an $(m+1) \times (n+1)$ matrix $A = (a_{ij})$ where a_{ij} is the minimum number of differences between the prefix $p_1 p_2 \cdots p_i$ of P and any substring of T ending at t_j, i.e., $a_{ij} = \min_{1 \leq l \leq j} D_{ed}(p_1 \cdots p_i, t_l \cdots t_j)$. We first need to show that $a_{i+1,j+1} - a_{i,j}$ equals either zero or one for any i and j. By dynamic programming, we have

$$a_{i,j} = \min \begin{cases} a_{i-1,j} + 1 \\ a_{i,j-1} + 1 \\ a_{i-1,j-1} & \text{if } p_i = t_j \text{ or } a_{i-1,j-1} + 1 \text{ otherwise} \end{cases}$$

It is then easy to see that $a_{i+1,j}$ equals either $a_{i,j}$ or $a_{i,j} + 1$; $a_{i,j+1}$ equals either $a_{i,j}$ or $a_{i,j} + 1$. Hence, $a_{i+1,j+1} - a_{i,j}$ equals either zero or one. That is, $\cdots, a_{i,j}, a_{i+1,j+1}, a_{i+2,j+2}, \cdots$ is a non-decreasing sequence. Therefore, from $G(P, j) \leq k$ we have $G(P_1, j - |P| + |P_1|) \leq k$. It then follows from Lemma 4 that $D_q(P_1, t_{j-|P|+1} \cdots t_{j-|P|+|P_1|}) \leq 2qk$.

2. From the fact that P occurs in T ending at position t_j with at most k differences implies that any suffix of P occurs in T ending at position t_j with at most k differences, we know that $G(P_2, j) \leq k$. It then follows from Lemma 4 that $D_q(P_2, t_{j-|P_2|+1} \cdots t_j) \leq 2qk$. $\qquad\square$

Lemma 7 *Let P_1 be any prefix of P and P_2 any suffix of P such that $|P_1| + |P_2| \geq |P|$. For any $1 \leq i \leq j \leq n$,*

$$D_q(P_1, t_i \cdots t_{i+|P_1|-1}) + D_q(P_2, t_{j-|P_2|+1} \cdots t_j) \leq 2q(D_{ed}(P, t_i \cdots t_j) + |P_1| + |P_2| - |P|).$$

Proof Observe that

$$D_{ed}(P, t_i \cdots t_j) \geq \min_{i \leq x < j} D_{ed}(P_1, t_i \cdots t_x) + \min_{x < y \leq j} D_{ed}(P_2, t_y \cdots t_j) - (|P_1| + |P_2| - |P|).$$

Thus, by Lemma 4, we have

$$D_{ed}(P, t_i \cdots t_j) \geq D_q(P_1, t_i \cdots t_{i+|P_1|-1})/2q + D_q(P_2, t_{j-|P_2|+1} \cdots t_j)/2q - (|P_1| + |P_2| - |P|).$$

□

Corollary 1 follows immediately from Lemma 7.

Corollary 1 *Let P_1 be any prefix of P and P_2 any suffix of P such that $|P_1| + |P_2| \geq |P|$. If the edit distance between P and a substring $t_i \cdots t_j$ of T is at most k, then*

$$D_q(P_1, t_i \cdots t_{i+|P_1|-1}) + D_q(P_2, t_{j-|P_2|+1} \cdots t_j) \leq 2q(k + |P_1| + |P_2| - |P|).$$

If we choose $|P_1| = |P_2| = l$, where $l \geq |P|/2$ is an integer to be stated later, then Lemma 6 says that the necessary condition for P to occur in T with at most k differences is that there exists a position j in T such that $D_q(p_1 \cdots p_l, t_{j-|P|+1} \cdots t_{j-|P|+l}) \leq 2qk$ and $D_q(p_{|P|-l+1} \cdots t_{|P|}, t_{j-l+1} \cdots t_j) \leq 2qk$. We call such a position j a candidate position of an (approximate) occurrence of P in T.

Let m_{min}/m_{max} denote the length of the shortest/longest pattern in the dictionary D respectively. We can expect that the lengths of the patterns in D are distributed randomly in the range $[m_{min}, m_{max}]$.

Let $l_0 = m_{min}$; $l_1 = 2l_0$; $\cdots$; $l_r = 2l_{r-1}$ where $r = \lfloor \log_2 \frac{m_{max}}{m_{min}} \rfloor$. We partition D into $r + 1$ equivalence classes $D_0, D_1, \cdots$ and D_r based on the lengths of the patterns as follows:

$$D_i = \{P \in D | l_i \leq |P| < l_{i+1}\} \text{ for } i = 0, 1, \cdots, r.$$

Algorithm B

1. Preprocessing the dictionary
   (a) Partition the dictionary D into $r + 1$ equivalence classes

   $$D_i = \{P \in D | l_i \leq |P| < l_{i+1}\} \ (\ i = 0, 1, \cdots, r)$$

   (b) For $i := 0$ to r Do
   compute the compact q-gram profile of the prefix and the suffix of length l_i of all patterns $P \in D_i$ (If $|P| = l_i$, we regard P as its prefix as well as its suffix). Let Γ_i denote the set of all such q-gram profiles.
   (c) For $i := 0$ to r Do
   construct the monotonous bisector tree MBT_i for the set Γ_i under the q-gram distance.
2. Searching
   For $i := 0$ to r Do
   (a) compute the compact q-gram profiles τ of all substrings of length l_i of T;
   (b) for each τ obtained in Step 2(a), perform the fixed-radius near neighbor search on the MBT_i of D_i with τ being the center and $2qk$ the radius. Let S_i denote the output of this substep.
   (c) select candidate patterns from S_i using Lemmas 6 and 7.
   (d) check candidate patterns obtained in Step 2(c) one by one against T using some approximate string matching algorithm.
3. Updating the dictionary.

Let us look at Step 2(c) and 2(d) in a bit more detail. Each entry in S_i has the form (P, f, x) where P is a pointer to some pattern in D_i, $f \in \{0, 1\}$ and x is a position in T. (P, f, x) means that the q-gram distance between the prefix of P if $f = 0$ or the suffix of P if $f = 1$ of length l_i and the substring $t_{x-l_1+1} \cdots t_x$ is no more than $2qk$. By Lemma 6, an entry (P, f, x) in S_i is a candidate occurrence of P in T ending at position x only if $f = 1$ and there exists another entry $(P, 0, y) \in S_i$ such that $y = x - |P| + l_i$. As to Step 2(d), for each candidate entry $(P, 1, x)$ found in Step 2(c) we run some approximate pattern matching algorithm (e.g., Landau and Vishkin 's algorithm [LV86]) for the string P and the substring $t_{x-|P|-k+1} \cdots t_x$ of T to determine whether P really occurs in T ending at position x or not.

The complexity. Both Steps 1(a) and 1(b) take time $O(M)$ and use space $O(M)$. Let M_i denote the total length of the pattern strings in D_i and z_i the number of elements (patterns) in D_i. Step 1(c) needs time $\sum_{i=0}^{r} M_i \log z_i$ $< M \log z_{max}$ where z_{max} denotes the maximal number of patterns in any sub-dictionary D_i $(i = 0, \cdots r)$ and storage $O(M)$. Hence, the algorithm uses time $O(M \log z_{max})$ and storage $O(M)$ to preprocess D into a set of $r+1$ monotonous bisector trees where $r = \lfloor \log_2 \frac{m_{max}}{m_{min}} \rfloor$.

Step 2(a) takes time $O(n)$. In Step 2(b), let $|S_i|$ denote the number of entries in the output set S_i. Then this substep needs time $O(n \log z_i + |S_i|)$. Recall that an entry $(P, f, x) \in S_i$ is a candidate occurrence of P in T ending at position x only if $f = 1$ and there exists another entry $(P, 0, y) \in S_i$ such that $y = x - |P| + l_i$. Step 2(c) can be carried out by first sorting the entries (P, f, x) in S_i then binary-searching the sorted set S_i and thus needs time $O(|S_i| \log |S_i|)$. Let S_i' denote the set of such candidate occurrences. Step 2(d) needs time $O(l_i(l_i + k)|S_i'|)$ if Landau and Vishkin's algorithm is used.

Therefore, the searching step (Step 2) needs expected time $O(n \log \frac{m_{max}}{m_{min}} \log z + m_{max}(m_{max} + k)cp)$ where n dotes the length of the text, z the number of patterns in the dictionary, m_{max} (m_{min}) the length of the longest (shortest) pattern in the dictionary and k the maximal number of differences allowed in a match and cp the number of candidate patterns found (in the loop of Step 2(c)).

Like in Algorithm A, inserting /deleting a pattern of length m into/from the dictionary needs time $O(m \log z)$.

6 Concluding remarks

Efficient algorithms for approximate dictionary matching problem are of enormous importance in molecular biology. While our algorithms are expected to answer different dictionary matching queries with different texts quickly, it remains to be seen what the performance will be for realistic data , such as those in the EMBL Data Library and in the Genbank.

Acknowledgements

Helpful discussions with Peter Widmayer are greatly appreciated.

References

[AC75] A. V. Aho and M. Corasick, Efficient string matching: an aid to bibliographic search. *Communications of the ACM*, June 1975, Vol. 18, No. 6, pp. 333 - 340

[AF91] A. Amir and M. Farach, Adaptive dictionary matching. *Proc. of the 32nd IEEE Annual Symposium on Foundation of Computer Science*, 1991, pp. 760 -766

[AFIPS93] A. Amir, M. Farach, R. Indury, J. A. Poutre and A. Schaeffer, Improved dynamic dictionary matching, *Proc. of the fifth Annual ACM-SIAM Symposium on Discrete Algorithms*, 1993, pp. 392 - 400

[AFM92] Amir, M. Farach and Y. Matias, efficient randomized dictionary matching algorithms. *Proc. of the 3rd Ann. Symp. on Combinatorial Pattern Matching*, 1992

[BRSWW93] E. Bugnion, T. Roos, F. Shi, P. Widmayer and F. Widmer, Approximate multiple string matching using spatial indexes. in *Proc. of the 1st South American Workshop on String Processing*, (eds.) R. Baeza-Yates and N. Ziviani, pp. 43 - 53, 1993

[FKS84] M. L. Fredman, J. Komlos and E. Szemeredi, Storing a sparse table with $O(1)$ worst case access time. *Journal of the ACM*, 31, 3(1984), 538 - 544

[IS92] R. Indury and A. Schaeffer, Dynamic dictionary matching with failure functions. in *Proc. of the 3rd Annual Symposium on Combinatorial Pattern Matching*, 1992

[KR78] T. Kohonen and E. Reuhkala, A very fast associative method for the recognition and correction of misspellt words, based on redundant hash-addressing. *Proc. 4th Joint Conf. on Pattern Recognition*, 1978, Kyoto, Japan, 807-809

[LV86] G. M. Landau and U. Vishkin, in *Proc. 18th ACM Symposium on Theory of Computing*, 1986, pp. 220 - 250

[NVZ92] H. Noltmeier, K. Verbarg and C. Zirkelbach, Monotonous bisector* trees - a tool for efficient partitioning of complex scenes of geometric objects. In *Data Structures and Efficient Algorithms: Final Report on the DFG Special Joint Initiative* Vol. 594 of L.N.C., Spring-Verlag, 1992

[OM88] O. Owolabi and D. R. McGregor: Fast approximate string matching. *Software - Practice and Experience* 18(4) (1988), 387-393

[Sha48] C. E. Shannon, A mathematical theory of communications. *The Bell Systems Techn. Journal* 27 (1948), 379-423

[Ukk91] E. Ukkonen, Approximate string matching with q-grams and maximal matches. Report, Department of Computer Science, University of Helsinki, Finland, 1991

[Ver92] K. Verbarg, Räumliche Indizes-Celltrees: Analyse und Vergleich mit Monotonen Bisektorbäumen, Diploma Thesis, Department of Computer Science, University of Würzburg, Germany, 1992

[WC76] C. K. Wong and A. K. Chandra, Bounds for the string editing problem. *Journal of the ACM*, vol.23, No.1, January 1976, pp. 13-16

[Zir90] C. Zirkelbach, Monotonous bisector trees and clustering problems, Report, Department of Computer Science, University of Würzburg, Germany, 1990

[Zir92] C. Zirkelbach, Geometrisches Clustern - ein metrischer Ansatz, Dissertation, Department of Computer Science, University of Würzburg, Germany, 1992

Undirected Vertex-Connectivity Structure and Smallest Four-Vertex-Connectivity Augmentation (Extended Abstract)

Tsan-sheng Hsu*

Inst. of Information Science, Academia Sinica, Nankang 11529, Taipei, Taiwan, ROC
tshsu@iis.sinica.edu.tw

Abstract. In this paper, we give an $O(n \cdot \log n + m)$-time algorithm to solve the problem of finding a smallest set of edges whose addition 4-vertex-connects an undirected graph, where n and m are the number of vertices and edges in the input graph, respectively. We also show a formula to compute this smallest number in $O(n \cdot \alpha(n,n) + m)$ time, where α is the inverse of the Ackermann function.

1 Introduction

The problem of augmenting a graph to reach a given connectivity requirement by adding edges has important applications in network reliability [3] and in statistical data security [4]. One version of the augmentation problem is to to satisfy the given requirement by adding as few edges as possible. We refer to this problem as the *smallest augmentation* problem. Linear-time algorithms and efficient parallel algorithms are known for solving the smallest augmentation problem on an undirected graph to reach k-vertex-connectivity, $k \leq 3$ [2, 10, 11, 12, 15, 16, 18]. Hsu [8] gave an almost linear-time algorithm to four-connect (i.e., 4-vertex-connect) a triconnected (i.e., 3-vertex-connected) graph by adding the smallest number of edges.

In this paper, we study the structure of an undirected graph that is not four-connected detailing on the parts we need to solve the smallest four-connectivity augmentation problem. We are unaware of any polynomial time algorithm for this problem when the input graph is not triconnected. In this paper, we give an $O(n \cdot \log n + m)$-time algorithm for solving this problem, where n and m are the number of vertices and edges in the input graph, respectively. We also show a formula to compute this smallest number in $O(n \cdot \alpha(n, n) + m)$ time, where α is the inverse of the Ackermann function.

Due to space limitation, we only sketch main ideas in this extended abstract.

2 Definitions

Vertex-Connectivity. The graph G with at least $k + 1$ vertices is *k-vertex-connected*, $k \geq 2$, if G is a complete graph with $k + 1$ vertices or G remains connected by removing any set of vertices with cardinality less than k.

* Research supported in part by NSC under the Grant No. 84-2213-E-001-005.

An equivalent characterization of k-vertex-connected graphs is due to Menger [14]. Two vertices are k-*vertex-connected*, if there are at least k internally vertex-disjoint paths between them. A set of vertices U is k-*vertex-connected*, if either (1) $|U| = 1$ and the degree of the vertex in U is at least k or (2) $|U| > 1$ and every two distinct vertices are k-vertex-connected.

Separating Set. Given a subset of vertices S in G, S *separates* two vertices u and v in G if $u \notin S$, $v \notin S$, and u and v are connected in G, but are disconnected in $G - S$. Let $\mathrm{COM}(G)$ be the number of connected components in G. A set of vertices S in G is a *separating set* if there are two distinct vertices $u \notin S$ and $v \notin S$ in G such that (1) there are $|S|$ internally vertex-disjoint paths between every two distinct vertices in $S \cup \{u, v\}$, and (2) u and v are separated in $G - S$. If the cardinality of S is k, then S is a separating k-set. For the case of $k = 3$, it is a *separating triplet*.

Wheel. A set of at least three separating ℓ-sets, $\ell \geq 2$, with a possible common intersection is a *wheel* [13]. A wheel can be represented by the set of vertices $C \cup \{W_0, W_1, \ldots, W_{q-1}\}$ which satisfies the following conditions: (1) $q > 2$, (2) $|W_i| = |W_j|$, (3) $\forall i \neq j$, $C \cup W_i \cup W_j$ is a separating ℓ-set unless in the case that $j = ((i + 1) \bmod q)$, (4) each vertex in C is adjacent to a vertex in each of the connected components created by removing any separating ℓ-set in the wheel, and (5) $\forall j \neq (i + 1) \bmod q$, $C \cup W_i \cup W_j$ is a degree-2 separating ℓ-set. The set of vertices C is the *center* of the wheel. Each set of vertices W_i is a *side* of the wheel. We define the *unit size* of the wheel to be ℓ.

We define a *wheel component* for a wheel W with unit size k as follows. It is either (1) a connected component in $G - W$, or (2) a side with degree $k - 1$. We define the *wheel constraint* of a wheel $W = C \cup \{W_0, W_1, \ldots, W_{q-1}\}$ for reaching k-vertex-connectivity to be $\lceil w(W)/2 \rceil + a_k(C, G)$ and is denoted as $\mathrm{WC}_k(W, G)$, where $w(W)$ is the number of wheel components in W. The wheel constraint of a wheel is the smallest number of edges needed to add to eliminate all separating sets represented by the wheel.

The wheel constraint contributes non-trivially in smallest four-connectivity augmentation. Though a graph that is not triconnected may contain a wheel, however, we can prove that wheels contribute no non-trivial information when a graph is not triconnected. Owing to space limitation, details are omitted. When a graph is triconnected, the effect of the wheel constraint is discussed in [8].

Block. Given a proper subset of vertices B in $G = (V, E)$, the *neighbor* of B in G, $\mathrm{NEIGHBOR}(B, G)$, is $\{u \mid u \in V \setminus B, u$ is adjacent to a vertex in $B\}$. An ℓ-*block* in G is a maximal set of vertices B such that there are at least ℓ internally vertex-disjoint paths between every two distinct vertices in B. By definition, a vertex with degree ℓ in G is an x-block, for all $x > \ell$.

A subset of vertices B in $G = (V, E)$ is a *special block* for reaching k-vertex-connectivity if (1) $B \subseteq S$, where S is a separating set with $|S| < k$, (2) $|\mathrm{NEIGHBOR}(B, G)| < k$, (3) there is no $Q \subset B$ with $|\mathrm{NEIGHBOR}(Q, G)| < k$, and (4) $\mathrm{NEIGHBOR}(B, G) \cup B \neq V$.

Block Graph. Given a graph $G = (V, E)$, we construct its 4-block graph,

4-BLK(G) as follows. We first find the 2-block graph, which is a forest, for G [16]. The set of vertices in the 2-block graph are the set of cut edges, separating 1-sets and 2-blocks in G. For each 2-block $\mathcal{B}_2$, we find its 3-block tree [7] (see also [10]). In the 3-block graph, a separating 2-set is either represented as a vertex or is represented as two vertices in a polygon. *Tutte component* [17] is triconnected. For each non-trivial 3-block, we find the 4-block tree for its Tutte component [8]. In the 4-block tree for $\mathcal{B}_3$, each separating triplet in $\mathcal{B}_3$ is represented (1) as a vertex, (2) as three vertices in a wheel, or (3) as the neighbor of a special block. The collection and relation of the above 2-block graph, the 3-block forest, and the 4-block forest is 4-BLK(G).

Augmentation Number. Given G, the smallest number of edges needed to add to G such that the resulting graph is k-vertex-connected is the *smallest k-vertex-connectivity augmentation number* AUG$_k(G)$.

3   Properties of Blocks

An ℓ-block $\mathcal{B}$ and a separating set $\mathcal{S}$ are *adjacent* if (1) $|\mathcal{S}| < \ell$, (2) every vertex $u \in \mathcal{S}$, it is either the case that $u \in \mathcal{B}$ or the case that u is adjacent to a vertex in $\mathcal{B}$, and (3) there is no separating set $\mathcal{S}'$ with $|\mathcal{S}'| < \ell$, $\mathcal{S} \neq \mathcal{S}'$, and $\mathcal{S}'$ separates a vertex in $\mathcal{S} \setminus \mathcal{S}'$ and a vertex in $\mathcal{B}$. Given a subset of vertices $\mathcal{B}$ in a graph, $|\mathcal{B}| > 1$, k is an *order* of $\mathcal{B}$ if there are k internally vertex-disjoint paths between every two vertices in $\mathcal{B}$.

Degree and Demand of a Block. Let $\mathcal{B}$ be a subset of vertices in $G = (V, E)$ and let $\ell_{\mathcal{B}}$ be the largest order of $\mathcal{B}$ if $|\mathcal{B}| > 1$. Then VDEG($\mathcal{B}, G$) = NEIGHBOR($\mathcal{B}, G$) if $|\mathcal{B}| = 1$ and VDEG($\mathcal{B}, G$) = $\{u \mid u$ is in a separating set that is adjacent to $\mathcal{B}\}$ if $|\mathcal{B}| > 1$. The **degree** of $\mathcal{B}$ in G is DEGREE($\mathcal{B}, G$) = $|$VDEG($\mathcal{B}, G$)$|$.

Definition 3.1 *Let $\mathcal{B}$ be a subset of vertices in $G = (V, E)$ where $|V| > k$ and let $\ell_{\mathcal{B}}$ be the largest order of $\mathcal{B}$ if $|\mathcal{B}| > 1$. The **demand** of $\mathcal{B}$ such that G can reach k-vertex-connectivity is*

$$
\text{DEMAND}_k(\mathcal{B}, G) = \begin{cases} \max\{k', 0\} & \text{if } |\mathcal{B}| = 1 \text{ or } \mathcal{B} \\ & \text{is a special block,} \\ \sum_{\forall R} \text{DEMAND}_k(R, G) & \text{if } 1 < |\mathcal{B}| < k \text{ and} \\ & \mathcal{B} \text{ is a separating set,} \\ 0 & \text{if } \mathcal{B} = V \text{ and} \\ & \mathcal{B} \text{ is a } k\text{-block,} \\ \max\{k' + \sum_{\forall Q} \text{DEMAND}_k(Q, G), 0\} & \text{if } \mathcal{B} \neq V \text{ and} \\ & \mathcal{B} \text{ is a } k\text{-block,} \\ \max\{k' + \sum_{\forall Q} \text{DEMAND}_k(Q, G), \\ \sum_{\forall H} \text{DEMAND}_k(H, G) + & \text{otherwise,} \\ \sum_{\forall T} \text{DEMAND}_k(T, G)\} \end{cases}
$$

where $k' = k - $ DEGREE($\mathcal{B}, G$), R is a special block in $\mathcal{B}$, Q is a special block in $\mathcal{B}$ that is contained in a separating r-set, $r < \ell_{\mathcal{B}}$, H is an $(\ell_{\mathcal{B}} + 1)$-block in $\mathcal{B}$

that is not contained in separating $\ell_\mathcal{B}$-sets, and T is a special block in $\mathcal{B}$ that is not contained in any $(\ell_\mathcal{B} + 1)$-block in $\mathcal{B}$.

Intuitively, given a subset of vertices $\mathcal{B}$ with $\mathrm{VDEG}(\mathcal{B}, G) \cup \mathcal{B} \subset V$, $\mathrm{VDEG}(\mathcal{B}, G)$ separates a vertex in $\mathcal{B} \setminus \mathrm{VDEG}(\mathcal{B}, G)$ from a vertex in $V \setminus (\mathcal{B} \cup \mathrm{VDEG}(\mathcal{B}, G))$. The demand of $\mathcal{B}$ is the minimum number of edges needed to add to $\mathcal{B}$ such that the degree of $\mathcal{B}$ is at least k in the resulting graph, i.e., the demand of $\mathcal{B}$ becomes zero.

Demanding Vertices in a Block. We now give a definition to identify vertices in an ℓ-block $\mathcal{B}_\ell$ to whom adding new edges increases the degree of $\mathcal{B}_\ell$ (i.e., decreases the demand of $\mathcal{B}_\ell$). Let $\mathcal{B}$ be a subset of vertices in $G = (V, E)$ with $\mathrm{DEMAND}_k(\mathcal{B}, G) > 0$. Let u be a vertex in $\mathcal{B}$. If there is a vertex v such that $\mathrm{DEMAND}_k(\mathcal{B}, G) > \mathrm{DEMAND}_k(\mathcal{B}, G \cup \{(u, v)\})$, then u is a *demanding vertex* in $\mathcal{B}$ for reaching k-vertex-connectivity.

Claim 3.2 *Let $\mathcal{B}_1$ and $\mathcal{B}_2$ be two ℓ-blocks, $\ell \leq 4$. Then $\mathcal{B}_1 \cap \mathcal{B}_2$ is a subset of a separating set that is adjacent to both $\mathcal{B}_1$ and $\mathcal{B}_2$ and whose cardinality is less than ℓ. Furthermore, $\mathcal{B}_1 \cap \mathcal{B}_2$ is in every separating set which separates a vertex in $\mathcal{B}_1 \setminus \mathcal{B}_2$ and a vertex in $\mathcal{B}_2 \setminus \mathcal{B}_1$.*

Claim 3.3 *Let $\mathcal{B}_1$ and $\mathcal{B}_2$ be two ℓ-blocks, $\ell \leq 4$. Let $\mathcal{I} = \mathcal{B}_1 \cap \mathcal{B}_2$. For reaching k-vertex-connectivity, $k \leq 4$, there is no special block in $\mathcal{I}$.*

Claim 3.4 *Let $\mathcal{B}_1$ and $\mathcal{B}_2$ be two distinct ℓ-blocks in G, $\ell \leq 4$. For making G four-connected, if $\mathcal{B}_2 \not\subseteq \mathcal{B}_1$ and $\mathcal{B}_1 \not\subseteq \mathcal{B}_2$, then there is no intersection between the set of demanding vertices in $\mathcal{B}_1$ and the set of demanding vertices in $\mathcal{B}_2$.*

Corollary 3.5 *Let u_1 and u_2 be two vertices in $G = (V, E)$. Then $\mathrm{DEMAND}_4(V, G \cup \{(u_1, u_2)\}) \geq \mathrm{DEMAND}_4(V, G) - 2$.*

Proof. (Sketch) Let w be an integer that is either one or two.

Case 1: u_w is not a demanding vertex for any block (special or non-special) in the graph. Increasing the degree of u_w does not help in decreasing the demand.

Case 2: u_w is a demanding vertex for a special block. By definition of a special block and Claim 3.3, u_w cannot be in any other special block. Also by the way the demand of the graph is computed, the demand of the graph is decreased by at most one by increasing the degree of u_w.

Case 3: u_w is a demanding vertex for an ℓ-block. By Claim 3.4, u_w cannot be a demanding vertex for any other ℓ-block. Thus the demand of the graph is decreased by at most one by increasing the degree of u_w.

Note that by adding an edge, several x-blocks that are in the same $(x - 1)$-block $\mathcal{B}$ can be merged into an x-block for all $1 \leq x \leq 4$. However, the demand of $\mathcal{B}$ can be decreased by at most two according to the way the demand is computed and the above discussion. The demands of those blocks not containing an endpoint of the added edge or that are not merged stay the same. For reaching four-connectivity, the demand of the graph is decreased by at most two by adding the edge (u_1, u_2).

Lemma 3.6 *We need to add at least* $\lceil \text{DEMAND}_4(V, G)/2 \rceil$ *edges to* $G = (V, E)$ *in order to raise the vertex-connectivity of* G *to four.*

Claim 3.7 *Let* ℓ *be the vertex-connectivity of a connected graph* $G = (V, E)$ *and let* u_1 *and* u_2 *be two vertices in* G. *Let* $G' = G \cup \{(u_1, u_2)\}$, *where* $(u_1, u_2) \notin E$. *Then* $\text{DEMAND}_k(V, G') = \text{DEMAND}_k(V, G) - 2$ *if* (1) $4 \geq k > \ell + 1$, (2) u_1 *and* u_2 *are in different* $(\ell + 1)$*-blocks with degree* ℓ, *and* (3) *for any* $\ell < x \leq k$, u_i *is a demanding vertex in an* x*-block,* $i \in \{1, 2\}$.

4 Properties of Separating Sets

4.1 Fundamental Properties

Given a set of separating sets $\{\mathcal{S}_1, \ldots, \mathcal{S}_w\}$, an x-*component* H, is a connected component in $G - \cup_{i=1}^{w} \mathcal{S}_i$, with the property that H is also a connected component in $G - \mathcal{S}'$, where (1) $\mathcal{S}' \subseteq \mathcal{S}_j$, for some $1 \leq j \leq w$, (2) H is created by removing $\mathcal{S}'$, (3) $x = |\mathcal{S}'|$, and (4) x is the smallest integer satisfying the above conditions.

Let H be a y-component created by removing a separating x-set, $x \geq y$, $\mathcal{S}$ from $G = (V, E)$ and let V_H be the set of vertices in H. Let $U \subseteq \mathcal{S}$ where $u \in U$ implies that there is a $(y + 1)$-block $\mathcal{B} \subseteq V_H \cup \mathcal{S}$ and $u \in \mathcal{B}$. We define $\Phi(H) = U \cup V_H$. It is denoted as H *contributes* $\text{DEMAND}_k(\Phi(H), G)$ in computing $\text{DEMAND}_k(V, G)$. Note that $\text{DEMAND}_k(\Phi(H), G) \geq \text{DEMAND}_{k-|\mathcal{S}|}(V_H, G - \mathcal{S})$. The following claims and corollary state the relation between the demand of the graph and each of the components created by removing a separating set.

Claim 4.1 *Given a separating set* $\mathcal{S}$ *in* $G = (V, E)$, *each* ℓ*-component,* $\ell \leq |\mathcal{S}|$, *in* $G - \mathcal{S}$ *contributes at least* $k - \ell$ *in computing* $\text{DEMAND}_k(V, G)$, $k \leq 4$.
Proof. Let H be an ℓ-component in $G - \mathcal{S}$. Let V_H be the set of vertices in H. Assume that H is created by removing $\mathcal{S}'$ where $\mathcal{S}' \subseteq \mathcal{S}$ and $|\mathcal{S}'| = \ell$. By properties of the 4-block graph, there is an $(\ell + 1)$-block with degree ℓ in $V_H \cup \mathcal{S}'$. Thus it contributes at least $k - \ell$ in computing $\text{DEMAND}_k(V, G)$.

Separating Degree and Dividing Degree. Given a subset of vertices $\mathcal{S}$, the *separating degree* $\text{SD}(\mathcal{S}, G)$ is the number of $|\mathcal{S}|$-components in $G - \mathcal{S}$ if $\mathcal{S}$ is a separating set. For convenience, if $\mathcal{S}$ is not a separating set, $\text{SD}(\mathcal{S}, G) = 1$. The *dividing degree* $\text{DD}(\mathcal{S}, G)$ is $\text{COM}(G - \mathcal{S}) - \text{COM}(G) + 1$.

We now state a claim with regard to the creation of new separating sets when adding edges. We will show that by adding edges properly, no separating set with undesirable properties (e.g., whose separating degree is greater than two) is created.

Claim 4.2 *Let* κ *be the vertex-connectivity of* $G = (V, E)$. *Let* u *and* v *be two vertices in* V, *and let* $G' = G \cup \{(u, v)\}$. (1) *Every separating* κ*-set in* G' *is also a separating* κ*-set in* G, *i.e., adding an edge creates no new separating* κ*-set.* (2) *Adding an edge creates only new separating* ℓ*-sets,* $\ell > \kappa$, *whose separating degrees are two.* (3) *If* u *and* v *are both not in any separating* ℓ*-set, then the separating degree of any separating* ℓ*-set in* G *does not increase by adding* (u, v).

Proof. All separating sets in G and G' must be of cardinality at least κ.

Part (1): Adding an edge in a graph does not decrease the number of internally vertex-disjoint paths between every two vertices. If S is a separating κ-set in G', let G_1 be a connected component in $G' - S$. If both u and v are in G_1, then S is a separating κ-set in G. The above is also true if both u and v are in S. If u is in G_1 and v is in S (or vice versa), then S is a separating set with cardinality at most κ in G. Thus the claim holds.

Part (2): Let S be a separating ℓ-set in G', $\ell > \kappa$, but not in G. If $\text{SD}(S, G') > 2$, then $\text{SD}(S, G) \geq 2$. Thus S is also a separating set in G.

Part (3): Without connecting a connected component to a separating set S, the separating degree of S does not increase.

4.2 Separation Constraint

Augmenting within a Separating Set. Let S be a subset of vertices in $G = (V, E)$. Then $a_k(S, G)$ is the minimum number of edges needed to add such that in the resulting graph $|\text{NEIGHBOR}(F, G)| \geq \min\{k, |V \setminus F|\}$ for every $F \subseteq S$. If S is a separating set with cardinality less than k, then $a_k(S, G)$ is the minimum number of edges to add such that there is no special block in S for making G k-vertex-connected.

Let S be a subset of vertices in $G = (V, E)$. If $|S| < k$, the *separation constraint* of S for making G k-vertex-connected is $\text{SC}_k(S, G) = a_k(S, G) + \text{AUG}_{k-|S|}(G - S)$. The separation constraint for S is the number of edges needed to add to G such that in the resulting graph (1) there is no separating set $S' \subseteq S$, (2) there is no separating set S' with cardinality less than k and $S \subset S'$, and (3) for any $F \subseteq S$, $\text{DEGREE}(F, G) \geq k$ unless $F \cup \text{NEIGHBOR}(F, G) = V$. The separation constraint of a graph for reaching k-vertex-connectivity is the largest separation constraint among all separating sets with cardinality less than k.

Lemma 4.3 (1) *We need to add at least* $\text{SC}_k(S, G)$ *edges to* G *in order to make* G k-vertex-connected. (2) $\text{SC}_k(S, G \cup \{e\}) \geq \text{SC}_k(S, G) - 1$, *where e is an edge not in G.*

Proof. Note that if G is k-vertex-connected, then $G - S$ must be $(k - |S|)$-vertex-connected. Thus we must add at least $\text{AUG}_{k-|S|}(G - S)$ edges. It is also true that for all $F \subset S$, either $F \cup \text{NEIGHBOR}(F, G) = V$ or $|\text{NEIGHBOR}(F, G)| > k$. Thus we must add at least $a_k(S, G)$ edges. Observing that if $a_k(S, G) > a_k(S, G \cup \{e\})$, then one of the endpoints of e must be in S. Thus adding e does not decrease $\text{AUG}_{k-|S|}(G - S)$. Observing also that if $\text{AUG}_{k-|S|}(G - S)$ is decreased by adding an edge e', then both of its endpoints must not be in S. Thus adding e' does not decrease $a_k(S, G)$. This proves the lemma.

The following claim states the way to reduce the separation constraint and the demand of a disconnected graph.

Claim 4.4 *For making a disconnected graph k-vertex-connected, $2 \leq k \leq 4$, we can add an edge such that the demand of the graph is reduced by two and at the same time the separating constraint of the graph is reduced by one.*

Proof. (Sketch) Let G_1 and G_2 be two distinct connected components in G. We can find a k-block $\mathcal{B}_i$, for all $i \in \{1, 2\}$, in G_i with degree less than k and $\mathcal{B}_i$ is properly contained in a w-block with degree less than w, for all $1 \leq w < k$. Let u_i, for all $i \in \{1, 2\}$, be a demanding vertex in $\mathcal{B}_i$. Note that u_i is not part of any separating set with cardinality less than k. From the definition of the demand of a graph, $\text{DEMAND}_k(V, G \cup \{(u_1, u_2)\}) = \text{DEMAND}_k(V, G) - 2$.

Let $G' = G \cup \{(u_1, u_2)\}$. Since $\text{COM}(G') = \text{COM}(G) - 1$, the separation constraint of any separating $(k-1)$-set is reduced by one. Let $\mathcal{S}$ be a separating set with cardinality less than $k - 1$. By the way the demand is defined, $\text{DEMAND}_{k-|\mathcal{S}|}(V \setminus \mathcal{S}, G' - \mathcal{S}) = \text{DEMAND}_{k-|\mathcal{S}|}(V \setminus \mathcal{S}, G - \mathcal{S}) - 2$. Using the above and the fact that $\text{COM}(G') = \text{COM}(G) - 1$, $\text{AUG}_{k-|\mathcal{S}|}(G' - \mathcal{S}) = \text{AUG}_{k-|\mathcal{S}|}(G - \mathcal{S}) - 1$. Thus the separation constraint of $\mathcal{S}$ is decreased by one.

4.3 Critical, Massive, and Balanced

Let $\mathcal{S}$ be a separating set with cardinality less than k in $G = (V, E)$. For making G k-vertex-connected, if $|\mathcal{S}| < k$ and $\text{SC}_k(\mathcal{S}, G) > \lceil \text{DEMAND}_k(V, G)/2 \rceil$, then $\mathcal{S}$ is *massive*. If $|\mathcal{S}| < k$ and $\text{SC}_k(\mathcal{S}, G) = \lceil \text{DEMAND}_k(V, G)/2 \rceil$, then $\mathcal{S}$ is *critical*. A graph with no critical and massive separating set is *balanced*. For a balanced graph, it seems that the "most important job" in augmenting a graph is to make sure that the demand of the graph becomes zero. For an unbalanced graph, "taking care of" critical and massive separating sets seems to be more important than "taking care of" the demand of the graph.

Structure of Massive Separating Sets.

Claim 4.5 *Let $\mathcal{S}$ be a massive separating ℓ-set, $\ell < 3$, for making a connected graph $G = (V, E)$ four-connected. (1) $a_4(\mathcal{S}, G) > 0$. (2) There is no other massive separating h-set, $h \leq 3$.*

Claim 4.6 *Let $\mathcal{S}$ be a massive separating triplet for making a connected graph $G = (V, E)$ four-connected. (1) If $a_4(\mathcal{S}, G) = 0$, then $\text{SD}(\mathcal{S}, G) \geq 4$ and there are at least four 3-components in $G - \mathcal{S}$ which each contributes exactly one in computing $\text{DEMAND}_k(V, G)$. (2) If $a_4(\mathcal{S}, G) > 0$, then G is triconnected, $\text{SD}(\mathcal{S}', G) = 2$ for every separating triplet $\mathcal{S}' \neq \mathcal{S}$, and there is no special block that is not in $\mathcal{S}$. (3) There is no other massive separating triplet.*

Proof. (Sketch) We only sketch the proof of part part (1). Let q_i be the number of i-components in $G - \mathcal{S}$. $\text{SC}_4(\mathcal{S}, G) = \sum_{i=1}^{3} q_i - 1$. Let $q_3 = q'_3 + q''_3$ where q'_3 is the number of 3-components which contributes exactly one in computing $\text{DEMAND}_k(V, G)$ and q'_3 is the number of other 3-components contributing at least two in computing $\text{DEMAND}_k(V, G)$. By Claim 4.1, $\text{DEMAND}_4(V, G) \geq q'_3 + 2 \cdot q''_3 + 2 \cdot q_2 + 3 \cdot q_3$. In order for $\text{SC}_4(\mathcal{S}, G) > \lceil \text{DEMAND}_4(V, G)/2 \rceil$, $q'_3 \geq 4$.

Corollary 4.7 *Let G be a connected graph that is not triconnected. For making G four-connected, if there is a massive separating set, then we can add an edge to reduce the separation constraint of G by one.*

Proof. Let S be the massive separating set. We first assume that $|S| < 3$. By Claim 4.5, $a_4(S, G) > 0$ and there is no other massive separating set. If $a_4(S, G) = \text{DEMAND}_4(S, G)$, we add an edge (u, v) such that u is a demanding vertex in a special block in S. By doing so, $a_4(S, G \cup \{(u, v)\}) = a_4(S, G) - 1$. Thus the separation constraint of the graph is reduced. If $a_4(S, G) < \text{DEMAND}_4(S, G)$, then we can find two special blocks B_1 and B_2 in S with $u_i \in B_i$ and $(u_1, u_2) \notin G$. $a_4(S, G \cup \{(u, v)\}) = a_4(S, G) - 1$. Thus the separation constraint of the graph is reduced.

We now assume that $|S| = 3$. By Claim 4.6, and there are at least four 3-components in $G - S$ and there is no other massive separating set. By adding an edge (u, v) such that u and v are in different 3-components in $G - S$, we reduce the dividing degree of S. Thus the separation constraint of the graph is reduced.

Structure of Critical Separating Sets.

Claim 4.8 *Let S be a separating set with cardinality less than three in $G = (V, E)$. For making G four-connected, if S is critical, then we can add an edge to reduce the separation constraint of every critical separating set by one.*

A set of separating sets $S = \{S_1, \ldots, S_k\}$ has a *sample set* $\{Q_1, \ldots, Q_r\}$ if S can be disjointedly partitioned into r nonempty partitions $W_1, \ldots, W_r$, where $Q_i \subseteq R$ for all $R \in W_i$ and Q_i is either a special block or a separating set. A sample set with the minimum cardinality is a *minimum sample set*.

Claim 4.9 *For making a connected graph $G = (V, E)$ four-connected, let $\mathcal{Y}$ be the set of critical separating triplets in G. The size of a minimum sample set for $\mathcal{Y}$ is at most two.*

4.4 Reducing the Separation Constraint

Claim 4.10 *Let u_1 and u_2 be two vertices in a connected graph $G = (V, E)$ that are not four-connected and $(u_1, u_2) \notin E$. Let S be a separating triplet in G. Then $\text{SC}_4(S, G \cup \{(u_1, u_2)\}) = \text{SC}_4(S, G) - 1$ if (1) $\text{SC}_4(S, G) > 2$, u_i is a demanding vertex in an x-block, for all $4 - \ell \leq x \leq 4$, and S separates u_1 and u_2, (2) u_1 and u_2 are demanding vertices in two distinct special blocks in S, or (3) u_1 is a demanding vertex in S, $u_2 \notin S$, and $a_4(S, G) = \text{DEMAND}_4(S, G)$.*

Lemma 4.11 *Let G be a connected graph that is not triconnected. For making G four-connected, if there is a critical or massive separating set, we can find two vertices u and v in G such that the separation constraint of G is reduced by one by adding the edge (u, v). Furthermore, the demand of G is reduced by two if there is a critical separating set.*

Proof. (Sketch) This lemma follows from Claims 4.6 and 4.5, and Corollary 4.7 when there is a massive separating set. If there is a critical separating set with cardinality less than three, we use Claim 4.8 to reduce the separation constraint of the graph by one. If there is a critical separating triplet, we use Claim 4.9. By Claim 3.7, Claims 4.8 and 4.9 also makes sure that the demand of the graph is reduced by two.

```
graph function aug0to4(graph G); {* The input graph G contains at least five vertices. *}
if G has exactly five vertices, then return a complete graph wit!. five vertices;
let T be 4-blk(G);
let s be the separation constraint of G; let d be the demand constraint of G;
while G is not triconnected do
        if G is not connected, then use Claim 4.4 to find a pair of vertices u and v
1.   else if d > 1 and s > d then
            if there is a massive separating set with cardinality less than three then
                use Claim 4.5 to find a pair of vertices u and v
            else if there is a massive separating triplet then
                use Claim 4.6 to find a pair of vertices u and v
            fi fi s := s - 1
2.   else if d > 1 and s = d then
            else if there is a critical separating set with cardinality less than three then
                use Claim 4.8 to find a pair of vertices u and v
            else if there is a critical separating triplet then
                use Claim 4.9 to find a pair of vertices u and v
            fi fi s := s - 1; d := d - 1
3.   else find two vertices u and v using Claim 3.7; d := d - 1
        fi fi fi
4.   G := G ∪ {(u, v)}; update T
end while;
return aug3to4(G) {* Function aug3to4 finds a smallest four-connectivity
augmentation for a triconnected G [8]. *}
end aug0to4;
```

Algorithm 1 An algorithm for finding a smallest set of edges whose addition four-connects an undirected graph with at least five vertices.

5   An Algorithm for Four-Connectivity Augmentation

A Simple Lower Bound for the Augmentation Number. Given $G = (V, E)$, the *demand constraint* for making G k-vertex-connected is $\lceil \text{DEMAND}_k(V, G)/2 \rceil$. The *wheel constraint* of G is the largest wheel constraint among all wheels with unit size less than k. Recall that the separation constraint of G is the largest separation constraint among all separating sets with cardinality less than k. We now give a theorem stating a lower bound of $\text{AUG}_k(G)$. Let $\text{LOW}_k(G) = \max\{\lceil \text{DEMAND}_k(V, G)/2 \rceil, \max_{\forall S}\{\text{SC}_k(S, G)\}, \max_{\forall W}\{\text{WC}_k(W, G)\}\}$, where S is a separating set with cardinality less than k, and W is a wheel with unit size less than k in G.

Theorem 5.1 $\text{AUG}_k(G) \geq \text{LOW}_k(G)$.

The Algorithm. Our algorithm is based on the following approach. Using the lower bound on the number of edges needed (Theorem 5.1) as a guideline, we make sure that each time we add an edge, this lower bound is decreased by one. We keep adding an edge until the graph is triconnected. Let $G = (V, E)$ be the resulting triconnected graph. We know that the lower bound given in Theorem 5.1 equals to $\text{AUG}_4(G)$ if G is triconnected [8]. Thus we can apply the algorithm given in [8] to wrap up the computation. By doing this, we guarantee that the number of edges added is minimum in four-connecting the original input graph. We describe our algorithm in Algorithm 1.

Lemma 5.2 *Let u and v be the two vertices found in step 4 of algorithm* aug0to4. *If G is not triconnected, then* $\text{LOW}_4(G \cup \{(u,v)\}) = \text{LOW}_4(G) - 1$.

Theorem 5.3 *Algorithm* aug0to4 *finds a smallest four-connectivity augmentation for G and* $\text{AUG}_4(G) = \text{LOW}_4(G)$.

Lemma 5.4 *The number* $\text{AUG}_4(G)$ *can be computed in* $O(n \cdot \alpha(n,n) + m)$ *time.*

To implement algorithm aug0to4, we are required to perform the following computations once an edge is added. (1) The algorithm must maintain and update the 4-block structure once a new edge is added. (2) The identification of critical and massive separating sets under the conditions that new separating sets may be created and the separation constraint of an existing separating set may decrease. Each of the above operations takes $O(\log n)$ time. Note that $\text{AUG}_4(G) = O(n)$. Thus we have the following lemma.

Lemma 5.5 *Algorithm* aug0to4 *runs in* $O(n \cdot \log n + m)$ *time.*

Acknowledgment. The author thanks helpful comments from referees.

References

1. G. Di Battista and R. Tamassia. On-line graph algorithms with SPQR-trees. In *Proc. 17th ICALP*, volume LNCS # 443, pages 598–611. Springer-Verlag, 1990.
2. K. P. Eswaran and R. E. Tarjan. Augmentation problems. *SIAM J. Comput.*, 5(4):653–665, 1976.
3. H. Frank and W. Chou. Connectivity considerations in the design of survivable networks. *IEEE Trans. on Circuit Theory*, CT-17(4):486–490, December 1970.
4. D. Gusfield. A graph theoretic approach to satistical data security. *SIAM J. Comput.*, 75:552–571, 1989.
5. F. Harary. The maximum connectivity of a graph. *Proc. Nat. Acad. Sci.*, 48:1142–1146, 1962.
6. F. Harary. *Graph Theory*. Addison-Wesley, Reading, MA, 1969.
7. J. E. Hopcroft and R. E. Tarjan. Dividing a graph into triconnected components. *SIAM J. Comput.*, 2:135–158, 1973.
8. T.-s. Hsu. On four-connecting a triconnected graph (extended abstract). In *FOCS'92*, pages 70–79, 1992.
9. T.-s. Hsu. *Graph Augmentation and Related Problems: Theory and Practice*. PhD thesis, University of Texas at Austin, October 1993.
10. T.-s. Hsu and V. Ramachandran. A linear time algorithm for triconnectivity augmentation. In *FOCS'91*, pages 548–559, 1991.
11. T.-s. Hsu and V. Ramachandran. On finding a smallest augmentation to biconnect a graph. *SIAM J. Comput.*, 22(5):889–912, 1993.
12. T. Jordán. Optimal and almost optimal algorithms for connectivity augmentation problems. In *Proc. 3rd IPCO Conference*, pages 75–88, 1993.
13. A. Kanevsky, R. Tamassia, G. Di Battista, and J. Chen. On-line maintenance of the four-connected components of a graph. In *FOCS'91*, pages 793–801, 1991.
14. K. Menger. Zur allgemeinen kurventheorie. *Fund. Math.*, 10:96–115, 1927.
15. J. Plesnik. Minimum block containing a given graph. *ARCHIV DER MATHEMATIK*, XXVII:668–672, 1976.
16. A. Rosenthal and A. Goldner. Smallest augmentations to biconnect a graph. *SIAM J. Comput.*, 6(1):55–66, March 1977.
17. W. T. Tutte. *Connectivity in Graphs*. University of Toronto Press, 1966.
18. T. Watanabe and A. Nakamura. A minimum 3-connectivity augmentation of a graph. *J. Comp. System Sci.*, 46:91–128, 1993.

Searching for a monotone function by independent threshold queries

Peter Damaschke

FernUniversität, Theoretische Informatik II

58084 Hagen, Germany

peter.damaschke@fernuni-hagen.de

Abstract

Let us be given an unknown monotone discrete function f with domain and range of size m and n, respectively, where $m \leq n$. A threshold query has the form "$f(x) \geq y$?" for a pair (x, y).

We give an optimal EREW PRAM algorithm that finds f in $O(\log n)$ time using $O(n)$ threshold queries and arithmetic operations, in such a way that simultaneous queries are independent, i.e. they call mutually different x and y. Our algorithm composes two distributed binary search strategies by accelerated cascading.

The result can be applied to certain segmentation problems, arising e.g. in digital image processing.

1 Problem and background

Let us be given a matrix M with n rows and m columns and with entries 0,1 satisfying the following condition: $M[i,j] \geq M[i, j+1]$ for all valid j in each row i, and $M[i,j] \geq M[i+1,j]$ for all valid i in each column j. We may ask each matrix entry whether it is 0 or 1. Our aim is to determine the matrix completely by such queries in an efficient way.

Obviously, the problem is equivalent to the search for a monotone discrete function f by threshold queries, i.e. for any pair (x, y) we may ask whether $f(x) \geq y$. The set of threshold queries asked during the search process is called *independent* if simultaneous queries (asked in the same unit of time) always call mutually distinct x and y, respectively. In the matrix formulation that means: Simultaneous queries must concern entries in mutually distinct rows and columns, respectively. The sense of this definition will become clear soon.

Some work has been done on related problems. In [6], the exact sequential worst-case complexity of a similar search problem was determined; in that paper, a query at argument x returned $f(x)$. The computation of the maximum of a set of numbers by threshold queries has been studied in [5]. Our problem can be reformulated as a special case of row minima search in totally monotone matrices (see [1] [2] [7]), but we aim at more efficient parallel algorithms for just this special case. A problem being very similar to ours has been solved in [8] in $O(\log \log n)$ time, but there arbitrary sets of simultaneous queries are allowed. In case of independent queries, a trivial argument shows that $\Omega(\log n)$ is a lower time bound, and hence our strategy is optimal.

Our interest in the problem is motivated by a companion paper [4] where we present an $O(\log n)$ time/ $O(n)$ work CREW PRAM algorithm that finds all maximal digital line segments in a digital curve (and therefore improves our first approach in [3]). In fact, we solved the problem in a more general setting where digital curve segmentation is only a special instance. A crucial part of our algorithm is the following result:

Theorem 1 *[4] For $m \leq n$, a discrete monotone function can be determined in $O(\log n)$ time by $O(n)$ threshold queries.*

As one easily sees, this result is asymptotically optimal on a CREW PRAM.

A closer inspection of [4] shows that all ingredients of the segmentation algorithm run even on an EREW PRAM; we need concurrent read operations only due to the way of application of Theorem 1 to segmentation. Here we do not want to repeat the whole segmentation algorithm, but at least we make clear this point.

Our general segmentation problem is defined as follows. Let S be the free semigroup over some alphabet A, and T another semigroup having a zero element 0. Furthermore, let $h : S \longrightarrow T$ be a homomorphism such that for each $a \in A$, $h(a)$ is computable in $O(1)$ sequential time, and each semigroup computation in T needs $O(1)$ sequential time. The problem is: Given an element of S, that is a word w over the alphabet A, find all inclusion-maximal connected subwords v of w with $h(v) \neq 0$.

In [4] we have shown that digital curve segmentation satisfies these presumptions, and we reduced the problem to triples of words $a = a_1 \ldots a_m$, $b = b_1 \ldots b_n$, and c (possibly empty), such that the h-images of a, b, c are nonzero, $h(acb) = 0$, and all output words have the form $a_x \ldots a_m \, c \, b_1 \ldots b_y$. For finding them, we first produce $h(c)$ and all $h(a_x \ldots a_m)$ and $h(b_1 \ldots b_y)$ by parallel prefix computations, and after that we search for the function $f(x) := \max\{y : \ h(a_x \ldots a_m \, c \, b_1 \ldots b_y) \neq 0\}$. Due to our presumptions, each threshold query "$f(x) \geq y$?" can be answered in $O(1)$ time. Hence the complexity bounds for segmentation follow from Theorem 1.

Unfortunately, an arbitrary number of simultaneous queries may access the same $h(a_x \ldots a_m)$ or $h(b_1 \ldots b_y)$. We can produce $O(n)$ copies of $h(c)$ in $O(\log n)$ time in advance, but it is too expensive to copy also the h-values of all suffixes of a and prefixes of b: If we do it in advance then the total work becomes superlinear, and if we produce the copies just when they are needed, the time bound can increase. The way out is obvious: Concurrent read operations are avoided if we allow only independent queries.

So we wish to strengthen Theorem 1 by the independence condition:

Problem 1 *Is there an EREW PRAM algorithm that determines a monotone discrete function in $O(\log n)$ time using $O(n)$ work with independent threshold queries?*

In this paper we give an affirmative answer. As stated above, this immediately yields an optimal EREW PRAM algorithm for segmentation. Our approach and analysis differs significantly from the search algorithm without independence condition in [4]; apparently a slight modification does not make it.

The above motivation may appear complex and specific; for the details we refer to [4] and the forthcoming full paper. However, the problem seems to be natural and interesting for itself.

2 A fast but non-linear algorithm

The following lemma is the most laborious part of our solution, whereas the remainder is not difficult.

Lemma 2 *A monotone discrete function can be found in $O(\log mn)$ time using $O(m \log n)$ independent threshold queries. The corresponding search algorithm runs on an EREW PRAM within these complexity bounds.*

Proof. We focus on the combinatorial side of the algorithm; the implementation on an EREW PRAM is straightforward then. The algorithm we shall give maintains a sequence of search intervals which fulfills the following invariants. (For convenience we use the matrix formulation.)

(1) Every interval is either a consecutive part of a column, or it is degenerated to the space between two neighbored entries of the column. The set of rows occupied by an interval is called its range. Similarly, the range of a degenerated interval is the space between two rows.

(2) All entries of the column below/above the interval are 0/1.

(3) Different intervals are in different columns.

(4) The sequence of intervals is ordered with respect to the columns they belong to.

(5) Two neighbored intervals in the sequence have either equal or disjoint range, and in the latter case, the range of the left interval is below the range of the right interval.

(6) At most two neighbored non-degenerated intervals have the same range. Such pairs of intervals are called twins.

Clearly, the search problem is solved if a non-degenerated interval is established in each column of the matrix. Initially we set two intervals, namely the entire first and last column. The algorithm runs in a number of rounds, each transforming the sequence of intervals into a new sequence. For specifying the algorithm we have to describe a single round. In the following presentation we omit some tedious details of non-essential case distinctions and of singular cases.

Step 1. For all non-degenerated intervals I, ask in parallel the central entry of I. (If I has even length then choose e.g. the lower of the two central entries.) If the answer is 0/1 then remove the lower/upper half of I.

Due to (3),(5) all queries are independent, except in the twin intervals. Due to (6) we can ask the queries in twin intervals in two units of time, thus guaranteeing independence. Note that all invariants remain true for the halved intervals.

Step 2. Insert degenerated intervals between any degenerated twins.

Step 3. For all pairs of neighbored intervals I, J with disjoint ranges do the following in parallel: If there are still columns between I and J then set a new interval K in the central column between I and J, and insert K into the sequence. (If the gap has an even length then choose e.g. the left of the two central columns.)

Initialize K as follows. First let the bottom of K be the bottom of I, and the top of K be the top of J. Due to monotonicity, K need not exceed this range. Now K can be partitioned into (possibly degenerated) subintervals I', L', J' where I' and J' has the same range as I and J, respectively, and the range of L' lies between them. Ask in succession the entries at the top of I', the bottom of L', the top of L', the bottom of J' (if existing). Depending on the anwers, K can be restricted to one of the three subintervals. Note that (3) and (5) ensure that the queries are independent. Due to the construction, (1)–(5) remain true, but (6) can be violated.

Step 4. In the new sequence, at most four consecutive intervals can have the same range. Whereever this happened, delete the inner interval(s) from the sequence. This restores (6) and finishes the round.

Time. Since each round requires constant time, we have to show that the algorithm stops after $O(\log mn)$ rounds. For this we introduce the notion of search area. For any pair of neighbored intervals I, J the search area is the rectangle (submatrix) between the columns of I, J, bounded by the rows of the bottom of I and the top of J. The search area of the whole sequence of intervals is the union of search areas of neighbored pairs.

Consider a pair of twins I, J. In step 1, the search area between I, J is reduced to a factor $1/2$, unless there remains the lower half of I and the upper half of J. But then the insertion of K in step 3 reduces the search area to $3/4$.

Consider a pair I, J with disjoint ranges, and assume that step 1 is already executed. One easily verifies the following claim: If the insertion of K in step 3 deletes a share of less than $1/6$ from the search area between I, J then halving K in the next round deletes a share of at least $1/6$.

Altogether, two consecutive rounds reduce the search area to at most some constant factor. Hence the algorithm terminates after $O(\log mn)$ rounds.

Work. Our next goal is to give an upper bound for the total number of queries. First of all, note that a search interval, once established, is never destroyed in a later round: Certainly, some intervals may be deleted in step 4, but this only happens if corresponding new intervals with the same veertical range have been initialized in the same round. So we can take another view of step 4: Instead of saying that some intervals are generated and others are deleted, we say that some existing intervals have been shifted horizontally into new columns, and the new intervals in these columns disappear straight after initialization.

Queries are only asked when a new interval is initialized or a current interval is halved. Whenever an interval I is shifted to the place of a new interval K, we charge I with the costs of initialization of K before K disappears. Further note that an interval which is shifted has been halved immediately before.

Altogether, in the life time of each interval we have $O(\log n)$ queries. Hence the total number of queries is $O(m \log n)$. $\square$

A more careful analysis would even give a bound of $O(m \log \log n + n)$ work, but this is needless for us.

3 A fast and linear algorithm

The next lemma provides a linear but still slow strategy.

Lemma 3 *For $m \leq n$, a monotone discrete function can be found in $O(\log^2 n)$ time using $O(n)$ independent threshold queries and work on an EREW PRAM.*

Proof. A very simple recursive algorithm is sufficient here. First find the 0/1 boundary in the central column by binary search. Then partition the problem into two subproblems concerning the left and right half of the matrix, and solve them recursively.

One easily sees that the queries asked by this algorithm are independent. Each level of the recursion needs $O(\log n)$ time. Since the number of levels is $\log m \leq \log n$, we need $O(\log^2 n)$ time.

The total number Q of queries is estimated as follows. In level $i \geq 0$ we have 2^i binary search processes with total range of at most n. Let s_{ij} be the length of the j-th search interval in the i-th level. So we have $Q = \sum_{i=0}^{\log m} \sum_{j=1}^{2^i} \log s_{ij}$ with $\sum_j s_{ij} = n$ for every i. Since the log function is concave, the inner sum is maximized if $s_{ij} = n/2^i$. Hence $Q = \sum_{i=0}^{\log m} 2^i \log(n/2^i) = O(n)$. $\square$

Finally we combine the advantages of both previous algorithms, using the technique of accelerated cascading.

Theorem 4 *For $m \leq n$, a monotone discrete function can be found optimally in $O(\log n)$ time using $O(n)$ independent threshold queries and work on an EREW PRAM.*

Proof. W.l.o.g. let be $m = n$; in case $m < n$ we may extend the matrix by uniform dummy columns.

We partition the matrix vertically into $n/\log n$ submatrices each consisting of approximately $\log n$ consecutive columns. First search for the unknown monotone function restricted to the borders of these submatrices. This can be done by Lemma 2 in $O(\log n)$ time using $O(m \log n) = O(n)$ work, since m is replaced by $n/\log n$.

After that we apply the same procedure horizontally, that is, we partition the matrix into roughly $n/\log n$ slices with $\log n$ rows, and search for the discrete inverse f^{-1} restricted to the borders of these slices.

Obviously, this breaks the problem into $O(n/\log n)$ independent subproblems of size not exceeding $\log n \times \log n$. They can be solved by Lemma 3 in $O(\log n)$ time using a total of $O(n)$ work. For this we may even use the sequential version of Lemma 3. $\square$

References

[1] M.J.Atallah, S.R.Kosaraju: An efficient parallel algorithm for the row minima of a totally monotone matrix, *J. of Algorithms* 13 (1992), 394-413

[2] P.G.Bradford, R.Fleischer, M.Smid: A polylog-time and $O(n\sqrt{\log n})$-work parallel algorithm for finding the row minima in totally monotone matrices, 1995, submitted

[3] P.Damaschke: Line segmentation of digital curves in parallel, *12th Symposium on Theoretical Aspects of Computer Science STACS'95*, Munich 1995, *LNCS* 900, Springer 1995, 539-549

[4] P.Damaschke: An optimal parallel algorithm for digital curve segmentation using Hough polygons and monotone function search, *3rd European Symposium on Algorithms ESA'95*, Corfu (Greece) 1995

[5] F.Gao, L.J.Guibas, D.G.Kirkpatrick, W.T.Laaser, J.Saxe: Finding extrema with unary predicates, *Algorithmica* 9 (1993), 591-600

[6] R.Hassin, N.Megiddo: An optimal algorithm for finding all the jumps of a monotone step-function, *J. of Algorithms* 6 (1985), 265-274

[7] R.Raman, U.Vishkin: Optimal randomized parallel algorithms for computing the row minima of a totally monotone matrix, *5th ACM-SIAM Symposium on Discrete Algorithms SODA '94*, 613-621

[8] R.Sarnath, X.He: Efficient parallel algorithms for selection and searching in sorted matrices, *6th Int. Parallel Processing Symposium IPPS'92*, IEEE Press 1992, 108-111

A Fast and Simple Algorithm for Identifying 2-Monotonic Positive Boolean Functions

Kazuhisa MAKINO and Toshihide IBARAKI

Department of Applied Mathematics and Physics,
Graduate School of Engineering, Kyoto University, Kyoto 606, Japan.
Email address: makino@kuamp.kyoto-u.ac.jp, ibaraki@kuamp.kyoto-u.ac.jp

Abstract. Consider the problem of identifying $\min T(f)$ and $\max F(f)$ of a positive (i.e., monotone) Boolean function f, by using membership queries only, where $\min T(f)$ $(\max F(f))$ denotes the set of minimal true vectors (maximal false vectors) of f. As the existence of a polynomial total time algorithm (i.e., polynomial time in the length of input and output) for this problem is still open, we consider here a restricted problem: given an unknown positive function f of n variables, decide whether f is 2-monotonic or not, and if f is 2-monotonic, output both $\min T(f)$ and $\max F(f)$. For this problem, we propose a simple algorithm, which is based on the concept of maximum latency, and show that it uses $O(n^2 m)$ time and $O(n^2 m)$ queries, where $m = |\min T(f)| + |\max F(f)|$. This answers affirmatively the conjecture raised in [3, 4], and is an improvement over the two algorithms discussed therein: one uses $O(n^3 m)$ time and $O(n^3 m)$ queries, and the other uses $O(nm^2 + n^2 m)$ time and $O(nm)$ queries.

1 Introduction

Consider the problem of identifying $T(f)$ (set of true vectors) and $F(f)$ (set of false vectors) of a given Boolean function (or a function in short) f by asking membership queries to an oracle whether $f(u) = 0$ or 1 holds for some selected vectors u [2]. In the terminology of computational learning theory [1, 14], this is the exact learning of a Boolean theory f by membership queries only. It is also a process of forming a theory that explains a certain phenomenon by collecting positive and negative data (in the sense of causing and not causing that phenomenon) [5]. In particular, we are interested in the case where f is known to be positive (or monotone). If f is a positive function, $T(f)$ and $F(f)$ can be compactly represented by $\min T(f)$ (the set of minimal true vectors) and $\max F(f)$ (the set of maximal false vectors). Therefore our problem is stated as follows.

Problem IDENTIFICATION
Input: a membership oracle for a positive function f.
Output: $\min T(f)$ and $\max F(f)$.

It is known [2] that the existence of a *polynomial total time algorithm* [9] (i.e., polynomial time in the length of input and output) for this problem is equivalent to the existence of polynomial algorithms for many other interesting problems encountered in various fields such as hypergraph theory, theory of coteries (used in distributed systems) [8], artificial intelligence and Boolean theory [2]. Unfortunately, the complexity of these problems is still open (see, e.g., [2, 9]), though the recent result by M. Fredman and L. Khachiyan [6] shows that it is unlikely to be NP-hard.

To partially answer this question, we consider here the following restricted problem:

Problem IDENTIFICATION-2M

Input: a membership oracle for a positive function f.

Output: If f is 2-monotonic, then $\min T(f)$ and $\max F(f)$; otherwise, "no".

We propose in this paper a simple algorithm for this problem, which is based on the concept of maximum latency [10] (see Section 2 for its definitions), and show that it uses $O(n^2 m)$ time and $O(n^2 m)$ queries, where $m = |\min T(f)| + |\max F(f)|$. In previous papers [3, 4], two algorithms were proposed for this problem; one uses $O(n^3 m)$ time and $O(n^3 m)$ queries, while the other uses $O(nm^2 + n^2 m)$ time and $O(nm)$ queries. Since $m \gg n$ can be expected in usual cases, our algorithm improves upon the previous algorithms, and also answers affirmatively to the conjecture raised in [3, 4].

2 Definitions and basic properties

2.1 2-monotonic positive Boolean functions

A *Boolean function*, or a *function* in short, is a mapping $f : \{0, 1\}^n \mapsto \{0, 1\}$, where $v \in \{0, 1\}^n$ is called a *Boolean vector* (a *vector* in short). If $f(v) = 1$ (resp. 0), then v is called a *true* (resp. *false*) vector of f. The set of all true vectors (resp. false vectors) is denoted by $T(f)$ (resp. $F(f)$). Two special functions with $T(f) = \phi$ and $F(f) = \phi$ are respectively denoted by $f = \perp$ and $f = \top$. Also we use notations $ON(v) = \{j \mid v_j = 1, j = 1, 2, \ldots, n\}$ and $OFF(v) = \{j \mid v_j = 0, j = 1, 2, \ldots, n\}$. A function f is *positive* if $v \le w$ (i.e., $v_j \le w_j, j = 1, 2, \ldots, n$) always implies $f(v) \le f(w)$. A positive function is also called *monotone*. A true vector v of f is *minimal* if there is no other true vector w such that $w < v$ (i.e., $w \le v$ and $w \ne v$), and let $\min T(f)$ denote the set of all minimal true vectors of f. A *maximal* false vector is symmetrically defined and $\max F(f)$ denotes the set of all maximal false vectors of f.

If f is positive, it is known that f has the unique disjunctive normal form (DNF) consisting of all prime implicants. There is one-to-one correspondence between prime implicants and minimal true vectors. For example, a positive function $f = x_1 x_2 \vee x_2 x_3 \vee x_3 x_1$ has prime implicants $x_1 x_2, x_2 x_3, x_3 x_1$ which correspond to minimal true vectors (110), (011), (101), respectively. In other words, the input length to describe a positive function f is $O(n|\min T(f)|)$ if it is represented in this manner. Sets $\min T(f)$ and $\max F(f)$ respectively define

$T(f)$ and $F(f)$ by

$$T(f) = \{v \mid v \geq a \text{ for some } a \in \min T(f)\}$$
$$F(f) = \{v \mid v \leq b \text{ for some } b \in \max F(f)\}.$$

If functions f and g (not necessarily positive) on the same set of variables satisfy $g(v) \leq f(v)$ for all $v \in \{0,1\}^n$, then we denote $g \leq f$. If $g \leq f$ and $g \neq f$, we denote $g < f$. An assignment A of binary values 0 or 1 to k variables $x_{i_1}, x_{i_2}, \ldots, x_{i_k}$ is called a k-*assignment*, and is denoted by

$$A = (x_{i_1} \leftarrow a_1, x_{i_2} \leftarrow a_2, \ldots, x_{i_k} \leftarrow a_k),$$

where each of $a_1, a_2, \ldots a_k$ is either 1 or 0. Let the complement of A, denoted by $\bar{A}$, represent the assignment obtained from A by complementing all the 1's and 0's in A. When a function f of n variables and a k-assignment A are given,

$$f_A = f_{(x_{i_1} \leftarrow a_1, x_{i_2} \leftarrow a_2, \ldots, x_{i_k} \leftarrow a_k)}$$

denotes the function of $(n-k)$ variables obtained by fixing variables $x_{i_1}, x_{i_2}, \ldots,$ x_{i_k} as specified by A.

Let f be a function of n variables. If either $f_A \leq f_{\bar{A}}$ or $f_A \geq f_{\bar{A}}$ holds for every k-assignment A, then f is said to be k-*comparable*. If f is k-comparable for every k such that $1 \leq k \leq m$, then f is said to be m-*monotonic*. (For more detailed discussion on these topics, see [12] for example.) In particular, f is 1-monotonic if $f_{(x_i \leftarrow 1)} \geq f_{(x_i \leftarrow 0)}$ or $f_{(x_i \leftarrow 1)} \leq f_{(x_i \leftarrow 0)}$ holds for any $i \in \{1, 2, \ldots, n\}$. A function f is positive if and only if f is 1-monotonic and $f_{(x_i \leftarrow 1)} \geq f_{(x_i \leftarrow 0)}$ holds for all i.

Now consider a 2-assignment $A = (x_i \leftarrow 1, x_j \leftarrow 0)$. If

$$f_A \geq f_{\bar{A}} \quad (\text{resp. } f_A > f_{\bar{A}})$$

holds, this is denoted $x_i \succeq_f x_j$ (resp. $x_i \succ_f x_j$). Variables x_i and x_j are said to be *comparable* if either $x_i \succeq_f x_j$ or $x_i \preceq_f x_j$ holds. When $x_i \succeq_f x_j$ and $x_i \preceq_f x_j$ hold simultaneously, it is denoted as $x_i \approx_f x_j$. If f is 2-monotonic, this binary relation $\succeq_f$ over the set of variables is known to be a total preorder [12]. A 2-monotonic positive function f of n variables is called *regular* if

$$x_1 \succeq_f x_2 \succeq_f \cdots \succeq_f x_n.$$

Any 2-monotonic positive function becomes regular by permuting variables.

The 2-monotonicity and related concepts have been studied in various names in the field such as threshold logic, game theory and hypergraph theory. The 2-monotonicity was originally introduced in conjunction with threshold functions (e.g., [12]), where a positive function f is *threshold* if there exist $n+1$ nonnegative real numbers $w_1, w_2, \ldots, w_n$ and t such that:

$$f(x) = \begin{cases} 1, & \text{if } \sum w_i x_i \geq t \\ 0, & \text{if } \sum w_i x_i < t. \end{cases}$$

As $w_i \geq w_j$ implies $x_i \succeq_f x_j$ and $w_i = w_j$ implies $x_i \approx_f x_j$, a threshold function is always 2-monotonic, although the converse is not true [12].

2.2 Maximum latency of 2-monotonic positive functions

Now let MT and MF respectively denote the partial knowledge of $\min T(f)$ and $\max F(f)$ currently at hand, i.e.,

$$MT \subseteq \min T(f) \text{ and } MF \subseteq \max F(f). \tag{1}$$

Define

$$T(MT) = \{v \mid v \geq w \text{ for some } w \in MT\}$$
$$F(MF) = \{v \mid v \leq w \text{ for some } w \in MF\}.$$

By assumption (1), $T(MT) \subseteq T(f)$ and $F(MF) \subseteq F(f)$ hold, and hence

$$T(MT) \cap F(MF) = \phi$$

holds. A vector u is called *unknown* if

$$u \in \{0,1\}^n \setminus (T(MT) \cup F(MF)),$$

since it is not known at the current stage whether u is a true vector or a false vector of f. If there is no unknown vector, then $T(MT) \cup F(MF) = \{0,1\}^n$ holds, i.e., $MT = \min T(f)$ and $MF = \max F(f)$ for some positive function f.

Definition 1. Given incomparable sets of vectors MT and MF $(\subseteq \{0,1\}^n)$ (i.e., no two vectors v and w in MT satisfies $v \leq w$ or $w \leq v$; similarly for MF) such that $MT \cup MF \neq \emptyset$ and $T(MT) \cap F(MF) = \emptyset$, a partial function g is defined by

$$g(v) = \begin{cases} 1, & v \in T(MT) \\ 0, & v \in F(MF) \\ \text{unknown}, & \text{otherwise.} \end{cases}$$

If MT and MF of g satisfy $MT \subseteq \min T(f)$ and $MF \subseteq \max F(f)$ for some (complete) positive function f, then g is called a *partial function of f*. If g is a partial function of f and $g \neq f$, then g is called a *proper* partial function of f. Denote the set of unknown vectors of g by $U(g)$, i.e.,

$$U(g) = \{0,1\}^n \setminus (T(MT) \cup F(MF)).$$

The *k-neighborhood* of g is defined by

$$N_k(g) = \{v \mid \| v - a \| \leq k \text{ for some } a \in MT \cup MF\},$$

where $\| w \|$ denotes $\sum_{i=1}^n |w_i|$. The *latency* of g, $\lambda(g)$, is the integer k satisfying

$$N_{k-1}(g) \cap U(g) = \phi \text{ and } N_k(g) \cap U(g) \neq \phi.$$

As a special case, if $U(g) = \phi$ (i.e., $g = f$), then $\lambda(g)$ is defined to be 0. $\lambda(g)$ is equivalently given by

$$\lambda(g) = \min\{\| u - a \| \mid u \in U(g), a \in MT \cup MF\}. \qquad \square$$

Definition 2. Let C_X be a subclass of positive functions. $C_X(n)$ denotes the set of functions in C_X with n variables. For $C_X(n)$, the *maximum latency* is defined by

$$\Lambda_X(n) = \max\{\lambda(g) \mid g \text{ is a partial function of } f \in C_X(n)\}. \qquad \square$$

We have the next results.

Proposition 3 [10]. *Let g be a partial function of $f \in C_X(n)$. If $N_{\Lambda_X(n)}(g) \cap U(g) = \phi$, then there is no unknown vector, i.e., $g = f$ holds.* $\qquad \square$

That is, in order to find an unknown vector, we only need to search $\Lambda_X(n)$-neighborhood of g.

Proposition 4 [10]. *Class C_{2M} of 2-monotonic positive functions satisfies, for any n,*

$$\Lambda_{2M}(n) = 1. \qquad \square$$

This theorem says that, if g is a proper partial function of $f \in C_{2M}$, then there is an unknown vector in the 1-neighborhood of g. However, we cannot directly apply Proposition 4 to problem IDENTIFICATION-2M, since an unknown positive function f might not be 2-monotonic.

3 Outline of the algorithm

This section gives an outline of our algorithm for problem IDENTIFICATION-2M. The details of each step and the analysis of its time complexity will follow in the subsequent sections.

Let MT and MF respectively denote the current partial knowledge of min $T(f)$ and max $F(f)$, i.e., $MT \subseteq \min T(f)$ and $MF \subseteq \max F(f)$. In our algorithm, we maintain four disjoint subsets of vectors MT_1, MT_2, MF_1 and MF_2 such that

$$MT_1 \cup MT_2 = MT \,(\subseteq \min T(f)),$$
$$MF_1 \cup MF_2 = MF \,(\subseteq \max F(f)),$$
$$N_1(MT_2 \cup MF_2) \subseteq T(MT) \cup F(MF),$$

where

$$N_k(S) = \{v \mid \| v - a \| \leq k \text{ for some } a \in S\}.$$

Let e_j denote the unit vector defined by $ON(e_j) = \{j\}$. At each iteration, we choose a vector $v \in MT_1 \cup MF_1$, and proceed in the following manner.

    (i) If $v \in MT_1$, then let $MT_1 := MT_1 \setminus \{v\}$ and $MT_2 := MT_2 \cup \{v\}$. Then find $y^j \in \max F(f)$ such that $y^j \geq v - e_j$ for each $j \in ON(v)$. If $y^j \notin MF$ for some $j \in ON(v)$, then update $MF_1 := MF_1 \cup \{y^j\}$.

    (ii) If $v \in MF_1$, then $MF_1 := MF_1 \setminus \{v\}$ and $MF_2 := MF_2 \cup \{v\}$. Then find $y^j \in \min T(f)$ such that $y^j \leq v + e_j$ for each $j \in OFF(v)$. If $y^j \notin MT$ for some $j \in OFF(v)$, then update $MT_1 := MT_1 \cup \{y^j\}$.

(iii) If $MT_1 \cup MF_1 = \phi$ holds, then we test whether the positive function g defined by $\min T(g) = MT$ is 2-monotonic. If g is 2-monotonic, then we output $MT_2 \, (= \min T(f))$ and $MF_2 \, (= \max F(f))$, and halt. Otherwise, we conclude that f is not 2-monotonic, and halt.

Initially MT_1, MT_2, MF_1 and MF_2 are appropriately prepared (see subsection 4.1), and then the loop of steps (i) (ii) is repeated until it halts in (iii).

The key points in implementing these steps are how to execute the following steps.

1. Prepare the initial sets MT_1, MT_2, MF_1 and MF_2.
2. Given an unknown vector u, compute a new maximal false (resp. minimal true) vector y such that $y \geq u$ (resp. $y \leq u$) which is used in the above steps (i) and (ii).
3. Check whether g is 2-monotonic or not in step (iii).

These will be separately discussed in the subsequent sections. Finally we show the main lemma which can be used to check the validity of step (iii).

Lemma 5. *Assume that given sets $MT \subseteq \min T(f)$ and $MF \subseteq \max F(f)$ of a positive function f satisfy $N_1(MT \cup MF) \subseteq T(MT) \cup F(MF)$, and let g be the positive function defined by $\min T(g) = MT$. Then f is 2-monotonic if and only if g is 2-monotonic.*

Proof. (1) The if-part: Assume that g is regular without loss of generality, and let $U = \{0,1\}^n \setminus (T(MT) \cup F(MF))$. Assuming that $U \neq \phi$, we derive a contradiction. Take a $u \in \max U$, where $\max U$ is the set of maximal unknown vectors (i.e., $u + e_j \in T(MT)$ for all $j \in OFF(u)$). Let $j = \max\{i \,|\, i \in OFF(u)\}$. Since $u + e_j \in T(MT)$, there exists $a \in MT$ such that $a \leq u + e_j$. Note that $a_j = 1$ holds since otherwise $a \leq u$, contradicting $u \in U$. Then $a - e_j \in F(MF)$ by assumption $N_1(MT \cup MF) \subseteq T(MT) \cup F(MF)$. Therefore, there exists $b \in MF$ such that $b \geq a - e_j$. This b satisfies $b_j = 0$ since otherwise $b \geq a$, a contradiction. For any $l \in OFF(u) \setminus \{j\}$,

$$a - e_j + e_l \in T(g) \subseteq T(MT) \cup U$$

by regularity of g, and hence $b \not\geq a - e_j + e_l$, i.e., $b_l = 0$ for all $l \in OFF(u)$ and hence $b \leq u$. (i) If $b = u$, then $u \in F(MF)$, which is a contradiction. (ii) If $b < u$, then $u \in T(MT)$ by $N_1(MT \cup MF) \subseteq T(MT) \cup F(MF)$, which is also a contradiction. Therefore, $U = \phi$, i.e., $g = f$ holds, and then f is 2-monotonic.

(2) The only if-part: If f is 2-monotonic, then Proposition 3 and Proposition 4 imply $f = g$. That is, g is 2-monotonic. □

4 Implementing the key steps

4.1 Initialization of MT_1, MT_2, MF_1 and MF_2

It may be convenient first to ask the oracle about $u = (111\ldots1)$. If $f(u) = 0$, then $f = \perp$ by the positivity of f (i.e., f is identified). Therefore assume that $f(u) = 1$ and obtain a minimal true vector $a \in \min T(f)$ by algorithm MINIMAL, which will be described in Section 4.3. Then initialize MT_1, MT_2, MF_1 and MF_2 as

$$MT_1 := \{a\}, \quad \text{and} \quad MT_2 := MF_1 := MF_2 := \phi.$$

4.2 Data structures of MT_1, MT_2, MF_1 and MF_2

As data structures of MT_1, MT_2, MF_1 and MF_2, we respectively use binary trees $B(MT_1)$, $B(MT_2)$, $B(MF_1)$ and $B(MF_2)$ of height n, in which the left edge (resp. right edge) from a node in depth $j - 1$ represents the case $x_j = 0$ (resp. $x_j = 1$). A leaf node t of $B(S)$ in depth n stores the vector $v \in S$, the components of which correspond to the edges of the path from the root to t. In order to have a compact representation, edges with no descendants are removed from $B(S)$. An example for such a binary tree is shown in Figure 1 corresponding to a set $S = \{011, 101, 110\}$. With this data structure, it is easy to see that we can apply

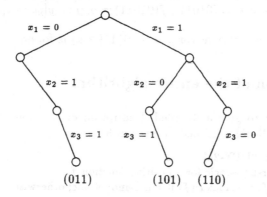

Fig. 1. A binary tree $B(S)$.

operations MEMBER (i.e., check if $v \in S$), INSERT (i.e., update $S := S \cup \{v\}$) and DELETE (i.e., update $S := S \setminus \{v\}$) in $O(n)$ time, respectively.

4.3 Construction of a minimal true vector or a maximal false vector of f

Given an unknown vector u for a positive function f, a vector y, satisfying either $y \in \min T(f)$ and $y \leq u$, or $y \in \max F(f)$ and $y \geq u$, can be computed as follows [2, 7, 14]. First ask the oracle whether $f(u) = 0$ or $f(u) = 1$ holds. We assume for simplicity that $f(u) = 0$ holds, and present an algorithm to find a $y \in \max F(f)$ such that $y \geq u$. Let $y[y_i = 1]$ denote the vector y with its i-th element y_i fixed to 1.

Algorithm MAXIMAL
Input: a vector $u \in F(f)$, and a membership oracle for a positive function f.

Output: a vector $y \in \max F(f)$ such that $y \geq u$.

1. $y := u$.

2. For $i = 1, 2, \ldots, n$, let $y := y[y_i = 1]$ if $f(y[y_i = 1]) = 0$ (membership queries are used here).

3. Output y. □

The case of $f(u) = 1$ can be treated analogously and the corresponding algorithm is called MINIMAL. Both algorithms issue $O(n)$ membership queries before outputting the final y.

As an example, consider a positive function f of five variables with $\max F(f) = \{11100, 01011\}$. Now let $u = (00010)$ be tested. Because $f(u) = 0$ (recall that $u < (01011)$), MAXIMAL tests the following sequence.

$$\begin{aligned}
y &= (00010), f(00010) = 0, \\
y[y_1 = 1] &= (10010), f(10010) = 1, \\
y[y_2 = 1] &= (01010), f(01010) = 0, y := y[y_2 = 1], \\
y[y_3 = 1] &= (01110), f(01110) = 1, \\
y[y_4 = 1] &= (01010), f(01010) = 0, y := y[y_4 = 1], \\
y[y_5 = 1] &= (01011), f(01011) = 0, y := y[y_5 = 1].
\end{aligned}$$

From this, a maximal false vector $y = (01011)\,(\geq u)$ is found.

5 Description of the entire algorithm

Now it is possible to give a detailed description of the algorithm for solving IDENTIFICATION-2M, outlined in Section 3.

Algorithm IDENTIFY-2M
Input: a membership oracle for a positive function f.
Output: $\min T(f)$ and $\max F(f)$ if f is 2-monotonic; otherwise, "no".

1 **begin**

2   Initialize $MT_1 := \{a\}$, $MT_2 := MF_1 := MF_2 := \phi$, as described in Subsection 4.1 ($f = \perp$ may be concluded here.);

3   **while** $MT_1 \cup MF_1 \neq \phi$ **do**

4   **begin**

5   Take a $v \in MT_1 \cup MF_1$ arbitrarily;

6   **if** $v \in MT_1$ **then**

7   **begin**

8         $MT_1 := MT_1 \setminus \{v\}$ and $MT_2 := MT_2 \cup \{v\}$;

9   Compute a maximal false vector y^j such that $y^j \geq v - e_j$ for every $j \in ON(v)$, by using MAXIMAL of Subsection 4.3;

10   **for** all the generated y^j **do**

11   **if** $y^j \notin MF_1 \cup MF_2$ **then** $MF_1 := MF_1 \cup \{y^j\}$

12   **end**

13   **else** $\{v \in MF_1\}$

14   **begin**

15       $MF_1 := MF_1 \setminus \{v\}$ and $MF_2 := MF_2 \cup \{v\}$;

16   Compute a minimal true vector y^j such that $y^j \leq v + e_j$ for every $j \in$ $OFF(v)$, by using MINIMAL of Subsection 4.3;

17   **for** all the generated y^j **do**

18   **if** $y^j \notin MT_1 \cup MT_2$ **then** $MT_1 := MT_1 \cup \{y^j\}$

19   **end**

20   **end**

21   Let g be the positive function satisfying $\min T(g) = MT_2$. If g is 2-monotonic, output $MT_2 (= \min T(f))$ and $MF_2 (= \max F(f))$; otherwise output "no".

22 **end**

Theorem 6. *Given a membership oracle of a positive function f of n variables, algorithm IDENTIFY-2M decides whether f is 2-monotonic or not, and if f is 2-monotonic, it outputs $\min T(f)$ and $\max F(f)$. The time required is $O(n^2 m)$ and the number of queries to the oracle is $O(n^2 m)$, where $m = |\min T(f)| + |\max F(f)|$.*

Proof. The correctness of algorithm IDENTIFY-2M follows from Lemma 5. We now analyze its time complexity. First, consider the while-loop of lines 3-20. The case of $v \in MT_1$ in line 6 occurs $|\min T(f)|$ times. For each v, line 8 (i.e., applying operations DELETE and INSERT to binary trees MT_1 and MT_2, respectively) is executed in $O(n)$ time, and at most n vectors y^j are constructed in line 9, each of which is computed by MAXIMAL in $O(n)$ time. Furthermore, for each such vector y, line 11 requires $O(n)$ time. Then the total time here is $O(n^2 |\min T(f)|)$ time. Analogously, in the case of $v \in MF_1$ (line 13), the total time required is $O(n^2 |\max F(f)|)$ time. Therefore, the while-loop requires

$$O(n^2 (|\min T(f)| + |\max F(f)|)) = O(n^2 m)$$

time.

Finally, the line 21 (i.e., checking the 2-monotonicity of g) can be done in

$$O(n^2 |\min T(g)|) = O(n^2 |\min T(f)|)$$

time by applying the algorithm proposed in [13] to check the regularity of a positive function (for details, see [11]).

Summing the above terms, we see that the time complexity of IDENTIFY-2M is $O(n^2 m)$.

Next, we count the number of queries to oracle. These queries are used only in lines 9 and 16; i.e., in the calls to MAXIMAL and MINIMAL. For each v and j, MAXIMAL (or MINIMAL) requires $O(n)$ queries. The total number of queries is $O(n^2 m)$ since the number of v considered in lines 9 and 16 is m and the number of j for each v is at most n. $\square$

6 Discussion

In this paper, we proposed a simple algorithm for IDENTIFICATION-2M, which is based on the concept of maximum latency for 2-monotonic positive functions. This algorithm uses $O(n^2 m)$ time and $O(n^2 m)$ queries, where $m = |\min T(f)| + |\max F(f)|$, an improvement upon the previously known complexities in [3, 4].

Another more ambitious goal is to develop a polynomial total time algorithm for general positive functions (or to disprove its existence). In this case, however, the maximum latency is not powerful enough as noted in [10], and some new tools would be necessary.

References

1. M. Anthony and N. Biggs, *Computational Learning Theory*, Cambridge University Press, 1992.
2. J. C. Bioch and T. Ibaraki, Complexity of identification and dualization of positive Boolean functions, RUTCOR Research Report RRR 25-93, Rutgers University, 1993; to appear in *Information and Computation*.
3. E. Boros, P. L. Hammer, T. Ibaraki and K. Kawakami, Identifying 2-monotonic positive Boolean functions in polynomial time, *ISA '91 Algorithms*, edited by W. L. Hsu and R. C. T. Lee, LNCS 557 (1991) 104-115.
4. E. Boros, P. L. Hammer, T. Ibaraki and K. Kawakami, Polynomial time recognition of 2-monotonic positive Boolean functions given by an oracle, RUTCOR Research Report RRR 10-95, Rutgers University, 1995; to appear in *SIAM J. Computing*.
5. Y. Crama, P. L. Hammer and T. Ibaraki, Cause-effect relationships and partially defined boolean functions, *Annals of Operations Research*, 16 (1988) 299-326.
6. M. Fredman and L. Khachiyan, *On the complexity of dualization of monotone disjunctive normal forms*, Technical Report LCSR-TR-225, Department of Computer Science, Rutgers University, 1994.
7. D. N. Gainanov, On one criterion of the optimality of an algorithm for evaluating monotonic Boolean functions, *U.S.S.R. Computational Mathematics and Mathematical Physics*, 24 (1984) 176-181.
8. T. Ibaraki and T. Kameda, A Theory of coteries: Mutual exclusion in distributed systems, *IEEE Trans. on Parallel and Distributed Systems*, 4 (1993) 779-794.
9. D. S. Johnson, M. Yannakakis and C. H. Papadimitriou, On generating all maximal independent sets, *Information Processing Letters*, 27 (1988) 119-123.
10. K. Makino and T. Ibaraki, The maximum latency and identification of positive Boolean functions, *ISAAC'94 Algorithms and Computation*, edited by D. Z. Du and X. S. Zhang, LNCS 834 (1994) 324-332.
11. K. Makino and T. Ibaraki, A fast and simple algorithm for identifying 2-monotonic positive Boolean functions, Technical Report of IEICE, COMP94-46 (1994) 11-20.
12. S. Muroga, *Threshold Logic and Its Applications*, Wiley-Interscience, 1971.
13. J. S. Provan and M. O. Ball, Efficient recognition of matroids and 2-monotonic systems, *Applications of Discrete Mathematics*, edited by R. Ringeisen and F. Roberts, SIAM, Philadelphia (1988) 122-134.
14. L. G. Valiant, A theory of the learnable, *Communications of the ACM*, 27 (1984) 1134-1142.

Deciding Bisimulation and Trace Equivalences for Systems with Many Identical Processes *

HSU-CHUN YEN   SHI-TSUEN JIAN   TA-PANG LAO

Dept. of Electrical Engineering, National Taiwan University
Taipei, Taiwan, R. O. C.

Abstract. In this paper, we study the complexity and decidability issues of checking trace and bisimulation equivalences for the model of *systems with many identical processes* with respect to various interprocess communication structures. In our model, each system consists of an arbitrary number of identical finite state processes using Milner's *Calculus of Communicating Systems* (CCS) as the style of interprocess communication. As it turns out, checking trace and bisimulation equivalences are undecidable for *star-like* and *linear* systems, whereas the two problems are complete for PSPACE and PTIME, respectively, for *fully connected* systems.

1 Introduction

In concurrency theory, various notions of *equivalence* have been proposed for capturing the essence of two concurrent systems being behaviorally equivalent [13]. The decidability and complexity issues of the *equivalence problem* (i.e., that of determining whether two systems are 'behaviorly equivalent') with respect to various equivalence notions are relatively well-studied for *finite-state systems*. See, e.g., [10, 11]. It is not surprising that for finite-state systems, the equivalence problem is decidable for all the equivalences proposed in the literature. Recently, considerable efforts have been directed to the study of the decidability and complexity issues of the equivalence problem for *infinite-state systems* including *Basic Process Algebra* (BPA, for short), *Basic Parallel Processes* (BPP, for short), *context-free processes*, and *Petri nets* [7, 9]. The reader is referred to [1] for a nice survey of such results.

In a recent article [5], checking bisimulation equivalence has been shown to be undecidable for *systems with indefinite number of identical processes* with respect to *propositional linear temporal logic* even without the next-time operator. (Two systems are said to be bisimulation equivalent if for every formula f written in propositional linear temporal logic, the sets of computations of the two systems satisfying f are identical.) This work can be thought of as an extension of that of [4, 12] in which the problem of determining whether a given system satisfies a specification given in *propositional linear temporal logic* without next-time operator (i.e., the *model checking problem*) for systems with many identical processes has been investigated.

* This work was supported in part by the National Science Council of the Republic of China under Grant NSC-84-2221-E-002-039.

In this paper, we study the decidability and complexity issues of checking bisimulation and trace equivalences for systems with many identical processes with respect to the following structures of interprocess communication: *fully-connected* topology, *star-like* topology, and *linear* topology. In our setting, each system consists of an arbitrary number of finite state processes using *Milner's CCS* as the style of interprocess communication. Given two such systems, our main interest lies in deciding whether the behaviors of two finite-state processes (taken from the two given systems) are *bisimulation equivalent* (or *trace equivalent*). Our results are summarized in Table 1. Despite the similarity in the under-

	Bisimulation Equivalence	Trace Equivalence
Fully-connected	PTIME-complete	PSPACE-complete
Star-like	Undecidable	Undecidable
Linear	Undecidable	Undecidable

Table 1. Complexities of the equivalence problem for a variety of systems with many identical processes.

lying model, our work differs from that of [5] in many aspects. What follows are the primary differences. First, all the processes in our model are identical. In [5], however, a so-called *synchronizer* is in presence whose structure may differ from the remaining *user* processes. Second, our underlying notions of bisimulation and trace equivalences are identical to those defined in the literature [1] (tailored to the model of systems with many identical processes), whereas the 'bisimulation' defined in [5] was built upon linear temporal logic (in their setting, two systems are 'bisimulation equivalent' iff they 'witness' the same set of formulas). Finally, Our results reveal an additional dimension of complexity regarding systems with many identical processes, namely, the structure through which processes communicate. Our results suggest that even in the absence of temporal logic, checking bisimulation (as well as trace) equivalence is undecidable for star-like and linear systems. In contrast, it is not clear whether the hardness (more precisely, undecidability) result of [5] results from the model of many identical processes itself, or from the expressiveness of temporal logic. For more about systems with many identical processes, the interested reader is referred to [2, 3, 4, 5, 6, 12, 14].

2 Definitions

We begin with the definition of *systems with many identical processes* (or simply *systems* if they are clear from the context). Our model is essentially the one proposed in [4]. A *process* is a 6-tuple $P=(Q, \delta, \Sigma, q_0, \Gamma, L)$, where Q is a finite set of *states*, $q_0(\in Q)$ is the *initial state*, δ ($\subseteq Q \times Q \times \{+, -\} \times \Sigma$) is the *transition relation*, Σ is a finite set of *messages*, Γ: a finite set of *labels*, and

303

L: $(\delta \to \Gamma)$ a labeling function which assigns labels to transitions. In our model, interprocess communication is based on the "hand-shaking" notion. Given a "c" in Σ, we can think of $-c$ as the action of *sending* message c and $+c$ as the action of *receiving* message c. For convenience, we denote (q, q', c) as $q \xrightarrow{c} q'$. We write $\bar{c}$ to denote the complement of c, i.e., $\overline{+c} = -c$ and $\overline{-c} = +c$.

Given two systems S_1 and S_2 and two designated processes P_1 and P_2 (in S_1 and S_2, respectively), our main concern in this paper is to determine whether P_1 and P_2 are bisimulation (or trace) equivalent in the presence of arbitrary numbers of identical processes in their respective systems with respect to the following interprocess communication structures: (1) fully-connected topology (see Fig. 1.(a)), (2) star-like topology (see Fig. 1.(b)), and (3) linear topology (see Fig. 1.(c)). The designated process is referred to as the *target* process throughout this paper. Notice that the target and the remaining processes are identical. Given a process P, we write P^F (P^S and P^L, respectively) to represent the system consisting of an arbitrary number of process P connected in a fully-connected (star-like and linear, respectively) fashion, or P^∞ if the underlying interconnection topology is not important.

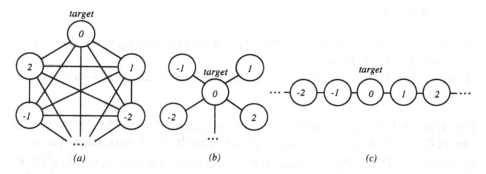

Fig. 1. System topology.

A *global state* s of a system P^∞, where $P = (Q, \delta, \Sigma, q_0, \Gamma, L)$, is a mapping from Z to Q such that $s(i)$ represents the current state of the process labeled i. Initially, the system is in its *initial state* s_0 with $s_0(i) = q_0, \forall i \in Z$. For convenience, we write $S(P^\infty)$ to denote the set of all global states of system P^∞. Given two global states s, s', and an action symbol $c \in \Sigma$, we say processes i and j, for some $i, j \in Z$, can communicate through the exchange of action symbol c in state s if the following hold:

(1) Processes i and j are connected to each other,
(2) $s(i) \xrightarrow{+c} q$ (resp., $s(i) \xrightarrow{-c} q$) and $s(j) \xrightarrow{-c} q'$ (resp., $s(j) \xrightarrow{+c} q'$) are defined in states $s(i)$ and $s(j)$, respectively, for some q and q', and
(3) $s'(l) = \begin{cases} q, & l = i \\ q', & l = j \\ s(l), & \text{otherwise} \end{cases}$

In this case, we write $s \mapsto^c_{\{i,j\}} s'$, or simply $s \overset{c}{\mapsto} s'$ if process names i and j are not important. A (global) transition $s \mapsto^c_A s'$ is said to be of *type 1* if $0 \in A$ (i.e., the target process is involved); otherwise, it is of *type 2*. A (*global*) *computation* is a sequence $\sigma : s_0 \mapsto^{a_1}_{A_1} s_1 \mapsto^{a_2}_{A_2} \cdots \mapsto^{a_n}_{A_n} s_n$, where $s_0, s_1, \cdots, s_n$ are global states and $\forall i, a_i \in \Sigma$. (We sometimes write $s_0 \overset{\sigma}{\mapsto} s_n$ as a shorthand for the above.) Global state s_n is said to be *reachable* through computation σ. A local state q is said to be *reachable* if there exists a computation σ such that $s_0 \overset{\sigma}{\mapsto} s$ and $s(i) = q$, for some $i \in Z$. Given a global computation $\sigma : s_0 \mapsto^{a_1}_{A_1} s_1 \mapsto^{a_2}_{A_2} \cdots \mapsto^{a_n}_{A_n} s_n$, we define $\rho(\sigma)$ to be a string $b_1 b_2 \cdots b_n \in \Gamma^*$ such that

$$
b_i = \begin{cases} L(t), & \text{if } 0 \in A_i \text{ and } t \text{ is the transition performed by} \\ & \text{the target process (i.e., process 0) at } s_{i-1}. \\ \lambda \text{ (the empty string)}, & \text{otherwise.} \end{cases}
$$

Intuitively, $\rho(\sigma)$ is the projection of transition labels of σ on the target process. Given a system P^∞, where $P = (Q, \delta, \Sigma, q_0, \Gamma, L)$, the *trace set* of P^∞, denoted by $Trace(P^\infty)$, is $\{\rho(\sigma) \mid \sigma \text{ is a global computation in } P^\infty\}$.

Given two systems P_1^∞ and P_2^∞, where $P_1 = (Q_1, \delta_1, \Sigma_1, q_0^1, \Gamma_1, L_1)$ and $P_2 = (Q_2, \delta_2, \Sigma_2, q_0^2, \Gamma_2, L_2)$, a "bisimulation" is an equivalence relation R over $S(P_1^\infty) \times S(P_2^\infty)$ such that for every $(s_1, s_2) \in R$,

1. for each $s_1 \overset{\sigma_1}{\mapsto} s_1'$ and $\rho(\sigma_1) = a \ (\in \Gamma_1)$, there exists an s_2', $s_2 \overset{\sigma_2}{\mapsto} s_2'$, $\rho(\sigma_2) = a$, and $(s_1', s_2') \in R$, <u>and</u>
2. for each $s_2 \overset{\sigma_2}{\mapsto} s_2'$ and $\rho(\sigma_2) = a \ (\in \Gamma_2)$, there exists an s_1', $s_1 \overset{\sigma_1}{\mapsto} s_1'$, $\rho(\sigma_1) = a$, and $(s_1', s_2') \in R$.

Two systems P_1^∞, P_2^∞ are said to be *trace equivalent*, denoted by $P_1^\infty \overset{T}{\sim} P_2^\infty$, iff $Trace(P_1^\infty) = Trace(P_2^\infty)$. P_1^∞ and P_2^∞ are *bisimilar* (or *bisimulation equivalent*), denoted by $P_1^\infty \overset{B}{\sim} P_2^\infty$, iff there exists a bisimulation R such that $(s_0^1, s_0^2) \in R$, where s_0^1 and s_0^2 are the initial global states of P_1^∞ and P_2^∞, respectively. The *bisimulation equivalence* (resp., *trace equivalence*) problem is that of, given two systems with many identical processes P_1^∞ and P_2^∞, determining whether $P_1^\infty \overset{B}{\sim} P_2^\infty$ (resp., $P_1^\infty \overset{T}{\sim} P_2^\infty$).

3 Fully-connected systems

In this section, we derive the complexities of deciding trace and bisimulation equivalences for fully-connected systems with many identical processes. The idea behind our derivation relies on showing that checking trace and bisimulation equivalences for fully-connected systems with many identical processes can be equated with that for finite state processes. As a result, our complexity results follow immediately from known results concerning the latter [11]. Given a system P^F, where $P = (Q, \delta, \Sigma, q_0, \Gamma, L)$, we establish an ordering on Q by letting $Q = \{q_0, q_1, \cdots, q_k\}$, for some k. Throughout the rest of this paper, we assume the existence of such an ordering.

To prove our main results, a few lemmas are required. Due to space limitation, their proofs are omitted here. The reader is referred to [15] for details.

Lemma 1. *Given a system P^F, the set of all reachable local states of P, denoted as $RS(P^F)$, can be constructed in polynomial time.*

Proof. What follows is a procedure that generates the set of all reachable local states in a greedy fashion. The procedure was originally proposed in [14]; see also [4].

Procedure Reachable_Set;
/* Given the description of a process $P = (Q, \delta, \Sigma, q_0, \Gamma, L)$, */
/* the output S consists of all reachable local states. */
$\quad S := \{q_0\}$;
$\quad$ **For** $i := 0$ **to** $|Q| - 1$ **do**
$\quad\quad$ **begin**
$\quad\quad\quad$ **If** $\exists p, q \in S$, $p \overset{+c}{\to} p'$, $q \overset{-c}{\to} q' \in \delta$, for some $c \in \Sigma$, and p'(or q') $\notin S$,
$\quad\quad\quad\quad$ **then** $S := S \cup \{p', q'\}$;
$\quad\quad\quad\quad$ **else return** (S);
$\quad\quad$ **end**;
end procedure. $\square$

As mentioned earlier, the key idea behind our derivation is to reduce the checking equivalence between fully-connected systems with many identical processes to that between finite state processes. Consider a system P^F, where $P = (Q, \delta, \Sigma, q_0, \Gamma, L)$, the associated finite state process $\tilde{P} = (\tilde{Q}, \tilde{\delta}, \tilde{\Sigma}, \tilde{q}_0)$ is constructed as follows: $\tilde{Q} = RS(P^F)$, $\tilde{q}_0 = q_0$, $\tilde{\Sigma} = \Gamma$, (q, q', l) (i.e., $q \overset{l}{\to} q'$), where $q, q' \in \tilde{Q}$ and $l \in \tilde{\Sigma}$, is defined in $\tilde{\delta}$ if there exist a $c \in \Sigma, r, r' \in RS(P^F)$, such that $(q, q', +, c), (r, r', -, c) \in \delta$ (or $(q, q', -, c), (r, r', +, c) \in \delta$), and $L((q, q', +, c)) = l$ (or $L(q, q', -, c) = l$). The idea is to retain transition $(q, q', +, c)$ (or $(q, q', -, c)$) that can be paired with a transition emanating from a state $r \in RS(P^F)$. Furthermore, the label of such a transition is $L((q, q', +, c))$ (or $L((q, q', -, c))$).

Lemma 2. $Trace(P^F) = Trace(\tilde{P})$, *where $\tilde{P}$ is the reduced finite state process of a system P^F.*

Lemma 3. *Given two systems P_1^F and P_2^F, let $\tilde{P}_1$ and $\tilde{P}_2$ be their reduced finite state processes, respectively. Then $P_1^F \overset{B}{\sim} P_2^F$ iff $\tilde{P}_1 \overset{B}{\sim} \tilde{P}_2$.*

Lemma 4. *Given two finite state processes $\tilde{P}_1$ and $\tilde{P}_2$, we can construct in polynomial time two systems P_1^F and P_2^F such that $\tilde{P}_1 \overset{T}{\sim} \tilde{P}_2$ (resp., $\tilde{P}_1 \overset{B}{\sim} \tilde{P}_2$) iff $P_1^F \overset{T}{\sim} P_2^F$ (resp., $P_1^F \overset{B}{\sim} P_2^F$).*

Using the above lemmas and the fact that checking trace and bisimulation equivalence for finite state processes are PSPACE-complete [11] and PTIME-complete [11], respectively, we have the following result:

Theorem 5. *Given two systems P_1^F and P_2^F, determining $P_1^F \overset{T}{\sim} P_2^F$ and $P_1^F \overset{B}{\sim} P_2^F$ are PSPACE-complete and PTIME-complete, respectively.*

4 Star-like systems

To show the bisimulation equivalence problem with a "star–like" structure (see Fig. 1.) to be undecidable, we reduce the halting problem for *two counter machines* (which is known to be undecidable) [7] to our problem. (See [8, 9] for similar proofs for the models of Petri nets and BPPs.) In a 2–counter machine, there are three operations associated with a counter: *increment, decrement,* and *test for zero.* More precisely, a transition of a 2-counter machine is of one of the following forms:

(1) $C := C + 1$; goto q (increment counter C)
(2) if $C > 0$ then $C := C - 1$; goto q (decrement counter C)
(3) if $C = 0$ then goto q (test counter C for zero)

Theorem 6. *The bisimulation equivalence problem is undecidable for star-like systems.*

Proof. It suffices to show that, given a 2-counter machine M, we can construct two systems P_1^S and P_2^S as shown in Fig. 2 in a way that M **doesn't halt** iff P_1^S **and** P_2^S **are bisimilar.** Let C_1 and C_2 be the two counters which are zero initially. In P_1^S, the values of the two counters are represented by the numbers of processes in states q_1^1 and q_2^1, respectively. (States q_1^2 and q_2^2 play the same role in P_2^S.) The simulation of incrementing and decrementing the two counters are easy; in what follows, we focus on how we can enforce the "test for zero" operation through the use of two systems on which the bisimilarity requirement is imposed.

As shown in Fig. 2, process P_1 consists of two copies of the finite state control of M. (For convenience, they are labeled $M_{1,1}$ and $M_{1,2}$.) Let states $\overline{h_1}, \overline{r_1}, \overline{r_1}'$, and $\overline{r_1}''$ be the "images" of states h_1, r_1, r_1', and r_1'', respectively. (The "h_1" represents the halt state of M.)

Suppose "if $C_1=0$ then goto r_1''" is defined in state r_1 of M, then $M_{1,1}$ and $M_{1,2}$ have transitions $r_1 \overset{a}{\rightarrow} r_1''$ and $\overline{r_1} \overset{a}{\rightarrow} \overline{r_1}''$, respectively. (These transitions can be paired with the "$\overset{+a}{\rightarrow}$" transition defined in state q_0^1.) In addition, we add a transition from r_1 to $\overline{r_1}''$ labeled $-e$ (see Fig. 2). Also we let $L(r_1 \overset{a}{\rightarrow} r_1'') = L(r_1 \overset{e}{\rightarrow} \overline{r_1}'')$. P_2 is identical to P_1 except with an additional transition labeled $-a$ from h_2 to z_2. We make $L(\overline{h_1} \overset{a}{\rightarrow} \overline{z_1})$ $(= L(h_2 \overset{a}{\rightarrow} z_2) = L(\overline{h_2} \overset{a}{\rightarrow} \overline{z_2}))$ be a unique symbol not used elsewhere in P_1^S and P_2^S.

A computation in P_1^S is said to be "valid" if during the course of the computation, the following hold: (1) the target process utilizes only those transitions belonging to $M_{1,1}$, and (2) when taking a transition like the $r_1 \overset{a}{\rightarrow} r_1''$ (in Fig. 2) which simulates a "test for zero" for counter C_1 (resp., C_2), none of the remaining processes is in state q_1^1 (resp., q_2^1) at the moment. (What it says is that any "test for zero" transition can only be taken while the corresponding counter of M is indeed zero.) A "valid" computation for P_2^S can be defined similarly.

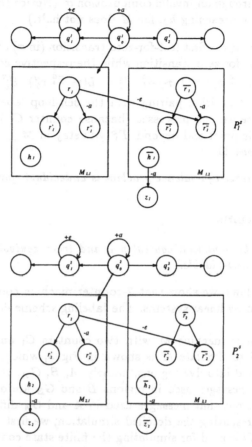

Fig. 2. The P_1^s and P_2^s used in the proof of Theorem 6.

We are in a position to argue that M does not halt iff $P_1^S \overset{B}{\sim} P_2^S$. Consider two cases:

(Case 1) Suppose M halts. To disprove $P_1^S \overset{B}{\sim} P_2^S$, first notice that F_2^S can do whatever F_1^S does. Since M halts, there exists a "valid" computation in P_2^F reaching h_2 using only those transitions belonging to $M_{2,1}$. To "keep pace with" P_2^S (intuitively, this is exactly what bisimilarity is all about), P_1^S must follow exactly the same trace as that of P_2^S in a step by step fashion. (Here, it is important to point out that P_1^S cannot "cheat" by entering $M_{1,2}$, for none of the remaining processes can exchange message "e" with the target process – condition (2) of a valid computation.) In the end, P_1^S and P_2^S end up in states h_1 and h_2, respectively. Hence, P_1^S and P_2^S cannot be bisimilar.

(Case 2) Suppose M does not halt. To prove $P_1^S \overset{B}{\sim} P_2^S$, consider the following subcases:

- P_2^S does not enter h_2. In this case, P_1^S can always follow suit.
- P_2^S enters h_2 through an invalid computation σ. (Notice that there exists no valid computation reaching h_2, for M does not halt.)

Suppose $r_2 \overset{-a}{\rightarrow} r_2''$ is the first test-for-zero transition (in σ) at which P_2^S cheats, i.e., executing a test-for-zero transition while the respective counter is non-zero. In our design, $L(r_1 \overset{-e}{\rightarrow} \overline{r_1''}) = L(r_1 \overset{-a}{\rightarrow} r_1'') = L(r_2 \overset{-a}{\rightarrow} r_2'')$. P_1^S, in this case, can take the $r_1 \overset{-e}{\rightarrow} \overline{r_1}''$ transition pairing with the self-loop labeled "$+e$" in state q_1^1 of some of the remaining processes (because counter C_1 is not zero at this moment). From this point and beyond, P_1^S will stay in $M_{1,2}$, and therefore can do whatever P_2^S does. □

Theorem 7. *The trace equivalence problem is undecidable for star-like systems.*

5 Linear systems

Theorem 8. *The bisimulation equivalence and trace equivalence problems for linear systems are undecidable.*

Proof. In what follows, we show that 2-counter machine computations can be faithfully simulated by linear systems. The labeling scheme depicted in Fig. 1 is assumed here.

Given a 2-counter machine M with two counters C_1 and C_2, the process structure designed to simulate M is shown in Fig. 3, which, for ease of explanation, is partitioned into five regions, namely, A, B, G_A, G_B, and D. For the sake of simplicity, messages used in regions B and G_B are omitted. Here, it is important to point out that messages used in A and G_A differ from that in B and G_B. Before presenting the detailed simulation, we first give the intuition. As before, region D is used for simulating the finite state control of M. (In particular, transitions involving "$p - p'$", "$r - r'$", and "$r - r''$" are examples of "increment C_1", "decrement C_1", and "test C_1 for zero", respectively.) Regions A and B are designed to simulate the two counters. Region G_A and G_B serve as two "gateways" through which unwanted messages can be "filtered out" for performing counter operations correctly. More will be said about this as our discussion progresses. To simulate M, consider the following computation:

1. By exchanging messages "g_a" and "g_b" with its two neighbors, process 0 enters region D and starts simulating M's computations. Depending on the order in which P_0 communicates with P_{-1} and P_1, we have the following two cases: (i) P_{-1} and P_1 enter regions G_A and G_B, respectively, or (ii) P_{-1} and P_1 enter regions G_B and G_A, respectively. In case (i), process P_i, $i \leq -2$ (respectively, $i \geq 2$), takes part in the simulation of counter C_1 (respectively, C_2). Case (ii) is symmetric. Without loss of generality, we assume case (i) throughout the rest of the proof.

309

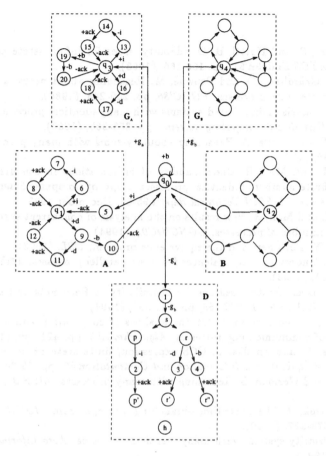

Fig. 3. The structure of the process used in the proof of Theorem 8.

2. At any time, the number of processes P_i, $i \leq -2$ (resp., $i \geq 2$), whose current states reside in region A (resp., B) represents the value of counter C_1 (resp., C_2). Furthermore, if the state of P_i, for some $i \leq -2$ (resp., $i \geq 2$), is in region A (resp., B), so are $P_{-2}, \cdots, P_i$ (resp., $P_2, \cdots, P_i$).

What we have discussed so far is based on the assumption that exactly one process (i.e., P_0) enters region D. Furthermore, M halts iff there exists a computation leading P_0 to state h. To complete the proof, it remains to show that even in the presence of two or more processes entering region D, state h can only be reached through "valid" computation. Due to space limitation, the detailed simulation is omitted. The reader is referred to [15] for details. □

Acknowledgments: We would like to thank the anonymous referees for their valuable comments.

References

1. Christensen, S. and Hüttel, H. Decidability issues for infinite-state processes – a survey. *EATCS Bulletin* 51, pp. 156-166, (1993).
2. Clark, E., Grümberg, O. and Browne, M. Reasoning about networks with many identical finite-state processes. *PODC'86*, pp. 240-248, (1986).
3. Garg, V. Analysis of distributed systems with many identical processes. *IEEE Int. Conf. on Distributed Computing Systems*, pp. 358-365, (1988).
4. German, S. and Sistla, A. Reasoning about systems with many processes. *JACM* 39, 675-735, (1992).
5. Girkar, M. and Moll, R. Undecidability of bisimulations in concurrent systems with indefinite number of identical processes. *Dept. of Computer Science, Univ. of Massachusetts, CMPSCI Technical Report* 93-86, December 1993.
6. Girkar, M. and Moll, R. New results on the analysis of concurrent systems with an indefinite number of processes. *CONCUR'94*, (1994).
7. Hirshfeld, Y. Petri nets and the equivalence problem. *CSL*, (1993).
8. Hüttel, H. Undecidable equivalences for basic parallel processes. *13th FST&TCS*, pp. 454-464, (1993).
9. Jančar, P. Decidability questions for bisimilarity of Petri nets and some related problems. *STACS'94, LNCS* 755, pp. 581-592, (1994).
10. Kanellakis, P. and Smolka, S. On the analysis of cooperation and antagonism in networks of communicating processes, *Algorithmica* 3, pp. 421-450, (1988).
11. Kanellakis, P. and Smolka, S. CCS expression, finite state processes, and three problems of equivalence. *Information and Computation* 86, pp. 43-68, (1990).
12. Sistla, A. and German, S. Reasoning with many processes. *LICS'87*, pp. 138-152, (1987).
13. van Glabbeek, R. The linear time –branching time spectrum. *CONCUR'90, LNCS* 458, pp. 278-297, (1990).
14. Yen, H. Priority systems with many identical processes. *Acta Informatica* 28, pp. 681-692 (1991).
15. Yen, H., Jian, S., and Lao, T. Deciding Bisimulation and Trace Equivalences for Systems with Many Identical Processes. Tech. Report, Dept. of Electrical Engineering, National Taiwan Univ., 1995.

Should Amdahl's Law Be Repealed?
(Invited Presentation)

Franco P. Preparata

Department of Computer Science, Brown University,
Providence, RI 02912, U.S.A.
franco@cs.brown.edu

Abstract

An appropriate observation made by Gene Amdahl in 1967 about data management housekeeping overhead was subsequently construed by his commentators as embodying a fundamental limitation to parallel computation and was elevated to to the rank of "Amdahl's Law". It is argued that Amdahl's Law, as formulated, has no fundamental character, but refers to specific technological choices, such as programs and input/output modes. Indeed, algorithmic research has shown that most problems are parallelizable. It is also argued here, on the basis of VLSI's area-time theory, that the I/O bandwidth is not a basic physical limitation. The seemingly inherently serial P-complete problems, as characterized by algorithmic research, constitute the only class to whose programs Amdahl's Law trivially applies.

Moreover, in a realistic computational model that fully accounts for the finiteness of the speed of light, it can be existentially shown that, with respect to multiprocessors, uniprocessors incur not only the obvious slowdown due to loss of parallelism but also a more subtle slowdown due to loss of locality. This makes the case for the multiprocessor and for the repeal of Amdahl's Law.

Embeddings of Hyper-Rings in Hypercubes

Yukihiro Hamada, Aohan Mei, Yasuaki Nishitani and Yoshihide Igarashi

Department of Computer Science, Gunma University,
Kiryu, 376 Japan
Email: igarashi@comp.cs.gunma-u.ac.jp

Abstract. A graph $G = (V, E)$ with N nodes is called an N-hyper-ring if $V = \{0, \cdots, N-1\}$ and $E = \{(u, v) \mid (u-v) \text{ modulo } N \text{ is a power of 2}\}$. We study embeddings of the 2^n-hyper-ring in the n-dimensional hypercube. We show a greedy embedding with dilation 2 and congestion $n+1$ and a modified greedy embedding with dilation 4 and congestion 6.

1 Introduction

Load balancing, communication locality, communication congestion, and node utility are important factors to achieve high performance on a parallel computer. These factors can be abstractly studied as the problem of embedding a process graph G in a process graph H[5, 6, 12, 13, 14]. In a process graph, the nodes are processes comprising a parallel program and the edges represent communication between processes. In order to simulate the parallel computation on a process graph G by another process graph H, we should consider a mapping (i.e., embedding) from G in H so that the maximum number of processes of G placed on any process of H, the maximum distance between any pair of processes of H corresponding to a pair of neighbor processes of G, and the maximum number of edges of G placed on any edge of H are small. These factors are called *load*, *dilation* and *congestion*, respectively. The ratio of the node set size of H to that of G is called *expansion* under consideration of an embedding of G in H. The expansion can be considered as a measure of node utility.

The embedding problem in the area of parallel algorithms and architectures is to find embeddings with small loads, dilations and congestions. Embeddings of various families of graphs in hypercubes have been much studied[1, 7, 8, 9, 14, 15]. Hyper-rings have appeared in literature under some different names, including an optimal broadcasting scheme[2, 10] and binary jumping networks[11]. Altman *et al.* showed that for any n, the n-dimensional hypercube can be embedded in the hyper-ring with 2^n nodes as its subgraph[3]. However, as far as we know, any nontrivial embedding of hyper-rings in hypercubes has not appeared in published literature. In this paper, we propose two types of embeddings of the hyper-ring with 2^n nodes in the n-dimensional hypercube. The first one is an embedding with dilation 2 and congestion $n+1$, and the other is an embedding with dilation 4 and congestion 6.

2 Preliminaries

A graph $G = (V, E)$ with N nodes is called the N-*hyper-ring* (N-HR for short) if $V = \{0, \cdots, N-1\}$ and $E = \{(u, v) \mid (u - v) \text{ modulo } N \text{ is a power of } 2\}$. The *n-dimensional hypercube* (n-HC for short) is a graph with 2^n nodes labeled by all n-bit binary numbers, where there exists an edge between two nodes if and only if their binary representations differ in exactly one bit position.

Let $G = (V, E)$ and $H = (W, F)$ be graphs. Let P be the set of paths in H. An embedding of G (a guest-graph) in H (a host-graph) is a pair of two mappings $\sigma : V \to W$ and $\rho : E \to P$ such that for each edge (u, v) in E, $\rho(u, v)$ connects $\sigma(u)$ and $\sigma(v)$. The efficiency of an embedding is measured by its *load, dilation, congestion* and *expansion*. The load of an embedding is the largest number of nodes mapped to a single node. The dilation of an edge e in G under an embedding is the length of path $\rho(e)$ in H. The dilation of an embedding is the maximum dilation among the dilations of all edges in G. The congestion of an edge f in H under an embedding is the number of edges e in G such that $\rho(e)$ contains f. The congestion of an embedding is the maximum congestion among the congestions of all edges in H. The expansion is the ratio of the cardinality of the node set of H to one of G.

We will use the Gray code for a one-to-one mapping between the nodes of the 2^n-HR and the nodes of the n-HC. The n-bit *Gray code* is an ordering of all the n-bit binary numbers such that a pair of any consecutive numbers (including a pair of the last number and the first number) differ in exactly one bit position. We denote the ith codeword of the Gray code by $x(i)$. That is,

$$x(i) = \begin{cases} 0 & \text{if } i = 0 \\ 2^t \oplus x(2^{t+1} - 1 - i) & \text{if } i > 0, \end{cases}$$

where $t = \lfloor \log_2 i \rfloor$, and $\oplus$ denotes the *bitwise exclusive-or* operation.

We denote the least significant bit of the binary representation of i by $b_0(i)$, and in general the $(k+1)$st least significant bit of the binary representation of i by $b_k(i)$. For a pair of integers, a and b, $[a]_b$ denotes a *modulo* b. For a nonnegative integer $i < 2^n$,

$$r_k^n(i) = \begin{cases} n-1 & \text{if } k \geq n-1 \\ k & \text{if } k < n-1 \text{ and } b_k(i) = 0 \\ r_{k+1}^n(i) & \text{otherwise.} \end{cases}$$

3 Greedy Embeddings

An embedding is determined by a bijection σ from the node set of the 2^n-HR to the node set of the n-HC together with an injection ρ from the edge set of the 2^n-HR to the set of paths connecting any two nodes in the n-HC. Let σ be the Gray code mapping x (i.e., $x(i)$ is the ith codeword of the Gray code), and let ρ be a mapping such that $\rho(i, j)$ is one of the shortest paths connecting $\sigma(i)$ and $\sigma(j)$ in the n-HC for any edge (i, j) in the 2^n-HR. We call an embedding specified by these σ and ρ a greedy embedding. Note that greedy embeddings of the 2^n-HR in the n-HC are not unique. We omit the proofs of the following three lemmas.

Lemma 1. *Let i be an integer such that $0 \leq i < 2^n$. Then, $x(i) \oplus x([i+1]_{2^n}) = 2^{r_0^n(i)}$.*

Lemma 2. *Let i and k be a pair of integers such that $0 \leq i < 2^n$ and $1 \leq k < n$. Then, $x([i+2^k]_{2^n}) = x(i) \oplus 2^{k-1} \oplus 2^{r_k^n(i)}$.*

Lemma 3. *Let i and k be a pair of integers such that $0 \leq i < 2^n$ and $0 \leq k < n$. Then, $x(i) \oplus x(i \oplus (2^{k+1} - 1)) = 2^k$.*

Let ρ be a mapping such that for each edge (i,j) in the 2^n-HR, (i,j) is mapped to one of the shortest paths connecting $x(i)$ and $x(j)$ in the n-HC. Then from Lemma 1 and Lemma 2 the greedy embedding specified by $\sigma = x$ and ρ has dilation 2. We therefore have the next theorem.

Theorem 4. *For $n \geq 2$ the dilation of any greedy embedding of the 2^n-HR in the n-HC is 2.*

Let u and v differ just in bit positions i and j. Then the shortest paths from u to v are $(u, u \oplus 2^i, (u \oplus 2^i \oplus 2^j) = v)$ and $(u, u \oplus 2^j, (u \oplus 2^j \oplus 2^i) = v)$. The former is called the $\oplus i$-path from u and the latter is called the $\oplus j$-path from u.

Let ρ map any edge (i,j) in the 2^n-HR to one of the shortest paths connecting nodes $x(i)$ and $x(j)$ in the n-HC. Then from Lemma 2, for any k $(1 \leq k < n)$, $\rho(i, [i+2^k]_{2^n})$ is the $\oplus(k-1)$-path from $x(i)$ or the $\oplus r_k^n(i)$-path from $x(i)$. When $k = n - 1$, for technical simplicity we always denote edge $e = (i, [i + 2^{n-1}]_{2^n})$ so that $i < [i + 2^{n-1}]_{2^n}$. If for any i $(0 \leq i < 2^n)$ and any k $(1 \leq k < n)$, $\rho(i, [i + 2^k]_{2^n})$ is the $\oplus(k-1)$-path from $x(i)$, then the embedding specified by $\sigma = x$ and ρ is called the K-type greedy embedding. If for any i $(0 \leq i < 2^n)$ and any k $(1 \leq k < n)$, $\rho(i, [i + 2^k]_{2^n})$ is the $\oplus r_k^n(i)$-path from $x(i)$, then the embedding specified by $\sigma = x$ and ρ is called the R-type greedy embedding.

Lemma 5. *For any i $(0 \leq i < 2^n)$ and any d $(0 \leq d < n)$, the congestion of edge $e = (x(i), x(i) \oplus 2^d)$ under the K-type greedy embedding or the R-type greedy embedding is at most $n + 1$.*

Proof. We first consider the K-type greedy embedding specified by x and ρ. From Lemma 1 edge $(i, [i + 2^0]_{2^n})$ in the 2^n-HR is mapped to edge $f = (x(i), x(i) \oplus 2^{r_0^n(i)})$ in the n-HC. Such an edge mapped from $(i, [i+1]_{2^n})$ is called a path of type 1 containing f in the n-HC. For $1 \leq k < n$ edge $(i, [i+2^k]_{2^n})$ in the 2^n-HR is mapped to path $p = (x(i), x(i) \oplus 2^{k-1}, x(i) \oplus 2^{k-1} \oplus 2^{r_k^n(i)})$. The path of length 2, p is called a path of type 2 containing edge $(x(i), x(i) \oplus 2^{k-1})$, and p is also called a path of type 3 containing edge $(x(i) \oplus 2^{k-1}, x(i) \oplus 2^{k-1} \oplus 2^{r_k^n(i)})$. These three types of paths containing e are shown in Fig. 1. Let $e = (x(i), x(i) \oplus 2^d)$ be an edge in the n-HC and $i' = i \oplus (2^{d+1} - 1)$. Then from Lemma 3, $x(i')$ is another end point of e (i.e., $x(i') = x(i) \oplus 2^d$).

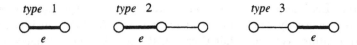

Fig. 1. 3 types of paths appeared in the greedy embedding.

Case 1 $(d < n - 1)$: We first consider paths of *type* 1 containing e. Since $e = (x(i), x(i'))$, $r_0^n(i) \neq r_0^n(i')$, the number of paths of *type* 1 containing e is at most 1.

We next consider paths of *type* 2 containing e. There are exactly two paths of *type* 2 containing e. From Lemma 2 and Lemma 3 these are $\rho(i, [i + 2^{d+1}]_{2^n})$ and $\rho(i', [i' + 2^{d+1}]_{2^n})$, where $i' = i \oplus (2^{d+1} - 1)$. The former is a path emanating from $x(i)$ and the latter is a path emanating from $x(i')$.

We now consider paths of *type* 3 containing e. Let p be a path of *type* 3 containing e and ending with $x(i')$. Then from Lemma 2 and Lemma 3, $p = \rho(i \oplus (2^k - 1), i') = (x(i \oplus (2^k - 1)), x(i), x(i'))$, where $i' = [(i \oplus (2^k - 1)) + 2^k]_{2^n}$, $d = r_k^n(i \oplus (2^k - 1))$ and $1 \leq k \leq d$. Let q be a path of *type* 3 containing e and ending with $x(i)$. Then $q = \rho(i' \oplus (2^j - 1), i) = (x(i' \oplus (2^j - 1)), x(i'), x(i))$, where $i = [(i' \oplus (2^j - 1)) + 2^j]_{2^n}$, $d = r_j^n(i' \oplus (2^j - 1))$ and $1 \leq j \leq d$. For $k = j$ at most one of p and q can exist, since $r_k^n(i \oplus (2^k - 1)) \neq r_k^n(i' \oplus (2^k - 1))$. Hence, the number of paths of *type* 3 containing e is at most d.

Summing up all possible paths of these three types, we can say that in this case at most $d + 3 \leq n + 1$ paths mapped from the edges in the 2^n-HR use edge e in the n-HC.

Case 2 $(d = n - 1)$: By the same argument as in the previous case, there is at most one path of *type* 1 containing e. From the definition of the K-type greedy embedding there is no path of *type* 2 containing e.

Let p be a path of *type* 3 containing e and ending with $x(i')$. Then $p = \rho(i \oplus (2^k - 1), i') = (x(i \oplus (2^k - 1)), x(i), x(i'))$, where $r_k^n(i \oplus (2^k - 1)) = n - 1$ and $1 \leq k \leq n - 1$. Let q be a path of *type* 3 containing e and ending with $x(i)$. Then $q = \rho(i' \oplus (2^j - 1), i) = (x(i' \oplus (2^j - 1)), x(i'), x(i))$, where $r_j^n(i' \oplus (2^j - 1)) = n - 1$ and $1 \leq j \leq n - 1$. Since for $1 \leq k \leq n - 2$, $r_k^n(i \oplus (2^k - 1)) \neq r_k^n(i' \oplus (2^k - 1))$, for each $1 \leq k = j \leq n - 2$, at most one of p and q can exist. Hence, the number of paths of *type* 3 containing e is at most $n - 2 + 2 = n$.

Summing up all possible paths of these three types, we can also say that in this case at most $n + 1$ paths mapped from the edges in the 2^n-HR use edge e in the n-HC.

We can also prove in a similar way that for the R-type greedy embedding the assertion of the lemma holds true. □

Theorem 6. *For $n \geq 2$ the congestion of each of the K-type greedy embedding and the R-type greedy embedding of the 2^n-HR in the n-HC is $n + 1$.*

Proof. From Lemma 5 the congestion of each of the K-type greedy embedding and the R-type greedy embedding is at most $n+1$. We first consider the K-type greedy embedding. We show that there exists an edge with congestion $n+1$ in the n-HC. Let $i = 2^{n-2} - 1$ and $i' = 2^{n-2}$. Then $e = (x(i), x(i'))$ is an edge in the n-HC. There are exactly one path of *type* 1 and exactly two paths of *type* 2 containing e. One of the paths of *type* 2 emanates from $x(i)$ and the other emanates from $x(i')$. There are exactly $n-2$ paths of *type* 3 containing e. These paths are $(x(i \oplus (2^k - 1)), x(i), x(i'))$ for $1 \leq k \leq n-2$. All these paths containing e are distinct. Hence, the congestion of e is $n+1$.

We next consider the R-type greedy embedding. Let $j = 2^{n-1} + 2^{n-2} - 1$ and $j' = 2^{n-1} + 2^{n-2}$. Then $e' = (x(j), x(j'))$ is an edge in the n-HC. There are exactly one path of *type* 1 and exactly $n-2$ paths of *type* 2 containing e'. There are exactly two paths of *type* 3 containing e'. These are $(x(2^{n-2}), x(j), x(j'))$ and $(x(2^{n-2} - 1), x(j'), x(j))$. All these paths containing e are distinct. Hence, the congestion of e' is $n+1$.

Therefore, from Lemma 5 and the congestions of the two edges given above, the assertion of the theorem holds. □

We do not know at present whether there exists a greedy embedding with congestion less than $n+1$. We can show that the congestion of any greedy embedding of the 2^n-HR in the n-HC is at most $2n$, but we omit the proof here.

4 An Embedding with Constant Dilation and Congestion

As shown in the previous section, the dilation of each greedy embedding of the 2^n-HR in the n-HC is optimal, but its congestion is not satisfactory. We adopt again the Gray code mapping $\sigma = x$ as a bijection from the node set of the 2^n-HR to the node set of the n-HC. To describe a mapping ρ from the edge set of the 2^n-HR to the set of paths in the n-HC, we introduce the following notation. For the n-HC, $\langle v; d_1, d_2, \cdots, d_m \rangle$ denotes a path $(v, v \oplus 2^{d_1}, v \oplus 2^{d_1} \oplus 2^{d_2}, \cdots, v \oplus 2^{d_1} \oplus 2^{d_2} \oplus \cdots \oplus 2^{d_m})$.

We define ρ as follows. For each edge $e = (i, [i + 2^k]_{2^n})$ in the 2^n-HR,

$$\rho(e) = \begin{cases} \langle x(i); r_0^n(i) \rangle & \text{if } k = 0 \\ \langle x(i); k-1, k \rangle & \text{if } k \neq 0, r_k^n(i) = k \\ \langle x(i); k-1, k, r_k^n(i), k \rangle & \text{if } k \neq 0, r_k^n(i) > k, \end{cases}$$

where for $k = n-1$ we consider that $i < 2^{n-1}$. Note that if $i < 2^{n-1}$ and $j = i + 2^{n-1}$ then (i, j) and $(j, [j + 2^{n-1}]_{2^n})$ denote the same edge.

We call the embedding defined above the modified greedy embedding.

Lemma 7. *Let $e = (x(i), x(i) \oplus 2^d)$ be an edge in the n-HC. Then the congestion of e under the modified greedy embedding is at most 6.*

Proof. From the definition of mapping ρ, there are 7 possible types of paths containing e in the n-HC. We refer to these types as *type* 1, $\cdots$, *type* 7 as shown in Fig. 2, where each of the bold edges stands for e. If an edge of the 2^n-HR is mapped by ρ to e in the n-HC, then e itself is said to be a path of *type* 1 containing e. Suppose that an edge $(u, [u + 2^k]_{2^n})$ of the 2^n-HR is mapped by ρ to a path of length 2 in the n-HC. If e is the first edge emanating from $x(u)$ of $\rho(u, [u+2^k]_{2^n})$, then $\rho(u, [u+2^k]_{2^n})$ is said to be a path of *type* 2 containing e. If e is the second edge ending at $x([u+2^k]_{2^n})$ of $\rho(u, [u+2^k]_{2^n})$, then $\rho(u, [u+2^k]_{2^n})$ is said to be a path of *type* 3 containing e. Suppose that an edge $(v, [v + 2^k]_{2^n})$ of the 2^n-HR is mapped by ρ to a path of length 4 in the n-HC. If e is the first (the second, the third, the fourth) edge of $\rho(v, [v + 2^k]_{2^n})$ in the direction from $x(v)$ to $x([v + 2^k]_{2^n})$, then $\rho(v, [v + 2^k]_{2^n})$ is said to be a path of *type* 4 (*type* 5, *type* 6, *type* 7, respectively) containing e.

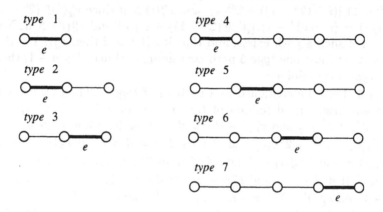

Fig. 2. 7 types of paths appeared in the modified greedy embedding.

Let $e = (x(i), x(i) \oplus 2^d)$ and $i' = i \oplus (2^{d+1} - 1)$. Then from Lemma 3, $x(i) \oplus 2^d = x(i')$ (i.e., $x(i')$ is another end point of e). In the following, each of the 7 types of paths containing $e = (x(i), x(i) \oplus 2^d) = (x(i), x(i'))$ can be characterized, and possible paths of each type containing e are stated.

type 1: If e is contained in a path p_1 of *type* 1, then $\rho^{-1}(p_1) = (i, [i + 1]_{2^n})$, $r_0^n(i) = d$ and $r_d^n(i) = d$, or $\rho^{-1}(p_1) = (i', [i' + 1]_{2^n})$, $r_0^n(i') = d$ and $r_d^n(i') = d$. If $r_0^n(i) = d$ then $r_0^n(i')$ cannot be d. Hence, there exists at most one path of *type* 1 containing e.

type 2: If e is contained in a path p_2 of *type* 2, then $\rho^{-1}(p_2) = (i, [i+2^{d+1}]_{2^n})$, $d \neq n - 1$ and $r_{d+1}^n(i) = d + 1$, or $\rho^{-1}(p_2) = (i', [i' + 2^{d+1}]_{2^n})$, $d \neq n - 1$ and $r_{d+1}^n(i') = d + 1$. Note that $r_{d+1}^n(i) = r_{d+1}^n(i')$. Hence, there exist at most two paths of *type* 2 containing e. When $d = n - 1$, there are no paths of *type* 2 containing e.

type 3: If e is contained in a path p_3 of *type* 3, then by Lemma 3, $\rho^{-1}(p_3) = (i \oplus (2^d - 1), [(i \oplus (2^d - 1)) + 2^d]_{2^n})$, and $r_d^n(i) = d$ since $r_d^n(i \oplus (2^d - 1)) = d$, or $\rho^{-1}(p_3) = (i' \oplus (2^d - 1), [(i' \oplus (2^d - 1)) + 2^d]_{2^n})$ and $r_d^n(i') = d$. If $d \neq n - 1$ and

$r_d^n(i) = d$, then $r_d^n(i')$ cannot be d. Hence, when $d \neq n-1$, there exists at most one path of *type* 3 containing e, and when $d = n-1$, there exist at most two paths of *type* 3 containing e. By a more careful and complicated analysis we can show that for $d = n-1$ there exists at most one path of *type* 3 containing e, but we omit the proof since for our purpose the result stated above is sufficient.

type 4: If e is contained in a path p_4 of *type* 4, then $\rho^{-1}(p_4) = (i, [i+2^{d+1}]_{2^n})$ and $r_{d+1}^n(i) > d+1$, or $\rho^{-1}(p_4) = (i', [i' + 2^{d+1}]_{2^n})$ and $r_{d+1}^n(i') > d+1$. Note that $r_{d+1}^n(i) = r_{d+1}^n(i')$, and that $d \neq n-1$ if such a path exists. If $r_{d+1}^n(i) > d+1$ then obviously $r_{d+1}^n(i) \neq d+1$. Hence, if there exist paths of *type* 4 containing e then there are no paths of *type* 2 containing e. Thus the sum of the number of paths of *type* 2 containing e and the number of paths of *type* 4 containing e is at most 2. When $d = n-1$, there are no paths of *type* 4 containing e.

type 5: If e is contained in a path p_5 of *type* 5, then by Lemma 3, $\rho^{-1}(p_5) = (i \oplus (2^d - 1), [(i \oplus (2^d - 1)) + 2^d]_{2^n})$, and $r_d^n(i) > d$ since $r_d^n(i \oplus (2^d - 1)) > d$, or $\rho^{-1}(p_5) = (i' \oplus (2^d - 1), [(i' \oplus (2^d - 1)) + 2^d]_{2^n})$ and $r_d^n(i') > d$. Note that $d \neq n-1$ if such a path exists, and that if $r_d^n(i) > d$ then $r_d^n(i') = d$. Hence, there exists at most one *type* 5 path containing e. When $d = n-1$, there is no path of *type* 5 containing e.

type 6: Let e be contained in a path p_6 of *type* 6. Then by using Lemma 3 twice and from the definition of *type* 6, for some $k > 0$, $\rho^{-1}(p_6) = (i \oplus 2^k, [(i \oplus 2^k) + 2^k]_{2^n})$ and $r_k^n(i \oplus 2^k) = d > k$, or for some $k' > 0$, $\rho^{-1}(p_6) = (i' \oplus 2^{k'}, [(i' \oplus 2^{k'}) + 2^{k'}]_{2^n})$ and $r_{k'}^n(i' \oplus 2^{k'}) = d > k'$. If $r_k^n(i \oplus 2^k) = d > k$ then $r_d^n(i) = d$ and $r_0^n(i) < d$, and if $r_{k'}^n(i' \oplus 2^{k'}) = d > k'$ then $r_d^n(i') = d$ and $r_0^n(i') < d$. If $d \neq n-1$ and $r_k^n(i \oplus 2^k) = d$ then $r_{k'}^n(i' \oplus 2^{k'})$ cannot be d. If $d \neq n-1$ and $r_{k'}^n(i' \oplus 2^{k'}) = d$ then $r_k^n(i \oplus 2^k)$ cannot be d.

Let us first consider a path of *type* 6 containing e, $p_6 = \rho(i \oplus 2^k, [(i \oplus 2^k) + 2^k]_{2^n})$. At this stage of the argument it is still not clear whether for a given i, such k is uniquely determined. Assume, to the contrary, that there are two paths of *type* 6 containing e, $\rho(i \oplus 2^k, [(i \oplus 2^k) + 2^k]_{2^n})$ and $\rho(i \oplus 2^t, [(i \oplus 2^t) + 2^t]_{2^n})$, where $r_k^n(i \oplus 2^k) = d > k$ and $r_t^n(i \oplus 2^t) = d > t$. Without loss of generality we may assume that $k < t$. For $i \oplus 2^k$ and $i \oplus 2^t$, they are bitwise equal except for the kth bit and tth bit. Hence, $b_t(i \oplus 2^k) \neq b_t(i \oplus 2^t)$. However, $b_t(i \oplus 2^k) = 1$ and $b_t(i \oplus 2^t) = 1$ since $r_k^n(i \oplus 2^k) = d > t > k$ and $r_t^n(i \oplus 2^t) = d > t$. This is a contradiction. Hence, it is not possible that there exist two distinct such paths. For the symmetry between i and $i' = i \oplus (2^{d+1} - 1)$, it is also not possible that there are two distinct paths of *type* 6 containing e, $\rho(i' \oplus 2^k, [(i' \oplus 2^k) + 2^k]_{2^n})$ and $\rho(i' \oplus 2^t, [(i' \oplus 2^t) + 2^t]_{2^n})$. Therefore, when $d \neq n-1$ there exists at most one path of *type* 6 containing e, and when $d = n-1$ there exist at most two paths of *type* 6 containing e.

Comparing the conditions for a path of *type* 1 containing e with the conditions for a path of *type* 6 containing e, if a path of the former type exists and $d \neq n-1$ then no path of the latter type exists. When $d = n-1$, if a path of the former type exists then only one path of the latter type exists. Thus, the number of paths of either *type* 1 or *type* 6 containing e is at most one when $d \neq n-1$, and at most two when $d = n-1$.

type 7: Let e be contained in a path p_7 of *type* 7. From the definition of mapping ρ and Lemma 3, for some u, $\rho^{-1}(p_7) = (u, [u + 2^d]_{2^n})$, $u = i \oplus 2^d \oplus (2^{r_d^n(u)+1} - 1)$ and $r_d^n(u) > d$, or for some v, $\rho^{-1}(p_7) = (v, [v + 2^d]_{2^n})$, $v = i' \oplus 2^d \oplus (2^{r_d^n(v)+1} - 1)$ and $r_d^n(v) > d$. Note that d cannot be $n - 1$ if such a path exists.

We first consider the former path. Since $r_d^n(i \oplus 2^d \oplus (2^{r_d^n(u)} - 1)) > d$ and $r_d^n(u) > d$, $r_d^n(i) > d$. It is still not clear at this stage of the argument whether for a given i such u can be uniquely determined. Assume, to the contrary, that there exists another $u' = i \oplus 2^d \oplus (2^{r_d^n(u')+1} - 1)$ satisfying $r_d^n(u') > d$. If $r_d^n(u') = r_d^n(u)$ then $u' = u$. Suppose that $u \neq u'$, and without loss of generality we may assume that $r_d^n(u') > r_d^n(u)$. Then for any s $(0 \leq s \leq r_d^n(u))$ the sth bit of u and the sth bit of u' should be equal. In particular, $b_{r_d^n(u)}(u) = b_{r_d^n(u)}(u') = 0$. However, since $r_d^n(u') > r_d^n(u) > d$, $b_{r_d^n(u)}(u') = 1$. This is a contradiction. Thus, u and u' cannot be distinct. Hence, there exists at most one u such that $\rho(u, [u + 2^d]_{2^n})$ is a path of *type* 7 containing e. By the same argument, $r_d^n(i') > d$ and there exists at most one v such that $\rho(v, [v + 2^d]_{2^n})$ is a path of *type* 7 containing e.

If $r_d^n(i) > d$ then $r_d^n(i') = d$. Hence, it is not possible that both $\rho(u, [u+2^d]_{2^n})$ and $\rho(v, [v+2^d]_{2^n})$ are paths of *type* 7 containing e. That is, there exists at most one path of *type* 7 containing e.

As discussed above, for $d \neq n - 1$, there are at most one path of either *type* 1 or *type* 6 but not both, two paths of either *type* 2 or *type* 4 but not both, one path of *type* 3, one path of *type* 5, and one path of *type* 7 containing e in the n-HC. That is, for $d \neq n - 1$ at most 6 paths mapped from the edges in the 2^n-HR use e in the n-HC. For $d = n - 1$, there are no paths of *type* 2, *type* 4, *type* 5 and *type* 7, at most two paths of either *type* 1 or *type* 6, and at most two paths of *type* 3 (by a complicated analysis we can show at most one of *type* 3) mapped from the edges in the 2^n-HR use e in the n-HC. Thus, the assertion of the lemma holds.

(In order to help the reader to understand the conditions that for each type a path containing e exists, we prepare Table 1 where each entry is "*yes*", "*no*", "—" or "*no/yes*". The meanings of "*yes*", "*no*" and "—" are "*the corresponding condition holds*", "*the corresponding condition does not hold*" and "*the corresponding condition is not cared*", respectively. Furthermore, "*no/yes*" is used to mean that "*no*" holds if $d \neq n - 1$ and "*yes*" holds if $d = n - 1$.) □

Theorem 8. *For $n \geq 3$, the dilation and congestion of the modified greedy embedding of the 2^n-HR in the n-HC are 4 and 6, respectively.*

Proof. From the definition of the modified greedy embedding its dilation is obviously 4. From Lemma 7 its congestion is not more than 6. Let the modified greedy embedding be specified by the mappings x and ρ. There are many edges with congestion 6 in the n-HC. We show one of them. Let $i = 0 \cdots 0001$ and $j = 0 \cdots 0010$. Let consider an edge $e = (x(i) = 0 \cdots 0001, x(j) = 0 \cdots 0011)$ in the n-HC. Then there are the following paths containing e in the n-HC:

type 1: Edge (i, j) in the 2^n-HR is mapped to path $(x(i), x(j))$ that is e itself.

Table 1. Conditions for existence of a path of each type.

	conditions					
	$r_0^n(i) = d$	$r_0^n(i') = d$	$r_d^n(i) = d$	$r_d^n(i') = d$	$r_{d+1}^n(i) = d+1$	$r_{d+1}^n(i') = d+1$
type 1	yes	no	yes	no/yes	—	—
	no	yes	no/yes	yes	—	—
type 2*†	—	—	—	—	yes	yes
type 3	—	—	yes	no/yes	—	—
	—	—	no/yes	yes	—	—
type 4*†	—	—	—	—	no	no
type 5*	—	—	yes	no	—	—
	—	—	no	yes	—	—
type 6	no	—	yes	no/yes	—	—
	—	no	no/yes	yes	—	—
type 7*	—	—	yes	no	—	—
	—	—	no	yes	—	—

* This type is possible only when $d \neq n - 1$.
† There may be two paths of this type.

type 2: Edge $(i, 0 \cdots 0101)$ in the 2^n-HR is mapped to path $(x(i), x(j), x(0 \cdots 0101))$ that contains e as its first edge. Edge $(j, 0 \cdots 0110)$ in the 2^n-HR is mapped to path $(x(j), x(i), x(0 \cdots 0110))$ that contains also e as its first edge.

type 3: Edge $(0 \cdots 0000, j)$ in the 2^n-HR is mapped to path $(x(0 \cdots 0000), x(i), x(j))$ that contains e as its second edge.

type 4: There are no paths of *type* 4 containing e.

type 5: Edge $(0 \cdots 0011, 0 \cdots 0101)$ in the 2^n-HR is mapped to path $(x(0 \cdots 0011), x(j), x(i), x(0 \cdots 0110), x(0 \cdots 0101))$ that contains e as its second edge.

type 6: There is no path of *type* 6 containing e.

type 7: Edge $(1 \cdots 1111, i)$ in the 2^n-HR is mapped to path $(x(1 \cdots 1111), x(1 \cdots 1110), x(1 \cdots 1101), x(j), x(i))$ that contains e as its fourth edge.

As shown above, the congestion of e in the n-HC is 6. Hence, the congestion of the modified greedy embedding is 6. □

5   Concluding Remarks

Since the dilation and the congestion of each of the K-type greedy embedding and the R-type greedy embedding are 2 and $n+1$, for $n \leq 5$ these greedy embeddings are better than the modified greedy embedding. However, the congestion of any greedy embedding grows as a linear function in n, whereas the congestion of the modified greedy embedding is constant 6. This property of the modified greedy embedding is a great advantage over greedy embeddings.

The edges in the 2^n-HR are roughly twice as many as the edges in the n-HC, and for an edge $(i, [i + 2^k]_{2^n})$ in the 2^n-HR, the Hamming distance between $x(i)$

and $x([i + 2^k]_{2^n})$ in the n-HC is 2 unless $k = 0$. Hence, the congestion of any embedding of the 2^n-HR using the Gray code mapping in the 2^n-HC cannot be better than 4. However, we do not know at present whether the congestion of the modified greedy embedding can be reduced to be 4 or 5 by a further modification. Another interesting problem is to find an efficient embedding of N-HR in an n-HC for an arbitrary N.

References

1. B. Aiello and F. T. Leighton, "Coding theory, hypercube embedding, and fault tolerance", *The 3rd Annual ACM Symp. on Parallel Algorithms and Architectures*, pp. 125–136, 1991.
2. N. Alon, A. Barak and U. Mauber, "On disseminating information reliably without broadcasting", *The 7th International Conference on Distributed Computing Systems*, pp. 74–81, 1987.
3. T. Altman, Y. Igarashi and K. Obokata, "Hyper-ring connection machines", *IEEE Region 10's 9th Annual International Conference*, Vol. 1, pp. 290–294, 1994.
4. S. N. Bhatt and J-Y. Cai, "Take a walk, grow a tree", *The 29th Annual Symp. on Foundations of Computer Science*, pp. 469–478, 1988.
5. S. N. Bhatt, F. R. K. Chung, J-W. Hong, F. T. Leighton and A. L. Rosenberg, "Optimal simulation by butterfly networks", *The 20th Annual ACM Symp. on Theory of Computing*, pp. 192–204, 1988.
6. S. N. Bhatt, F. R. K. Chung, F. T. Leighton and A. L. Rosenberg, "Optimal tree machines", *The 27th Annual Symp. on Foundations of Computer Science*, pp. 274–282, 1986.
7. S. N. Bhatt, F. R. K. Chung, F. T. Leighton and A. L. Rosenberg, "Efficient embeddings of trees in hypercubes", *SIAM J. Computing*, Vol. 21, pp. 151–162, 1992.
8. M. Y. Chan, "Embedding of d-dimensional grids into optimal hypercubes", *The 1st Annual ACM Symp. on Parallel Algorithms and Architectures*, pp. 52–57, 1989.
9. K. Efe, "Embedding mesh of trees in the hypercube", *J. Parallel and Distributed Computing*, Vol. 11, pp. 222–230, 1991.
10. Y. Han and R. Finkel, "An optimal scheme for disseminating information", *Proceedings of 1988 International Conference on Parallel Processing*, pp. 198–203, 1988.
11. Y. Han, Y. Igarashi, K. Kanai and K. Miura, "Fault-tolerance broadcasting in binary jumping networks", *The 3rd International Symp. on Algorithms and Computation, Lecture Notes in Computer Science*, Vol. 650, pp. 145–154, 1992.
12. R. Koch, F. T. Leighton, B. Maggs, S. Rao and A. L. Rosenberg, "Work-preserving emulations of fixed-connected networks", *The 21st Annual ACM Symp. on Theory of Computing*, pp. 227–240, 1989.
13. F. T. Leighton, *Introduction to Parallel Algorithms and Architectures: Arrays. Trees. Hypercubes*, Morgan Kaufman Publishers, San Mateo, 1992.
14. F. T. Leighton, M. Newman, A. G. Ranade and E. Schwabe, "Dynamic tree embeddings in butterflies and hypercubes", *The 1st Annual ACM Symp. on Algorithms and Architectures*, pp. 224–234, 1989.
15. M. Röttger, U-P. Schroeder and W. Unger, "Embedding 3-dimensional grids into optimal hypercubes", *Technical Report*, Dept. of Mathematics and Computer Science, Univ. of Paderborn, Paderborn, Germany, 1994.

A Fast Algorithm for Computing Optimal Rectilinear Steiner Trees for Extremal Point Sets*

EXTENDED ABSTRACT

Siu-Wing Cheng and Chi-Keung Tang

Department of Computer Science
The Hong Kong University of Science and Technology
Clear Water Bay, Hong Kong
Contact author e-mail: scheng@cs.ust.hk.

Abstract. We present a fast algorithm to compute an optimal rectilinear Steiner tree for extremal point sets. A point set is extremal if each point lies on the boundary of a rectilinear convex hull of the point set. Our algorithm can be used in homotopic routing in VLSI layout design and it runs in $O(k^2 n)$ time, where n is the size of the point set and k is the size of its rectilinear convex hull.

1 Introduction

Routing in VLSI layout design calls for connecting terminals in the same plane with horizontal and vertical wiring. The terminals are points that sit on the boundaries of isothetic modules which are impenetrable rectilinear polygons. The terminals are divided into groups known as *nets* [9] and the terminals in each net are to be wired together in the routing region outside the modules. The connection within a net is a *rectilinear Steiner tree* (RST). The total length of wiring is often required to be minimized which corresponds to finding a *minimal* (optimal) rectilinear Steiner tree (MRST). In *homotopic routing* [5, 6], the routing region outside the modules is automatically partitioned into rectilinear polygons without holes, such that each polygon contains exactly one net and the connection for each net must lie within its polygon. Thus, homotopic routing requires the solution to a special MRST problem where the terminal points lie on the boundary of a rectilinear polygon. It is commonly assumed that the input specifies the polygon and the terminal points in cyclic order on the boundary of this polygon [2, 8].

Let n be the size of S. Related results appear in [1, 2, 4, 5, 7, 8]. The best running time for terminal points on a rectilinear polygon is $O(n^4 \log n)$ time [5]. Given the large number of nets that need to connected, a faster algorithm is desired. For the special case where the terminal points in S lie on the boundary

---

* This work is supported in part by the RGC CERG grant HKUST 190/93E.

of a rectilinear convex hull of S, Richards and Salowe [8] present an $O(k^4 n)$-time algorithm, where k is the size of the rectilinear convex hull. Cheng *et al* [2] improve upon the worst-case and give an $O(n^3)$-time algorithm.

In this paper, we present an $O(k^2 n)$-time algorithm for the MRST problem when the terminal points in S lie on the boundary of a rectilinear convex hull of S. We combine the topological results in [8] and a modification of the dynamic programming framework in [2]. To obtain the speedup, we define new partial solutions and fast algorithms to compute them.[2] Section 2 defines the terminologies. Section 3 defines the partial solutions and the dynamic programming process. Sections 4–6 provide more details of the algorithm. All geometric objects in this paper are rectilinear. For brevity, we omit 'rectilinear' for the rest of this paper. Details and proofs omitted in this abstract can be found in the technical report [3].

2 Terminologies

Given a terminal point set S, we denote the boundary of the convex hull of S by $RC(S)$. Ignoring degeneracy, $RC(S)$ consists of four *boundary staircases* B_i, $1 \leq i \leq 4$, which can be obtained by cutting $RC(S)$ at the highest, lowest, leftmost, and rightmost vertices. B_1 is at the lower right; B_2 at the lower left; B_3 at the upper left; and B_4 at the upper right. An *inner boundary corner* is a reflex corner on $RC(S)$. An *outer boundary corner* is a convex corner on $RC(S)$. Note that an outer boundary corner vertex must be a terminal point. A *boundary edge* is a maximal straight line segment on $RC(S)$. S is *extremal* if every terminal point in S lies on $RC(S)$.

A *node* is either a terminal point or the vertex at an inner boundary corner. A *grid line* is a line segment joining two points on $RC(S)$ such that at least one of them is a node. The endpoints of a grid line are called *exterior vertices*. A *grid graph* is the union of $RC(S)$ and all the grid lines. The intersection of two grid lines inside $RC(S)$ is a *grid vertex*. Given exterior vertices c and d, $B(c, d)$ is the portion of $RC(S)$ traversed from c to d in the counterclockwise order. Define $[c, d]$ to be the counterclockwisely ordered sequence of exterior vertices on $B(c, d)$. Define $S[c, d]$ to be subsequence of $[c, d]$ that contains c, d, and all terminal points in $[c, d]$.

A Steiner tree vertex is either a terminal point, the intersection of two orthogonal line segments contained in the tree, or the vertex at an inner boundary corner. A tree edge is a line segment in the tree incident to two adjacent vertices. A Steiner point is a vertex that is not a terminal point and has degree at least three. An *interior edge* is a tree edge whose interior lies inside $RC(S)$. An *interior line* is a maximal sequence of adjacent and collinear interior edges. A *complete interior line* is an interior line whose endpoints lie on $RC(S)$. If c and d are endpoints of an interior line, then we denote the interior line by (c, d)

---

[2] Some partial solutions are variants of partial solutions defined in [2]. We also develop faster algorithm to compute them, which requires new insights into the structure of a MRST.

or (d, c). A degree two vertex incident to two orthogonal tree edges is a *corner-vertex*. An *interior corner* is a corner whose corner-vertex lies inside $RC(S)$. The two interior lines incident to an interior corner-vertex are the *legs* of the interior corner. If both legs of an interior corner intersect $RC(S)$, it is called a *complete interior corner*. Three interior edges incident to the same vertex form a T and the vertex is a *T-vertex*. The interior line that contains the two collinear interior edges is the *head* of the T or T-head. The interior line containing the other interior edge is the *body* of the T or T-body. Yang and Wing [10] prove that there is an optimal Steiner tree which is a subgraph of the grid graph. Therefore, we will assume that all Steiner trees considered in this paper are subgraphs of the grid graph. For any exterior vertices c and d, whenever we refer to any 'interior edge/line/corner' of a Steiner tree for $S[c, d]$, we always refer to the interior of $RC(S)$ (not $RC(S[c, d])$). Let $\tau(S[c, d])$ be the set of all Steiner trees for $S[c, d]$ which are subgraphs of the grid graph.

3 Partial solutions and dynamic programming

We shall color $O(k)$ grid lines blue, where k is the size of $RC(S)$. A line/edge is blue if it lies on a blue grid line. A complete interior corner is blue if its two legs are blue. An exterior vertex is blue if it is the endpoint of a blue grid line or it is the vertex at an outer boundary corner. Our coloring procedure is an extension of that in [8]. For each inner boundary corner, the corner-vertex and a constant number of terminal points are colored blue in [8], while our procedure also colors the grid lines incident to the corner vertex and these terminal points blue. We want to point out that grid lines are colored blue only once by inspecting $RC(S)$. We may refer to the colors of vertices/edges/lines of a Steiner tree in $\tau(S[c, d])$ for some exterior vertices c and d. Nevertheless, the coloring is always done based on $RC(S)$ (not $RC(S[c, d])$).

It is proved in [8] that by sliding interior lines and flipping interior corners, there exists a subset of optimal Steiner trees in $\tau(S)$ that satisfy what we call the *blue topological properties* :

B1 If v_1 and v_2 are two adjacent Steiner points in the interior of an interior line ℓ, then both v_1 and v_2 have degree three. Moreover, the two interior lines properly incident[3] to ℓ at v_1 and v_2 must be on opposite sides of ℓ. That is, the interior lines properly incident to ℓ *alternate* between the two sides of ℓ.

B2 A T-body is an interior edge incident to a node on the boundary.

B3 Every interior corner must be a complete interior corner. Moreover, at most one of the two legs can have more than one interior edge properly incident to it. Each such properly incident edge is a T-body and is incident to a node on the boundary by B2. Among the interior edges properly incident to a leg ℓ, the interior edge closest to the corner-vertex must lie on the side of ℓ opposite to the side of ℓ containing the other leg.

---

[3] An interior line or edge is *properly incident* to another line ℓ if the interior line or edge is incident to the interior of ℓ.

B4 No complete interior corner can be flipped to overlap with $RC(S)$.

B5 All legs of complete interior corners are blue. Among all interior edges properly incident to a leg, the one closest to the corner-vertex is blue.

B6 All complete interior lines whose interior contain Steiner point(s) are blue.

The blue topological properties imply that every interior line is either a complete interior line, a leg of a complete interior corner, or the body of a T. We define below the partial solutions needed in the dynamic programming process. Let b_r and b_t be two exterior vertices such that $[b_r, b_t]$ lie on at most three boundary staircases (i.e., $[b_r, b_t] \subseteq B_1 \cup B_2 \cup B_3$, or $B_2 \cup B_3 \cup B_4$, or $B_3 \cup B_4 \cup B_1$, or $B_4 \cup B_1 \cup B_2$.)

A τ-structure is denoted by $Tree(b_r, b_t)$. $Tree(b_r, b_t) \in \tau(S[b_r, b_t])$ and either at least one of b_r and b_t is blue, or (b_r, b_t) is a non-blue grid line.

A β-structure is denoted by $G(b_r, b_t)$. $G(b_r, b_t) = T_1 \cup T_2$, $T_1 \in \tau(S[b_r, a_i])$ and $T_2 \in \tau(S[a_{i+1}, b_t])$ such that a_i, a_{i+1} are neighboring points in $S[b_r, b_t]$ that minimizes the total length of $G(b_r, b_t)$. At least one of b_r and b_t must be blue.

A α-structure is denoted by $C(b_r, b_t, (b_r, v))$. Given an interior line $\ell = (c, d)$, let $openspan(\ell)$ be the closed half plane that contains c and bounded by an infinite line through d orthogonal to ℓ.[4] $C(b_r, b_t, (b_r, v))$ is defined when v is a grid vertex or exterior vertex, (b_r, v) is a blue interior line and $[b_r, b_t] \subseteq openspan((b_r, v))$. $C(b_r, b_t, (b_r, v))$ is the minimal Steiner tree that connects (b_r, v) and $S[b_r, b_t]$, and satisfies the following requirement: every interior edge incident to v or properly incident to (b_r, v) is orthogonal to (b_r, v) and incident to an exterior vertex in $[b_r, b_t]$. We can similarly define a 'clockwise' version $C_c(b_r, b_t, (b_r, v))$ to connect (b_r, v) and $S[b_t, b_r]$.

A $\alpha\beta$-structure is denoted by $CG(b_r, b_t, (b_r, v))$. $CG(b_r, b_t, (b_r, v))$ is defined when b_t is blue, v is a grid vertex or an exterior vertex, (b_r, v) is a blue interior line, and $[b_r, b_t] \subseteq openspan((b_r, v))$ or $b_t \notin openspan((b_r, v))$. $CG(b_r, b_t, (b_r, v))$ equals $C(b_r, b_x, (b_r, v)) \cup G(b_x, b_t)$ where b_x minimizes the total length among exterior vertices in $[b_r, b_t] \cap openspan((b_r, v)) - \{b_t\}$. We can similarly define a 'clockwise' version $CG_c(b_r, b_t, (b_r, v))$.

We call the collection of the α, β, τ, and $\alpha\beta$-structures for all $[b_r, b_t]$ on at most three boundary staircases *the complete set of partial solutions*. Given the complete set of partial solutions, we can then construct an optimal Steiner tree:

Theorem 1. *For an extremal point set S of size n, an optimal Steiner tree can be computed in $O(k^2 n)$ time given the complete set of partial solutions.* □

The proof will be sketched in the appendix. The computation of partial solutions is based on dynamic programming which has $O(kn)$ iterations. In iteration j, we compute the above structures for all $[b_r, b_t]$ such that $|[b_r, b_t]| = j$. In each iteration, due to dependency, we must compute β-, $\alpha\beta$-, τ-, and α-structures in this order. More details of their construction are described in Sections 4–6.

---

[4] Note that $openspan((c, d)) \neq openspan((d, c))$.

4 β- and $\alpha\beta$-structures

β-structure. If both b_r and b_t are blue, $G(b_r, b_t)$ is the minimal $Tree(b_r, a_i) \cup Tree(a_{i+1}, b_t)$ for some neighboring points $a_i, a_{i+1} \in S[b_r, b_t]$. This takes $O(n)$ time. For the case where either b_r or b_t is blue, we assume that $G(b_r, b_t)$ satisfies the blue topological properties; otherwise, $G(b_r, b_t)$ cannot be part of an optimal Steiner tree that satisfies the blue topological properties. W.l.o.g., assume that b_r is blue and so b_t is non-blue. In $G(b_r, b_t)$, there are three possible topologies: (1) b_t has degree zero, (2) b_t has degree one and b_t is not incident to any interior line, (3) b_t is incident to an interior line. In case (1), $G(b_r, b_t)$ is $G(b_r, a_j) \cup \{b_t\}$, where a_j is the terminal point in $S[b_r, b_t]$ closest to b_t. In case (2), $G(b_r, b_t)$ is $G(b_r, b_{t-1}) \cup B(b_{t-1}, b_t)$. In case (3), let $b_z \in [b_r, b_t]$ such that (b_z, b_t) is a grid line and let b_x be the blue exterior vertex in $[b_r, b_t]$ closest to b_r. It can be proved that the 'junction' between the two trees in $G(b_r, b_t)$ is either in $[b_r, b_z]$, or $[b_r, b_x]$, or $[b_x, b_t]$. So $G(b_r, b_t)$ equals $G(b_r, b_z) \cup Tree(b_z, b_t)$, or $G(b_r, b_x) \cup Tree(b_x, b_t)$, or $B(b_r, b_x) \cup G(b_x, b_t)$. Thus, it takes $O(1)$ to return the minimal tree.

$\alpha\beta$-structure. Let $\ell = (b_r, c)$ be a blue complete interior line. Let $v_0, v_1, v_2 \cdots$ be the sequence of grid vertices on ℓ at increasing distances from b_r and $v_0 = b_r$. We show how to compute $CG(b_r, b_t, \ell_j)$ for all $\ell_j = (b_r, v_j)$ such that $[b_r, b_t] \subseteq openspan(\ell_j)$ or $b_t \notin openspan(\ell_j)$ in $O(n)$ time. The clockwise versions can be computed similarly. The computation is inductive on j. $CG(b_r, b_t, \ell_0)$ equals $G(b_r, b_t)$. Consider $j \geq 1$. If $[b_r, b_t] \not\subseteq openspan(\ell_j)$ and $b_t \in openspan(\ell_j)$, then we can stop the inductive computation. Otherwise, either v_j has degree one, or v_j has degree two in $CG(b_r, b_t, \ell_j)$. In the first case, $CG(b_r, b_t, \ell_j)$ equals $CG(b_r, b_t, \ell_{j-1}) \cup \{(v_{j-1}, v_j)\}$. In the second case, v_j is incident to an interior edge orthogonal to ℓ_j and which intersects $B(b_r, b_t)$ at an exterior vertex, say b_x. So $CG(b_r, b_t, \ell_j)$ equals $C(b_r, b_x, \ell_j) \cup G(b_x, b_t)$.

5 τ-structures

We first define four subsets of $\tau(S[b_r, b_t])$ (possibly non-disjoint and empty) when b_r or b_t is blue. W.l.o.g., we assume throughout this section that $[b_r, b_t]$ lies on $B_1 \cup B_2 \cup B_3$. The other placements of $[b_r, b_t]$ can be handled symmetrically. In the following, let $\{b_r, b_{r+1}, \cdots, b_{t-1}, b_t\}$ be the sequence $[b_r, b_t]$. Let $\tau^*(S[b_r, b_t]) \subseteq \tau(S[b_r, b_t])$ be the set of all trees that satisfy the blue topological properties.

Class-1. Let b_x be the blue exterior vertex in $[b_r, b_t]$ closest to b_r. Let b_y be the blue exterior vertex in $[b_r, b_t]$ closest to b_t. If b_r is blue, then the Class-1 trees are $Tree(b_r, b_{t-1}) \cup B(b_{t-1}, b_t)$ and $B(b_r, b_x) \cup Tree(b_x, b_t)$. If b_t is blue, then the Class-1 trees are $B(b_r, b_{r+1}) \cup Tree(b_{r+1}, b_t)$ and $Tree(b_r, b_y) \cup B(b_y, b_t)$. If both b_r and b_t are blue, then the Class-1 trees are the above four trees.

Class-2. A Steiner tree in $\tau^*(S[b_r, b_t])$ that contains a blue complete interior line incident to b_r or b_t.

Class-3. A Steiner tree in $\tau^*(S[b_r, b_t])$. If there exists $b_x \in [b_r, b_t] - \{b_r, b_t\}$ such that (b_r, b_x) or (b_t, b_x) is a grid line, then $Tree(b_r, b_x) \cup Tree(b_x, b_t)$ is a Class-3 tree. Note that there are at most two such choices for b_x for any $[b_r, b_t]$.

Class-4. A Steiner tree in $\tau^*(S[b_r, b_t])$ that contains a blue complete interior corner such that one of its legs is incident to b_r or b_t.

Theorem 2. *Let $E(Class-1) \subseteq \tau(S[b_r, b_t])$ be the set of all trees that can be transformed into a Class-1 tree by sliding edge(s) and/or flipping corner(s). If b_r or b_t is blue and $Tree(b_r, b_t) \in \tau^*(S[b_r, b_t]) - E(Class-1)$, then $Tree(b_r, b_t) \in \bigcup_{j=2}^{4} Class-j$.* □

Theorem 2 means that if b_r or b_t is blue, then it suffices to construct a Steiner tree whose total length is not greater than that of the minimal Steiner tree in Class-j, for $1 \leq j \leq 4$. In the rest of this section, we first describe how to do this in $O(k)$ time, and then describe how to compute $Tree(b_r, b_t)$ in $O(1)$ time when (b_r, b_t) is a non-blue grid line.

Class-1. We compute at most four candidate trees as described in the definition of Class-1 and return the minimal one. It is clear that $O(1)$ time is sufficient.

Class-2. Assume that the minimal Class-2 tree contains (b_r, b_x). The case for (b_t, b_x) is symmetric. If $b_t \in openspan((b_x, b_r))$, the tree is $CG(b_r, b_x, (b_r, b_x)) \cup C(b_x, b_t, (b_x, b_r))$. Otherwise, $RC(S[b_r, b_t]) = RC(S[b_r, b_{t-1}]) \cup B(b_{t-1}, b_t)$ and so the tree equals $Tree(b_r, b_{t-1}) \cup B(b_{t-1}, b_t)$ which is in Class-1 and has been handled before. So $O(1)$ time is sufficient.

Class-3. The tree is the minimal $Tree(b_r, b_x) \cup Tree(b_x, b_t)$, where $b_x \in [b_r, b_t]$ and (b_r, b_x) or (b_t, b_x) is a grid line (not necessarily blue). There are at most two such grid lines for any $[b_r, b_t]$. So this can be done in $O(1)$ time.

Class-4. There are four possible configurations depending on the orientation of the blue complete interior corner. Configuration A: the horizontal leg is incident to b_t and the vertical leg points downward. Configuration B: the vertical leg is incident to b_r and the horizontal leg points leftward. Configuration C: the vertical leg is incident to b_r and the horizontal leg points rightward. Configuration D: the horizontal leg is incident to b_t and the vertical leg points upward. We compute four candidate trees in $O(k)$ time for these four configurations and return the minimal tree. Configurations A and B are symmetric and configurations C and D are symmetric. Therefore, we only discuss configurations A and C.

Configuration A. Let c be the corner-vertex and let the vertical leg be incident to $b_s \in [b_r, b_t]$. Let (b_t, c, b_s) denote the blue complete interior corner. So b_s and b_t are blue but the position of b_s is unknown. It can be proved that in a tree T of configuration A and $T \notin E(Class-1)$, b_r has degree one in T and b_r is the endpoint of an interior edge properly incident to (b_t, c). Since we know how to handle Class-1, we will make the above assumptions about b_r. By B3, either (1) there is exactly one interior edge properly incident to (b_t, c) (with b_r as one endpoint), or (2) there is at most one interior edge properly incident to (b_s, c). Refer to Fig.1(a) and 1(b). In Fig.1(a), b_r is blue by B5 and so the tree equals $CG_c(b_s, b_r, (b_s, c)) \cup CG(b_s, b_t, (b_s, c)) \cup \{(b_r, d), (b_t, c)\}$. We try all $O(k)$ positions for b_s and return the tree with minimal total length. In Fig.1(b), let b_x be the

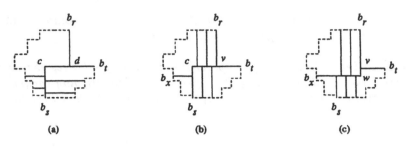

Fig. 1.

left endpoint of the interior edge properly incident to (b_s, c). By B5, b_x is blue. It can be proved that there is one less interior edge properly incident to the lower side of (b_t, c) than its upper side, as shown in Fig.1(b). Thus, we can convert Fig.1(b) to Fig.1(c) and so the tree equals $C_c(b_x, a_p, (b_x, w)) \cup CG(b_x, b_t, (b_x, w)) \cup \{(b_r, w), (b_t, v)\}$, where a_p is the terminal point in $S[b_r, b_t]$ closest to b_r. We try all $O(k)$ positions for b_x and return the tree with minimal total length.

Configuration C. Let c be the corner-vertex and let the horizontal leg be incident to $b_s \in [b_r, b_t]$. Let (b_r, c, b_s) denote the blue complete interior corner. So b_r and b_s are blue but the position of b_s is unknown. If $b_s = b_t$, then by B3, (b_r, c) or (b_t, c) has zero or one interior edge properly incident to it. For the case where no interior edge is properly incident to (b_t, c), we have $CG(b_r, b_t, (b_r, c)) \cup \{(b_t, c)\}$. Consider the case where exactly one interior edge f is properly incident to (b_t, c). Let b_x be the endpoint of f on the boundary which is blue by B5. Then we have $CG(b_r, b_x, (b_r, c)) \cup G(b_x, b_t) \cup \{(b_t, c), f\}$ and we try all $O(k)$ positions for b_x to minimize the total length. The two cases where at most one interior edge is properly incident to (b_r, c) is handled similarly. Suppose that $b_s \neq b_t$. It can be proved that b_t is the endpoint of an interior edge properly incident to (b_r, c). See Fig.2. Either same number of interior edges are properly incident to both sides

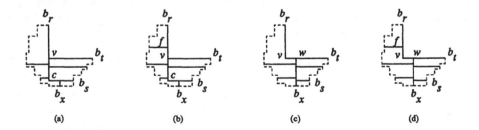

Fig. 2.

of (b_r, c) (see Fig.2(a)), or there is one more interior edge properly incident to the left side of (b_r, c) than the right side (see Fig.2(b)). For Fig.2(a), we convert Fig.2(a) to Fig.2(c) and so the tree equals $CG_c(b_x, b_r, (b_x, w)) \cup C(b_x, b_t, (b_x, b_t)) \cup$

$\{(b_r, v), (v, w)\}$. We try all $O(k)$ positions for b_x to minimize the total length. In Fig.2(b), there is one interior edge, say f, properly incident to (b_r, v). Let b_z be the exterior vertex in $[b_r, b_t]$ such that (b_z, b_t) is a horizontal grid line. Let b_p be last node in the sequence $[b_r, b_z]$ (recall that a node is a terminal point or an inner boundary corner vertex). Let b_q be the lower endpoint of the vertical boundary edge that contains b_z. Let a_p (a_q) be the terminal point closest to b_p (b_q) in $S[b_p, b_x]$ ($S[b_q, b_x]$). Let g be the interior edge immediately below f and properly incident to (v, c). It can be proved that f can be slid downward until f contains b_p or b_z. Thus, we can convert Fig.2(b) to Fig.2(d) by sliding f downward and (v, c) to the right. The resulting tree equals $G(b_r, b_p) \cup C_c(b_x, a_q, (b_x, w)) \cup C(b_x, b_t, (b_x, w)) \cup \{f, (b_r, v), (v, w)\}$ or $G(b_r, b_z) \cup B(b_z, b_q) \cup C_c(b_x, a_q, (b_x, w)) \cup C(b_x, b_t, (b_x, w)) \cup \{f, (b_r, v), (v, w)\}$. We try all $O(k)$ positions for b_x to minimize the total length.

(b_r, b_t) **is a non-blue grid line.** $Tree(b_r, b_t)$ is assumed to satisfy the blue topological properties; otherwise, it cannot be part of an optimal Steiner tree that satisfies the blue topological properties. Since (b_r, b_t) is a non-blue grid line, b_r and b_t lie in the interior of two opposite boundary edges. In $Tree(b_r, b_t)$, either no interior edge is incident to b_r or b_t, or an interior line ℓ is incident to b_r or b_t. In the first case, the tree equals $B(b_r, b_{r+1}) \cup Tree(b_{r+1}, b_{t-1}) \cup B(b_{t-1}, b_t)$. In the second case, ℓ must be (b_r, b_t) and ℓ does not contain any Steiner point by B6. Let a_q be the terminal point in $S[b_r, b_t]$ closest to b_t. Let (b_s, a_q) be a grid line parallel to ℓ. Let b_p be the blue exterior vertex in $[b_r, b_t]$ closest to b_r. If b_p (b_s) is closer to b_r, then the tree equals $B(b_r, b_p) \cup Tree(b_p, a_q) \cup \{\ell\}$ ($B(b_r, b_s) \cup Tree(b_s, a_q) \cup \{\ell\}$).

6 α-structure

Let $\ell = (b_r, v)$ be a blue complete interior line such that $[b_r, b_t] \subseteq openspan(\ell)$. It suffices to compute $C(b_r, b_t, \ell)$ as for any grid vertex w on ℓ such that $[b_r, b_t] \subseteq openspan((b_r, w))$, whenever $C(b_r, b_t, (b_r, w))$ is needed, it can be constructed in $O(1)$ time as $(C(b_r, b_t, \ell) - \{\ell\}) \cup \{(b_r, w)\}$. Thus, we only need to compute a total of $O(kn)$ α-structures. Let $[b_r, b_t]$ be $\{b_r, b_{r+1}, \cdots, b_{t-1}, b_t\}$. W.l.o.g., we assume that ℓ is vertical and b_r is the upper endpoint of ℓ. There are two possible topologies in $C(b_r, b_t, \ell)$: (1) no interior edge is incident to b_t and ℓ, (2) an interior edge f is incident to b_t and ℓ. In case (1): Either no interior edge is properly incident to ℓ, or some interior edge is properly incident to ℓ. In the first case, the tree equals $Tree(b_r, b_t) \cup \{\ell\}$. In the second case, the tree equals $C(b_r, b_{t-1}, \ell) \cup B(b_{t-1}, b_t)$ or $C(b_r, b_x, \ell) \cup Tree(b_x, b_t)$, where (b_x, b_t) is a vertical grid line. In case (2), there are three possible topologies: (2.1) b_t has degree one, (2.2) b_t has degree more than one and b_t is not incident to a tree edge perpendicular to f, (2.3) b_t is incident to a tree edge g perpendicular to f. In case (2.1), the tree equals $C(b_r, a_i, \ell) \cup \{f\}$, where a_i is the terminal point in $S[b_r, b_t]$ closest to b_t. In case (2.2), the tree equals the tree constructed in case (2) when computing $C(b_r, b_{t-1}, \ell)$. In case (2.3), f can be slid upward and this reduces case (2.3) to case (1).

330

References

1. M. Bern, *Faster Exact Algorithms for Steiner Tree in Planar Networks*, Networks, 7 (1990), pp. 109–120.
2. S.W. Cheng, A. Lim, and C. Wu, *Optimal Rectilinear Steiner Tree for Extremal Point Sets*, in Proc. Int'l Symp. Alg. and Comput. (ISAAC), 1993, LNCS 762, pp. 523–532.
3. S.W. Cheng and C.K. Tang, *A Fast Algorithm for Computing Optimal Rectilinear Steiner Trees for Extremal Point Sets*, Technical Report HKUST-CS95-20, Department of Computer Science, HKUST.
4. R. E. Erickson, C. Monma, and A. F. Veinott, *Send-and-split Method for Minimum-cost Network Flows*, Mathematical Operation Research, 12 (1987), pp. 634–664.
5. M. Kaufmann, S. Gao, and K. Thulasiraman, *On Steiner Minimal Trees in Grid Graphs and Its Application to VLSI Routing*, in Proc. Int'l Symp. Alg. and Comput. (ISAAC), 1994, LNCS 834, pp. 351–359.
6. C.E. Leiserson and F.M. Maley, *Algorithms for Routing and Testing Routability of Planar VLSI Layouts*, Proc. 7th Ann. Symp. Theory of Comput., 1985, pp. 69–78.
7. J. Provan, *Convexity and the Steiner Tree Problem*, Networks, 18 (1988), pp. 55–72.
8. D. Richards and J. Salowe, *A Linear-Time Algorithm to Construct a Rectilinear Steiner Minimal Tree for k-Extremal Point Sets*, Algorithmica, 7 (1992), pp. 247–276.
9. J.D. Ullamn, *Computational Aspects of VLSI*, Computer Science Press, 1984.
10. Y. Y. Yang and O. Wing. Optimal and Suboptimal Solution Algorithms for the Wiring Problem. *Proc. IEEE Int'l Symp. Circuit Theory*, pp. 154–158, 1972.

A   Sketch of proof of Theorem 1

For $1 \le j \le 3$, we compute in $O(k^2 n)$ time a tree G_j with total length no greater than the minimal tree in Group-j. Thus, the minimal one is the desired solution. G_1 is $RC(S)$ minus the longest portion of the boundary between two adjacent

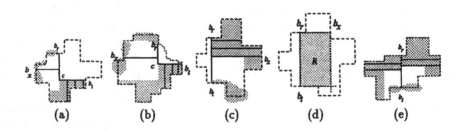

Fig. 3. G_2 and G_3.

terminal points on the boundary, which can be found in $O(n)$ time.

The minimal Group-2 tree contains a complete interior corner whose legs are incident to two blue exterior vertices b_r and b_t. Let c be the corner-vertex. By B3, there are $O(k^2)$ possible choices for b_r and b_t. There are three possible cases. Case (1.1): b_r *and* b_t *lie on opposite boundary staircases* (e.g., see Fig.3(a)). The subtree that connects $S[b_t, b_r]$ and the complete interior corner equals $C(b_t, a_i, (b_t, c)) \cup C_c(b_r, a_{i+1}, (b_r, c))$ for some neighboring points $a_i, a_{i+1} \in S[b_t, b_r]$, which can be found in $O(n)$ time. Consider the subtree T connecting $S[b_r, b_t]$ and the complete interior corner. If (b_r, c) has no interior edge properly incident to it, then $T = CG_c(b_t, b_r, (b_t, c)) \cup \{(b_r, c)\}$. If there is an edge f properly incident to (b_r, c) and a blue node b_x in $[b_r, b_t]$, then $T = G(b_r, b_x) \cup CG_c(b_t, b_x, (b_t, c)) \cup \{f, (b_r, c)\}$. We try all $O(k)$ choices for b_x to minimize the total length. The two cases where (b_t, c) has at most one interior edge properly incident to it can be handled similarly. Case (1.2): b_r *and* b_t *lie on the same boundary staircase* (e.g., Fig.3(b)). As in case (1.1), the subtree that connects $S[b_t, b_r]$ and the complete interior corner can be computed in $O(n)$ time. By B4, we cannot flip the complete interior corner to overlap with the boundary. Thus, there is exactly one interior edge properly incident to one of the legs. Suppose that this edge is properly incident to (b_r, c) and a blue node b_x. Then the subtree that connects $S[b_r, b_t]$ and the complete interior corner equals $G(b_r, b_x) \cup CG_c(b_t, b_x, (b_t, c)) \cup \{(b_r, c)\}$. We try all $O(k)$ positions for b_x to minimize the total length. Note that $[b_x, b_t]$ may not lie on three boundary staircases, but $CG_c(b_t, b_x, (b_t, c))$ can be computed in an $O(k^2 n)$-time preprocessing step. Case (1.3): b_r *and* b_t *lie on consecutive boundary staircases*. This case can be handled by a method similar to the one in case (1.2).

The minimal Group-3 tree contains a complete interior line (b_r, b_t). When (b_r, b_t) is blue, there are three cases. Case (2.1): b_r *and* b_t *lie on opposite boundary staircases* (e.g., Fig.3(e)). The tree equals $CG(b_r, b_t, (b_r, b_t)) \cup CG_c(b_r, b_t, (b_r, b_t))$. Case (2.2): b_r *and* b_t *lie on two consecutive boundary staircases* (e.g., Fig.3(c) and 3(d)). Suppose that some interior edge is properly incident to the right side of (b_r, b_t) as shown in Fig.3(c). Let $b_x \in [b_t, b_r]$ be the right endpoint of the lowest interior edge properly incident to the right side of (b_r, b_t). Then the tree equals $CG(b_r, b_t, (b_r, b_t)) \cup C_c(b_r, b_x, (b_r, b_t)) \cup G(b_t, b_x)$. We try all $O(n)$ positions for b_x to minimize the total length. Suppose that no interior edge is properly incident to the right side of (b_r, b_t) as shown in Fig.3(d). We can further assume that (b_r, b_t) is a longest complete interior line and it is the rightmost one among all longest vertical complete interior lines. It can be proved that the region R shown in Fig.3(d) is empty. Let b_x other than b_r and b_t be an exterior vertex at a corner of R as shown in Fig.3(d). If b_x is a terminal point, then a candidate tree is the minimal of $Tree(b_r, b_t) \cup G(b_t, b_x) \cup Tree(b_x, b_r)$ and $Tree(b_r, b_t) \cup Tree(b_t, b_x) \cup G(b_x, b_r)$. If b_x is not a terminal point, then we also need to compare the above two candidates with $Tree(b_r, b_t) \cup Tree(b_t, a_i) \cup Tree(a_{i+1}, b_r)$, where a_i and a_{i+1} are the two terminal points in $S[b_t, b_r]$ neighboring to b_x, in order to select the minimal one. For the case that (b_r, b_t) is non-blue, by B6, (b_r, b_t) cannot contain any Steiner point. This makes the configuration similar to Fig.3(d) and can be computed by a similar method.

Algorithms for Finding f-Colorings of Partial k-Trees
—— An Extended Abstract ——

Xiao Zhou* and Takao Nishizeki**

* Education Center for Information Processing
** Graduate School of Information Sciences
Tohoku University, Sendai 980-77, JAPAN

Abstract. In an ordinary edge-coloring of a graph $G = (V, E)$ each color appears at each vertex $v \in V$ at most once. An f-coloring is a generalized edge-coloring in which each color appears at each vertex $v \in V$ at most $f(v)$ times, where $f(v)$ is a positive integer assigned to v. This paper gives a linear-time sequential algorithm and an optimal parallel algorithm which find an f-coloring of a given partial k-tree with the minimum number of colors.

1 Introduction

This paper deals with a generalized edge-coloring, called an f-coloring, of partial k-trees. A partial k-tree, formally defined later, is a simple graph, that is, it has no multiple edges or self-loops. A partial 3-tree is depicted in Figure 1. Throughout the paper we consider only partial k-trees with bounded k, that is, graphs with bounded tree-width. The class of partial k-trees includes many interesting classes of graphs such as series-parallel graphs and outerplanar graphs.

An *edge-coloring* of a graph G is to color all the edges of G so that no two adjacent edges are colored with the same color. There are linear-time sequential and optimal parallel algorithms to edge-color partial k-trees with the minimum number of colors [12, 13, 14].

Let f be a function which assigns a positive integer $f(v)$ to each vertex $v \in V$. Then an f-*coloring* of G is to color all the edges of G so that, for each vertex $v \in V$, at most $f(v)$ edges incident to v are colored with the same color. Figure 1 illustrates an f-coloring of a partial 3-tree with three colors, where numbers next to vertices v are $f(v)$'s and three colors are indicated by thin, thick and dotted lines. Thus an f-coloring of graph $G = (V, E)$ is a partition of edge set E to several subsets so that the subgraph induced by each subset satisfies the degree-constraints expressed by f. The ordinary edge-coloring is a special case of an f-coloring for which $f(v) = 1$ for every vertex $v \in V$. We call the minimum number of colors needed for an f-coloring the f-*chromatic index* of G, and denote it by $\chi'_f(G)$. We call $\Delta_f(G) = \max_{v \in V} \lceil d(v)/f(v) \rceil$ the *maximum f-degree* of graph G, where $d(v)$ is the degree of vertex v. It is known that $\chi'_f(G) = \Delta_f(G)$ or $\Delta_f(G) + 1$ for any simple graph G [5]. The f-*coloring problem* is to find an *optimal f-coloring* of G, that is, an f-coloring using $\chi'_f(G)$

---

* Email: zhou@ecip.tohoku.ac.jp
** Email: nishi@ecei.tohoku.ac.jp

colors. The problem arises in many applications, including the scheduling of file transfers in computer networks [8].

Since the ordinary edge-coloring problem is NP-complete [6], the f-coloring problem is also NP-complete. Therefore it is very unlikely that there exists a sequential algorithm which solves the f-coloring problem for general graphs in polynomial time. Furthermore, the f-coloring problem is one of the typical "edge-partitioning problems" which, as mentioned in [2], does not appear to be efficiently solvable even for partial k-trees although many combinatorial problems of other types can be solved very efficiently for partial k-trees.

In this paper we give two efficient algorithms, a sequential algorithm and a parallel one, to solve the f-coloring problem for partial k-trees if all $f(v)$'s are bounded.

The sequential algorithm takes linear time. The idea behind it is threefold: (1) we show that a partial k-tree G can be f-colored with $\chi'_f(G)$ colors in linear time if $\Delta_f(G)$ is small, say $\Delta_f(G) < 6k$; (2) we prove that a partial k-tree can be f-colored with exactly $\Delta_f(G)$ colors if $\Delta_f(G)$ is large, say $\Delta_f(G) \geq 2k$; and (3) we show that a partial k-tree G of large $\Delta_f(G)$ can be decomposed into several subgraphs $G_1, G_2, \cdots, G_s$ of small $\Delta_f(G_i)$ such that $\Delta_f(G) = \sum_{i=1}^{s} \Delta_f(G_i)$ and $2k \leq \Delta_f(G_i) < 6k$ for each i, and hence an optimal f-coloring of G can be obtained simply by extending those of $G_1, G_2, \cdots, G_s$. Note that the optimal f-colorings of $G_1, G_2, \cdots, G_s$ can be found in linear time by (1).

The parallel algorithm is an optimal and the first NC parallel algorithm for the f-coloring problem on partial k-trees G. Given a tree-decomposition of G, the algorithm finds an optimal f-coloring of G in $O(\log n)$ time with $O(n/\log n)$ processors. It is known that a tree-decomposition of G can be found in $O(\log^2 n)$ time with $O(n)$ processors [1, 10]. Our idea is to partition the vertex set of G into $O(\log n)$ subsets so that the "forward degree" of each vertex is bounded. The parallel computation model we use is a concurrent-read and exclusive-write parallel random machine (CREW PRAM).

Thus our two algorithms for the f-coloring problem imply those for the ordinary edge-coloring problem in [12, 13, 14] as a special case where $f(v) = 1$ for every $v \in V$. The methods which we develop in this paper appear to be useful for many other problems, especially for the "edge-partition problem with respect to property π," which asks to partition the edge set of a given graph into the minimum number of subsets so that the subgraph induced by each subset satisfies the property π.

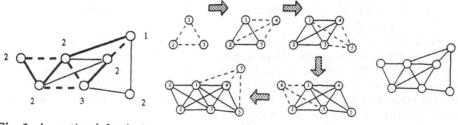

Fig. 1. An optimal f-coloring of a partial 3-tree.

(a)　　　　　　　　(b)

Fig. 2. (a) 3-trees and (b) a partial 3-tree.

2 Definition

In this section we give some definitions. Let $G = (V, E)$ denote a graph with vertex set V and edge set E. We often denote by $V(G)$ and $E(G)$ the vertex set and the edge set of G, respectively. We denote by n the number of the vertices in G. The paper deals with *simple* graphs without multiple edges or self-loops. An edge joining vertices u and v is denoted by (u, v). The *degree* of vertex $v \in V(G)$ is denoted by $d(v, G)$ or simply by $d(v)$. The *maximum degree* of G is denoted by $\Delta(G)$ or simply by Δ. The minimum number of colors needed for an edge-coloring is called the *chromatic index* of G and denoted by $\chi'(G)$. Vizing showed that $\chi'(G) = \Delta$ or $\Delta + 1$ for any simple graph G [3, 11]. The graph obtained from G by deleting all vertices in $V' \subseteq V(G)$ is denoted by $G - V'$. The graph obtained from G by deleting all edges in $E' \subseteq E(G)$ is denoted by $G - E'$. For $E' \subseteq E(G)$, $G[E']$ denotes the subgraph of G induced by the edges in E'.

The class of *k-trees* is defined recursively as follows:

(a) A complete graph with k vertices is a k-tree.
(b) If $G = (V, E)$ is a k-tree and k vertices $v_1, v_2, \cdots, v_k$ induce a complete subgraph of G, then $G' = (V \cup \{w\}, E \cup \{(v_i, w) | 1 \le i \le k\})$ is a k-tree where w is a new vertex not contained in G.
(c) All k-trees can be formed with rules (a) and (b).

A graph is called a *partial k-tree* if it is a subgraph of a k-tree. Thus a partial k-tree $G = (V, E)$ is a simple graph, and $|E| < kn$. Figure 2(a) illustrates a process of generating a 3-tree, and Figure 2(b) depicts a partial 3-tree. In this paper we assume that k is a constant.

Let f be a function which assigns a natural number $f(v)$ to each vertex $v \in V$. One may assume without loss of generality that $f(v) \le d(v)$ for each vertex $v \in V(G)$. Furthermore we assume in the paper that all $f(v)$'s are bounded. We call $d_f(v, G) = \lceil d(G, v)/f(v) \rceil$ the *f-degree* of vertex v, and often denote $d_f(v, G)$ simply by $d_f(v)$. We call $\Delta_f(G) = \max\{d_f(v) | v \in V(G)\}$ the *maximum f-degree* of G.

3 f-Chromatic Index

It is known that $\chi'_f(G) = \Delta_f$ or $\Delta_f + 1$ for any simple graph G [5]. In this section we first show that $\chi'_f(G) = \Delta_f$ holds for any partial k-tree G with $\Delta_f \ge 2k$, and then show that the f-chromatic index $\chi'_f(G)$ can be determined in linear time for any partial k-tree G.

Let $G = (V, E)$ be a graph. For an edge $(u, v) \in E$, we denote by $d_u^*(v)$ the number of v's neighbors other than u whose f-degrees are Δ_f. We call (u, v) an *f-eliminable edge* if either $d_f(u) + d_u^*(v) \le \Delta_f$ or $d_f(v) + d_v^*(u) \le \Delta_f$. Note that an "$f$-eliminable edge" is a generalization of an "eliminable edge" for the ordinary edge-coloring [3, 9, 11]. Then we have the following theorem, which is a generalization of Vizing's "adjacency lemma" [3, 9, 11].

Theorem 1. *Let e be an f-eliminable edge of a simple graph G. If $G - \{e\}$ has an f-coloring φ' with $\Delta_f(G)$ colors, then G has an f-coloring φ with $\Delta_f(G)$ colors. Furthermore, φ' can be updated to φ in time $O(n)$.*

Proof. Using generalized techniques of shifting a "fan-sequence" and switching an "alternating path," one can undate φ' to φ in time $O(n)$. The detail of the proof is omitted in this extended abstract. $\mathcal{Q.E.D.}$

For partial k-trees we have the following lemma.

Lemma 2. *If the maximum f-degree of a partial k-tree G is large, say $\Delta_f(G) \geq 2k$, then G has an f-eliminable edge.*

Proof. Since G is a partial k-tree, G has at least one vertex u such that $d(u) \leq k$. Let $S = \{u \in V(G) \mid d_f(u, G) \leq k\}$. Then $S \neq \phi$ since $d_f(u) = \lceil d(u)/f(u) \rceil \leq d(u)$. Furthermore $V - S \neq \phi$ since $\Delta_f \geq 2k > k$. Since G is a partial k-tree, $G' = G - S$ is a partial k-tree and is not empty. Therefore G' has at least one vertex v such that $d(v, G') \leq k$. Since $k + 1 \leq d_f(v, G) \leq d(v, G)$ and $d(v, G') \leq k$, G has an edge (u, v) joining v and a vertex $u \in S$. Since $u \in S$, $d_f(u) \leq k < \Delta_f$. On the other hand, none of v's neighbors in S has an f-degree Δ_f, and hence $d_u^*(v) \leq d(v, G') \leq k$. Therefore $d_f(u) + d_u^*(v) \leq 2k \leq \Delta_f$, and hence edge (u, v) is eliminable. $\mathcal{Q.E.D.}$

We are now ready to prove the following theorem.

Theorem 3. *If a partial k-tree G has $\Delta_f(G) \geq 2k$, then $\chi'_f(G) = \Delta_f$.*

Proof. By Lemma 2 G has an f-eliminable edge e_1. Since $G - \{e_1\}$ is also a partial k-tree, $G - \{e_1\}$ has an f-eliminable edge e_2 if $\Delta_f(G - \{e_1\}) \geq 2k$. Thus there exists a sequence of edges $e_1, e_2, \cdots, e_m$ such that

(a) $\Delta_f(G') = \Delta_f(G) - 1$ where $G' = G - \{e_1, e_2, \cdots, e_m\}$; and
(b) every edge e_i, $1 \leq i \leq m$, is f-eliminable in $G - \{e_1, e_2, \cdots, e_{i-1}\}$.

By the result in [5] $\chi'_f(G') \leq \Delta_f(G') + 1 = \Delta_f(G)$. Therefore, applying Theorem 1 repeatedly, we have $\chi'_f(G) = \Delta_f(G)$. $\mathcal{Q.E.D.}$

For partial k-trees with bounded degrees we have the following theorem.

Theorem 4. *An optimal f-coloring of a partial k-tree G with bounded degrees can be found in linear time.*

Proof. Using the dynamic programming method based on a tree-decomposition of G, one can find an optimal f-coloring of G. Since Δ is bounded, one can show that the size of a DP table is bounded and hence the DP algorithm takes $O(n)$ time. We omit the detail of the proof in this extended abstract. $\mathcal{Q.E.D.}$

From these two theorems we can derive the following theorem.

Theorem 5. *The f-chromatic index $\chi'_f(G)$ of a partial k-tree $G = (V, E)$ can be determined in linear time.*

Proof. If Δ_f is large, say $\Delta_f \geq 2k$, then by Theorem 3 $\chi'_f(G) = \Delta_f(G)$ and hence $\chi'_f(G)$ can be computed in linear time. Thus we may assume that $\Delta_f(G)$ is small, say $\Delta_f(G) < 2k$. Then Δ is bounded since $\Delta \leq \Delta_f \cdot \max\{f(v) \mid v \in V\}$, and we assume that $f(v) = O(1)$ for every $v \in V$. Therefore, by Theorem 4, $\chi'_f(G)$ can be computed in linear time. $\mathcal{Q.E.D.}$

4 f-Coloring

In the preceding section we have shown that the f-chromatic index $\chi'_f(G)$ of a given partial k-tree G can be determined in linear time. In this section we give a linear-time algorithm which actually obtains an optimal f-coloring of G. If $\Delta_f(G)$ is small, say $\Delta_f(G) < 6k$, then $\Delta(G)$ is bounded and hence by Theorem 4 one can obtain an optimal f-coloring of G in linear time. Therefore it suffices to give a linear-time algorithm only for the case $\Delta_f(G) \geq 6k$.

The proofs in the preceding section do not yield a linear-time algorithm for the case $\Delta_f(G) \geq 6k$, as follows. Lemma 2 implies that a partial k-tree G with $\Delta_f(G) \geq 6k$ necessarily has an f-eliminable edge e. If an f-coloring of $G - \{e\}$ with $\Delta_f(G)$ colors is known, then, by Theorem 1 one can obtain an f-coloring of G with $\chi'_f(G) = \Delta_f(G) \, (> 2k)$ colors in $O(n)$ time. By Lemma 2 there exists an edge-sequence $e_1, e_2, \cdots, e_m$ such that

$$\Delta_f(G - \{e_1, e_2, \cdots, e_m\}) = 6k$$

and e_i is an f-eliminable edge in $G - \{e_1, e_2, \cdots, e_{i-1}\}$ for every i, $1 \leq i \leq m$. By Theorem 4 one can obtain an f-coloring of $G' = G - \{e_1, e_2, \cdots, e_m\}$ with $\chi'_f(G') = 6k \, (> 2k)$ colors in $O(n)$ time. Add edges $e_m, e_{m-1}, \cdots, e_2, e_1$ to G' in this order, and modify the f-coloring of G' to an f-coloring of G with $\Delta_f(G)$ colors by repeatedly using the generalized techniques of "shifting a fan sequence" and "switching an alternating path." Such a repetition of recoloring would take time $O(n^2)$. (However, such a method provides a practical algorithm of time complexity $O(n^2)$ with a small coefficient which finds an optimal f-coloring of G with $\Delta_f(G) \geq 2k$.)

Our idea is to decompose $G = (V, E)$ to several edge-disjoint subgraphs $G_1, G_2, \cdots, G_s$ such that $\chi'_f(G) = \sum_{j=1}^s \chi'_f(G_j)$ when $\Delta_f(G) \geq 6k$. Let c be a fixed positive integer, and let $E_1, E_2, \cdots, E_s$ be a partition of E. We say that $E_1, E_2, \cdots, E_s$ is a (Δ_f, c)-partition of E if $G_j = G[E_j]$, $1 \leq j \leq s$, satisfies

(a) $\Delta_f(G) = \sum_{j=1}^s \Delta_f(G_j)$; and
(b) $\Delta_f(G_j) = c$ for each j, $1 \leq j \leq s - 1$, and $c \leq \Delta_f(G_s) < 2c$.

Since G is a partial k-tree, subgraphs $G_1, G_2, \cdots, G_s$ of G are partial k-trees, too. Theorem 3 implies that $\chi'_f(G) = \Delta_f(G)$ since $\Delta_f(G) \geq 6k$. Choose $c = 3k$, then $\Delta_f(G_j) \geq c = 3k$ and hence $\chi'_f(G_j) = \Delta_f(G_j)$ for each j, $1 \leq j \leq s$. Since $\Delta_f(G_j) < 2c = 6k$ for $1 \leq j \leq s$, by Theorem 4 an optimal f-coloring of G_j with $\Delta_f(G_j)$ colors can be found in linear time. On the other hand, the condition (a) above implies

$$\chi'_f(G) = \sum_{j=1}^s \chi'_f(G_j).$$

Therefore optimal f-colorings of G_j, $1 \leq j \leq s$, can be immediately extended to an optimal f-coloring of G. Thus we have the following algorithm.

Algorithm f-Coloring(G)
begin

    if $\Delta_f(G) < 6k$ **then**
1   find an optimal f-coloring of G by a DP method   { Theorem 4 }

else begin

2   find a $(\Delta_f, 3k)$-partition $E_1, E_2, \cdots, E_s$ of E;

3   find an optimal f-coloring of G_j for each j, $1 \leq j \leq s$, by a DP method;

                                                         { Theorem 4 }

4   extend these optimal f-colorings of $G_1, G_2, \cdots, G_s$

         to an optimal f-coloring of G

   end

end.

We have the following lemma.

Lemma 6. *If a partial k-tree $G = (V, E)$ has $\Delta_f(G) \geq 6k$, then a $(\Delta_f, 3k)$-partition of E can be found in linear time.*

An algorithm EdgePartition to find a $(\Delta_f, 3k)$-partition will be given later although the proof of Lemma 6 is omitted in this extended abstract.

From Theorem 3 and Lemma 6 we have the following theorem.

Theorem 7. *An optimal f-coloring of a partial k-tree $G = (V, E)$ can be founded in linear time.*

Proof. By Theorems 3, 4 and Lemma 6 the algorithm above correctly finds an optimal f-coloring of a partial k-tree G. Therefore it suffices to prove the time complexity. By Lemma 6 line 2 can be done in linear time. By Theorem 4 line 1 can be done in linear time, since

$$\Delta(G) \leq \Delta_f(G) \cdot \max\{f(v) \mid v \in V\} < 6k \cdot \max\{f(v) \mid v \in V\} = O(1).$$

Similarly one can obtain an optimal f-coloring of G_j, $1 \leq j \leq s$, in linear time since $\Delta_f(G_j) < 6k$. Hence line 3 can be done in linear time. At line 4 one can immediately extend these optimal f-colorings of $G_1, G_2, \cdots, G_s$ to an optimal f-coloring of G in linear time since

$$\chi'_f(G) = \Delta_f(G) = \sum_{j=1}^{s} \Delta_f(G_j) = \sum_{j=1}^{s} \chi'_f(G_j).$$

$$Q.\mathcal{E}.\mathcal{D}.$$

In the remaining of this section we give the algorithm EdgePartition. We construct a (Δ_f, c)-partition of edge set E from a partition $S_1, S_2, \cdots, S_l$ of vertex set V. For each $v \in S_i$, $1 \leq i \leq l$, let

$$BW(v, G) = \{(v, w) \in E \mid w \in S_j \text{ and } j < i\},$$
$$FW(v, G) = \{(v, w) \in E \mid w \in S_j \text{ and } j \geq i\},$$
$$bw(v, G) = |BW(v, G)|, \text{ and}$$
$$fw(v, G) = |FW(v, G)|.$$

Thus $d(v, G) = bw(v, G) + fw(v, G)$. We call the edges in $BW(v, G)$ *backward edges*, those in $FW(v, G)$ *forward edges*, $bw(v, G)$ the *backward degree* of v, and $fw(v, G)$ the *forward degree* of v. (See Figure 3.) For a fixed positive integer c we say that $S_1, S_2, \cdots, S_l$ is a (fw, c)-partition of V if $fw(v, G) \leq cf(v)$ for every vertex $v \in V$. Then we have the following lemma.

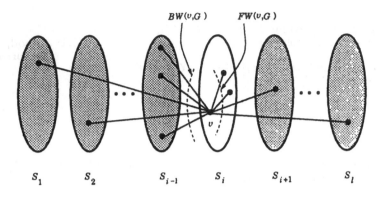

Fig. 3. Illustration for notations.

Lemma 8. *A $(fw, 3k)$-partition $S_1, S_2, \cdots, S_l$ of V can be found in linear time for every partial k-tree $G = (V, E)$.*

Proof. One can easily find a $(fw, 3k)$-partition of V as follows. Since G is a partial k-tree, G has a vertex of degree $\leq k$. Let S_1 be the set of vertices u having degree $\leq 3kf(u)$, and delete all vertices in S_1 from G. The resulting graph is a partial k-tree since G is a partial k-tree. Let S_2 be the set of vertices v having degree $\leq 3kf(v)$ in the resulting graph, and delete all vertices in S_2. Repeating the same operation, one can find a $(fw, 3k)$-partition $S_1, S_2, \cdots, S_l$ of V. Clearly the partition can be found in linear time. $\qquad$ Q.E.D.

We then present the algorithm EdgePartition to find a $(\Delta_f, 3k)$-partition $E_1, E_2, \cdots, E_s$ of E, where $s = \lfloor \Delta_f / 3k \rfloor$. We now outline the algorithm before describing it formally. We first find a $(fw, 3k)$-partition $S_1, S_2, \cdots, S_l$ of V. We then construct $E_1, E_2, \cdots, E_{s-1}$ as follows, and finally let $E_s = E - E_1 \cup E_2 \cup \cdots \cup E_{s-1}$. For each vertex v, we number the backward edges incident to v from 1 to $bw(v, G)$; we then insert the first $3kf(v)$ backward edges to E_1, the second $3kf(v)$ backward edges to E_2, and so on whenever there remain $3kf(v)$ or more backward edges; and we finally insert all the remaining backward edges to $E_{p(v)+1}$ if $p(v) + 1 \leq s - 1$, where $p(v) = \lfloor bw(v, G)/3kf(v) \rfloor \leq s$. For each j, $1 \leq j \leq s - 1$, we have

$$fw(v, G[E_j]) \leq 3kf(v),$$
$$d(v, G[E_j]) = fw(v, G[E_j]) + bw(v, G[E_j]),$$

and hence

$$d(v, G[E_j]) - 3kf(v) \leq bw(v, G[E_j]).$$

If $d(v, G[E_j]) > 3kf(v)$, then we adjust the degree $d(v, G[E_j])$ to exactly $3kf(v)$ by deleting $d(v, G[E_j]) - 3kf(v)$ backward edges incident to v from E_j. More precisely, we first adjust the degrees of vertices in S_l, then vertices in S_{l-1}, and so on. Since a deleted backward edge incident to $v \in S_i$ is not incident to any vertex $w \in S_{i'}$, $i' \geq i + 1$, the deletion does not change $d(w, G[E_j])$ which has been adjusted so that it does not exceed $3kf(v)$. Thus, when the algorithm terminates, we have $\Delta_f(G[E_j]) \leq 3k$ for every j, $1 \leq j \leq s - 1$. Since $s = \lfloor \Delta_f / 3k \rfloor$ and $E_s = E - E_1 \cup E_2 \cup \cdots \cup E_{s-1}$, we have $3k \leq \Delta_f(G[E_s])$.

We now formally describe the algorithm EdgePartition.

Algorithm EdgePartition(G)
{ $G = (V, E)$ is a partial k-tree with $\Delta_f \geq 6k$ }
begin
1   find a $(fw, 3k)$-partition $S_1, S_2, \cdots, S_l$ of V;   { Lemma 8 }
2   $s := \lfloor \Delta_f / 3k \rfloor$;
3   **for each** $v \in V$ **do**
4   **begin**
5   number the backward edges in $BW(v, G)$ from 1 to $bw(v, G)$;
6   let $BW(v, G) = \{e_{v_q} | 1 \leq q \leq bw(v, G)\}$
7   **end**;
8   **for each** j, $1 \leq j \leq s$, **do** $E_j := \phi$;   { initialize edge-sets E_j }
9   **for each** $v \in V$ **do** $p(v) := \lfloor bw(v, G)/3kf(v) \rfloor$;
   { construct $E_1, E_2, \cdots, E_{s-1}$ }
10 **for each** $v \in V$ and each j, $1 \leq j \leq \min\{p(v), s-1\}$, **do**
11   $E_j := E_j \cup \{e_{v_q} \in BW(v, G) | 3kf(v)(j-1) < q \leq 3kf(v)j\}$;
        { insert to E_j exactly $3kf(v)$ backward edges incident to v, and hence
              $bw(v, G[E_j]) = 3kf(v)$ although probably $fw(v, G[E_j]) > 0$
              and $d(v, G[E_j]) = fw(v, G[E_j]) + bw(v, G[E_j]) > 3kf(v)$}
12 **for each** $v \in V$ with $p(v) \leq s-2$ **do**
13   $E_{p(v)+1} := E_{p(v)+1} \cup \{e_{v_q} \in BW(v, G) | 3kf(v)p(v) < q \leq bw(v, G)\}$;
        { insert to $E_{p(v)+1}$ the remaining backward edges incident to v,
              and hence $bw(v, G[E_{p(v)+1}]) < 3kf(v)$ }
   { adjust $d(v, G[E_j])$ so that it does not exceed $3kf(v)$ }
14 **for** $i := l$ **downto 1 do**
15   **for each** $v \in S_i$ and each j, $1 \leq j \leq s-1$, **do**
                        { $d(v, G[E_j]) - 3kf(v) \leq bw(v, G[E_j])$ }
16   **if** $d(v, G[E_j]) > 3kf(v)$ **then**
17   delete $d(v, G[E_j]) - 3kf(v)$ backward edges in $BW(v, G[E_j])$ from E_j;
                          { $d(v, G[E_j]) = 3kf(v)$ }
                          { $\Delta_f(G[E_j]) \leq 3k$ }
   { construct E_s }
18 $E_s := E - E_1 \cup E_2 \cup \cdots \cup E_{s-1}$
end

5   Parallel Algorithm

In this section we give an optimal parallel algorithm which finds an optimal f-coloring of a partial k-tree G. Our algorithm runs in $O(\log n)$ parallel time with $O(n/\log n)$ processors if a tree-decomposition of G with treewidth $\leq k$ is given.

If Δ is bounded, say $\Delta < 6k$, then the sequential algorithm in Theorem 4 can be modified to a parallel algorithm as in the following theorem.

Theorem 9. *Let G be a partial k-tree with bounded degrees given by its decomposition tree. Then there is a parallel algorithm which solves the f-coloring problem for G in $O(\log n)$ time using a total of $O(n)$ operations.*

If $\Delta_f(G)$ is bounded, then $\Delta(G)$ is bounded and hence the algorithm in Theorem 9 is an optimal parallel algorithm for such a case. Therefore it suffices to give an optimal parallel algorithm only for the case when $\Delta_f(G) \geq 6k$.

We have the following lemma.

Lemma 10. *Every partial k-tree $G = (V, E)$ has a $(fw, 3k)$-partition $S_1, S_2, \cdots,$ S_l of V with $l = O(\log n)$ and such a partition can be found in $O(\log n)$ parallel time using a total of $O(n)$ operations.*

By the lemma above the algorithm EdgePartition can be modified to a parallel algorithm as in the following lemma.

Lemma 11. *If a partial k-tree $G = (V, E)$ has the maximum degree $\Delta_f(G) \geq 6k$, then a $(\Delta_f, 3k)$-partition of E can be found in $O(\log n)$ parallel time using a total of $O(n)$ operations.*

Zhou, Nakano and Nishizeki [13] have given the following lemma.

Lemma 12. *Let T be a tree-decomposition of a partial k-tree G with treewidth $\leq k$, and let $E_1, E_2, \cdots, E_s$ be any partition of E with $s = O(n)$. Then one can find tree-decompositions T_j of $G_j = G[E_j]$ with treewidths $\leq k$ for all j, $1 \leq j \leq s$, in $O(\log n)$ parallel time using a total of $O(n)$ operations.*

From Theorem 9 and Lemmas 11 and 12, we have the following theorem.

Theorem 13. *The f-coloring problem can be solved in $O(\log n)$ parallel time using a total of $O(n)$ operations for a partial k-tree $G = (V, E)$ given by its tree-decomposition with treewidth $\leq k$.*

Proof. It suffices to show that the algorithm f-$Coloring$ can be optimally done in parallel. Since a tree-decomposition is given, Theorem 9 implies that line 1 of f-$Coloring$ can be done in $O(\log n)$ parallel time using a total of $O(n)$ operations. By Lemma 11 line 2 can be done in $O(\log n)$ parallel time using a total of $O(n)$ operations. Since $\Delta(G_j) < 6k$, by Theorem 9 one can obtain an f-coloring of G_j with $\Delta(G_j)$ colors in $O(\log n_j)$ parallel time using a total of $O(n_j)$ operations where n_j is the number of vertices in G_j. Since G_j is a subgraph of G induced by E_j, $n_j \leq 2|E_j|$. Therefore

$$\sum_{j=1}^{s} n_j \leq \sum_{j=1}^{s} 2|E_j| \leq 2|E| < 2kn.$$

Hence line 3 can be done in $O(\log \max_{1 \leq j \leq s} n_j) = O(\log n)$ parallel time using a total of $O(n)$ operations. Since $\Delta_f(G) = \sum_{j=1}^{s} \Delta_f(G_j)$, one can immediately extend these optimal f-colorings of $G_1, G_2, \cdots, G_s$ to an optimal f-coloring of G with $\Delta_f(G)$ colors in $O(\log n)$ parallel time using a total of $O(n)$ operations. $\mathcal{Q.E.D.}$

The following general lemma is well-known [4, 7].

Lemma 14. *Let A be a given algorithm with $O(\log n)$ parallel computation time. Suppose that A involves a total of m operations. Then A can be implemented using p processors in $O(m/p + \log n)$ parallel time.*

If there is an algorithm A which solves the f-coloring problem in $O(\log n)$ parallel time using a total of $m = O(n)$ operations, then by adapting Lemma 14 with choosing $p = n/\log n$ one can know that A can be implemented using $O(n/\log n)$ processors in $O(\log n)$ parallel time. Thus by Theorem 13 and Lemma 14 we have the following corollary.

Corollary 15. *The f-coloring problem can be solved in $O(\log n)$ parallel time with $O(n/\log n)$ processors for a partial k-tree $G = (V, E)$ given by its tree-decomposition with treewidth $\leq k$.*

Remark. If not all $f(v)$'s are bounded, then our sequential algorithm takes polynomial-time and our parallel algorithm is not always optimal but is an NC parallel algorithm.

References

1. H. L. Bodlaender. A linear time algorithm for finding tree-decompositions of small treewidth. In *Proc. of the 25th Ann. ACM Symp. on Theory of Computing*, pp. 226–234, San Diego, CA, 1993.
2. R. B. Borie, R. G. Parker, and C. A. Tovey. Automatic generation of linear-time algorithms from predicate calculus descriptions of problems on recursively constructed graph families. *Algorithmica*, 7, pp. 555–581, 1992.
3. S. Fiorini and R. J. Wilson. *Edge-Colourings of Graphs*. Pitman, London, 1977.
4. A. Gibbons and W. Rytter. *Efficient Parallel Algorithms*. Cambridge Univ. Press, Cambridge, 1988.
5. S. L. Hakimi and O. Kariv. On a generalization of edge-coloring in graphs. *Journal of Graph Theory*, 10, pp. 139–154, 1986.
6. I. Holyer. The NP-completeness of edge-colouring. *SIAM J. Comput.*, 10, pp. 718–720, 1981.
7. J. JáJá. *An Introduction to Parallel Algorithms*. Addison-Wesley, New York, 1992.
8. S. Nakano and T. Nishizeki. Scheduling file transfers under port and channel constraints. *Int. J. Found. of Comput. Sci.*, 4(2), pp. 101–115, 1993.
9. T. Nishizeki and N. Chiba. *Planar Graphs: Theory and Algorithms*. North-Holland, Amsterdam, 1988.
10. B.A. Reed. Finding approximate separators and computing tree-width quickly. In *Proc. of the 24th Ann. ACM Symp. on Theory of Computing*, pp. 221–228, 1992.
11. V. G. Vizing. On an estimate of the chromatic class of a p-graph. *Discret Analiz*, 3, pp. 25–30, 1964.
12. X. Zhou, S. Nakano, and T. Nishizeki. A linear algorithm for edge-coloring partial k-trees. In *Proc. of the First Europian Symposium on Algorithms, Lect. Notes in Computer Science, Springer-Verlag*, 726, pp. 409–418, 1993.
13. X. Zhou, S. Nakano, and T. Nishizeki. A parallel algorithm for edge-coloring partial k-trees. In *Proc. of the Fourth Scandinavian Workshop on Algorithm Theory, Lect. Notes in Computer Science, Springer-Verlag*, 824, pp. 359–369, 1994.
14. X. Zhou and T. Nishizeki. Optimal parallel algorithms for edge-coloring partial k-trees with bounded degrees. *IEICE Trans. on Fundamentals of Electronics, Communication and Computer Sciences*, E78-A, pp. 463–469, 1995.

Spanning Closed Trail and Hamiltonian Cycle in Grid Graphs *

Hwan-Gue, Cho[1] and Alexander Zelikovsky[2]

[1] Dept. of Computer Science, Pusan National University,
Pusan 609-735, Korea, hgcho@hyowon.cc.pusan.ac.kr
[2] Institute of Mathematics, Academiei 5
Kishinev, Moldova, 17azz@mathem.moldova.su

Abstract. In this paper we study a trail routing and a hamiltonian cycle in a class of grid graphs, *polycube* and *polymino*. A *Spanning closed trail* is an eulerian subgraph containing all vertices of a given graph. For general grid graphs we prove that the problem of finding that trail is NP-complete and for a wide subclass of grid graphs, called *polymino*, we give an an optimal algorithm if it exists. For polycube graphs we prove that every polycube has a spanning closed trail. Finally we show that a graph product G to a simple path with length n, $G \times P_n$, is hamiltonian for all $n \geq 2$, if G is a polymino with a perfect matching.

1 Introduction

Spanning closed trail(for brevity trail) can be considered as one relaxation between two cycles, hamiltonian and eulerian cycle. While hamiltonian routing is to visit all vertices(at most once) with some edges(at most once) and eulerian routing is to visit all edges(at most once) with all vertices(at least once), this trail routing is to visit all vertices(at least once) with some edges(at most once).

We know the spanning trail can be applied in practice such as *one-way street problem* and *optimal page retrieval problem*[10, 8]. It is obvious that if a graph G has a trail, it is *super eulerian* graph[4]. There are interesting charterizations on super eulerian graphs[1, 2, 3, 4]. They gave some sufficient conditions for super eulerian graphs, but did not show any algorithms to find a trail.

2 Spanning Closed Trail in Polyminos

In the following a *path* means an alternating sequence of vertices and edges and a *trail* is a path with all distinct edges[5]. The planar *rectangular grid* consists of vertices $\{(x, y)| \ x, y \in Z\}$ and edges $\{(u, v)| \ u, v \in V, |u - v| = 1\}$. A *grid graph* $R = (V, E)$ is a finite subgraph of the planar rectangular grid [7]. A bounded face F of R with exactly four vertices on its boundary is called a *cell*, all other

---

* Parts of this work were done while authors visited in Max-Planck-Institut für Informatik, Saarbrücken, Germany.

bounded faces are called *holes*. A grid graph is a *polymino* if every edge is one of boundary edges of a unit cell. A *simple polymino* is a 2-connected grid graph without holes. This section is devoted to the following.

Problem 1. **The Trail Problem:** Given a grid graph R, find a trail for R if it exists.

Note that if a hamiltonian cycle exists, then it is the shortest trail. The Hamilton circuit problem is NP-complete for grid graphs, but it is still unknown whether this problem is NP-complete for simple polyminos (see [6]).

We show that the Trail Problem can be solved in linear time for simple polyminos and so called *cell connected* polyminos with holes. From the other side we prove that the Trail Problem for arbitrary polyminos is NP-complete.

2.1 Trails in Simple Polyminos

Let R be a simple polymino bounded by a simple polygon P and let T be its trail. Since T is a planar eulerian graph, the dual graph for T is bipartite. Color all faces of T into two colors black and white, such that the unbounded face is white. Any cell obtains the color of the face it belongs to. There is one to one correspondence between trails of R and *proper* colorings of cells such that

(1) the cells of the unbounded face are white;
(2) the edges of black cells form a connected subgraph;
(3) each vertex belongs to at least one black and one white cell.

We need the following definitions. The *cell graph* of R is a graph on cells of R such that two cells are adjacent if they have an edge in common. So the cell graph is a dual graph for R without a vertex corresponding to the unbounded face. R is called *cell-connected* if its cell graph is connected. A *corner* of a polymino is a vertex of degree 2. An edge e of R is a *neck* if $e \notin P$ and the both ends of e belong to P. *Inner* vertices of R do not belong to its boundary P. An inner vertex is an *eye* if all its four neighbors belong to the boundary. A *block* is a polymino bounded by a square with a side of size 2. A *dragon* is a simple polymino built from blocks such that any two blocks are disjoint or may have in common exactly one edge (Fig. 1). Inner vertices of a dragon are *eyes*. A *head* of a polymino R is a block with one edge in common with the rest of R. Heads are darkened in Fig. 1. The following lemma is obvious.

Lemma 2. *Any dragon except a block has at least two heads.*

Lemma 3. *Dragons have no trails.*

Proof. In a proper coloring all corner cells of a head should be black and the rest cells of a head should be white. Therefore if we remove a head from a dragon, then the rest dragon should have a proper coloring also (Fig. 2 (a)). Removing heads one by one leads to a block which has no proper coloring, since all its cells should be black and eye cannot belong to a white cell.

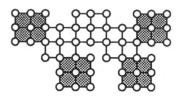

Fig. 1. A dragon with dark heads.

The main result of this section says that the converse statement is also true.

Theorem 4. *A simple polymino has a closed spanning trail if and only if it is not a dragon.*

To prove Theorem 4 we show the way how to find a trail if it exists.

Theorem 5. *A spanning closed trail for a simple polymino, if it exists, can be found in linear time.*

Proof. First we consider the case of a polymino R without eyes and necks. We show that the *chess* coloring of R defined below is proper. Initially all cells of the grid are white. We color black a cell c of R if (i) c contains a corner of R and/or (ii) the sum of coordinates of the left low corner of c is even.

The chess coloring satisfies the condition (2) since R does not contain necks and the cell graph of R is connected. Note that each vertex belongs to at least one black cell. Any boundary vertex obviously belong to at least one white cell of the unbounded face. Any inner vertex has at least one inner neighbor. These two inner vertices belong to two cells with an inner edge in common. One of these two cells is white. Thus we proved that the chess coloring is proper.

Now assume that R is not a dragon. Below we present several linear-time procedures which transform R to a set of polyminos without eyes and necks. Cut_{head} replaces a head H of R by one edge (Fig. 2 (a)). Walking along the boundary polygon P we apply Cut_{head} to the head met first. If as a result a new head appears, then cut this new one and so on. If a new head does not appear, then continue our walk along P until we return to the initial vertex.

Cut_{neck} (Fig. 2 (b)) gives two simple polyminos from the given polyminos such that both contain the neck edge and call Cut_{head}.

Still there are two cases when R contains eyes. A *bump* is a union of three corners slightly different from a head (Fig. 3 (a)). A *tumor* looks like a bump inside the polymino). Cut_{bump} and Cut_{tumor} remove a block adding two or four edges to P, respectively (Fig. 3(a) and 3(b)).

It is not difficult to see that any proper coloring of polyminos obtained after application of procedures above can be extended to a proper coloring of the initial polymino. Thus we apply to R Cut_{heads}, Cut_{neck} for each neck every time removing heads appeared, and finally Cut_{bumps} and Cut_{tumors} for each bump and tumor, respectively.

Simple polyminos resulted do not contain eyes and necks, so the chess coloring is proper for them. Then we can easily obtain a proper coloring of the initial simple polymino running cut procedures in the reverse order.

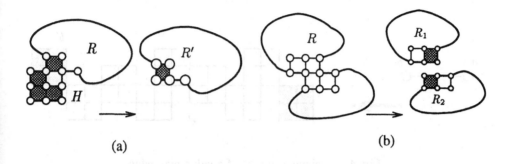

(a)   (b)

Fig. 2. Procedures for (a) Cut_{head} and Cut_{neck}.

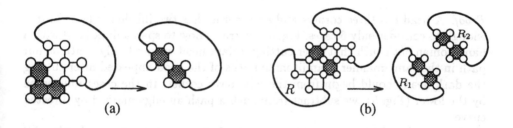

(a)   (b)

Fig. 3. Procedures for (a) Cut_{bump} and (b) Cut_{tumor}.

2.2   The Trail Problem in Polymino with Holes

In this subsection we will prove a negative result on Trail Problem (TP) for grid graphs with holes. This result for TP is somehow stronger than the negative result for the Hamilton circuit problem since it holds not for arbitrary grid graphs but for polyminos. We also show that the Trail Problem for cell-connected polyminos can be solved in linear time.

Theorem 6. *The Trail Problem in polymino with holes is NP-complete.*

Proof. Our proof is similar to the proof in [6]. We transform the Hamilton Circuit Problem (HCP) in planar cubic maps to Trail Problem. For each vertex of a cubic map we assign points of a block. For each edge of a cubic map M we replace it by a dragon with two heads, i.e. *a chain-dragon*. Let an edge e be incident to a vertex a in a cubic map. Then the corresponding block a' and chain-dragon e' have a common corner (Fig. 4).

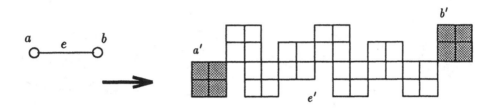

Fig. 4. A polymino image of a cubic map edge.

Lemma 7. *Any chain-dragon has a Hamiltonian path P which connects any pair a and b of corners from different heads. Moreover P may go along arbitrary two edges incident to a and b, respectively.*

Proof. A head has three corners and each corner has two incident edges. So it is enough to consider only 6 cases. Fig. 5 illustrates how to span a dark head with a hamiltonian path such that after cutting a dark head we need only continue our path in the same manner. Hamiltonian paths of the head appeared after cutting the dark head should begin from the left lower corner. In the cases illustrated by the lower (Fig. 5), we subtract from such a path an edge marked by directed curve.

Assume that M has a hamiltonian cycle. In the corresponding polymino M' we can span all images of vertices and Hamiltonian edges with the corresponding cycle. The rest of vertices, i.e vertices of chain-dragons, we can span with the corresponding trails, since we can omit one corner of such chain-dragon.

Suppose we have a trail T for the polymino M'. This trail T cannot span one chain-dragon unless it contains a hamiltonian path between the corresponding two corners of its heads. The chain-dragons which are spanned by paths form some cycle. This cycle should correspond to some hamiltonian cycle H in the original map M, since all other chain-dragons cannot connect images of vertices and T should be disconnected if H does not cover all points of M.

Still we can solve the Trail Problem in the *tree* polymino in which any 2-connected component is cell-connected.

Theorem 8. *The Trail Problem for tree polyminos can be solved in linear time. A tree polymino has a trail if and only if it does not contain a dragon as its 2-connected component.*

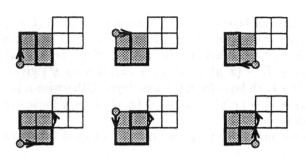

Fig. 5. A dark head spanned with a thick hamiltonian path.

Proof. Let R be a cell-connected polymino with at least one hole. We apply the algorithm for simple polyminos to R. A dragon may appear only while application of the procedure *cut necks*. Let R' be a dragon obtained after cutting a neck from R. R' can have only two heads and R should be a chain-dragon. By Lemma 7 we can span all points of R' by a hamiltonian path P connecting corners of two heads of R' which correspond to two vertices of the neck of R, Moreover we may lead P through two edges which correspond to the neck of R. This hamiltonian path corresponds to some hamiltonian cycle of R with a doubled neck.

In the following section we will consider Trail Problem in 3-dimensional grids, called *polycube* graph.

3 Spanning closed trail in polycube graphs

In this section we give an optimal algorithm for the trail in a class of *cubic grid* graphs. The cubic grid graph consists of vertices $\{ (x, y, z) | x, y, z \in Z \}$ and edges $\{ (p, q) | p, q \in V \text{ and } |p - q| = 1 \}$. A *polycube graph* $R = (V, E)$ is a finite subgraph of the cubic grid graph. If a polycube does not have any empty space in its internal body, then we call it *simple polycube*.

Previous sections proved that every simple polymino has a trail if it is not a dragon polymino. This section shows that every simple polycube has a trail, even though a polycube graph has a *dragon polymino* in one of its layers. In the following i-th layer of a polycube G denotes the plane polymino subgraph which is horizontally embedded in G with height i.

Lemma 9. *Let G_p be a polycube. If $(i + 1)$-th layer of G_p is not a dragon and there is a closed trail which spans upto i-th layer from the bottom, then there is a trail spanning upto $(i + 1)$-th layer.*

Proof. Let t_{i+1} be a trail in the $(i + 1)$-th layer. t_{i+1} can be found easily since $(i + 1)$-th layer is not a dragon. And let T_i be a closed trail upto i-th layer from 0-th layer. Since G_p is a polycube structure, at least one unit polycube is shared by every two adjacent layers. So there is an edge (x, y) in $(i + 1)$-th layer which has a corresponding edge (x', y') in i-th layer, where $(x, x'), (y, y') \in E(G_p)$.

Suppose that (x, y) is included in t_{i+1}. If (x', y') is not included in T_i, we know that $t_{i+1} + T_i - (x, y) + (x', y') + (x, x') + (y, y')$ gives a bigger closed trail for upto $(i + 1)$-th layer from the base layer. Otherwise if (x', y') is included in T_i, then $t_{i+1} + T_i - (x, y) - (x', y') + (x, x') + (y, y')$ also gives another trail upto $(i + 1)$-th layer.

If (x, y) is not included in t_{i+1}, we can get a closed trial similarly.

Lemma 10. *If D is a dragon polymino, then there is a spanning trail(not closed) which starts with x and ends at y, for every edge (x, y) of D.*

Proof. Let $[x, y]$-trail denote a spanning trail (not closed) which starts with x and ends at y in a layer. We construct $[x, y]$-trail by an constructive procedure on the number of blocks. In a single block B, it is easy to see that there is $[x, y]$-trail for every edge (x, y) of B.

We show how to find the $[x, y]$-trail for a given (x, y) edge in D. Let B_1 be a single block of D which contains (x, y). The whole D can be constructed by attaching blocks to B_1 one by one. Let $\{B_i\}$ denote a sequence block attached to rebuild D from B_1.

Now we consider the step of attaching B_2 to one of boundary edges in B_1. Let (p, q) be the neck edge shared by two heads B_1 and B_2. Since there is $[p, q]$-trail in B_2 for any boundary edge (p, q), we can merge these two trails, $[x, y]$-trail of B_1 and $[p, q]$-trail of B_2 into a bigger $[x, y]$-trail which spans two headed dragon $B_1 + B_2$. By this inductive procedure we can make $[x, y]$-trail for D by attaching each B_i to a boundary edge of $[x, y]$-trail.

Lemma 11. *Let G_P be a polycube. Suppose that its $(i + 1)$-th layer is a dragon. If there is a closed trail which spans upto i-th layer, then we can make a closed trail upto $(i + 1)$-th layer.*

Proof. According to the proof of Lemma 9, we know that there exists at least one edge (x, y) in $(i + 1)$-th layer such that $(x, x'), (y, y') \in E(G_p)$ and (x', y') is an edge of i-th layer. Then we can find $[x, y]$-trail in $(i + 1)$-th layer by Lemma 10. Let t_{i+1} be $[x, y]$-trail of $(i + 1)$-th layer and T_i be the closed trail upto i-th layer. If (x, y) is not included in T_i, then $t_{i+1} + T_i + (x, x') + (y, y') + (x', y')$ gives a closed trail spanning upto $(i + 1)$-th layer from base layer. Otherwise if (x, y) is included in T_i, then by the similar way of proof in Lemma 9, we can construct another trail for upto $(i + 1)$-layer.

Lemma 12. *Let G_p be a polycube graph and both i-th and $(i+1)$-th layer are dragons. Then there is a closed trail in both layers.*

Proof. There are a pair of corresponding edges such that (x, y) and (x', y') is an edge of i-th and $(i + 1)$-th layer, respectively and $(x, x'), (y, y') \in E(G_p)$ by Lemma 9. By Lemma 10 we can find $[x, y]$-trail in i-th layer and $[x', y']$-trail in $(i + 1)$-th layer. Then adding two additional edges such that $[x, y]$-trail + $[x', y']$-trail + $(x, x') + (y, y')$ gives a closed trail which spans both dragon layers.

The following theorem gives the main result of this section.

Theorem 13. *Every simple polycube has a spanning closed trail. And we can find it in linear time.*

Proof. In order to find a trail of a simple polycube, we construct it by merging trails of each layer one by one from the base layer. Combining all above lemmas in this section completes this theorem.

For merging trails, firstly we have to find edge (x, y) in $(i + 1)$-th layer with the corresponding edge (x', y') in i-th layer. It is easy to find such an edge by looking all edges in one layer. This work can be done within $O(n_i)$ steps, where n_i is the number of edges in i-th layer.

To merge two trails into one trail, it is enough to consider at most four edges by Lemma 9 and 10. Therefore the total number of edges examined to find a closed trail for a polycube G_p is $O(|E(G_p)|)$.

4 Hamiltonian cycle in polycube graphs

Till now there are a little results on the hamiltoncity of grid graphs. They gave a necessary and sufficient condition for the existence of the hamiltonian path in rectangular grids [6]. Also they have shown that the hamiltonian problem for general grids is NP-complete. But it should be noted that their grid graph is different from our polymino graphs in that our polymino model does not allow any degenerated faces or holes.

This section introduces *n-layer graph* as one extension of planar grids. For a given planar graph G, the n-layer graph of G is defined as the graph product $G \times P_n$ denoted as $(G \times n)$, where P_n is a linear path with length n.

A polymino is said to have an even(odd) parity, if it is (not) a bipartite graph with even partition of vertices. We know that if a polymino has an even parity then it has a perfect matching.

Though little was known about the properties of polymino, we conjecture that deciding hamiltoncity of a simple polymino is NP-complete even though that graph has an even parity without holes. The main result of this section is stated.

Lemma 14. *If G is a simple polymino, $(G \times 2n)$ is hamiltonian for $n \leq 1$.*

Proof. It is trivial that a rectangular grid with an even length side is hamiltonian. So we cut G into vertical layers, find hamiltonian cycles for each layer and merge all hamiltonian cycles into one.

Theorem 15. *If G is a simple polymino with a perfect matching M, then n-layer $(G \times n)$ is hamiltonian for $n \geq 2$.*

Proof. It is enough to prove $(G \times 3)$ is hamiltonian. At first we find a perfect matching M. Then we get Depth First Search tree with matching M, DFS_M, which is a spanning tree including all M. This DFS_M searching is different to the original DFS in that DFS_M prefers the matched edge in searching procedure.

Assume DFS_M procedure just has visited vertex p and there are several unvisited edges incident to p. If there is an edge $(p, q) \in M$ and it is not still included in the current search tree, then we include (p, q) rather than visiting other unmatched edges. If there is no matched edges incident to p, then we take an arbitrary unmatched and unvisited edge.

Let S_M be a spanning tree obtained by DFS_M searching. It should be noted that S_M tree has no vertex whose degree is greater than 4 since G is a 2-connected polymino. If a vertex x is of degree 4 in S_M, then one of its adjacent vertex should have been visited previously along the path from the other adjacent vertices of p in S_M due to the 2-connectivity of G.

Fig. 6 shows that x and y must be in the same 2-connected component, so there is a path connecting x and y. In S_M p can not be of degree 4 by the property of DFS_M. Note that thick edges (solid thin edges) denote M (S_M edges not included in M).

Now we construct a hamiltonian cycle with S_M. It is obvious that if $|M| = 2$ then $(S_M \times 3)$ has a hamiltonian cycle. As an induction base we assume that if $|M| \leq i$ then $(S_M \times 3)$ is hamiltonian. Suppose that $|M| = i$ and S_M is given, then we make another spanning tree S'_M by deleting a leaf edge $m = (u, v)$ of S_M. Let u be a leaf vertex in S_M, then it is obvious that $m = (u, v)$ should be a matched edge.

By induction hypothesis we can construct a hamiltonian cycle in $(S'_M \times 3)$. Now we restore m to S'_M to make a bigger hamiltonian cycle for $(S_M \times 3)$. In order to restore (u, v) with preserving hamiltoncity, we prepare L and T-type components. Note that the degree of v of m is 2 or 3 in S_M. We can easily select L or T-type path to extend the hamiltonian cycle of S'_M in each edge restoring step. Illustrations of degree 2, 3 connections in Fig.6 proves its validity and correctness. This completes the proof.

Corollary 16. *If G is a 2-connected planar graph with a perfect matching, then $(G \times n)$ is hamiltonian.*

5 Concluding Remarks

In this paper we proved that all polymino graphs except dragons are super eulerian and every simple polycube is super eulerian. And our proof procedure implies an optimal linear time algorithm. And we gave one class of hamiltonian graph such as $(G \times n)$ where G is 2-connected planar graph with a perfect matching. Now we are trying to find the longest trail in a simple polymino. We hope to see other applications for trails and a hamiltonian routing in a polycube.

References

1. P. A. Catlin, *A reduction method to Find Spanning Eulerian Subgraphs*, J. Graph theory, Vol.12, No. 1, pp.29-44, 1988.

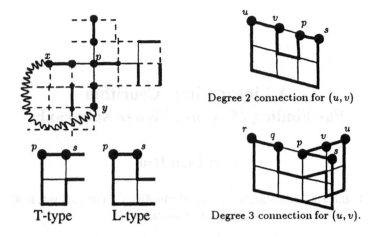

Degree 2 connection for (u, v)

T-type   L-type   Degree 3 connection for (u, v).

Fig. 6. (p, s), (r, q) are the matched edges. $m = (u, v)$ is restored to S'_M

2. P. A. Catlin, *Spanning trails* , J. Graph Theory, vol.11, No.2, pp.161-167, 1987.

3. P. A. Catlin, *Spanning Eulerian subgraphs and matching* , Discrete Math., 76, pp. 95-116, 1989

4. Z. H. Chen, *Spanning closed trails in graphs.* Discrete Math. 117, pp. 57-71, 1993.

5. F. Harary, *Graph Theory*, Addison-Wesley Pub., 1972.

6. A. Itai, C. H. Papadimitriou and J.L. Szwarcfiter, *Hamiltonian Paths in Grid Graphs.*, SIAM, Comp., Vol., 11, No.4, Nov., pp.676-686, 1982.

7. M. Kaufmann and K. Mehlhorn, *A linear-time algorithm for the Homotopic routing in grid graph* SIAM J. Comput. vol. 23, No. 2. pp. 227-246, Apr. 1994.

8. T. H. Lai and L. S. Wei, *The edge hamiltonian path problem is NP-complete for bipartite graphs* Infom. Process. Lett., 46, pp. 21-26, 1993.

9. C.H. Papadimitriou and U.V. Vazirani, *On two geometric problems related to the traveling salesman problem*, J. of Algorithms, vol.5, pp.231-246, 1984.

10. F. S. Roberts, *Graph Theory and Its applications to Problems of Society*, CBMS-NSF Regional Conference Series in Applied Mathematics, Vol. 29, SIAM, 3rd, pr. 1993.

A Linear Time Algorithm
For Finding Maximal Planar Subgraphs

Wen-Lian Hsu[1]

Institute of Information Science, Academia Sinica, Taipei, Taiwan, ROC
email: hsu@iis.sinica.edu.tw

ABSTRACT. Given an undirected graph G, the maximal planar subgraph problem is to determine a planar subgraph H of G such that no edge of G-H can be added to H without destroying planarity. Polynomial algorithms have been obtained by Jakayumar, Thulasiraman and Swamy [6] and Wu [9]. O($m\log n$) algorithms were previously given by Di Battista and Tamassia [3] and Cai, Han and Tarjan [2]. A recent O($m\alpha(n)$) algorithm was obtained by La Poute [7]. Our algorithm is based on a simple planarity test [5] developed by the author, which is a vertex addition algorithm based on a depth-first-search ordering. The planarity test [5] uses no complicated data structure and is conceptually simpler than Hopcroft and Tarjan's path addition and Lempel, Even and Cederbaum's vertex addition approaches.[1]

1. Introduction

Given an undirected graph, the *planarity testing problem* is to determine whether there exists a clockwise edge ordering around each vertex such that the graph can be drawn in the plane without any crossing edges. Linear time planarity testing algorithm was first established by Hopcroft and Tarjan [4] based on a "*path addition* approach". A "*vertex addition* approach", originally developed by Lempel, Even and Cederbaum [8], was later improved by Booth and Lueker [1] to run in linear time using a data structure called a "PQ-tree". These approaches are quite involved. We developed a very simple linear time algorithm [5] based only on a depth-first search tree. The key to our approach is to add vertices according to a postordering obtained from a depth-first-search tree. Given an undirected graph G, the *maximal planar subgraph problem* is to determine a planar subgraph H of G such that no edge of G-H can be added to H without destroying planarity. Polynomial time algorithms have previously been given by [2,3,6,9] and a recent O($m\alpha(n)$) algorithm was obtained by La Porte [7]. We solved the MPS problem in linear time by modifying the algorithm in [5]. In the next section we briefly describe the idea in [5]. Our MPS algorithm is discussed in Sections 3 and 4. Finally, the time complexity is discussed in Section 5.

---

[1]The research of this author was supported in part by the National Science Council of the Republic of China.

2. A Review of Our Modified Vertex Addition Approach for Planarity Testing

In the two previous approaches [4,8] of planarity test, the partial subgraph constructed at each iteration is always connected. The st-numbering of Lempel et al's approach further requires that those vertices not added induce a connected subgraph. We adopted a "vertex addition" approach which only requires that those vertices "not added" induce a connected subgraph. Thus, a simple postordering of a depth-first search tree of G suffices.

Let G_i be the subgraph at the i-th iteration consisting of the first i vertices and those edges among them. In our approach G_i may be disconnected, but the embedding for each biconnected component of G_i, once determined, is never changed.

Assume the given graph G is biconnected and the degree of each vertex is at least 3. To simplify the discussion we use a *generalized forest representation F_i* for each G_i. Each node in the forest F_i represents either

(a) an original vertex of G (denoted by a *v-node*) adjacent to vertices not in G_i.
(b) a biconnected component of G_i whose planar embedding has already been determined (denoted by a *c-node*).

Let T^* be a rooted depth-first search tree of G. All edges in T^* are called *tree edges* and the other edges of G are called *back edges*. Each tree edge is directed away from the root and each back edge is directed form a node to one of its ancestors. Assume the vertices of G are labeled by a postordering of T^*. Thus, the label of a parent is always larger than those of its children. Sort the neighbors of every vertex in $O(m)$ time.

The vertices are added one by one according to their ascending label. Consider the beginning of the i-th iteration of the algorithm at which point vertex i is to be added to G_{i-1}. We need to determine an ordering of edges emanating from vertices in G_{i-1} to i. If a vertex in the current forest has a back edge pointing to i, then there must exist a tree edge directed from i to its root. List the trees of F_{i-1} whose roots are adjacent to i in the current generalized forest as $T_1, T_2, ..., T_r$. Let S be a subset of vertices. Denote by *G[S]* the subgraph of G induced on S. Since i is an articulation vertex of the induced subgraph $H_i = G[\{i\} * T_1 * ... * T_r]$, H_i is planar if and only if each $G[\{i\} * T_j]$ is planar, j = 1, ..., r. Hence, it suffices to consider the embedding problem of each $G[\{i\} * T_j]$. We shall find an embedding of $G[\{i\} * T_j]$ that corresponds to a partial embedding of G. Denote the root of T_j by r_j.

Define the *external degree* of a vertex at iteration i to be the the number of its neighbors among $\{i+1,...,n\}$. Since some of those vertices in G_{i-1} with 0 external degree do not have to be examined for future embedding, we shall apply a **vertex contraction** procedure to eliminate them. The contraction procedure replaces a connected subgraph by a contracted edge, which allows us to recognize and embed the planar graph more efficiently. Let T_j' denote the tree after the vertex contraction from T_j.

At the beginning of the i-th iteration reduce the external degrees of all neighbors of vertex i in F_i by 1. The contraction procedure starts by marking all vertices of T_j that are adjacent to i. Note that a vertex which is not marked at this stage can become marked later through a contracted edge. There are two types of contraction:

contracting vertices on a path and contracting vertices on a cycle. We shall consider the former contraction first. Scan all marked vertices in ascending order. If a marked vertex u is a leaf with external degree 0, then contract u by deleting u and marking parent(u). The edge from parent(u) to i now represents the path: **parent(u)-u-i.**

'During the contraction process, if an internal vertex u of T_j becomes a leaf, then all of its descendants must have external degree 0. If several (but not all) children of a vertex u have been contracted when u is scanned, then the edge from u to i represents the connection between i and every vertex in those contracted children subtrees. Each back edge from a vertex u in T_j' to i actually represents the edge connection between i and all contracted vertices in a partial subtree of T_j rooted at u.

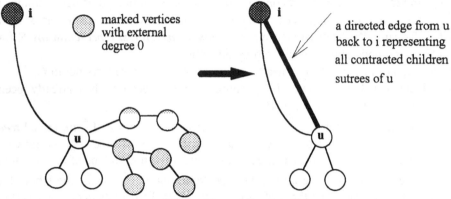

marked vertices with external degree 0

a directed edge from u back to i representing all contracted children sutrees of u

Fig. 2.1 Vertex Contraction

Let u be a marked v-node in T_j' none of its descendents is a contracted c-node. Vertex u is said to be a **terminal node** if none of its descendants is marked. We can show that, if G is planar, then there exist at most two terminal nodes in T_j'.

Now consider the embedding at the i-th iteration. Consider two cases: (1) there is only one terminal node u. Let P be the path from r_j to u in T_j'; (2) there are two terminal nodes u and u'. Let P be the unique path from u to u' plus the edge (u',i). We shall describe how to embed path P and those contracted edges in two cases.

2.1 A Special Case of the Embedding

Consider the special case that path P contains only v-nodes. The two cases of path P in the previous section are analogous. Hence, we shall only describe the first case. The back edge (u,i) together with the tree edge (i,r_j) and edges in P form a cycle **C**. Let u_0 (= u), u_1, ..., u_t be the marked nodes in P. We first describe how to construct an internal planar embedding of the biconnected component **H** composed of C and those contracted vertices represented by (u_1,i), ..., (u_t,i). Embed the contracted edges (u_1,i), ..., (u_t,i) on the same side of C. Convert each edge (u_s,i) back to its uncontracted partial subtree in T_j and connect all vertices in the subtree to i. There are many different ways to achieve this. One way is to lay the subtree down on the plane (namely, specify a left-right children relationships) and to arrange a connecting order based on a preorder traversal.

To represent the biconnected component H in the forest after the embedding of path P, let $k_1, ..., k_s$ be vertices (ordered along the cycle) of C which are adjacent to vertices outside H (namely, either incident to a tree edge or adjacent to a vertex in $\{i+1,...,n\}$). Contract all vertices in $C-\{k_1, ..., k_s\}$ one by one by deleting such a vertex and connect its two neighbors in C. Store $k_1, ..., k_s$ in a circular linked list as the *representative cycle C(H)* for H. C' will be denoted by a *c-node* in the forest F_i. For future iterations, one can concentrate only on the representative cycle for H.

To maintain the forest structure when more than one T_j is considered, make a copy i_j of i in each T_j and change all back edges (u,i) from vertices in T_j to (u,i_j). Connect i_j with i using a **virtual tree edge**. Hence, i_j (instead of i) is placed into the cycle C. A tree edge of T_j, once becomes an edge in H, is no longer considered as a tree edge.

2.2 The General Embedding

Now, consider the general case in which path P contains some c-nodes. A node in P which is not the end node is called an *intermediate node* of P. In a path P the representative cycle of each intermediate c-node H contains exactly two vertices (called *high P-vertex h_1* and *low P-vertex h_2*, respectively, according to their labels; in case H is an end c-node, then only the high vertex h_1 exists) adjacent to the two tree edges of P. Each v-node in P and each high, low P-vertex of a c-node in P is referred to as a P-vertex. For every non-P-vertex u in the representative cycle define the *upward* (resp. *downward*) *direction* as the direction which leads u first to the high (resp. low) P-vertex along the cycle.

In addition to the above vertex contraction procedure, we need to contract vertices in the c-node as follows. Let u be a marked non-P-vertex in a c-node which is not incident to any tree edge nor adjacent to vertices in $\{i+1,...,n\}$ (a vertex satisfying the latter condition is said to be *saturated*). If u has only one neighbor in the c-node (which must be the high P-vertex), then delete u and mark the high P-vertex; otherwise, let u_1, u_2 be the two neighbors of u in the representative cycle of H. If one of u_1 or u_2 is a marked non-P-vertex, then contract u by deleting u and connecting u_1, u_2 by an edge. This edge now represents the path: u_1-u-u_2. It is easy to verify that the contracted graph is planar iff the uncontracted one is. Note that an entire c-node could be contracted to a single vertex through this operation. Also, after the contraction, no two saturated vertices can be adjacent to each other.

A c-node is marked if one of the vertices in its representative cycle is marked. A marked c-node is a *terminal c-node* if none of its descendants is marked. A marked v-node none of whose descendants are marked is called a *terminal node*. A marked non-P-vertex u in a c-node is a *terminal node* if either it has non-zero external degree or it is incident to a tree edge. A marked P-vertex in a c-node is a *terminal node* if there exists a non-P-vertex which is a terminal node by the above definition.

Obviously, only "one-half" of the c-node can be embedded inside the cycle formed by path P and the edge from the last marked vertex of P to i. It is quite easy to determine which half should be embedded inside by the following Theorem 2.3.

Theorem 2.2. *If G is planar, then there exist at most two terminal nodes in T_j'.*

Consider the contracted T_j'. Let P be a path connecting r_j to a terminal v-node.

Theorem 2.3. *Each marked vertex in a c-node H of P must be adjacent to the high P-vertex h_1. If H is an intermediate c-node in P, then h_1 and h_2 must either be adjacent to each other or to a saturated vertex in C.*

We now describe the embedding of the current graph at iteration i. Refer to the "outside face" as the one containing all vertices in $\{i+1,...,n\}$. The reason that these vertices must be contained in one face is that they form a connected induced subgraph of G. Consider two cases:

Case 1. There is only one terminal v-node u.

Let P be the path from r_j to u. Unlike the special case that P contains no c-node (where a unique cycle is formed), we now form two cycles based on P. The **inner cycle** C_1 is used to construct an embedding of the resulting biconnected subgraph; the **outer cycle** C_2 is used to construct the representative cycle for the corresponding c-node. For each intermediate c-node H of P, if $h_1(H)$ is adjacent to $h_2(H)$, define $P_1(H)$ to be the edge $(h_1(H),h_2(H))$, define $P_2(H)$ to be the other half of C(H) from $h_1(H)$ to $h_2(H)$; otherwise, let u(H) be a saturated vertex adjacent to both $h_1(H)$ and $h_2(H)$ (by Theorem 2.3 such a vertex must exist), define $P_1(H)$ to be the set of edges $\{(h_1(H),u(H)),(u(H),h_2(H))\}$, define $P_2(H)$ to be the other half of C(H) from $h_1(H)$ to $h_2(H)$.

Case 2. There are two terminal vertices u and u'.

In case u and u' are on the same c-node, say m, then the outer cycle C_2 consists of the edges (u,i), (u',i) and the path on C(M) from u to u' not passing through $h_1(M)$; the inner cycle C_1 consists of the edges (u,i), (u',i) and the path on C(M) from u to u' passing through $h_1(M)$. Hence, assume u and u' are not on the same c-node. Let m be the least common ancestor of u and u' in T_j'. If m is a v-node then consider the unique path from u to u' and, similar to Case 1, we can form the outer cycle and inner cycle. If m is a c-node, then let v and v' be the vertices connected to u and u' respectively. Then we can form the cycles similar to the above description except that on the c-node M, the outer cycle uses the path on C(M) from v to v' not passing through $h_1(M)$; the inner cycle uses the path from v to v' passing through $h_1(M)$. A necessary condition for the graph to be embeddable is that all vertices in C_1-C_2 are saturated.

We now discuss the time complexity of the PT algorithm. The construction of a depth-first search tree and a postorder traversal takes $O(m+n)$ time. At each iteration, Let the degree of i in G_{i-1} be denoted by $\deg_{Gi}(i)$. Vertex contraction takes time proportional to $\deg_{Gi}(i)$. The terminal nodes can be identified by checking the smallest marked vertex in T_j'. In the embedding of a path P starting from a terminal vertex, each tree edge of P will become a non-tree edge. Summing over all iterations, the number of times a tree edge becomes a non-tree edge is at most $O(n)$.

The key saving in our algorithm is that, in each c-node h, the set of marked vertices with zero external degree in C(H) must be arranged consecutively and hence, can be easily determined. Furthermore, vertices in $P_2(H)$ never need to be traversed in composing the outer cycle. It suffices to find its two end vertices in the linked list.

Therefore, besides traversing the tree edges from the terminal vertices (which totaled to be $O(n)$), the amount of work involved in each iteration is proportional to $\deg_{G_i}(i)$. Hence, the complexity of the algorithm is $O(m+n)$.

3. The Maximal Planar Subgraph Algorithm

We now discuss our maximal planar subgraph (MPS) algorithm. It can be viewed as an extension of our planarity Testing (PT) algorithm . The basic idea is to emulate the steps in the PT algorithm. We shall describe one important property of planar graphs that froms the basis of the MPS algorithm. The terminal nodes in the PT algorithm can all be determined uniquely at each iteration. In fact, one can have the following theorem.

Theorem 3.1. *A graph G is planar if, in the above planarity test there are at most two terminal nodes at each iteration.*

By definition, terminal nodes are actually determined after the introduction of certain external edges (these are the edges inhibiting the contraction process). Since the appearance of any external edge in the final subgraph is still in question, we assume all external edges are likely to exist at any iteration. The algorithm will simply "reserve" two terminal node slots to be filled by some candidates later and proceed to the next iteration. This allows the MPS algorithm to proceed more or less like the PT algorithm. Note that as long as there are no more than two terminal nodes created, Theorem 2.2 will guarantee that the final subgraph is planar. The main difference now is that whenever a new edge is included by the MPS algorithm it will exclude a set of edges from further consideration. Such a set can be determined uniquely. Furthermore, no vertex contraction is applied.

The main idea of the MPS algorithm is to traverse the tree towards the root from those vertices adjacent to the new vertex. To accommodate for slot reservation, we need to label the vertices and biconnected components traversed in the algorithm so that each edge will be traversed a constant number of times.

Each tree in the current forest that contains vertices adjacent to the new vertex will produce at most one biconnected component containing that new vertex. The MPS algorithm will assume such a component is already formed (which is actually pending on the determination of terminal nodes) and each vertex of the boundary cycle is labeled with the new vertex. Thus, the next time any vertex, say u, of this component, say C, is traversed two things will happen: (1) vertex u will be identified to be lying on the boundary of the biconnected component formed by the current new vertex; (2) the traversal will continue with the new vertex of C, thus skipping all other vertices of component C.

Let i be the new vertex in the current iteration. Consider a tree with root x in the current forest that contains at least two vertices adjcent to i. Such a tree will create a biconnected component for i at this iteration. We shall associate the edge (i,x) with this component. The MPS algorithm traverses the vertices and edges of the tree as follows. It first picks the neighbor of i with the lowest order, say u, in the tree and traverses the unique path from u to x. Each vertex v on that path will be labeled with (i,x,d), where d is the distance between v and u measured by the number of edges along the path.

Since the postordering guarantees that vertices with lower order will be considered before those with higher orders, those neighbors of i along the path from u to i will be labeled and they will no longer be considered at this iteration. After the unique path from u has been traversed the algorithm will pick the next lowerest unlabeled vertex, say u', adjacent to i and traverse the unique path from u' to i. The traversal will terminate when it encounters a vertex already labeled. The algorithm then picks the next lowerest unmarked vertex and continues until all neighbors of i in this tree have been considered. Everytime a new edge is added to the subgraph some other edges could be deleted. We shall discuss the edge exclusion in the next iteration.

4. Determination of Terminal Nodes

As discussed in the last section, the determination of terminal nodes in any iteration depends on the inclusion of future edges. Thus, everytime a new edge is added to the subgraph G' some candidate edges in previous iterations will be excluded. This process is continued so that for each specific iteration the candidate edges are eliminated until no more than two terminal nodes can appear. Note that at the end of an iteration i all edges in G from i to its descendents have been considered and will be included in the subgraph G' if not already excluded in the process.

Each back edge can form a unique cycle with edges in the tree. Such a cycle is said to be a fundamental cycle.

Let i, j be two integers with i < j satisfying the following conditions:
(a) There is an edge (j,u) between j and a descendent u of i.
(b) Let x be the child of i on the unique tree path from u to i. A component associated with the edge (i,x) was created in the i-th iteration.
(c) i, j are the smallest such integers respectively.

We shall describe the process of edge addition with regard to those edges from j to the descendants of i. The MFS algorithm will traverse the unique tree path from u to i. Each vertex traversed will be labeled. Consider two cases: (i) there is an edge in G' from i to a descendent of u; (ii) no such edge exists in G'. As discussed in the last section each vertex encountered will be labeled with (i,x,d), where d is the distance between v and u measured by the number of edges along the path. Whenever a vertex, say v, encountered has m(v) = (i,x,d') we check m(i,x). In the discussion below we shall describe the exclusion of candidate edges. Consider the following three cases:

(1) m(i,x) = 0: (j,u) is the first external edge added to G' with u being a descendant of i. We shall give vertices u and v special names, u_1 and v_1, respectively and change m(i,x) to be 1. A biconnected component for i will be formed as depicted in the left diagram of Figure 4.1, where the triple thick lines indicate that there is a biconnected component for i whose outer boundary has not yet been fixed. The MPS algorithm will then traverse the unique tree path from v to i (thus, edges on this path are traversed twice).

(2) m(i,x) = 1: some biconnected component for i has already been discovered.

We change m(i,x) to 2 and continue the traversal until a vertex, say v', labeled with (j,d") is encountered, where j ≠ i. The MFS algorithm will then visit i, skipping vertices on the unique tree path between v' and i. Consider the following two subcases:

(a) If v' = v₁, the unique path from v₁ to i (respectively, v) is "on" the outer boundary of the biconnected component for i (respectively, for j). However, the remaining part is still not fixed as depicted in the middle diagram of Figure 4.1.

(b) If v' ≠ v₁, then the unique path from v to v₁ is on the outer boundary of the biconnected component for i as well as the component for j and all the remaining part is not fixed as depicted in the right diagram of Figure 4.1.

We shall give vertices u and v special names, u₂ and v₂. Note that if a part is wrapped inside the biconnected component for j, then it is immaterial which path in that part should be on the outer boundary of the component for i. Furthermore, only descendents of v will remain as viable candidates and we can eliminate the external edges for all vertices wrapped inside the biconnected component for j.

Finally, the path from u₁ through v, v₂, i to j has its vertices divided into two segments: those in between u₁ and v₂ and those in between v₂ and j (even though the exact path from v₂ to i has not yet been fixed). Such a partition has the following effect: if any vertex in a segment is first encountered through a tree traversal when an edge (j',u') is added where u' is a descendant of i and j' is an ancester of j, then the other segment will be "wrapped" inside by the edge (j,u₂).

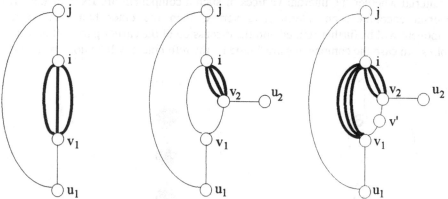

Fig. 4.1. The change of mark for the edge (i,x) from 0 to 2

(3) m(i,x) = 2: let the label of i be (j,d).

This is the situation that, at the j-th iteration some vertex (which is then named u₂) adjacent to j caused m(i,x) to be changed from 1 to 2. By the postordering of the depth-first-search tree, we can argue that v must be an ancestor of either v₁ or v₂. Hence v must be inserted into the segment between v₁ and v₂. Thus, this segment is futher divided. We call such a process a **refinement**. Futhermore, if there were similar vertices, say v', being inserted in between v₁ or v₂ before v, then v' cannot be an ancestor of v. Hence, the vertices inserted into the segment between v₁ and v₂ can be stored in two stacks, one for the ancestors of v₁ the other for the ancestors of v₂.

Thus, each such insertion takes only constant time. At the end of the insertion process, the MFS algorithm will then pick the next neighbor of j and continue with the traversing process.

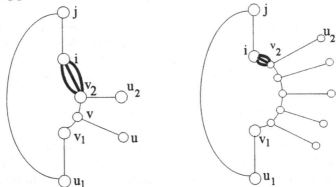

Fig. 4.2. The refinement of the outer boundary of a biconnected component

Each time such a refinement occurs, However, the argument at the end of (2) with regard to the wrapping effect of the edge (j,u) still holds except that there are more segments now to be wrapped. If a traversal does not run into any intermediate vertex of a biconnected component, then this component can be regarded as an "edge", one that connects the lower P-vertex to its high P-vertex. Hence, it is immaterial whether the internal vertices of such a component are fixed or not. If a traversal encounters an intermediate vertex, then the outer boundary of this component will be further refined and the discussion of the earlier part of this section applies. In case the components are "nested", the refinement will be done recursively.

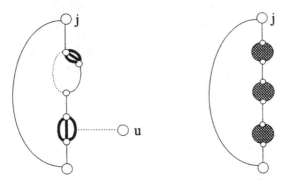

the outer boundary of each biconnected component can be regarded as an edge

Fig. 4.3. Nested biconnected components

5. Complexity Analysis

The argument for the linear complexity is mainly based on the fact that the number of times an edge is traversed is a constant. First of all, consider each tree edge. The first time a tree edge e is traversed is trickled by the addition of a back edge to the subgraph G'. Whenever edge e is wrapped inside a biconnected component, it is eliminated from further consideration. Otherwise, e will be traversed the second time when a potential biconnected component is formed for some edge (i,x). The third time the edge is traversed (if not already wrapped inside a component) is when it is on the outer boundary cycle for some component.

Now consider each back edge. The first time a back edge is added to the subgraph G' a biconnected component is formed and this edge is on the outer boundary cycle. Afterwards this edge will be traversed only when it is wrapped inside a component.

Finally, when a biconnected component is regarded as an "edge", the number of times such an edge will be traversed is also a constant, based on an argument similar to the beginning paragraph of this section. Hence, the time complexity of the MPS algorithm is linear.

REFERENCES

1. K. S. Booth and G. S. Lueker, *Testing the consecutive ones property, interval graphs, and graph planarity using PQ-tree algorithms*, **J. Comput. Syst. Sci.** 13(1976), 335-379.

2. J. Cai, X. Han and R. T. Tarjan, *An O(mlogn)-time algorithm for the maximal planar subgraph problem*, **SIAM J. Comput.** 22(1994), 1142-1162.

3. G. Di Battista and R. Tamassia [1989], *Incremental planarity testing*, in Proc. 30th Annual IEEE Symposium on Foundation of Computer Science, 436-411.

4. J. E. Hopcroft and R. E. Tarjan, *Efficient planarity testing*, **J. Assoc. Comput. Mach.** 21(1994), 549-568.

5. W. K. Shih and W. L. Hsu, *A simple test for planar graphs*, 1993, submitted.

6. R. Jakayumar, K. Thulasiraman and M. N. S. Swamy, *An $O(n^2)$ algorithms for graph planarization*, **IEEE Trans. CAD** 8(1989), 257-267.

7. J. A. La Poute, *Alpha-Algorithms for incremental planarity testing*, **STOC** 1994, 706-715.

8. A. Lempel, S. Even and I. Cederbaum, *An algorithm for planarity testing of graphs*, **Theory of Graphs**, ed., P. Rosenstiehl, Gordon and Breach, New York, 1967, 215-232.

9. W. Wu, *On the planar imbedding of linear graphs*, **J. Systems Sci. Math. Sci.** 5, 290-302, 1985.

Illumination with Orthogonal Floodlights[*]
(Extended Abstract)

James Abello[1], Vladimir Estivill-Castro[2], Thomas Shermer[3], Jorge Urrutia[4]

[1] Department of Computer Science, Texas A & M University,
College Station, Texas MS 3112, US. abello@cs.tamu.edu.
[2] Laboratorio Nacional de Informática Avanzada,
Rébsamen 80, Xalapa, Veracruz 91000, México. vlad@xalapa.lania.mx.
[3] School of Computer Science, Simon Fraser University,
Burnaby, B.C. V5A 1S6, Canada. shermer@cs.sfu.ca.
[4] Department of Computer Science, University of Ottawa,
Ottawa, Ontario K1N 6N5, Canada. jorge@csi.uottawa.ca.

Abstract. We provide the first tight bound for covering a polygon with n vertices and h holes with vertex guards. In particular, we provide tight bounds for the number of floodlights, placed at vertices or on the boundary, sufficient to illuminate the interior or the exterior of an orthogonal polygon with holes. Our results lead directly to simple linear, and thus optimal, algorithms for computing a covering of an orthogonal polygon.

1 Introduction

The question of guarding a polygonal art gallery has raised many problems ranging from polygon decomposition and problem complexity to combinatorial structure of visibility graphs [1, 11]. Moreover, the study of visibility in this type of geometric setting has not only been naturally motivated by many applications, but it has also been fundamental in developing many theoretical and practical results [13]. Despite the many variants of the problem, little regard has been placed to the assumption that guards can cover a complete 2π range of orientations around them. Rawlins [12] studied visibility along finitely oriented staircases and provided corresponding Art Gallery Theorems. However, only recently the question of studying visibility covers with *floodlights* (that is, static guards with restricted angle of vision) has been raised. The problem of covering a line with floodlights has been labeled the *stage illumination problem* [3, 5] while the problem of covering the plane has also been discussed [3, 14].

Observe that in O'Rourke's book [11] the first sentence of Chapter 5 says "One of the major open problems in the field of art gallery theorems is to establish a theorem for polygons with holes". In this paper we extend the tight bound of $\lfloor 3(n-1)/8 \rfloor$ [6] for the number of orthogonal floodlights sufficient to cover an orthogonal polygon with n vertices to the case of polygons with holes. In particular, we show that $\lfloor (3n + 4(h-1))/8 \rfloor$ orthogonal floodlights placed

[*] This work was partially carried out under grant CONACYT 3912-A9402 in México.

at vertices are always sufficient and sometimes necessary to cover an orthogonal polygon with n vertices and h holes. Moreover, we show that the above bound is reduced to $\lfloor (n+2h)/4 \rfloor$ if the orthogonal floodlights may be placed in the boundary of the polygon. Hoffman [8] proved the first tight bound previous to these results; namely, $\lfloor n/4 \rfloor$ guards are sometimes necessary and always sufficient to cover an orthogonal polygon independent of the number of holes. However, Hoffmann's results require that the guards be placed on points other than the vertices or the boundary. Hoffman also claims an $O(n^{1.5} \log^2 \log\log n)$ guard placement algorithm. Two groups independently [2] proved that $\lfloor (n+h)/3 \rfloor$ is the corresponding tight bound for general polygons; but again, guards must be allowed in the interior of the polygon. Other efforts to provide tight bounds for general polygons with holes and orthogonal polygons with holes have been successful for only special cases [7, 13]. Thus, our results provide the first theorem with a tight bound for vertex guarding polygons with holes, for all values of n and h. Moreover, our proofs lead to linear algorithms that are simple and practical since they avoid complex polygon decomposition (we do not need to triangulate or to find a quadrilateralization of the polygon). We observe that, for the case $h = 0$, the original art gallery theorem requires long proofs [10], however, our proofs are simple.

Moreover, from the point of view of total aperture used, the original art gallery theorems may require an aperture of $3\pi/2$ for each of the $\lfloor n/4 \rfloor$ floodlights resulting in a total aperture of $3n\pi/8$ (in fact, Fig. 2.12 of O'Rourke's book [11] illustrates a possible worst case where all floodlights that the algorithm finds must be of aperture $3\pi/2$). This can be regarded as unnecessarily inefficient, since for example, for polygons without holes our results demonstrate that only $\lfloor (3n-4)/8 \rfloor$ floodlights, each of aperture $\pi/2$, are needed even if we are required to place them at vertices. The total aperture obtained with our algorithms is half the naive use of the original art gallery theorem and corresponding algorithms.

In Section 2 we demonstrate that an aperture of $\pi/2$ is necessary for vertex floodlights to illuminate an orthogonal polygon. We also show that $\lfloor (3n + 4(h-1))/8 \rfloor$ orthogonal floodlights are sometimes necessary. Section 3 proves that $\lfloor (3n + 4(h-1))/8 \rfloor$ orthogonal floodlights placed at vertices are always sufficient and describes the linear algorithm to find the covering. Section 4 discusses the case where the floodlights are placed on the boundary. We show that $\lfloor (n+2h)/4 \rfloor$ floodlights are always sufficient and sometimes necessary. Section 5 provides an Art Gallery Theorem and a corresponding linear algorithm for illuminating the exterior of a simply-connected orthogonal polygon with holes; namely that $(n + 4)/2$ orthogonal floodlights are always sufficient and sometimes necessary. Section 6 provides some final remarks.

2 Necessity Bounds

Consider a polygonal art gallery given by an orthogonal polygon P in the plane with n vertices and h holes. There are n_0 vertices on the outer boundary and n_i vertices in the i-th hole, with $\sum_{i=0}^{h} n_i = n$. An α-floodlight is a lamp that shines light in a wedge of aperture α. We are interested in determining a set of α-floodlights that illuminate (cover) the interior or the exterior of P. If we require that the floodlights be placed at vertices, we will call them vertex floodlights, or v-floodlights for short. Note that no more than one floodlight is allowed at a point in P, otherwise, we have one floodlight of larger aperture. The following result demonstrates that an aperture of $\pi/2$ is the smallest aperture necessary for orthogonal polygons.

Theorem 1. *For all $\epsilon > 0$, there is an orthogonal polygon that can not be illuminated with vertex floodlights in each vertex with aperture $\frac{\pi}{2} - \epsilon$.*

Proof. Let λ be a line segment with endpoints a and b and middle point in the origin O. Let the slope of λ be $\delta < \epsilon$ and $\theta = \frac{\pi}{2} - \delta$. Let λ' be a segment orthogonal to λ with the same length as λ and also with middle point in the origin, but with endpoints c and d. We construct an orthogonal polygon with 12 vertices in the shape of an helix and whose reflex vertices are precisely a, b, c and d; refer to Fig. 1. If the prongs of P are large enough, the line of slope

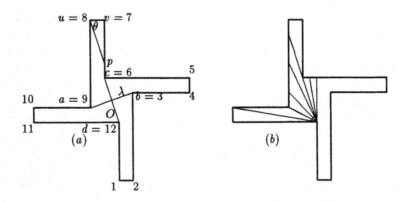

Fig. 1. This orthogonal helix requires four orthogonal v-floodlights.

$\theta + \pi/2$ through u intersects the side cv in a point p far enough from c; see Fig. 1 again. The reader can now verify that P can not be illuminated with floodlights of aperture $\frac{\pi}{2} - \epsilon$ in each vertex. $\square$

In what follows we concentrate on floodlights with aperture $\pi/2$; we will call them orthogonal floodlights. The helix of Fig. 1 will provide the building block to prove a bound of the number of orthogonal floodlights necessary to illuminate an orthogonal polygon with n vertices and h holes. First, we will prove the bound for the case $h = 0$, that is polygons without holes, and later the general case.

Consider again the orthogonal polygon in Fig. 1, where the prongs are long enough that a floodlight as shown in part (b) can illuminate at most one prong.

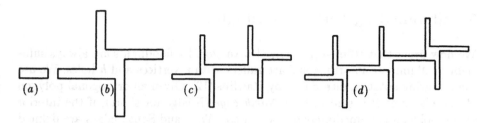

Fig. 2. Orthogonal polygons that require one $\pi/2$-floodlight for each prong.

Clearly, at least 4 orthogonal floodlights are required to cover this polygon. Now, consider the progression illustrated by Fig. 2. At each new stage, we merge a copy of the polygon in Fig. 1 by its left prong. Vertices 10 and 11 are identified with the two right-most vertices in the previous figure. It is not hard to see that at each stage, three more orthogonal v-floodlights are needed but only eight vertices are added. It now follows that $\lfloor (3n - 4)/8 \rfloor$ floodlights are necessary.

We now provide polygons that demonstrate the necessity of $\lfloor \frac{3n+4(h-1)}{8} \rfloor$ orthogonal floodlights. Consider the polygon P_{32} of Fig. 3 (a). It has 32 vertices and one hole. This polygon is constructed with four copies of the helix of Fig. 1 joined to form a polygon with one hole. Moreover, P_{32} requires 12 orthogonal floodlights, one for each of the eight prongs and one for each of the four alleys. This polygon provides the basic building block. For larger values of h, we join the the right-most edge of one copy of P_{32} with the left-most edge of another. For example, Fig. 3 (b) is a polygon with 60 vertices and two holes that requires 23 orthogonal floodlights. The process can be repeated to generate polygons with $32 + 28(h-1)$ vertices and h holes that need $12 + 11(h-1)$ orthogonal floodlights. Other values of n can be obtained eliminating the extra prongs.

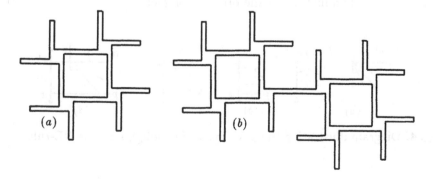

Fig. 3. Polygons that require one $\pi/2$-floodlight for each prong or alley.

3   Illuminating Vertex Floodlights

We now prove that $\lfloor (3n + 4(h-1))/8 \rfloor$ orthogonal v-floodlights are always sufficient to illuminate an orthogonal art gallery with n vertices and h holes. We use the following notation introduced by Rawlins [12]. Given an orthogonal polygon P, an edge e of P is said to be a *North edge* (N-edge for short), if the interior of the polygon is immediately below e. East, West and South edges are defined analogously. A vertex is said to be a *North-East vertex* (NE-vertex for short), if the polygonal edges that intersect at the vertex are an N-edge and an E-edge. NW-vertices, SE-vertices and SW-vertices are defined similarly. Since a vertex may also be convex or reflex, there are eight possible types of vertices in an orthogonal polygon. We define the following floodlight placement rule.

Definition 2. *North-East rule* (NE-rule): For each North edge e of the polygon, place a floodlight aligned with e at the East vertex of e. For each East edge e of the polygon, place a floodlight aligned with e at the North vertex of e. For a diagram of the NE-rule see Fig. 4 (a).

Lemma 3. *Let P be an orthogonal polygon with holes. The NE-rule produces an assignment of floodlights that illuminates the interior of P.*

Proof. Let p be a point in the interior of P. Let x be the first point in the border of P visible by a horizontal ray from p to the East. Clearly, x is in an E-edge e and p is visible from x. Consider a point x' in e just above x and consider the rectangle R with extreme points at x' and p; see Fig. 4 (b). Clearly, if x' is close enough to x, the rectangle R is contained in P. Consider moving x' North until it cannot be moved further without R leaving P. This happens because
- x' has reached the North vertex of e, in which case, p is illuminated by a floodlight at this point; see Fig. 4 (c), or
- the upper side of the rectangle R has coincided with a North edge, in which case, p is illuminated by a floodlight at the East point of this North edge; see Fig. 4 (d).

In both cases, p is illuminated and the proof is complete.   □

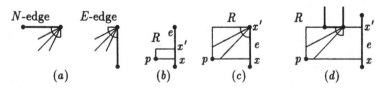

Fig. 4. Diagram illustrating the placement of floodlights by the NE-rule

Similarly, we can define a NW-rule, a SE-rule and a SW-rule, each illuminating the polygon. We are now ready to prove sufficiency.

Lemma 4. *Let P be an orthogonal polygon with n vertices and h holes. A total of $\lfloor(3n + 4(h-1))/8\rfloor$ orthogonal floodlights are sufficient to illuminate P.*

Proof. Illuminate the polygon P by each of the four rules proposed above. Let $\|X\|$ denote the number of floodlights used by the X rule. Note that each edge of the polygon receives at most two floodlights (for example, a N-edge receives a floodlight at its E-vertex in the NE-rule and at its W-vertex in the NW-rule) and the sets of floodlights of any pair of rules is disjoint. Moreover, in the NE-rule, a NE-convex vertex receives only one floodlight. Thus, the number $\|NE\|$ of floodlights used by the NE-rule is given by

$$\|NE\| = \|SE\|_r + \|NW\|_r + \|NE\|_c,$$

where $\|SE\|_r$ is the number of SE-reflex vertices, $\|NW\|_r$ is the number of NW-reflex vertices, and $\|NE\|_c$ is the number of NE-convex vertices. Thus, the total number of orthogonal floodlights used by the four rules is given by $\|NE\| + \|NW\| + \|SE\| + \|SW\| = 2r + c$, where r is the number of reflex vertices in the polygon P and c is the number of convex vertices. Since for an orthogonal polygon with no holes $c = (n + 4)/2$ [11] and $r = (n - 4)/2$ [11], and also, for a polygon with holes, the convex vertices on a hole are reflex vertices for the hole, while the reflex vertices on a hole are convex vertices for the hole, we have that the covering rule that uses the minimum number of floodlights uses

$$\left\lfloor \frac{2r + c}{4} \right\rfloor = \left\lfloor \frac{2(r_0 + c_1 + \ldots + c_h) + c_0 + r_1 + \ldots + r_h}{4} \right\rfloor$$

floodlights, where $c_i + r_i = n_i$ and c_i is the number of convex vertices in the i-th hole. Since $n = \sum_{n=0}^{h} n_i$, we have that $\lfloor(2r + c)/4\rfloor$ is

$$\left\lfloor \frac{2\left[\frac{n_0-4}{2} + \sum_{i=1}^{h} \frac{n_i+4}{2}\right] + \frac{n_0+4}{2} + \sum_{i=1}^{h} \frac{n_i-4}{2}}{4} \right\rfloor = \left\lfloor \frac{3n + 4(h-1)}{8} \right\rfloor.$$

This completes the proof. □

We have proved the following result.

Theorem 5. *If P is an orthogonal polygon with n vertices and h holes, then $\lfloor(3n + 4(h-1))/8\rfloor$ orthogonal floodlights are always sufficient and sometimes necessary to illuminate P.*

We claim that this result is significant despite the fact that it may suggest that more guards are required than in the original art gallery theorem. We support this claim with three observations:
1.- Using Theorem 5, the total aperture of $3\pi n/8$ proposed by the original version of the orthogonal art gallery theorem, for polygons without holes, has been reduced by half. Moreover, for polygons with holes, the total aperture using floodlights is always less than the $\pi n/2$ proposed by Hoffmann's result.

2.- The placing rules lead directly to a linear algorithm that is much simpler than the algorithms for guards that require trapezoidization, quadrilateralization or decomposition into L-shaped pieces [11]. The algorithm consists of a traversal of the boundary of the polygon that counts the types of vertices. It computes the number of vertex floodlights required by each of the four rules, determines which one uses the minimum and assigns floodlights in the corresponding vertices with a second traversal of the boundary.

3.- The algorithm may place many fewer floodlights that $\lfloor(3n + 4(h - 1))/8\rfloor$; for example, in a staircase polygon, only one floodlight is used.

4 Illuminating with Floodlights on the Boundary

An orthogonal art gallery P with no holes and r reflex vertices can be partitioned into $\lfloor r/2\rfloor + 1$ L-shaped pieces [11], and since each L-shaped piece can be illuminated with one floodlight, we have that $\lfloor n/4\rfloor$ orthogonal floodlights are sometimes necessary and always sufficient to illuminate an orthogonal polygon. This seems to contradict our previous results; however, the missing detail is that using O'Rourke's algorithm to partition P into L-shaped pieces, some of the floodlights will be placed in the interior of the polygon. This seems rather unsatisfactory.

In this section, we first show that we can illuminate a polygon P with no holes with $\lfloor n/4\rfloor$ orthogonal floodlights placed at points in the boundary of P. We prove this by showing that $\lfloor r/2\rfloor + 1$ orthogonal floodlights at points in the boundary are always sufficient to illuminate P, where r is the number of reflex vertices in P. Then, we demonstrate that any orthogonal polygon with n vertices and h holes can be illuminated with $\lfloor(n + 2h)/4\rfloor$ floodlights in its boundary.

For a polygon with no holes, necessity of $\lfloor n/4\rfloor$ floodlights is established by the well-known "comb" example [11, Figure 2.18]. Sufficiency follows an inductive argument similar to O'Rourke's proof of the orthogonal art gallery theorem [11, Sections 2.5 and 2.6].

A *horizontal cut* of an orthogonal polygon P is an extension of the horizontal edge incident to a reflex vertex through the interior of the polygon. A cut *resolves* a reflex vertex in the sense that the vertex is no longer reflex in either of the two pieces of the partition determined by the cut. Clearly, a cut does not introduce any reflex vertices. A horizontal cut is an *odd-cut* (also and H-odd-cut) if one of the halves contains an odd number of reflex vertices.

Lemma 6. *Let P be an orthogonal polygon and partition P into $P_1, P_2, \ldots, P_t$ by drawing all H-odd-cuts and all H-cuts that are visibility rays of two reflex vertices. Then, each P_i is in general horizontal position and can be covered with $\lfloor r_i/2\rfloor + 1$ floodlights in the boundary of P, where r_i is the number of reflex vertices in P_i.*

In this extended abstract we omit the proof. Since the H-graph can be constructed in linear time [11] and trapezoidization can be achieved in linear time [4], it is not hard to see that the above argument results in a linear algorithm. Thus, we have obtained the following result.

Theorem 7. *Let P be an orthogonal polygon with n vertices, then $\lfloor n/4 \rfloor$ orthogonal floodlights placed in the boundary of P are always sufficient and sometimes necessary to illuminate the interior of P. Moreover, such set of floodlights can be found in $O(n)$ time.*

We are now ready to discuss the case of orthogonal polygons with holes.

Theorem 8. *An orthogonal polygon with n vertices and h holes can be illuminated with at most $\lfloor \frac{n+2h}{4} \rfloor$ floodlights in its boundary. Moreover, this bound is tight, since there are orthogonal polygons with n vertices and h holes that require $\lfloor \frac{n+2h}{4} \rfloor$ floodlights.*

Proof. To prove sufficiency, let P be a polygon with n vertices and h holes. Resolve all holes of P and construct a polygon P' as follows. Let v_0 be a reflex vertex that belongs to a hole P_0. Cut P along a horizontal segment λ in the interior of P that joins v_0 with the boundary of P. Include λ as two segments of the boundary. By Theorem 7, the polygon P' can be illuminated with $\lfloor (n + 2h)/4 \rfloor$ floodlights in its boundary. However, the cuts to construct P' from P were always horizontal, and the $\lfloor (n + 2h)/4 \rfloor$ floodlights provided by Theorem 7 are placed on vertical edges of P'. Thus, these floodlights are on vertical edges of P. Moreover, it is not hard to see that the floodlights illuminate P. To prove necessity, we consider the generic polygon of Fig. 5. For each $m > 0$, this polygon

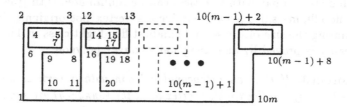

Fig. 5. A polygon with m holes and $10m$ vertices that requires $3m$ $\pi/2$-floodlights.

can be configured to have m holes and $10m$ vertices and requires $3m$ orthogonal floodlights. □

5 Illuminating the Exterior

Two other variants of the illumination problem are "The Fortress Problem" and the "Prison Yard Problem". The first asks for the number of floodlights to illuminate the exterior of the polygon, while the second asks for the number needed to see both the interior and the exterior.

In this section we demonstrate that $(n+4)/2$ orthogonal floodlights are sometimes necessary and always sufficient to illuminate the exterior of an orthogonal polygon P with n vertices and even with h holes.

We prove sufficiency first. Let P be an orthogonal polygon with n vertices and h holes. Again, let n_0 be the number of vertices in the outer boundary and n_i the number of vertices in the i-th hole. Illuminating the exterior of P requires us to illuminate the interior of the holes. Since P is simply-connected, the holes do not have holes of their own, thus, by the results of earlier sections, the i-th hole can be illuminated with fewer than $n_i/2$ orthogonal floodlights (recall that n_i must be even for $i = 0, \ldots, h$).

If we prove that the exterior of the polygon P' defined by the n_0 vertices of the outer boundary of P can be illuminated with $n_0/2 + 2$ floodlights, then we would have shown that $2 + \sum_{i=0}^{h} n_i/2 = 2 + n/2$ floodlights are sufficient.

Recall that a set S in the plane is orthogonally convex if any horizontal or vertical line segment intersects S in a connected set. The orthogonal convex hull $C(S)$ of a set S is the smallest orthogonally convex set containing S. Consider the orthogonal convex hull of P'. The boundary of $C(P')$ is composed of at most four staircases (and possibly fewer). Moreover, if this hull $C(P')$ has k vertices, its exterior can be illuminated with $k/2 + 2$ vertices by traveling around the boundary and placing one floodlight in every other vertex of each staircase and one on each of the North-most, South-most, East-most and West-most edges. Observe also that, for each new vertex introduced to the boundary of $C(P')$, a vertex is resolved from P' and no floodlights are assigned to a resolved vertex.

It only remains to illuminate the bays of P'; that is the regions exterior to P' but interior to the hull $C(P')$. We can not apply Theorem 5 directly and conclude that a bay with b vertices can be illuminated with $\lfloor (3b - 4)/8 \rfloor$ orthogonal floodlights since a bay may have a vertex not originally in P'. However, estimating the size of the second smallest floodlight class in Theorem 5 suffices. Necessity is proved by an orthogonally convex polygon.

Theorem 9. *If P is an orthogonal polygon with n vertices and h holes, then $n/2+2$ orthogonal floodlights (vertex or at the boundary) are sometimes necessary and always sufficient to illuminate the exterior of P.*

Observe that the proof of this theorem leads again to a linear algorithm that does not require polygon partitioning and thus is simple and practical.

For the Prison Yard Problem, a rectangle illustrates that there is no solution for vertex floodlights unless we allow two floodlights to be placed at one vertex.

6 Concluding Remarks

We have established combinatorial tight bounds for illuminating the interior or exterior of orthogonal polygons with holes by orthogonal floodlights. Our proofs are simple and lead to linear algorithms on the number n of vertices of the polygon. The algorithms compute a covering achieving the bounds, and since any algorithm must inspect all points of the input, the computational complexity of our algorithms is optimal.

We have shown that $\pi/2$ is the smallest aperture necessary and in fact, it is not hard to see that for $\alpha \in [\pi/2, 3\pi/2)$ all our results hold for α-floodlights. It is trivial to note that for $\alpha \geq 3\pi/2$, α-floodlights are general vertex guards in orthogonal polygons. However, several open problems remain. What bounds can be found for other classes of polygons? Is computing the minimum set of covering α-floodlights an NP-hard or NP-complete problem? If the floodlights are each allowed to have a different aperture α_k, what can be said about the problem of finding a cover that optimizes the total angle power given by $\sum_{i=1}^{k} \alpha_k$?

References

1. J. Abello, O. Egecioglu, and Kumar K. Visibility graphs of staircase polygons and the weak Bruhat order I: From polygons to maximal chains. *Discrete and Computational Geometry.* in press.
2. I. Bjorling-Sachs and D. L. Souvaine. A tight bound for guarding polygons with holes. Report LCSR-TR-165, Lab. Comput. Sci. Res., Rutgers Univ., New Brunswick, NJ, 1991.
3. P. Bose, L. Guibas, A. Lubiw, M. Overmars, D. Souvaine, and J. Urrutia. The floodlight problem. *Int. J. On Computational Geometry.* in press.
4. B. Chazelle. Triangulating a simple polygon in linear time. *Discrete Comput. Geom.*, 6:485–524, 1991.
5. J. Czyzowicz, E. Rivera-Campo, and J. Urrutia. Optimal floodlight illumination of stages. *Proc. of the 5th CCCG*, pages 393–398, Waterloo, Ontario, August 1993. Univ. of Waterloo.
6. V. Estivill-Castro and J. Urrutia. Optimal floodlight illumination of orthogonal art galleries. In *Proc. of the Sixth CCCG*, pages 81–86, Saskatoon, Canada, August 1994. University of Saskatchewan.
7. Györi. E., F. Hoffman, K. Kriegel, and T. Shermer. Generalized guarding and partitioning for rectilinear polygons (extended abstract). In *Proc. of the Sixth CCCG*, pages 302–307, Saskatoon, Canada, August 1994. U. of Saskatchewan.
8. F. Hoffmann. On the rectilinear art gallery problem. In *Proc. of the Int. Col. on Automata, Languages, and Programming 90*, pages 717–728. Springer-Verlag Lecture Notes on Computer Science 443, 1990.
9. F. Hoffmann, M. Kaufmann, and K. Kriegel. The art gallery theorem for polygons with holes. In *Proc. 32nd IEEE Sympos. Found. Comput. Sci.*, pages 39–48, 1991.
10. J. Kahn, M. Klawe, and D. Kleitman. Traditional galleries require fewer watchmen. *SIAM J. Algebraic Discrete Methods*, 4:194–206, 1980.
11. J. O'Rourke. *Art Gallery Theorems and Algorithms.* Oxford University Press, New York, 1987.
12. G. J. E. Rawlins. *Explorations in restricted-orientation geometry.* Ph.D. thesis, Univ. Waterloo, Waterloo, ON, 1987.
13. T. Shermer. Recent results in art galleries. *Proceedings of the IEEE*, 80:1384–1399, 1992.
14. W. Steiger and I. Streuni. Positive and negative results on the floodlight problem. In *Proc. of the Sixth CCCG*, pages 87–92, Saskatoon, Canada, August 1994. University of Saskatchewan.

No Quadrangulation is Extremely Odd[*]

Prosenjit Bose[1] and Godfried Toussaint[2]

[1] University of British Columbia
[2] McGill University

Abstract. Given a set S of n points in the plane, a quadrangulation of S is a planar subdivision whose vertices are the points of S, whose outer face is the convex hull of S, and every face of the subdivision (except possibly the outer face) is a quadrilateral. We show that S admits a quadrangulation if and only if S does not have an odd number of extreme points. If S admits a quadrangulation, we present an algorithm that computes a quadrangulation of S in $O(n \log n)$ time, which is optimal, even in the presence of collinear points. If S does not admit a quadrangulation, then our algorithm can quadrangulate S with the addition of one extra point, which is optimal. Finally, our results imply that a k-angulation of a set of points can be achieved with the addition of at most $k - 3$ extra points within the same time bound.

1 Introduction

In the study of *finite element methods* and *scattered data interpolation* ([8], [9]), it has recently been shown that quadrangulations may be more desirable objects than triangulations. A quadrangulation of a set of points S is a planar subdivision whose vertices are the points of S, whose outer face is the convex hull of S, and every face of the subdivision (except possibly the outer face) is a quadrilateral.

The above two applications provide new motivation for the study of quadrangulations of point sets from the computational geometry point of view. We remark that quadrangulations of polygons have been investigated in the computational geometry literature for some time in the context of guarding or illumination problems ([5], [7], [10], [13], [14]).

In this paper we characterize those sets of points that admit a quadrangulation. We show that S admits a quadrangulation if and only if S does not have an odd number of extreme points. If S admits a quadrangulation, we present an algorithm that computes a quadrangulation of S in $O(n \log n)$ time, which is optimal, even in the presence of collinear points. If S does not admit a quadrangulation, then our algorithm can quadrangulate S with the addition of one extra point, which is optimal. Our algorithm is conceptually simple, but to achieve the $O(n \log n)$ time complexity, we need to use some complicated data structures. However, from the conceptual description of the algorithm, a very simple $O(n^2)$

---

[*] research supported by Killam and NSERC postdoctorate fellowships and grants NSERC-OGP0009293 and FCAR-93ER0291. Email: jit@cs.ubc.ca and godfried@cs.mcgill.ca

time algorithm is implied which may be more desirable from a practical point of view. Finally, our results imply that a set of points S admits a k-angulation for any k with the addition of at most $k-3$ Steiner points. A k-angulation of a set of points S is a planar subdivision whose vertices are the points of S, whose outer face is the convex hull of S, and every face of the subdivision (except possibly the outer face) is a simple polygon with k vertices.

Some proofs and details are omitted in this extended abstract. For full detailed proofs, we refer the reader to the technical report [3].

2   Existence of Quadrangulation

Most of the geometric and graph theoretic terminology used is standard and for details, we refer the reader to O'Rourke [11], Bondy and Murty [2], and Preparata and Shamos [12].

In this section, we characterize the sets of points that admit a quadrangulation. The characterization is surprisingly simple. We first describe a necessary condition for a point set to admit a quadrangulation.

Lemma 1. *If a set of points S admits a quadrangulation then S has an even number of points on its convex hull.*

Proof. We use a counting argument to prove the lemma. Let $Q(S)$ represent a quadrangulation of S. Let $E(S)$ be the number of edges, $V(S)$ the number of vertices and $F(S)$ the number of faces of $Q(S)$.

Let CH represent the number of vertices on the outer face of $Q(S)$. Let $F_I(S)$ represent the number of internal faces of $Q(S)$. Since every internal face is a quadrilateral, we have that $4F_I(S) + CH = 2E(S)$ (see [2] for a detailed proof of this fact). Therefore, $CH = 2(E(S) - 2F_I(S))$. Since $2(E(S) - 2F_I(S))$ is an even number, CH must be even for the relation to hold. Therefore, we conclude that if S admits a quadrangulation then the convex hull of S must contain an even number of extreme points.

We now show that the above necessary condition is also sufficient. In this extended abstract, we only prove it for points in general position. Details on how to remove this condition can be found in the full paper [3].

Lemma 2. *If S has an even number of points on its convex hull then S admits a quadrangulation.*

Proof. We proceed by induction on the number of points inside the convex hull of S, $CH(S)$.

Basis: Let S be a set of points with an even number of points on $CH(S)$ and no points inside $CH(S)$. Let $1, 2, \ldots, n$ represent the n points as they occur in counter-clockwise order on the convex hull of S. We construct a quadrangulation of S in the following manner.

Join i to $i+1$ for all $i = 1, 2 \ldots, n$. Join n to 1. So far we have constructed a convex polygon. Finally, join 1 to j for all j satisfying the following: $j = 1, \ldots, n-2$, $j \neq 2$ and $j = 0$ mod 2. The result is a quadrangulation of S.

Inductive Hypothesis: A set of points S with an even number of points on $CH(S)$ and k points inside $CH(S)$ where $k \geq 0$ and $k \leq m$ for a fixed constant m admits a quadrangulation.

Inductive Step: Let S be a set of points with an even number of points on $CH(S)$ and $m+1$ points inside $CH(S)$. We now must show that S admits a quadrangulation.

Since $m+1 \geq 1$, S must have at least one point inside $CH(S)$. Let q be such a point. Let S' be the set of points S with q removed. By construction, $CH(S)$ is the same as $CH(S')$, which means that S' has an even number of vertices on the convex hull and at most m points inside its convex hull. Therefore, by the inductive hypothesis, S' admits a quadrangulation, denoted by $Q(S')$. If we re-insert q, then q must be contained in some quadrilateral of $Q(S')$, say D. The point q cannot lie on an edge of $Q(S')$ since we assumed that no three points in S are collinear. Let $D = a, b, c, d$ be the four vertices of the quadrilateral containing q. If D is convex, then we can adjust $Q(S')$ to be a quadrangulation of S by adding the edge $[qa]$ and $[qc]$. If D is not convex, without loss of generality, let us assume that c is a reflex vertex. Again, we can adjust $Q(S')$ to be a quadrangulation of S by adding the edge $[qa]$ and $[qc]$.

We conclude with the following theorem:

Theorem 3. *A set of points S in general position admits a quadrangulation if and only if the convex hull of S has an even number of extreme points.*

3 Algorithms

The existence proof in the previous section (Lemma 2) immediately implies the following Sequential Insertion (SI) Algorithm. Compute the convex hull of S in $O(n \log n)$ time with any of several algorithms available (see [11]). If the number of convex hull vertices is even, partition the convex hull into quadrilaterals in $O(n)$ time by joining one vertex to every other vertex of the polygon in a clockwise fashion. Subsequently, the remaining points are inserted one at a time. The insertion stage of the algorithm is the crucial part of the entire procedure. The obvious method of inserting the points results in an $O(n^2)$ time algorithm. However, an implementation of the SI Algorithm using a sweep-line approach (see [11] or [12] for examples of sweep-line algorithms) yields an $O(n \log n)$ time algorithm.

Theorem 4. *Let S be a set of n points in general position in the plane. If n is even, then S can be quadrangulated in $O(n \log n)$ time, using the SI algorithm.*

From a theoretical computational complexity point of view, the SI algorithm is optimal. However the main drawback of the sequential insertion algorithm, as

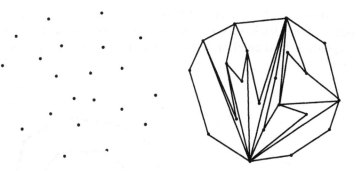

Fig. 1. A set of points S and the quadrangulation obtained with the Sequential Insertion Algorithm

Figure 1 illustrates, is that it yields quadrangulations that are not desirable in practice since they tend to yield long non-convex quadrangles when fat convex quadrangles are preferred. Therefore we omit the implementation details of the SI algorithm and instead describe another algorithm below which has much greater promise of yielding nice quadrangulations in practice and can handle collinearities with little effort.

Let $P = (p_1, p_2, ..., p_n)$ denote a simple polygonal chain spanning the set S of n planar points given. A triangulation of P, denoted by $T(P)$, is a triangulation of S, $T(S)$, such that it contains all the edges of P as a subset of the edges of $T(S)$. The edges in $T(P)$, other than the edges of P, are called the *diagonals* of $T(P)$. The dual graph of $T(P)$ is a graph $G(V, E)$ whose vertex set V corresponds to the set of triangles in $T(P)$ and two vertices are connected with an edge if, and only if, their corresponding triangles share a diagonal. A triangulation of P is called *serpentine* if its dual graph is a chain.

Our approach will be to show that a set of points S always admits a simple spanning polygonal chain (whose vertex set is precisely the set of points in S) that in turn always admits a serpentine triangulation. In such a triangulation the diagonals (and hence triangles) are ordered in accordance to the ordering of the vertices of the dual chain. By *removing* every other diagonal starting at one end of the chain, we will obtain the desired quadrangulation except perhaps a single triangle at the end. In such an eventuality we add one Steiner point outside the convex hull of S to convert the final triangle to a quadrangle. We then show that the Steiner point is necessary if, and only if, the number of points h on the convex hull of S is odd. We will first discuss this approach when the given points are in general position and then we will show the minor modifications that can be made in order to handle point sets with collinearities.

We should point out that in a different context concerned with fast rendering in computer graphics, Arkin et al. [1] proposed an $O(n \log n)$ time algorithm for obtaining a serpentine triangulation (Hamiltonian in their terminology) of a set of points. However, their algorithm is based on sequentially inserting triangles in a triangulation of the points and hence, like our SI algorithm, its application to our problem leads to very poor quadrangles similar to those in Figure 1. The serpentine triangulation algorithm that we propose yields much nicer quadrilaterals.

Fig. 2. A set of points S and the convex spiral of S.

A simple spanning polygonal chain most convenient for our purpose is the convex spiral of S. We define the convex spiral of S constructively as follows. Let $p_{x-\min}$ denote the point of S with the minimum x-coordinate. If more than one point satisfies this property select the point which also has minimum y-coordinate. Construct an infinite directed half-ray anchored at $p_{x-\min}$ and pointing in the $+y$ direction. Mark the point $p_{x-\min}$. This marked point is the first vertex of the spiral chain. Now rotate the half-ray in a clockwise direction until it coincides with an unmarked point of S. This point is marked, it becomes the second vertex of the spiral chain and the ray is translated in the direction it is pointing so that it is anchored at the newly marked vertex just found. Continue this process until no unmarked points remain in S. We will refer to this procedure as the *Spiraling Procedure*. For the set of points in Figure 1 the resulting convex spiral is illustrated in Figure 2. This structure has appeared in the literature before and is closely related to the *onion peeling* of a set or the convex layers [4]. In fact one can compute the spiral and the convex layers, one from the other, in $O(n)$ time [12]. This structure has also been used for example in statistics to define a generalization of the concept of the median to two-dimensional data. In this application the median is defined as the last vertex of the spiral chain. We now show that the convex spiral of a set S admits a serpentine triangulation. Refer to Figure 3.

Lemma 5. *Let S be a set of n points in the plane. The convex spiral of S admits a serpentine triangulation.*

Proof. If all the points of S lie on their convex hull then the required triangulation can be obtained trivially by connecting any vertex of the convex hull to

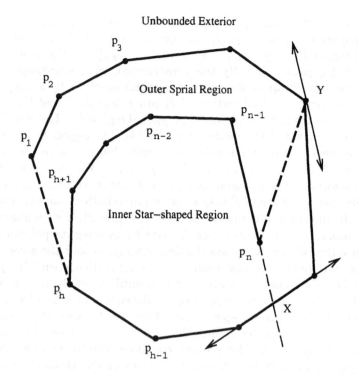

Fig. 3. Illustrating the partition of the region inside the convex hull of S into the *outer* spiral region and the *inner* star-shaped region.

all the others. Therefore we assume there are points of S in the interior of the convex hull. To simplify the proof we divide the spiral region to be triangulated into two regions: an outer (spiral) region and an inner (star-shaped) region. We then show that each region admits a serpentine triangulation in such a way that these can be concatenated into one final serpentine triangulation.

Without loss of generality, assume that the points in S are re-labelled so that they are ordered in accordance with the convex spiral, i.e., $p_1 = p_{x-\min}$, p_2 is the next vertex of the spiral and so on until the last vertex of the spiral is labelled p_n. Since S has h vertices on the convex hull there exists one edge of the convex hull, namely $[p_1, p_h]$, which is a diagonal in every triangulation of the convex spiral P. The union of this edge $[p_1, p_h]$ with P partitions its complement, i.e., that part of the plane (E^2) excluding P and $[p_1, p_h]$, into two connected regions, one unbounded (exterior to the convex hull) and one bounded (the spiral region). Now extend the edge $[p_{n-1}, p_n]$ of P in the direction of p_n and let X be the first intersection point of this extension with P. At this point X, construct a line locally tangent to P. Now rotate this line in a counter-clockwise direction until it meets the first vertex of P, say Y, such that the line is parallel to the edge $[p_{n-1}, p_n]$. Finally, insert the diagonal $[p_n, Y]$. This diagonal partitions the spiral polygonal region into two regions: the outer spiral polygonal region $P_o = [p_1, p_2, \ldots, Y, p_n, p_{n-1}, \ldots, p_h]$ and the inner polygonal region $P_i = [p_n, p_{n-1}, \ldots, Y]$. By this construction, P_i is star-shaped from p_n. Therefore we can obtain a serpentine triangulation of P_i by simply inserting diagonals between p_n and all vertices of P_i other than p_{n-1} and Y.

It remains to triangulate the spiral polygonal region P_o. This can be accomplished with a variant of the *rotating calipers* [15]. The region P_o can be viewed locally as a convex polygonal annulus with outer chain $C_o = [p_1, p_2, \ldots, Y]$ and inner convex chain $C_i = [p_h, p_{h+1}, \ldots, Y, \ldots, p_{n-1}, p_n]$. We place one supporting line of the *rotating calipers* tangent to C_o and the other parallel to it and tangent to C_i. Note that a portion of P may be common to both C_o and C_i when P has high enough winding number (as is the case in Figure 2). If we make two copies of this common portion we can view C_o and C_i as separate polygonal chains. To initialize the calipers we locate the first through p_1 and the second through p_h. At the start of the rotation both lines are set collinear with $[p_1, p_h]$. Next, we *rotate* the calipers in a clockwise direction until one of the lines meets a new vertex. This vertex will be either p_2 or p_{h+1} - depending on which has the smaller angle. This identifies a new co-podal pair of vertices between the chains C_o and C_i and this pair in turn determines the new diagonal to be inserted between them in the triangulation. This process is continued until the inner caliper line contains $[p_{n-1}, p_n]$. Since at each step one vertex on one chain is advanced, the insertion of the new diagonal creates one triangle in the triangulation. Since by this construction diagonals can never be inserted between two vertices of the same chain, we obtain a serpentine triangulation of the region P_o. Since the final triangle of the triangulation of P_o is adjacent to (shares a diagonal with) the first triangle of the triangulation of P_i, it follows that their concatenation is also serpentine.

By Lemma 5 the Spiraling Rotating Caliper (SRC) Algorithm produces a serpentine triangulation. Therefore the diagonals and triangles of $T(P)$ are ordered according to the order of the vertices comprising the dual chain of $T(P)$. Let $D = (d_1, d_2, \ldots, d_m)$ denote the diagonals in this order, where $d_1 = [p_1, p_h]$. If m is even, then if we delete from the triangulation every other diagonal $d_2, d_4, \ldots$ and so on, we obtain a quadrangulation. If m is odd we may delete every other diagonal starting from the last diagonal d_m. This will quadrangulate P except for the presence of triangle $\Delta(p_1, p_{h+1}, p_h)$ or $\Delta(p_1, p_2, p_h)$. Finally, by inserting one Steiner point just outside the convex hull of S near the edge $[p_1, p_h]$ we may convert this triangle to a quadrangle.

Consider now the complexity of the SRC algorithm. The first step of the algorithm is to compute the convex spiral of the given set of points. If the spiraling procedure, as described in the paragraph above the statement of Lemma 5, is used, then $O(n^2)$ time is needed for this step. However, we may compute the convex layers of S in $O(n \log n)$ time using the algorithm of Chazelle [4] or the algorithm of Hershberger and Suri [6]. This is the most difficult step in the algorithm, as both these algorithms are fairly involved. From an implementation point of view, it might be preferable to use the simpler algorithm. From the convex layers, we can compute a convex spiral P of S in $O(n)$ time with the procedure of Preparata and Shamos [12]. The spiraling rotating caliper algorithm for obtaining a serpentine triangulation of P described in the proof of Lemma 5 runs in $O(n)$ time since no backtracking is involved. Finally, triangulating the star-shaped interior, deleting the unwanted diagonals and inserting the Steiner point all require no more than $O(n)$ time. Therefore the entire algorithm runs in $O(n \log n)$ time. We conclude with the following theorem.

Theorem 6. *Let S be a set of n points in general position in the plane. The SRC algorithm quadrangulates S, with the addition of at most one Steiner point, in $O(n \log n)$ time.*

We have shown above that the Spiraling Rotating Caliper (SRC) Algorithm computes a quadrangulation for any set of n points and that this can be done with at most one Steiner point. The above theorem implies the following corollary:

Corollary 7. *A k-angulation of a set of points can be achieved with the addition of at most $k - 3$ extra points.*

Now we show that the Steiner point is necessary only if the number of convex hull vertices h is odd and when h is even the SRC-Algorithm always yields a quadrangulation without the need for Steiner points.

Theorem 8. *Let S be a set of n points in general position in the plane. The SRC algorithm quadrangulates S with at most one Steiner point if the number of vertices on the convex hull of S is odd.*

Proof. Let h denote the number of points on the convex hull of S and let k denote the number of points in the interior of the convex hull. Therefore $n = h + k$. From Euler's theorem it follows that the number of edges in any triangulation $T(S)$ equals $2h + 3k - 3$. Now, any spanning simple polygonal chain on n vertices contains $n - 1$ edges. Since the outer shell of the convex spiral contains $h - 1$ edges and the inner spiral contains k edges, it follows that the entire convex spiral contains $h + k - 1$ edges. Therefore the number of diagonals in $T(S)$ equals $(2h + 3k - 3) - (h + k - 1) = h + 2k - 2$. Since 2 is even, $2k$ is even for any value of k, it follows that the expression is even if and only if h is even. Since the diagonals are ordered it follows that by removing every other diagonal the SRC-Algorithm yields a quadrangulation without a Steiner point if, and only if, h is even. Furthermore, one Steiner point is needed if, and only if, h is odd.

Figure 4 illustrates the convex spiral and resulting quadrangulation obtained with the SRC-Algorithm. Note that the quadrangulation is much *nicer* than the quadrangulation (shown in Figure 1) obtained with the sequential insertion method. In fact for this example not only are the quadrangles obtained with the SRC-Algorithm fat but they are all convex.

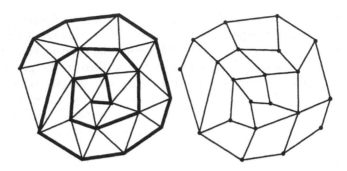

Fig. 4. Illustrating the serpentine triangulation and the resulting quadrangulation of the set of points in Figures 1 and 2.

Only a few minor modifications need to be made to convert the proof of Theorem 6 into one that applies to sets of points that have collinearities. We assume that the convex hull of the points is not a line segment. Furthermore, in the presence of collinear points, degenerate quadrangles may be necessary. We allow degenerate quadrangles which are three-sided figures (four edges) consisting of four vertices three of which are collinear such that edges only intersect at their endpoints.

We note that since by definition the quadrangulation of a point set contains the convex hull, it follows that computing a quadrangulation of a point set has an $\Omega(n \log n)$ lower bound. Therefore, the SRC algorithm is optimal.

Theorem 9. *A set of points S not all on a line may be quadrangulated with at most one Steiner point in $\Theta(n \log n)$ time with the SRC-Algorithm even in the presence of collinear points.*

4 Conclusions

As a side benefit of our quadrangulation algorithm, we have obtained an optimal algorithm for computing serpentine (Hamiltonian) triangulations of point sets that yield very nice triangles. Arkin et al. [1] have shown that such triangulations have applications in fast rendering algorithms in computer graphics. Our algorithm yields much nicer triangulations than the algorithm of Arkin et al.

It is easy to see that a set of points does not always yield a quadrangulation where every quadrangle is convex. This observation leads to the following interesting questions. Can one decide if a set of points admits a quadrangulation where every quadrangle is convex? Can one compute the quadrangulation of a set of points that has the maximum number of convex quadrilaterals? As Lubiw [10] has shown that the decision problem is NP-complete for simple polygons with holes, it seems reasonable to believe that the same might hold for point sets.

Finally, preliminary experiments confirm that for several measures of the quality of the quadrangles, the SRC-algorithm is statistically significantly superior to the sequential insertion (SI) algorithms. Furthermore, SRC yields quadrangulations in which 55% of the quadrangles are convex on average compared to only 30% for SI.

Acknowledgments: We thank Larry Schumaker, Bernard Chazelle, Joe Mitchell, Subhash Suri, Mike McAllister, T.S. Michael, and Anna Lubiw for interesting discussions on this topic.

References

1. Arkin, E., M. Held, J. Mitchell, and S. Skiena, Hamiltonian triangulations for fast rendering, *Algorithms-ESA'94*, J. van Leeuwen, ed., Utrecht, NL, LNCS 855, pp. 36-47, September 1994.
2. Bondy, J. and U.S.R. Murty, *Graph Theory with Applications*. Elsevier Science, New York, New York, 1976.
3. Bose, P. and G. Toussaint, *No Quadrangulation is Extremely Odd*. Tech. Rpt. 95-03, Dept. of Comp. Sci., University of British Columbia, Vancouver, Canada, 1995.
4. Chazelle, B., On the convex layers of a planar set, *IEEE Transactions on Information Theory*, **IT-31**, pp. 509-517, 1985.
5. Everett, H., W. Lenhart, M. Overmars, T. Shermer, and J. Urrutia, Strictly convex quadrilateralizations of polygons, in *Proceedings of the 4th Canadian Conference on Computational Geometry*, pp. 77-83, 1992.
6. Hershberger, J., and S. Suri, Applications of a semi-dynamic convex hull algorithm, in *Proceedings of the second S.W.A.T.*, *Lecture Notes in Computer Science 447*, Bergen, Sweden, pp. 380-392, 1990.
7. Kahn, J., M. Klawe, D. Kleitman, Traditional galleries require fewer watchmen, *SIAM J. Algebraic Discrete Methods*, **4**, pp.194-206, 1983.
8. Lai, M., *Scattered data interpolation and approximation by using C^1 piecewise cubic polynomials*, submitted for publication.
9. Lai, M., and L. Schumaker, Scattered data interpolation using C^2 piecewise polynomials of degree six, *Third Workshop on Proximity Graphs*, Mississippi State University, Starkville, Mississippi, December 1-3, 1994.
10. Lubiw, A., Decomposing polygonal regions into convex quadrilaterals, in *Proceedings of the 1st ACM Symposium on Computational Geometry*, pp.97-106, 1985.
11. O'Rourke, J., *Computational Geometry in C*, Cambridge University Press, 1994.
12. Preparata, F., and M. Shamos, *Computational Geometry: An Introduction*, Springer-Verlag, New York, 1985.
13. Sack, J., and G. Toussaint, A linear-time algorithm for decomposing rectilinear star-shaped polygons into convex quadrilaterals, *Proc. 19th Annual Conf. on Communications, Control and Computing*, Allerton, pp. 21-30, 1981.
14. Sack, J., and G. Toussaint, Guard placement in rectilinear polygons, in *Computational Morphology*, G. Toussaint, ed., North-Holland, pp. 153-175, 1988.
15. Toussaint, G., Solving geometric problems with the rotating calipers, *Proc. IEEE MELECON 83*, Athens, Greece, pp. A10002/1-4, 1983.

Finding the Medial Axis of a Simple Polygon in Linear Time

Francis Chin*   Jack Snoeyink†   Cao An Wang‡

Abstract

We give a linear-time algorithm for computing the medial axis of a simple polygon P, This answers a long-standing open question—previously, the best deterministic algorithm ran in $O(n \log n)$ time. We decompose P into pseudo-normal histograms, then influence histograms and xy monotone histograms. We can compute the medial axes for xy monotone histograms and merge to obtain the medial axis for P.

1 Introduction

The *medial axis* of a simple plane polygon P goes by many names, including *symmetric axis* or *skeleton*. One of the more picturesque is the *grassfire transform*: Imagine igniting all boundary points of P. If the flame burns inward at a uniform rate, then the *quench points* where the flame meets and extinguishes itself define the medial axis. Equivalently, the medial axis is the locus of all centers of circles inside P that touch the boundary of P in two or more points.

The medial axis was proposed and named by Blum [3] in 1967 article entitled, "A transformation for extracting new descriptors of shape." The pattern recognition literature uses it heavily as a one-dimensional structure that represents two-dimensional shape [7, 14, 16]; it has also been used in solid modelling [18], mesh generation [8], pocket machining [9], etc.

To a computational geometer, the medial axis of an n-gon P is a Voronoi diagram [2, 15] whose sites are the open edges and the vertices of the boundary. In 1982, Lee [13] developed an $O(n \log n)$ algorithm to compute the medial axis.Since that time, it has been an open question to determine the time required to compute the medial axis. Up to this year there were two significant milestones: In 1987 Aggarwal et al. [1] first published an algorithm that can compute the medial axis of a *convex* polygon P in linear time. In 1991 Devillers [5] first published a *randomized* algorithm for the medial axis that runs in $O(n \log^* n)$ expected time.

*Dept of Comp. Science, University of Hong Kong, Hong Kong. chin@csd.hku.hk

†Dept of Comp. Science, UBC, Vancouver, BC, Canada V6T 1Z4. snoeyink@cs.ubc.ca Partially supported by an NSERC grant and a BC ASI Fellowship.

‡Dept of Comp. Science, Memorial University of Newfoundland, St. John's, NFLD, Canada A1C 5S7 wang@garfield.cs.mun.ca Work partially supported by NSERC grant OPG0041629.

Decompositions of P into histograms have been applied to the *constrained Voronoi diagram* of vertices of P. Djijev and Lingas' [6] showed that the algorithm of Aggarwal et al. [1] could find the constrained Voronoi for an xy monotone histogram. Klein and Lingas [11] gave a randomized algorithm for the constrained Voronoi of a histogram and, by merging, computed the constrained Voronoi of P in $O(n)$ expected time. Wang and Chin [19] gave a deterministic algorithm by further decomposing histograms.

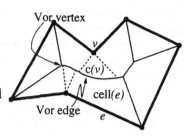

Fig. 1: Medial axis (solid) and Voronoi diagram (solid & dotted)

We extend Wang and Chin's decomposition to compute the medial axis of a simple polygon P. Our algorithm decomposes P into normal histograms, then into influence histograms, and xy monotone histograms. It computes the Voronoi diagrams of xy monotone histograms, and merges to obtain the medial axis of P. After reviewing definitions and known results about Voronoi diagrams and histogram decompositions in Section 2, we describe the new steps in reverse order: In Section 3, we extend the algorithm of Aggarwal et al.[1] to compute the Voronoi diagram of selected edges and vertices of an xy-monotone histogram. In the process, we simplify part of the analysis. In Section 4, we compute the Voronoi diagram of a histogram by decomposing it into influence histograms and xy monotone histograms. We conclude in Section 5.

Klein and Lingas have recently extended their work on histogram decompositions to obtain the medial axis of P in *expected* linear time [12]. Their algorithm adds edges to close of all histogram polygons and applies randomization twice: once to compute the medial axis of all edges of a histogram polygon, and again when non-P edges are removed and the medial axes are merged. The first is strongly predicated on the fact that all edges affect the medial axis, so it is unclear how to directly make it deterministic. Addition and deletion of edges also adds to the programming complexity.

2 Preliminaries

Let P be a simple polygon with n vertices, $\{p_1, p_2, \ldots, p_n\}$. The boundary ∂P consists of these vertices and the edges (open line segments) between consecutive vertices. We assume that the vertices and edges of P are in general position, which can be simulated by (actual or conceptual) perturbation of the input.

2.1 The Medial Axis and Voronoi Diagram of P

The *Voronoi diagram* [2, 15] of a set of *sites* is the partition of the plane into connected regions having the same set of closest sites. This partition consists of *Voronoi cells, edges,* and *vertices,* as in Figure 1. The *medial axis* of P is the locus of all centers of circles contained in P that touch ∂P in two or more points. Thus, the medial axis consists of Voronoi vertices and Voronoi edges. The only Voronoi edges that are not part of the medial axis are the bisectors of an edge and an incident vertex. We will, therefore, concentrate on computing the Voronoi diagram $V(P)$ in this abstract, and obtain the medial axis by removing these Voronoi edges.

The *constrained Voronoi diagram* of a set of sites on the boundary of P is the Voronoi diagram in which distance is measured along a shortest path inside P. All of our Voronoi diagrams should be considered constrained Voronoi diagrams, even though we typically omit the word "constrained." In our algorithms, the reflex vertices are always sites, so the shortest paths are line segments.

Algorithms to merge Voronoi diagrams have been important since Shamos and Hoey [17]. See also Kirkpatrick [10] or Klein and Lingas [11].

Lemma 2.1 *Let Q be a polygon that is divided into Q_1 and Q_2 by a diagonal e. Let subsets of vertices and edges S_1, S_2, and $S = S_1 \cup S_2$ be the sites in Q_1, Q_2, and Q respectively. Given the Voronoi diagrams of S_1 in Q_1 and S_2 in Q_2, one can obtain the Voronoi diagram of S in Q in time proportional to the number of Voronoi edges that intersect e and the number of new edges added.*

2.2 Histograms

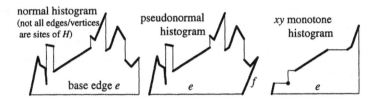

Fig. 2: Histograms

A *normal histogram* (NH) is a simple polygon H whose boundary consists of a *base edge e* and a chain that is monotone with respect to e Typically, we rotate normal histograms so that the base is along the x axis and the rest of the polygon is in the positive quadrant, as in Figure 2. The *base line* is the line through e. A *pseudo-normal histogram* (PNH), defined by Klein and Lingas [11], can be viewed as a normal histogram with a missing corner. Because of the merge lemma, an NH is as good as a PNHwith respect to computing a Voronoi diagram.

Corollary 2.2 *The constrained Voronoi diagram of selected sites of an n-vertex PNH can be obtained from the diagram of the corresponding NH in $O(n)$ time.*

An NH is an *xy-monotone histogram* if, after putting the base along the x axis, the y coordinates of non-base vertices are monotone increasing or monotone decreasing.

2.3 Decomposing P into Pseudo-Normal Histograms

Klein and Lingas' algorithm for the constrained Voronoi diagram [11] is based on decomposing a polygon P into PNHs, computing their Voronoi diagrams (via

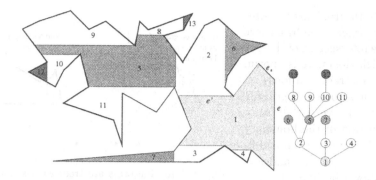

Fig. 3: A decomposition of P into a tree of PNH's

the corresponding NHs), and then merging. In the full paper, we note that only two calls to a linear-time trapezoidation algorithm are needed.

Figure 3 shows a polygon P decomposed into thirteen PNHs. PNH_1 is associated with the vertical base e missing its upper corner; PNH_2 with the horizontal base e' is missing its left corner, etc. The decomposition can be represented as a tree whose nodes are PNHs and whose edges represent adjacency of two PNHs.

Lemma 2.3 *An n-vertex simple polygon P can be decomposed into a set of PNHs having a linear number of vertices.*

It has been proved that the Voronoi diagrams of two PNH's do not interfere with each other as long as these two PNH's are (i) at the same depth not facing each other, or (ii) with their corresponding depths more than two apart. Although the proof given by Klein and Lingas is only for vertex sites, the proof can be easily extended to include edge sites.

Lemma 2.4 (Klein and Lingas [11, Thm 4.6]) *Given the Voronoi diagrams for the PNHs in a decomposition of an n-vertex simple polygon P, the Voronoi diagram for P can be computed in linear time.*

The task that remains is to compute the constrained Voronoi diagram of selected vertex and edge sites inside a normal histogram.

3 Computing the Voronoi of an xy Monotone Histogram and the Voronoi Extension

In this section, we provide efficient subroutines for two Voronoi diagram problems. See Figure 4.

Voronoi of xy monotone histogram Compute the constrained Voronoi diagram of selected sites in an xy monotone histogram H.

Voronoi extension Given the intersection of the Voronoi diagram of the sites in a histogram H with the base edge e, compute the Voronoi diagram in the halfspace below the base line.

In both problems the sites have an order given by a curve and the intersection of the Voronoi with the curve is known. In this abstract, we focus on the xy monotone histogram H, where the curve is the monotone portion of the boundary of H. We direct initial and final rays to $x = -\infty$ and $y = \infty$ as shown in figure 4.

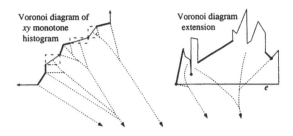

Fig. 4: The diagrams are trees with known leaves

Monotonicity in both x and y not only implies that the boundary of H does not leave the bounding box of the diagonals that join consecutive sites in x coordinate order, but also that the Voronoi diagram inside the box is completely determined by the endpoints of the diagonals. In fact, we replace portions of the boundary that are not sites with these diagonals. To avoid a degenerate case, if an edge and its lower endpoint are both sites, we consider them as a unit.

In the full paper, we sketch a randomized incremental construction that can solve both problems in expected linear time. For deterministic algorithms, we adapt the work of Aggarwal et al. [1].

3.1 Deterministic Algorithms

The basic idea of [1] is to identify a fraction of the Voronoi cells that are not adjacent to each other, remove their corresponding sites, recursively compute the Voronoi diagram of the remaining sites, and then independently merge in the non-adjacent cells. Identifying non-adjacent cells is complicated by the fact that one does not have the cell descriptions until the algorithm is done.

Let $s_1, s_2, \ldots, s_k$ be the list of sites in order along the curve. (Include sites at infinity as s_0 and s_{k+1}.) We mark sites red and blue to satisfy three rules:

1.) No two adjacent sites are *red*.
2.) No three adjacent sites are *blue*.
3.) If the portion of the boundary below 5 sites, $(s_{i-2}, s_{i-1}, s_i, s_{i+1}, s_{i+2})$, has a circle below that touches s_{i-1} and s_{i+1} and does not

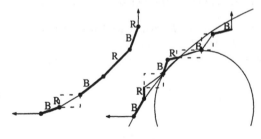

Fig. 5: Possible markings

contain any point of s_{i-2}, s_i, or s_{i+2}, then site s_i must be *red*.

Figure 5 shows two histograms with possible markings. Recall that if an edge and its lower endpoint are both sites, we consider them as a unit—otherwise symbolic perturbation would be needed so that two open edge sites with the same (non-site) endpoint could not be co-circular with two other sites. Marks on the concave chain at left are constrained only by rules 1 and 2, while the reds at right are forced by the empty circles.

Lemma 3.1 *A marking of sites can be computed in linear time.*

Proof: The first and third rules do not conflict because the third cannot apply to two adjacent sites. Thus, we can mark sites in linear time by initializing all sites to blue, then marking the sites that rule 3 says must be red. For each sequence of $i > 2$ blues that remain, we mark every other other site red, starting with the second and ending one or two before the last. ∎

This coloring has the following independence property:

Lemma 3.2 *Consecutive red sites cannot have Voronoi cells that are adjacent.*

Proof: Consider two sites s and s' that have adjacent Voronoi cells. There is a circle in H that touches s and s' and excludes all other sites.

If s and s' have a single site t between them, then by rule 3, t is marked red and by rule 1, both s and s' are blue. On the other hand, if two sites t and t' lie between s and s', then move the circle center along the bisector of s and s' toward t and t', as in figure 6. The new circle exits the old only between s and s'. Therefore, by monotonicity, the circle encounters t or t' before encountering any other. Suppose it encounters t', then t is red by rule 3 and s must be blue. Therefore, either s and s' are not both red or they are not consecutive. ∎

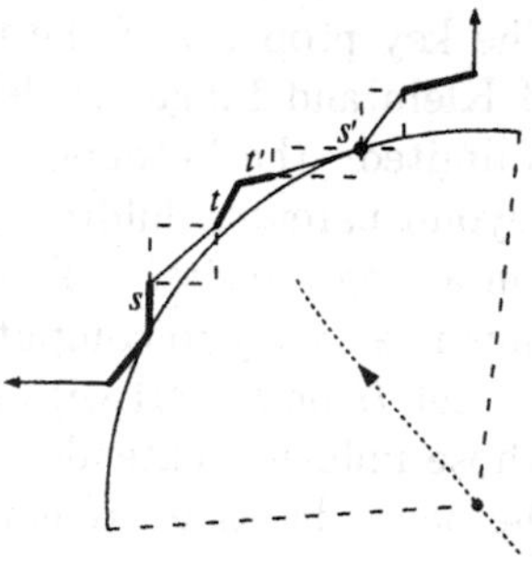

Fig. 6: s or s' is blue

These two lemmas are sufficient to use the algorithm of Aggarwal et al. [1], as we shall briefly describe. They prove the following combinatorial lemma.

Lemma 3.3 (Aggarwal et al. [1]) *Let T be a binary tree embedded in the plane. Each leaf of T has an associated "neighborhood," which is a connected subtree rooted at that leaf, and leaves adjacent in the topological order around the tree have disjoint neighborhoods. Then there are a fixed fraction of the leaves with disjoint, constant-size neighborhoods, and such leaves can be found in linear time (assuming that neighborhoods can be traced out in breadth-first order).*

The rest follows Aggarwal et al. [1]; see their paper for more detail.

Theorem 3.4 *The Voronoi diagram of an xy monotone histogram can be computed in linear time.*

Proof: By rule 2, a constant fraction of the sites are marked blue. We compute their Voronoi diagram recursively and let T be the tree of Voronoi edges that start between blues sites that are separated by reds. Now, the "neighborhood" of a leaf is the portion of the Voronoi edge that is farther from the blue sites that define it than from the red site that is being inserted. Lemma 3.2 says that adjacent neighborhoods are disjoint, so Lemma 3.3 says that a constant fraction of the red sites with disjoint, constant-size neighborhoods can be found. These red sites can be merged into the blue diagram in constant time apiece.

Finally, a constant fraction of the sites remain red; we again compute their Voronoi diagram recursively and merge it into the blue Voronoi diagram—we can do this in linear total time if we merge connected portions starting and ending on the histogram boundary. ∎

We can deterministically compute the Voronoi diagram extension by a similar algorithm.

Theorem 3.5 *Given the intersection Voronoi diagram of a histogram with the base edge, the extension below the base can be computed in linear time.*

4   Finding the Voronoi Diagram of a Normal Histogram

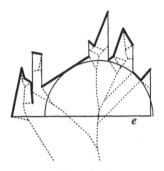

Fig. 7: The constrained Voronoi of selected sites

The key property of the histogram decomposition of Klein and Lingas is that the influence of a site is limited—the Voronoi cell of site does not extend beyond parents, children, or siblings of histograms containing the site. Following Wang and Chin [19], we show that limiting influence is also key to computing the Voronoi diagram of an NH.

Let H be an NH with horizontal base edge e. We can identify the sites of H whose influence extends across the base edge by considering circles centered at the base edge that are empty of sites.

Lemma 4.1 *In an NH H, the sites whose Voronoi cells extend below the base line are those that can be touched by a circle centered at the base edge and empty of other sites.*

We assume that H has been decomposed into a linear number of horizontal trapezoids, which is the result of our decomposition into PNHs or of running Chazelle's algorithm [4]. We call the horizontal segments introduced by trapezoidation *chords*. The dual graph of the trapezoidation of H is a tree rooted below the base edge whose edges correspond to chords, as illustrated in Figure 8.

The *influence region* of H is the union of all circles centered at the base edge whose interiors do not intersect a site. The *influence histogram*, IH, consists of all horizontal trapezoids that intersect the influence region.[1] We find the IH by exploring dual tree

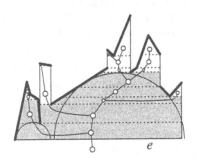

Fig. 8: Trapezoids in an IH

of the trapezoidation and maintaining stacks of sites whose Voronoi cells may intersect the base.

---

[1]This influence region and influence histogram are slightly different from Wang and Chin's corresponding ones for the constrained Voronoi diagram of vertices of P [19].

Lemma 4.2 *The influence histogram of H can be computed in time proportional to the number of its trapezoids.*

Proof: All circles in this algorithm will be centered on the base edge e. Any two such circles intersect in at most one point above the base line. Let C_s denote the largest circle that crosses a chord s and whose interior does not contain a site on or below s. If C_s exists, it either touches sites to the left and right of its center, or is centered at a (non-site) endpoint of e, or is infinite. The first case corresponds to the intersection of e with the bisector of the two sites touched, as can be seen in Figure 7.

We compute all such largest circles. Initialize empty stacks L and R, for left and right. If the left endpoint of the base edge is a site, insert it into L; similarly, insert the right endpoint into R. Let chord $s = e$.

Now, we maintain two invariants:

1. With respect to the histogram consisting of the trapezoids output so far, the stacks contain the sites whose Voronoi cells intersect e in order.

2. The largest circle C_s is determined by the sites at the tops of L and R.

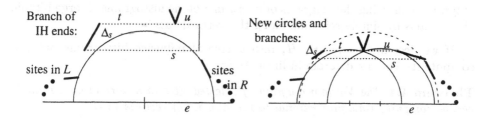

Fig. 9: The largest circle C_s and the trapezoid Δ_s

Let Δ_s be the trapezoid above s and let t and u be the at most two other chords of Δ_s as illustrated in Figure 9. Include Δ_s in the output. If the segment of C_s above s is contained in Δ_s then this branch of the IH is complete. If C_s crosses chord t but does not contain a site in Δ_s then let $s = t$ and continue. Otherwise, we need to re-establish the invariants, as at the right of Figure 9.

First, the new sites in Δ_s may be closer to e than sites in L or R—new Voronoi cells may crowd out old ones. If the largest circle determined by the next-to-top site in L and sites in Δ_s does not contain the top site in L, then pop the top of L. The first site to remain on the stack certifies that all sites beneath also remain, since stacked sites appear in the same order as their Voronoi cells intersect e. Handle R similarly.

Next, we look at the new empty circles crossing chords t and u—if there are none, then the branch of the IH ends here. If only one chord is crossed, then we push the new sites whose Voronoi cells intersect e on the L and/or R stacks and continue. Otherwise, both chords are crossed, as in Figure 9; some site, call it f, touches C_t on the right and C_u on the left. We continue building the

IH across t using circle C_t and stack L and a new stack R' that contains site f alone. Building IH across u uses C_u, R, and a new L' containing only f. Site f will not be popped off these stacks because among the empty circles between C_t and C_u there must be one that touches f directly above the center—this circle, tangent to t and u, certifies that the Voronoi cell of f in histogram H intersects the base edge e. Thus, the branches can be computed independently and the invariants can be maintained. ∎

The IH contains all sites whose Voronoi cells intersect the base edge. It may also contain other sites, but they can be grouped into xy monotone histograms.

Lemma 4.3 *The Voronoi diagram of an influence histogram H can be computed in time proportional to the number of vertices in H.*

Proof: Sketch: Each time the dual of the IH branches—and a circle is replaced by two—we know a vertical segment to the base edge contained in the Voronoi cell of the site forcing the branch. Cut the histogram along this segment. We are left with pairs, a histogram and a circle that intersect all of its horizontal trapezoids. Thus, the histograms are *bitonic* and can be cut into two xy monotone histograms. Compute the Voronoi diagrams of a bitonic histogram by merging the diagrams of its xy monotone histograms. Merge bitonic histograms by simply rejoining along the vertical cuts. ∎

If we decompose an NH, H, into a tree of influence histograms, we can compute its Voronoi diagram in linear time.

Theorem 4.4 *The Voronoi diagram of selected sites in a normal histogram H can be computed time proportional to the number of vertices in H.*

From this theorem, with the lemmas from Section 2.3, we obtain the final result.

Corollary 4.5 *The Voronoi diagram of the vertices and edges of a polygon P can be computed time proportional to the number of vertices in P.*

5 Conclusion

We have given an optimal linear-time algorithm for computing the Voronoi diagram or medial axis of a simple polygon. Several problems for simple polygons can be solved in linear time based on this result: computing the largest inscribed circle, building a query structure for the closest boundary point, and determining the buffer zone of all points within ε of a simple polygonal curve. This algorithm also applies to other L_p metrics and constant-complexity convex distance functions.

Acknowledgements

We thank Otfried Schwartzkopf and David Kirkpatrick for discussions on medial axis algorithms, and Bethany Chan, Siu-Wing Cheng, and Michael McAllister for their comments on drafts of this paper.

References

[1] A. Aggarwal, L. J. Guibas, J. Saxe, and P. W. Shor. A linear-time algorithm for computing the Voronoi diagram of a convex polygon. *Disc. & Comp. Geom.*, 4:591–604, 1989.

[2] F. Aurenhammer. Voronoi diagrams—A survey of a fundamental geometric data structure. *ACM Comp. Surveys*, 23(3):345–405, 1991.

[3] H. Blum. A transformation for extracting new descriptors of shape. In W. Whaten-Dunn, editor, *Proc. Symp. Models for Perception of Speech and Visual Form*, pages 362–380. MIT Press, 1967.

[4] B. Chazelle. Triangulating a simple polygon in linear time. *Disc. & Comp. Geom.*, 6:485–524, 1991.

[5] O. Devillers. Randomization yields simple $O(n \log^* n)$ algorithms for difficult $\Omega(n)$ problems. *Int. J. Comp. Geom. App.*, 2(1):97–111, 1992.

[6] H. Djidjev and A. Lingas. On computing the Voronoi diagram for restricted planar figures. In *WADS '91: Second Workshop on Data Structures and Algorithms*, number 519 in LNCS, pages 54–64. Springer-Verlag, 1991.

[7] R. O. Duda and P. E. Hart. *Pattern Classification and Scene Analysis*. John Wiley & Sons, 1973.

[8] H. N. Gürsoy and N. M. Patrikalakis. An automatic coarse and fine surface mesh generation scheme based on medial axis transform: Part I algorithm. *Engineering with Computers*, 8:121–137, 1992.

[9] M. Held. *On the Computational Geometry of Pocket Machining*. Number 500 in LNCS. Springer-Verlag, 1991.

[10] D. G. Kirkpatrick. Efficient computation of continuous skeletons. In *Proc. 18th FOCS*, pages 162–170, 1977.

[11] R. Klein and A. Lingas. A linear-time randomized algorithm for the bounded Voronoi diagram of a simple polygon. In *Proc. 9th Ann. ACM Symp. Comp. Geom.*, pages 124–132, 1993.

[12] R. Klein and A. Lingas. Fast skeleton construction. To appear, *Proc. 3rd Europ. Symp. on Alg. (ESA'95)*, 1995.

[13] D. T. Lee. Medial axis transformation of a planar shape. *IEEE Trans. Pat. Anal. Mach. Int.*, PAMI-4(4):363–369, 1982.

[14] S. Lu, H. Xu, and C. Wang. Detecting and eliminating false strokes in skeletons by geometric analysis. In *SPIE's OE/Technology'92, Vision Geometry*, Boston, Nov. 1992.

[15] A. Okabe, B. Boots, and K. Sugihara. *Spatial Tessellations: Concepts and Applications of Voronoi Diagrams*. John Wiley & Sons, 1992.

[16] A. Rosenfeld. Axial representation of shape. *Comp. Vis. Graph. Image Proc.*, 33:156–173, 1986.

[17] M. I. Shamos and D. Hoey. Closest point problems. In *Proc. 16th FOCS*, pages 151–162, 1975.

[18] P. Vermeer. Two-dimensional mat to boundary conversion. In *Proc. 2nd Symp. Solid Model. Appl.*, pages 493–494, 1993.

[19] C. A. Wang and F. Chin. Finding the constrained Delaunay triangulation and constrained Voronoi diagram of a simple polygon in linear time. To appear, *Proc. 3rd Europ. Symp. on Alg. (ESA'95)*, 1995.

The First Subquadratic Algorithm for Complete Linkage Clustering*

Drago Krznaric and Christos Levcopoulos

Department of Computer Science
Lund University, Box 118, S-221 00 Lund, Sweden.

1 Introduction

Let S be any set of n points (called vertices) in the plane. The complete linkage (c-link) clustering produces a hierarchy of clusters as follows: Initially each vertex constitutes a cluster. As long as there is more than one cluster, the two closest clusters are merged, where the distance between two clusters is defined as the distance between the two vertices, one from each cluster, that are farthest apart.

Hierarchical clustering algorithms are of great importance for structuring and interpreting data in domains such as biology, medicine, geographical information systems, and image processing [3]. Among the different methods for producing a hierarchy of clusters, the c-link clustering is one of the most well-known, and has thus been used for applications [1, 2, 3].

In this paper we show that a c-link hierarchy can be computed in $O(n \log^2 n)$ worst-case time using linear space. In addition we show that a hierarchy of clusters that approximate a c-link hierarchy can be computed in $O(n \log n)$ time (actually $O(n)$ time suffices if the Voronoi diagram is known). The techniques presented in this paper can also be generalized to other types of clusterings.

The paper is organized as follows. In Sect. 2 we define a hierarchy of well separated clusters, so-called 2-clusters, which are easy to compute and that are useful in order to organize the algorithm in a hierarchical fashion. In that section we also define more precisely in which sense we approximate a c-link hierarchy, and state a lemma that is crucial for our approach. An outline of the algorithm is given in Sect. 3, and in Sect. 4 we describe how it is possible to compute c-link distances between clusters in total time $O(n \log^2 n)$. Finally we describe in Sect. 5 how an approximation of a c-link hierarchy can be computed.

2 Preliminaries

We use the hierarchical clustering method from [7] to decompose S into subsets for which we compute a c-link hierarchy locally. The *rectangular diameter* of a vertex set D, abbreviated rdiam(D), is defined as the diameter of the smallest enclosing rectangle with sides parallel to the coordinate axes.

---

* This paper was partially supported by TFR. E-mail: {drago,christos}@dna.lth.se.

Definition 1. A subset D of S is a *2-cluster* if and only if the distance between vertices of D and vertices of $S - D$ is greater than $2 \cdot \mathrm{rdiam}(D)$ or D equals S.

In [7] it was shown that any two non-identical 2-clusters are either disjoint or one of them is a proper subset of the other. This property causes the 2-clusters to form in a natural way the following unique hierarchy.

Definition 2. The *2-cluster tree* of S is a rooted tree whose nodes correspond to distinct 2-clusters, where the root corresponds to S and the leaves to single vertices in S. Let a be any internal node and let A be its corresponding 2-cluster. Then the children of a correspond to all 2-clusters C such that $C \subset A$ and there is no 2-cluster B such that $C \subset B \subset A$.

In the continuation we will by a 2-cluster also refer to its corresponding node of the 2-cluster tree and vice versa. A simple algorithm that computes the 2-cluster tree in $O(n)$ time from a Euclidean minimum spanning tree of S was given in [7]. As a byproduct of that algorithm, for each 2-cluster D, we also receive a shortest diagonal such that it has one endpoint in D and the other in $S - D$. The following is another property about 2-clusters that we will use.

Fact 3 (Lemma 3.3 in [7]). *Let D be an arbitrary 2-cluster with $m \geq 2$ children. Among all diagonals with endpoints in distinct children of D, let l be the length of a shortest and let l' be the length of a longest such diagonal. Then the ratio between l' and l is less than 2^{2m-3}.*

We define an approximation of a c-link hierarchy as follows.

Definition 4. An *ϵ-procedure* for the c-link clustering produces a hierarchy of clusters by repeatedly merging two clusters in such a way that the following holds: If d is the distance between the two clusters next merged and d' is the distance between the two closest clusters, then $\frac{d}{d'} \leq 1 + \epsilon$ (all distances in this sentence are according to the c-link measure). We say that a hierarchy of clusters is an *ϵ-approximation* of a c-link hierarchy if it can be produced by an ϵ-procedure.

The algorithm in the next section merges two clusters C and C' only if the c-link distance between them is locally shortest, in the sense that no other cluster is within smaller c-link distance from C or C'. The main reason for this approach is that we can check efficiently whether the c-link distance between two clusters is locally shortest, because there are only a constant number of other clusters in their vicinity. This is formalized in the following lemma.

Lemma 5. *For any ϵ-procedure P with $\epsilon \leq 0.1$, and after an arbitrary number of mergings with P, the following holds: if the c-link distance of each pair of clusters is greater than l, then no circle of radius $O(l)$ contains more than a constant number of clusters.*

Proof. Is omitted in this short version but can be found in [5]. □

3 An Outline of the Algorithm

Let $\text{dist}(C, C')$ stand for the c-link distance between two clusters C and C' (the way in which we compute c-link distances will be discussed in Sect. 4).

3.1 Processing Each 2-cluster Separately and in Phases

It is easy to show that each 2-cluster is also a cluster in any c-link hierarchy. Thus we can compute a c-link hierarchy of S by a depth-first search from the root of the 2-cluster tree, and, for each 2-cluster D, clustering the children of D when backtracking from D. Therefore, in the continuation, it suffices to describe how we compute a c-link hierarchy of an arbitrary 2-cluster D, under the assumption that we have computed it within each child of D.

Let l be the length of a shortest diagonal such that it connects two vertices belonging to distinct children of D. We compute a c-link hierarchy of D in a sequence $p_0, p_1, \ldots$ of phases. The objective of a phase p_i is to merge each pair (C, C') of clusters such that $\text{dist}(C, C') \in [2^i l, 2^{i+1} l)$. We define the *parameter* of phase p_i, abbreviated as l_i, to be the number $2^i l$.

Now, let D' be a child of D, and let l' be the length of a shortest diagonal such that it has one endpoint in D' and the other in $S - D'$. Clearly there is no other cluster within c-link distance $< l'$ from D'. So D' can be ignored until we reach a phase at which we merge clusters such that the c-link distance between them is $\geq l'$. Therefore, if $l' \in [l_i, 2l_i)$, D' is assigned to phase p_i. In this way we assign each child of D to a phase, and the child is kept idle until we reach that phase. We observe that, if D has m children, then $O(m)$ time suffices by Fact 3 in order to assign the children of D to various phases.

3.2 Computing a Phase p_i

Let L be a list of all clusters that have been assigned to a phase p_i, either initially or during the previous phase. For each cluster C in L, we compute a set $N(C)$ including all other clusters C' in L such that $\text{dist}(C, C') < 2l_i$, but not including any cluster C'' such that the c-link distance of (C, C'') is greater than, say, $10l_i$.

By Lemma 5 a set $N(C)$ contains a constant number of clusters. We can therefore find a set $N(C)$ in constant time by using the range searching technique developed in [6]. That technique builds basically on the construction of a threaded quadtree for S, which can be done in $O(n)$ time given the Voronoi diagram of S (see [6] for more details).

Next we create a graph G whose nodes correspond to the clusters in L. For each pair (C, C') of nodes (clusters) we have an edge between them if and only if $\text{dist}(C, C') < 2l_i$. We define the weight of an edge (C, C') to be equal to $\text{dist}(C, C')$. Observe that it suffices to consider the constant number of clusters in $N(C)$ when calculating which edges that are to be incident to C. The phase p_i can now be computed by traversing G as below described.

We start at any edge e of G and perform a *shortest first search* from e in the following way. First we mark e as visited. If there is an edge incident to e

of less weight than e, we visit that edge and repeat the procedure recursively. The search ends when we are at an edge e' which is not adjacent to an edge of smaller weight, at which time we merge the two clusters that correspond to the endpoints of e'. This means that the edge e' is shrunk to a single node v of G. Further, the weight of every edge incident to v is updated appropriately, and those that become $\geq 2l_i$ are removed from G. Finally, we pick the edge of smallest weight incident to v and repeat the procedure from this edge. In case there is no edge incident to v, we backtrack to the latest visited edge of G, or if all edges of G are unvisited then we can repeat the procedure from an arbitrary edge.

Eventually the graph G will become edgeless. We are then left with a set of clusters, each cluster corresponding to a node in G. The phase p_i is completed after having assigned these clusters to the next phase p_{i+1}.

3.3 Analysis of the Run Time (Sketch)

We first observe that the number of edges in G is by Lemma 5 linear with respect to number of clusters in the list L. Thus, if we temporarily ignore the time used for computing c-link distances between clusters, the total time used for computing phase p_i is linearly bounded by the number of clusters in L, because we do not backtrack successively more than twice in G without merging two clusters.

Now, let C be an arbitrary cluster in L. We observe that there is some other cluster within c-link distance $O(l_i)$ from C (C would otherwise constitute a 2-cluster which is not possible because of the way in which we assign the children of D to various phases). Therefore, within a constant number of phases after p_i, either C participates in a merging, or a cluster within distance $O(l_i)$ from C participates in a merging, and so we can associate C with that merging. By Lemma 5 we associate in this way no more than a constant number of clusters to each merging. Thus, since we perform at most $n - 1$ mergings during the whole algorithm, we obtain

Lemma 6. *Let S be any set of n vertices in the plane. Given the Voronoi diagram of S, a complete linkage hierarchy of S can be computed in time $O(n)$ plus the time taken to compute c-link distances between clusters.*

4 Computing Complete Linkage Distances

In order to compute c-link distances between clusters efficiently, each cluster is kept partitioned into subsets for which we have a structure, that we call *Voronoi structure*. The Voronoi structure of a set V of vertices consists of the following pieces:

1. the farthest point Voronoi diagram of V,
2. a data structure for point location in this Voronoi diagram, and

3. the (at most) four extreme vertices of V in x and y direction, which we call *xy-extremes*.

We may, for example, compute the farthest point Voronoi diagram according to [9], and use the data structure from [4] for point location. The Voronoi structure of V can then be constructed in time $O(|V|\log|V|)$, and, using this Voronoi structure, we can for an arbitrary point p find the vertex in V that is farthest from p in time $O(\log|V|)$.

Initially each vertex in S is equipped with its trivial Voronoi structure. For a cluster C, a subset of C which is equipped with a Voronoi structure shall be called a *Voronoi child* (or *vorchild* for short). The vorchildren of a cluster are kept in a list for which we maintain the following invariants: The vorchildren appear in the list by increasing order with respect to their cardinalities, and the cardinalities of two consecutive vorchildren in the list differs at least by a factor 4. Thus the number of vorchildren of a cluster is logarithmic with respect to the cardinality of the cluster.

Now, in order to maintain these invariants, whenever we merge a pair (C_1, C_2) of clusters into a single cluster C, we impose the following operations: Let n_1 and n_2 be the cardinalities of C_1 and C_2, respectively, with $n_1 \leq n_2$. First all vorchildren of C_1 and those vorchildren of C_2 that have cardinality $\leq 4n_1$ are merged into a single vorchild V (their Voronoi structures are thrown away and we construct a new Voronoi structure of their union). Next we create a list L that consists of V behind which we hang on what remains of C_2's list of vorchildren (we remove a vorchild from a list when it participate in a merging). It is easy to show that $|V| < 7n_1$, so the vorchildren in L are sorted with respect to their cardinalities. Now, as long as the cardinality of the first (smallest) vorchild in L is larger than a quarter of the second vorchild in L, we merge these two vorchildren into a single vorchild and insert the newly created vorchild first in L. The list L that remains after this procedure comprise C's list of vorchildren.

It remains to describe how the c-link distance between two clusters C_1 and C_2 is computed at a phase p_i. Let V_1 and V_2 be two vorchildren of C_1 and C_2, respectively. First we compute the rectangular diameter, $\mathrm{rdiam}(V_1 \cup V_2)$, which we defined in Sect. 2. The rectangular diameter is easily obtained in constant time by using the xy-extremes of V_1 and V_2. If $\mathrm{rdiam}(V_1 \cup V_2) < l_i$ (the parameter of p_i) then also the c-link distance of (V_1, V_2) is $< l_i$, and so this pair of vorchildren is not relevant (the c-link distance of (C_1, C_2) is $\geq l_i$, otherwise they would have been merged at some previous phase). Therefore, only if $\mathrm{rdiam}(V_1 \cup V_2) \geq l_i$, we proceed and compute the c-link distance between V_1 and V_2 as follows. Suppose that $|V_1| \leq |V_2|$. For a vertex v in V_1 we compute $\mathrm{rdiam}(\{v\} \cup V_2)$. Again, if $\mathrm{rdiam}(\{v\} \cup V_2) < l_i$, v is not relevant. Otherwise, we use the Voronoi structure of V_2 to find the vertex in V_2 that lies farthest from v. By considering all vertices of V_1 in this way, we thus compute the c-link distance between V_1 and V_2 (if it is $\geq l_i$). Finally, by following this procedure for all pairs of vorchildren, one from C_1 and the other from C_2, we obtain the c-link distance between C_1 and C_2.

Theorem 7. *Let S be any set of n vertices in the plane. A complete linkage hierarchy of S can be computed in time $O(n\log^2 n)$.*

Proof. Throughout the proof, by $\log n$ we shall mean $\lceil \log_2 n \rceil$.

Lemma 8. *The total time for constructing Voronoi structures is $O(n \log^2 n)$.*

Proof. Let v be an arbitrary vertex. When v participates in a merging of vorchildren, we can differ between the following two cases.

(i) The vorchild that v belongs to after this merging has cardinality at least 1.25 times the cardinality of the vorchild that v belonged to before this merging. It is easy to show that v can participate in a merging of this type at most $\lceil 3.2 \log n \rceil$ times.
(ii) v belongs to a cluster which is being merged with a cluster of at least the same cardinality. It is easily seen that v can participate in a merging of this type at most $\log n$ times.

When constructing the Voronoi structure of a set V of vertices, each vertex in V contributes $O(\log |V|)$ time. Thus the total time for computing Voronoi structures is $O(n \log^2 n)$. □

Next we note that the total time spent for computing the rectangular diameter of pairs of vorchildren is $O(n \log^2 n)$, because each cluster has at most $\log n$ vorchildren and the complete linkage distance is computed for at most $O(n)$ pairs of clusters (see Sect. 3.3).

For the sake of analysis, when we use the point location structure for the (farthest point) Voronoi diagram of a vorchild V_2 to find its vertex that lies farthest from a query vertex v, we say that v makes a *query* on V_2. In particular, when a vertex v makes a query on a vorchild V_2 of cardinality in $[2^k, 2^{k+1})$, we say that v makes a *k-query* on V_2. Similarly, we say that a vertex v makes an *estimate* on vorchild V_2 when we compute $\mathrm{rdiam}(\{v\} \cup V_2)$ (just before deciding whether v shall make a query). When v makes an estimate on V_2 and V_2 has cardinality in $[2^k, 2^{k+1})$, we say that v makes a *k-estimate* on V_2.

To complete the proof of this theorem it suffices to show that if any vertex uses $O(\log n)$ time to make any query and $O(1)$ time to make any estimate, then the total time used by the algorithm is $O(n \log^2 n)$. For this purpose, imagine a bank containing coins, the coins having value so that $\log n$ coins suffice to make any query and one coin suffices to make any estimate. In the remainder, c is a sufficiently large constant (an upper bound on c can be derived from this proof, but we do not explicitly calculate it here).

Initially each vertex receives $c \log^2 n + c \log n$ coins from the bank. At any phase p_i with parameter l_i, when a vorchild V expires by some merging, if a vertex v belongs to a vorchild of cardinality $\leq 2|V|$ which is within c-link distance $\leq 10 l_i$ from V, then v receives $2c \log n$ coins from the bank. Similarly, when a vorchild V is created, each vertex in V receives $c \log n$ coins from the bank.

Lemma 9. *The total amount of coins that the bank spends is $O(n \log^2 n)$.*

Proof. Let V_1 be a vorchild which is created or that expires by some merging at a phase p_i, and let n_1 be the cardinality of V_1. First we recall that an arbitrary cluster C has at most one vorchild of cardinality in $[2^k, 2^{k+2})$ for any positive integer k. Thus the total number of vertices in vorchildren of C that have cardinality $\leq 2n_1$ is $< 3n_1$, and by Lemma 5 there are at most $O(1)$ clusters within distance $O(l_i)$ from V_1. Hence, when V_1 is created or expires by some merging, it causes the bank to spend a total amount of $O(n_1 \log n)$ coins. Thus each vertex in V_1 contributes $O(\log n)$ coins to this amount. But it follows from the proof of Lemma 8 that the sum of cardinalities of all vorchildren is $O(n \log n)$. So the total amount of coins that the bank spend is $O(n \log^2 n)$. $\square$

It suffices now to show that for an arbitrary vertex v it holds that v can afford to spend $\log n$ coins whenever v makes a k-query and one coin whenever v makes a k-estimate. For this purpose, imagine arrays Q and E with slots $q_1, q_2, \ldots, q_{\log n}$, respectively $e_1, e_2, \ldots, e_{\log n}$. From the coins that v received initially from the bank, a portion of $c \log^2 n$ and a portion of $c \log n$ coins are divided equally among the slots of Q and among the slots of E, respectively, so that each slot of Q contains $c \log n$ coins whereas each slot of E contains c coins. The vertex v spends $\log n$ coins in slot q_k whenever v makes a k-query, and one coin in e_k whenever v makes a k-estimate.

Let V be a vorchild such that v has made a k-query on V. When V expires by merging, v receives $2c \log n$ coins from the bank, and distributes them as follows: $c \log n$ coins are inserted in slot q_k, and c coins are inserted in each slot of the array E. On the other hand, when the vorchild that v belongs to expires and a new vorchild V is created which contains v, the vertex v receives $c \log n$ coins from the bank. At such a occasion, v inserts c coins in each slot of the array E.

Lemma 10. *For a sufficiently large c, no slot of array Q becomes empty.*

Proof. Suppose for contradiction that v inserts $c \log n$ coins in q_k at time t_1 and that v after that has made c k-queries without inserting any coin in q_k, and let t_2 be the time when the last of these c k-queries was made. If v becomes a member of a vorchild of cardinality $\geq 2^{k+1}$, then it cannot make a k-query again (a vertex makes queries only on vorchildren whose cardinality is at least as large as its own vorchild). Therefore we can assume in the continuation that v belongs to a vorchild of cardinality $< 2^{k+1}$ between t_1 and t_2.

Now, assume that v makes a k-query on a vorchild V' at some time t' between t_1 and t_2. First we observe that V' cannot expire by some merging at any time before t_2 (otherwise v would insert $c \log n$ more coins in q_k). We claim that v can make at most $c' = O(1)$ k-queries on V', where the constant c' does not depend on c. Indeed, by Lemma 5 it follows that, at any phase p_i, no more than $O(1)$ clusters which are within distance $O(l_i)$ from v are merged during phase p_i. Hence, as our algorithm works, v makes at most $O(1)$ k-queries on V' during any phase. The claim follows now easily, because within a constant number of phases after t', rdiam($\{v\} \cup V'$) becomes less than the parameter of a phase.

Let p_j be the phase which includes time t_2. From above we see that between time t_1 and t_2, v has made at least c/c' k-queries on different vorchildren. But no two of these vorchildren can ever belong to the same cluster (recall that a cluster can only have one vorchild of cardinality in $[2^k, 2^{k+1})$). Hence, at time t_2, there are at least c/c' clusters within c-link distance $\leq 10l_j$ from v, contradicting Lemma 5 if c is sufficiently large. □

It remains to show that our arbitrary vertex v can spend one coin in slot e_k whenever v makes a k-estimate. Suppose for contradiction that v inserts c coins in e_k at some time, call it t_1, and that v after that has made c k-estimates without inserting any coin in e_k, and let t_2 be the time when the last of these c k-estimates was made. Let V be the vorchild which v belonged to just after time t_1. First we observe that V cannot expire by a merging before time t_2, because otherwise a new vorchild containing v would be created, at which time v would insert c coins in e_k. Now, assume that v makes a k-estimate on a vorchild V' at some time t' between t_1 and t_2. Recall that V' cannot expire by some merging at any time before t_2. Therefore, since both V and V' remain between t' and t_2, within a constant number of phase after t' we reach a phase, say p_i, such that rdiam$(V \cup V')$ is smaller than l_i. So after p_i no vertex of V will make a k-estimate on V'. Let p_j be the phase which includes time t_2. As in the proof of Lemma 10, for some constant c' independent of c, we have that there are at time t_2 at least c/c' clusters within c-link distance $\leq 10l_j$ from v, contradicting Lemma 5 if c is sufficiently large. □

5 Computing an ϵ-Approximation

The accuracy of our approximation depends on how well we approximate the c-link distance between two clusters. For this purpose we keep track of the extreme vertices of each cluster in k directions (k is an integer ≥ 4 and the directions are uniformly chosen so that the angle between consecutive directions equals $2\pi/k$). Such extreme vertices shall be called k-extremes. We define the k-distance between two clusters C and C', denoted by k-dist(C, C'), as the distance between the two k-extremes that are farthest apart, one from C and the other from C'.

Lemma 11. *Using the k-distance in the algorithm of Section 3, we compute an ϵ-approximation of a complete linkage hierarchy with $\epsilon < \frac{5}{k^2-5}$.*

Proof. Is omitted in this short version but can be found in [5]. □

From Lemma 11 above it follows that the statement in Lemma 5 holds if $k \geq 8$ (ϵ is then < 0.1). For smaller k however (between 4 and 7) we can instead rely on the following lemma (note that we in this lemma do not consider an arbitrary ϵ-procedure as in Lemma 5).

Lemma 12. *Using the k-distance in the algorithm of Section 3, at any phase p_i, no circle of radius $O(l_i)$ contains more than a constant number of clusters (within the 2-cluster being processed).*

Proof. Is omitted in this short version but can be found in [5]. □

For two clusters C and C' we can compute k-dist(C, C') in $O(k)$ time by keeping track of the convex hull of the k-extremes of each cluster. Indeed, the diameter of a convex hull can be computed in linear time, and two convex hulls can be merged in linear time (see, for example, Sect. 3.3.5 and 4.2.3 in [8]). So the total time used for computing k-distances between clusters is $O(kn)$, since we compute it for at most $O(n)$ pairs of clusters (see Sect. 3.3). Hence, combining Lemmata 6 and 11, we obtain

Theorem 13. *Let S be any set of n vertices in the plane. Given the Voronoi diagram of S, we can compute an ϵ-approximation of a complete linkage hierarchy of S with the above method in time $O(\frac{n}{\sqrt{\epsilon}})$ for $0 < \epsilon < 5/11$.*

However, for very small values of ϵ we can use the ideas from Sect. 4 to compute the k-distances in total time $O(n \log^2 k)$. Indeed, we can use the algorithm in that section with only the following two modifications.

1. *Clean-up:* If a cluster C is created such that the total sum of cardinalities in all vorchildren of C is $\geq 2k$, we replace these vorchildren by a single vorchild V whose vertices correspond to the k-extremes of C (we insert V into C's list of vorchildren and remove all other vorchildren).
2. We define the *vorcardinality* of a cluster as the total cardinality in all its vorchildren, and we use the vorcardinality instead of the actual cardinality throughout the algorithm (this affects only the merging of vorchildren).

By the clean-ups we get that each cluster has at most $O(\log k)$ vorchildren, and that each of its k-extremes belongs to one of its vorchildren. To compute the k-distance between two clusters we use their vorchildren in the same way as in Sect. 4.

In this paragraph we consider the time used for computing Voronoi structures. For simplicity we suppose that each vertex receives initially $\lceil 8.4 \log 2k \rceil + 2$ bank notes, each note having value so that the vertex can participate in one clean-up or one merging of vorchildren (thus a bank note has value $O(\log k)$ since all vorchildren have now cardinality at most $2k$). Let v be an arbitrary vertex. When v participates in a merging of vorchildren, either the new resulting vorchild (that v becomes a member of) has cardinality at least 1.25 times greater than its previous vorchild, or its previous vorchild belongs to a cluster being merged with a cluster of at least the same vorcardinality. Thus v cannot participate in more than $\lceil 4.2 \log 2k \rceil$ mergings of vorchildren without participating in a clean-up. Further, when v participates in a clean-up, at least k vertices participating in this clean-up will never again belong to some vorchild. Therefore, if v is not one of these vertices, v can receive $\lceil 4.2 \log 2k \rceil + 1$ more bank notes from one of them. In this way, v has enough bank notes to participate in additional $\lceil 4.2 \log 2k \rceil$ mergings of vorchildren and one clean-up. Thus we can conclude that the total time used for computing Voronoi structures is $O(n \log^2 k)$.

It remains to consider the time used for computing k-distances between clusters. Recall that all vorchildren have now cardinality at most $2k$. Therefore, referring to the proof of Theorem 7, we can use an analogous analysis but having

$O(\log k)$ slots in arrays Q and E and using $O(\log k)$ coins when making a query. In this way we can show that the total time used for computing k-distances is $O(n \log^2 k)$, thus obtaining the following theorem.

Theorem 14. *Let S be any set of n vertices in the plane. Given the Voronoi diagram of S, we can compute an ϵ-approximation of a complete linkage hierarchy of S in time $O(n \log^2 \frac{1}{\epsilon})$ for $0 < \epsilon < 5/11$.*

6 Final Remarks and Further Research

The techniques in this paper can also be adapted for other measures of proximity between two clusters, for example:

1. the perimeter of the convex hull (or of the smallest enclosing rectangle, circle, etc.) of their union,
2. different sorts of weighted distances between the clusters (like the distance between their centroids).

For several such measures we can prove the analog of Lemma 5, so the same approach yields fast algorithms.

In addition, extending the approach to more than two dimensions, Lemmata 5 and 12 can still be adapted, although the constants grow with the number of dimensions.

Acknowledgment We are grateful to Dr. Günter Rote for his very helpful and encouraging comments, and for pointing out some errors in an earlier draft.

References

1. Franz Aurenhammer. *Voronoi diagrams - A survey of a fundamental geometric data structure*. ACM comp. Surveys 23(3), 1991, 345-405.
2. A. D. Gordon. *Classification*. Chapman and Hall, 1981.
3. A. K. Jain and R. C. Dubes. *Algorithms for clustering data*. Prentice Hall, New Jersey, 1988.
4. D. G. Kirkpatrick. *Optimal search in planar subdivisions*. SIAM Journal of Computing 12(1), 1983, 28-35.
5. D. Krznaric and C. Levcopoulos. *Fast algorithms for complete linkage clustering*. Tech. Rep. LU-CS-TR:95-143, Dep. Comp. Sci., Lund Univ., 1995.
6. D. Krznaric and C. Levcopoulos. *Computing a threaded quadtree (with links between neighbors) from the Delaunay triangulation in linear time*. 7th Canadian Conf. on Computational Geometry, 1995, 187-192.
7. D. Krznaric and C. Levcopoulos. *Computing hierarchies of clusters from the Euclidean minimum spanning tree in linear time*. 15th Conf. on FST & TCS, Lecture Notes in Comp. Sci., 1995.
8. F. Preparata and M. Shamos. *Computational geometry: an introduction*. Springer-Verlag, 1985.
9. M. I. Shamos. *Computational geometry*. Ph. D. thesis, Dep. Comp. Sci., Yale Univ., 1978.

Matching Nuts and Bolts Faster*

Phillip G. Bradford   Rudolf Fleischer

Max-Planck-Institut für Informatik, Im Stadtwald, 66123 Saarbrücken, Germany. E-mail:
{bradford,rudolf}@mpi-sb.mpg.de.

Abstract. The problem of matching nuts and bolts is the following : Given a collection of n nuts of distinct sizes and n bolts such that there is a one-to-one correspondence between the nuts and the bolts, find for each nut its corresponding bolt. We can *only* compare nuts to bolts. That is we can neither compare nuts to nuts, nor bolts to bolts. This humble restriction on the comparisons appears to make this problem very hard to solve. In fact, the best deterministic solution published to date is due to Alon *et al.* [2] and takes $\Theta(n \log^4 n)$ time. Their solution uses (efficient) graph expanders. In this paper, we give a simpler $O(n \log^2 n)$ time algorithm which uses only a simple (and not so efficient) expander.

1   Introduction

In [12], page 293, Rawlins posed the following interesting problem :

We wish to sort a bag of n nuts and n bolts by size in the dark. We can compare the sizes of a nut and a bolt by attempting to screw one into the other. This operation tells us that either the nut is bigger than the bolt; the bolt is bigger than the nut; or they are the same size (and so fit together). Because it is dark we are not allowed to compare nuts directly or bolts directly.
How many fitting operations do we need to sort the nuts and bolts in the worst case?

As a mathematician (instead of a carpenter) you would probably prefer to see the problem stated as follows ([2]) :

Given two sets $B = \{b_1, \ldots, b_n\}$ and $S = \{s_1, \ldots, s_n\}$, where B is a set of n distinct real numbers (representing the sizes of the bolts) and S is a permutation of B, we wish to find efficiently the unique permutation $\sigma \in S_n$ so that $b_i = s_{\sigma(i)}$ for all i, based on queries of the form compare b_i and s_j. The answer to each such query is either $b_i > s_j$ or $b_i = s_j$ or $b_i < s_j$.

The obvious information theoretic lower bound shows that at least $\Omega(n \log n)$ comparisons are needed to solve the problem, even for a randomized algorithm. In fact, there is a simple randomized algorithm which achieves an expected running time of $O(n \log n)$, namely Quicksort : Pick a random nut, find its matching bolt, and then split the problem into two subproblems which can be solved recursively, one consisting of the nuts and bolts smaller than the matched pair and one consisting of the larger ones. The standard analysis of randomized Quicksort gives the expected running time as stated above (see for example [5]).

Unfortunately, it is much harder to find an efficient deterministic algorithm. The only one known to us is the algorithm by Alon *et al.* [2] which is also based on Quicksort. To find a good pivot element which splits the problem into two subproblems of nearly the same size, they run $\log n$ iterations of a procedure which eliminates half of the nuts in each iteration while maintaining at least one good pivot; since there is only one nut left in the end, this one must be

---
* The authors were supported by the ESPRIT Basic Research Actions Program, under contract No. 7141 (project ALCOM II).

a good pivot. This procedure uses the edges of a highly efficient expander of degree $\Theta(\log^2 n)$ to define its comparisons. Therefore, finding a good pivot takes $\Theta(n \log^3 n)$ time, and the entire Quicksort takes $\Theta(n \log^4 n)$ time.

In this paper, we propose a simpler algorithm to find a good pivot (see Section 3 for details). First, we connect the set of nuts with the set of bolts via some expander of constant degree and compare each nut to all the bolts to which it is connected by an edge of the expander. We discard all nuts which are only connected to smaller bolts or only connected to larger bolts. Then we play a simple knockout tournament on the remaining nuts (where in each round half of the nuts are eliminated) which guarantees that the winner of the tournament is a good pivot. Since we can play each round of the tournament in $O(n)$ time, we can find a good pivot in $O(n \log n)$ time. Therefore, we can solve the nuts and bolts matching problem in $O(n \log^2 n)$ time.

And recently, Komlós, Ma and Szemerédi [7] have advised us that they have found an $O(n \log n)$ time algorithm for solving the nuts and bolts problem.

Alon *et al.* [2] mention two potential applications of this problem: the first is local sorting of nodes in a given graph [6], and the second is selection of read only memory with a little read/write memory [11].

In the next section, we describe the Quicksort algorithm more formally and recall some facts about expanders. In Section 3, we show how we can efficiently find a good pivot. And we conclude with some remarks in Section 4.

2 Basic Definitions

Let $S = \{s_1, \ldots, s_n\}$ be a set of nuts of different sizes and $B = \{b_1, \ldots, b_n\}$ be a set of corresponding bolts. For a nut $s \in S$ define $rank(s)$ as $|\{t \in B \mid s \geq t\}|$. The rank of a bolt is defined similarly. For a constant $c < \frac{1}{2}$, s is called a *c-approximate median* if $cn \leq rank(s) \leq (1-c)n$. Similarly, define the *relative rank* of s with respect to a subset $T \subseteq B$ as $rank_T(s) := \frac{|\{t \in T \mid s \geq t\}|}{|T|}$.

If T is a multiset then the relative rank of s with respect to T is defined analogously, where each $t \in T$ is counted according to its multiplicity in T.

The algorithm for matching nuts and bolts works as follows.

(1) Find a c-approximate median s of the n given nuts (we will determine c later).

(2) Find the bolt b corresponding to s.

(3) Compare all nuts to b and all bolts to s. This gives two piles of nuts (and bolts as well), one with the nuts (bolts) smaller than s and one with the nuts (bolts) bigger than s.

(4) Run the algorithm recursively on the two piles of the smaller nuts and bolts and the two piles of the bigger nuts and bolts.

In the next section, we will show how to find a c-approximate median in $O(n \log n)$ time, where c is a small constant. Then our main result follows immediately.

Theorem 1. *We can match n nuts with their corresponding bolts in $O(n \log^2 n)$ time.*

Proof. The correctness of the algorithm above follows immediately from the correctness of Quicksort. For the running time observe, that each subproblem has size at most $(1 - c)n$, hence the depth of the recursion is only $O(\log n)$, and in each level of the recursion we spend at most $O(n \log n)$ time to compute the c-approximate median and $O(n)$ time to split the problem into the two subproblems. □

We now recall some facts about expanders (see for example [8] if you want to learn more about expanders). Let $0 < \alpha \leq \frac{1}{2}$ and $\alpha c < 1$. An (n, k, α, c)-*expander* is a k-regular bipartite graph on vertices $\mathcal{I}$ (inputs) and $\mathcal{O}$ (outputs), where $|\mathcal{I}| = |\mathcal{O}| = n$, such that every subset $A \subset \mathcal{I}$ of size at most αn is joined by edges to at least $|A| \left(1 + c(1 - \frac{|A|}{n})\right)$ different outputs. The constant c is called the *expansion factor* of the graph.

Theorem 2 (Alon, Galil, and Milman [3], Cor. 2.3). *If $n = m^2$ for some integer m, then we can construct an $(n, 9, \frac{1}{2}, 0.41)$-expander in $O(n)$ time.*

Corollary 3. *Let $0 < \delta \leq \frac{1}{2}$ and $\gamma_\delta = \frac{3-5\delta}{5\delta(1-\delta)}$. Then there exists an integer q_δ such that for any n, where $n = m^2$ for some integer m, we can construct an $(n, q_\delta, \delta, \gamma_\delta)$-expander in $O(n)$ time. In such an expander, any subset of the inputs of size δn is connected to at least $\frac{3}{5}n$ different outputs.*

Proof. We take a series of the expanders of Theorem 2 and identify the outputs $\mathcal{O}_1$ of the first one with the inputs $\mathcal{I}_2$ of the second one, the outputs $\mathcal{O}_2$ of the second one with the inputs $\mathcal{I}_3$ of the third one, and so on. Then there is an integer k_δ (independent of n) such that any set of δn inputs of $\mathcal{I}_1$ is connected to at least $\frac{n}{2}$ different outputs of $\mathcal{O}_{k_\delta}$ and hence to at least $(1 + \frac{0.41}{2})\frac{n}{2} > \frac{3}{5}n$ different outputs of $\mathcal{O}_{k_\delta+1}$. We can easily calculate k_δ by computing the series defined by $a_0 := \delta$ and $a_{i+1} := a_i(1 + 0.41(1 - a_i))$; then k_δ is the smallest index i such that $a_i \geq \frac{1}{2}$.

Hence, to get the desired bipartite graph, we only have to connect each node v of $\mathcal{I}_1$ to all nodes w of $\mathcal{O}_{k_\delta+1}$ which can be reached from v by traversing a path which uses exactly one edge from each of the $k_\delta + 1$ expanders. Then the degree of any node is clearly at most $q_\delta := 9^{k_\delta+1}$. To make the graph q_δ-regular we can add arbitrary dummy edges without destroying the expansion property. Further, the expansion factor of the graph is at least γ_δ because $(1 + \gamma_\delta(1 - \delta))\delta n = \frac{3}{5}n$ (note that subsets of the inputs of size smaller than δn are even better expanded). $\square$

3 Finding a c-Approximate Median

Our algorithm to find a c-approximate median is based on a knockout tournament played on some subset of the nuts. We start with a subset $S_1 \subset S$ of the nuts where each nut $s \in S_1$ has a set $T_1(s)$ of two bolts associated with it; for all s, the sets $T_1(s)$ need not be disjoint, but every bolt may appear only in a constant number of them. We describe later how S_1 is constructed.

We then play $\lceil \log |S_1| \rceil$ rounds of the tournament, where in each round half the nuts survive for the next one. Intuitively, we take any two nuts together with their sets of associated bolts, determine which nut splits the union of both sets of bolts less equally, eliminate that nut, and give both sets of bolts to the surviving nut. Unfortunately, pairing the nuts arbitrarily does not quite work, i.e., the winner of the tournament would not necessarily be a c-approximate median, but there is a simple way to overcome that difficulty.

In general, let S_i be the set of nuts before we start round i. For each nut $s \in S_i$ let $T_i(s)$ be the multiset of bolts associated with s and let $r_i(s) := rank_{T_i(s)}(s)$ be the relative rank of s with respect to its set of bolts $T_i(s)$. Let $S_i^{high} := \{s \in S_i \mid r_i(s) \geq \frac{1}{2}\}$ be the nuts in S_i of high relative rank and $S_i^{low} := \{s \in S_i \mid r_i(s) < \frac{1}{2}\}$ be the set of nuts in S_i of small relative rank.

We play the *knockout tournament* as follows.

$i := 1;$
while $|S_i| > 2$ **do**

   (1) Pair the nuts of S_i^{high} arbitrarily. If $|S_i^{high}|$ is odd then we eliminate the single nut which did not get a partner.

(2) Let (s_1, s_2) be a pair of nuts from S_i^{high}. Compute the relative ranks of s_1 and s_2 with respect to the multiset $T := T_i(s_1) \cup T_i(s_2)$. Note that it is sufficient to compare s_1 with all bolts in $T_i(s_2)$ and s_2 with all bolts in $T_i(s_1)$, because $rank_T(s_j) = \frac{1}{2}(r_i(s_j) + rank_{T_i(s_{3-j})}(s_j))$, for $j = 1, 2$ (here we use Observation 4 (c)).

Whichever nut s has relative rank closer to $\frac{1}{2}$ survives in S_{i+1} and is associated with the multiset $T_{i+1}(s) := T$.

(3) Repeat steps (1) and (2) with S_i^{low} instead of S_i^{high}.

od

Let l be the value of i after the **while**-loop terminates, i.e., $|S_l| \leq 2$. We claim that if S_1 was sufficiently large then every nut in S_l is a c-approximate median, where c is a small constant (see Lemma 5). But first we make a few simple observations.

Observation 4. *Assume we play the tournament starting with some set S_1 of nuts. Then*

(a) $\lceil \log |S_1| \rceil - 1 \leq l \leq \lceil \log |S_1| \rceil$.

(b) $S_l \neq \emptyset$.

(c) *For $i = 1, \ldots, l$ and all $s \in S_i$, $|T_i(s)| = 2^i$. In particular, $|T_l(s)| \geq \frac{|S_1|}{2}$ for all $s \in S_l$.*

(d) *Each round needs $O(|S_1|)$ time.*

Proof.

(a) In each round, we eliminate half of the nuts which could be paired, and at most two unpaired nuts, i.e., $|S_{i+1}| \geq \frac{|S_i| - 2}{2}$. We stop if at most two nuts remain. It is easy to show by induction on $|S_1|$ that l must be at least $\log(|S_1| + 2) - 1$. This proves the first inequality. The second inequality follows directly from $|S_{i+1}| \leq \frac{|S_i|}{2}$.

(b) We never eliminate all nuts.

(c) By induction on i.

(d) Observe that in each round, every bolt is involved in at most one comparison (in step (2)). Since there are a total of $2|S_1|$ bolts in the first round and we never let additional bolts enter the game, we do at most $2|S_1|$ comparisons in each round. Further, pairing the nuts, computing the relative ranks, and merging two multisets of bolts does not increase the asymptotic complexity. $\qquad\square$

Lemma 5. *Let $S_1 \subseteq S$ and $\beta = \frac{|S_1|}{n}$. Suppose, each nut $s \in S_1$ lies between the two bolts in $T_1(s)$, i.e., $b_{\text{low}}(s) < s < b_{\text{high}}(s)$ if $T_1(s) = \{b_{\text{low}}(s), b_{\text{high}}(s)\}$, and every bolt appears at most q times in the sets $T_1(s)$. Then any $s \in S_l$ is a c-approximate median, where $c = \frac{\beta}{8q}$.*

Proof. Before the first round, we have $r_1(s) = \frac{1}{2}$ for all $s \in S_1$, and hence $\frac{1}{4} \leq r_2(s) \leq \frac{3}{4}$ for all $s \in S_2$. We now prove by induction, that this inequality holds after each round.

Assume we know that $\frac{1}{4} \leq r_i(s) \leq \frac{3}{4}$ for all $s \in S_i$. Let (s_1, s_2) be a pair from S_i^{high}, where w.l.o.g. $s_1 < s_2$. Let T be the multiset $T_i(s_1) \cup T_i(s_2)$. Since s_1 is larger than half of the bolts in $T_i(s_1)$, it must be larger than a quarter of the bolts in T. On the other hand, it is smaller than a quarter of the bolts in $T_i(s_1)$ and smaller than a quarter of the bolts in $T_i(s_2)$ (because it is smaller than s_2); hence it is smaller than a quarter of the bolts in T. Therefore, the inequality holds for s_1, and we only eliminate s_1 if the relative rank of s_2 with respect to T is even closer to $\frac{1}{2}$.

Let $s \in S_l$. Since $T_l(s)$ contains at least $\frac{\beta n}{2}$ bolts by Observation 4 (c), we conclude from the inequality above that s is larger than $\frac{\beta n}{8}$ bolts and smaller than another $\frac{\beta n}{8}$ bolts. Since each bolt may have up to q copies in $T_l(s)$, $\frac{\beta n}{8q} \leq rank(s) \leq (1 - \frac{\beta}{8q})n$, i.e., s is a $\frac{\beta}{8q}$-approximate median. $\qquad \square$

Now we can give our algorithm to find a c-approximate median.

(1) If n is not the square of an integer, then add at most $2\sqrt{n}$ dummy nuts and bolts to make n the square of an integer.

(2) Let $B = \mathcal{I}$ and $S = \mathcal{O}$ be the sets of vertices of the $(n, q, \frac{1}{20}, \frac{220}{19})$-expander of Corollary 3. We compare each bolt with all the nuts to which it is connected by an edge of the expander.
Let S_1 be the set of nuts which are compared to at least one smaller bolt and at least one larger bolt. For all $s \in S_1$, pick arbitrarily one of the smaller and one of the larger bolts and put them into the set $T_1(s)$.

(3) Now play the knockout tournament starting with S_1. Let S_l be the set of (at most two) winners of the tournament. Choose any $s \in S_l$ as a good pivot.

Theorem 6. *This algorithm computes in $O(n \log n)$ time a c-approximate median, where $c = \frac{1}{80q}$ is a small constant not depending on n.*

Proof. Construction of the expander and hence of set S_1 takes $O(n)$ time (note that enlarging n slightly in step(1) does not increase the asymptotic complexity of the following steps). And the tournament takes $O(|S_1| \log |S_1|) = O(n \log n)$ time by Observation 4.

It remains to show that we really compute a c-approximate median for some constant c. First, observe that $|S_1| \geq \frac{n}{10}$. To see this, let B_1 be the set of bolts with rank at most $\frac{n}{20}$ and B_2 be the set of bolts with rank at least $\frac{19}{20}n$. Since the expander connects any subset of B of size $\frac{n}{20}$ to at least $\frac{3}{5}n$ different nuts in S, there must be a set of at least $\frac{n}{5}$ nuts which are connected to bolts in both B_1 and B_2. But then at least $\frac{n}{5} - 2 \cdot \frac{n}{20} = \frac{n}{10}$ of the nuts must have rank between $\frac{n}{20}$ and $\frac{19}{20}n$, which means they are in the set S_1.

Next, observe that each bolt appears in at most q of the sets $T_1(s)$ because the expander connects each bolt to exactly q nuts. Hence by Lemma 5, any $s \in S_l$ is a c-approximate median, where $c = \frac{1}{80q}$. $\qquad \square$

4 Conclusions

We have presented an $O(n \log^2 n)$ time deterministic algorithm for matching nuts and bolts. This improves the first $O(n(\log n)^{O(1)})$-time solution of this problem, given by Alon *et al.* [2], by a factor of $\log^2 n$. As already mentioned in [2], the methods described in this (and their) paper seem not to be sufficient to reduce the complexity below $O(n \log^2 n)$.

Unfortunately, the constants hidden in our O-notation are incredibly large (far beyond the constant 10^8 in [2]). This is mainly due to the iterative construction in Corollary 3 which produces an expander of enormous, but still constant, degree (a short calculation shows that the degree is $q_{\frac{1}{20}} = 9^9$ because we must build the expander from nine copies of the simple expander of Theorem 2). On the other hand, the standard counting argument shows that there is a family of bipartite 24-regular graphs on $2n$ nodes which connect any subset of the inputs of size $\frac{n}{4}$ to at least $\frac{7}{8}n$ different outputs. Using these graphs in the construction of the set S_1 would give us an algorithm with fairly reasonable constants.

Noga Alon ([1]) found the following clever way of constructing S_1 using a 198-regular bipartite graph, which also improves drastically the running time of our algorithm.

Call an undirected graph G an (n, d, λ)-graph if it is a d-regular graph on n nodes, and the absolute value of each of the eigenvalues of its adjacency matrix, besides the largest, is at most λ. Call a sequence of integers *dense* if for every $\epsilon > 0$ there exists some $m_0 = m_0(\epsilon)$ so that for every $m > m_0$ the sequence contains a member between m and $(1 + \epsilon)m$. It is known that for every $d = p + 1$ where p is a prime congruent to 1 modulo 4 there is a dense sequence of integers so that for every member n of the sequence there is an explicit $(n, d, 2\sqrt{d-1})$-graph which can be constructed in time linear in the number of its edges ([9], [10]). For more details on such graphs see, e.g., [4].

From [4], page 122, Cor. 2.5., it follows that in an (n, d, λ)-graph, for any two subsets B and C of the nodes of cardinalities bn and cn respectively, where $bc > \frac{\lambda^2}{d^2}$, there is at least one edge between B and C. Choosing $d = 198$ and $\lambda = 2\sqrt{197}$, we can construct a bipartite graph $G = (T \cup S, E)$ by taking two copies V' and V'' of the nodes V of the explicit (n, d, λ)-graph and joining $u' \in V'$ to $v'' \in V''$ iff $uv \in E$. Since $\frac{1}{6} \cdot \frac{1}{8} = \frac{1}{48} > \frac{\lambda^2}{d^2}$, the graph G has the property that any subset of T of size $\frac{n}{6}$ is connected to any subset of S of size $\frac{n}{8}$ by at least one edge.

If we now put the nuts in S and the bolts in T and greedily choose a maximal family of pairwise disjoint triples t_i^1, s_i, t_i^2, where $s_i \in S$, $t_i^1, t_i^2 \in T$ and $t_i^1 < s_i < t_i^2$, then we choose at least $\frac{n}{12}$ such triples. This can be seen as follows. Let T_{low} be the $\frac{n}{3}$ bolts in T of smallest rank, T_{high} be the $\frac{n}{3}$ bolts in T of highest rank, and S_{med} be the $\frac{n}{3}$ nuts in S of medium rank. If the greedy procedure finds less than $\frac{n}{12}$ disjoint triples, then at least $\frac{n}{6}$ bolts in T_{low}, $\frac{n}{6}$ bolts in T_{high}, and $\frac{n}{4}$ nuts in S_{med} have not been chosen. But by the property of G, one of the unchosen nuts in S_{med} must be connected to an unchosen bolt in T_{low} and an unchosen bolt in T_{high}, a contradiction.

Using these $\frac{n}{12}$ triples as the start set S_1 in our tournament will give us an $\frac{1}{96}$-approximate median (Lemma 5).

Acknowledgements

We thank Noga Alon for contributing the efficient construction of the set S_1.

References

1. N. Alon. Personal communication, 1995.

2. N. Alon, M. Blum, A. Fiat, S. Kannan, M. Naor and R. Ostrovsky. *Matching nuts and bolts.* Proceedings of the 5th Annual ACM-SIAM Symposium on Discrete Algorithms (SODA'94), 1994, pp. 690–696.

3. N. Alon, Z. Galil and V.D. Milman. *Better expanders and superconcentrators.* Journal of Algorithms **8** (1987), pp. 337–347.

4. N. Alon and J. Spencer. *The Probabilistic Method.* John Wiley and Sons Inc., New York, 1992.

5. T.H. Cormen, C.E. Leiserson and R.L. Rivest. *Introduction to algorithms.* MIT Press, 1990.

6. W. Goddard, C. Kenyon, V. King and L. Schulman. *Optimal randomized algorithms for local sorting and set-maxima.* SIAM Journal on Computing 22, 1993, pp. 272–283.

7. J. Komlós, Y. Ma and E. Szemerédi. *Matching nuts and bolts in $O(n \log n)$ time.* Personal communication, Yuan Ma, 1995.

8. A. Lubotzky. *Discrete groups, expanding graphs and invariant measures.* Birkhäuser Verlag, 1994.

9. A. Lubotzky, R. Phillips and P. Sarnak. *Explicit expanders and the Ramanujan conjectures.* Proceedings of the 18th ACM STOC, 1986, pp. 240–246.
   See also: A. Lubotzky, R. Phillips and P. Sarnak. *Ramanujan graphs.* Combinatorica **8** (1988), pp. 261–277.

10. G.A. Margulis. *Explicit group-theoretical constructions of combinatorial schemes and their application to the design of expanders and superconcentrators.* Problemy Peredachi Informatsii **24** (1988), pp. 51–60 (in Russian).
    English translation in Problems of Information Transmission **24** (1988), pp. 39–46.

11. J.I. Munro and M. Paterson. *Selection and sorting with limited storage.* Theoretical Computer Science 12, 1980, pp. 315–323.

12. G.J.E. Rawlins. *Compared to what ? An introduction to the analysis of algorithms.* Computer Science Press, 1992.

Linear matching-time algorithm for the directed graph isomorphism problem

James Jianghai Fu

Department of Computer Science, University of Waterloo
Waterloo, ON, Canada, N2L 3G1
Email: jfu@neumann.uwaterloo.ca

Abstract. The directed graph isomorphism problem is a natural extension of the subtree isomorphism problem and has applications to type systems, functional languages and cyclic term graph rewriting systems. In this paper, we present an efficient algorithm for the ordered labeled directed graph isomorphism problem. Once the pattern graph is preprocessed, the matching process of our algorithm runs in time $O(|E_T|)$, where $|E_T|$ is the number of edges in the target graph. Our algorithm is particularly useful when the same pattern graph is used many times or the size of the pattern graph is small, which is typically the case in the above applications.

1 Introduction

A *rooted directed graph* is a directed graph in which there is a node designated as the root, from which there is a path to every other node. If there are cycles in a rooted directed graph, an infinite tree can be obtained by unfolding the graph. We consider the following problem: given a pattern directed graph P and a target directed graph T, are there subgraphs T' of T such that the tree obtained by unfolding T' is identical to the tree obtained by unfolding a subgraph of P? Informally, we call this problem the *directed graph isomorphism problem*.

Besides its theoretical interest, directed graph isomorphism has several applications. One application is Katzenelson, Pinter and Schenfeld's type-checking system [11] in which type expressions are represented by *type-graphs* which are *ordered labeled* rooted directed graphs, where an ordered labeled directed graph is a directed graph in which every node is associated with a label and the left-to-right order of sibling nodes is significant. Testing whether two type-graphs are equivalent is essentially a subproblem of the directed graph isomorphism problem. Other applications can be found in functional languages in which recursively typed languages are represented by directed graphs [10], in type inference and program analysis algorithms in which regular tree expressions are represented by directed graphs [2, 8] and in cyclic term graph rewriting systems [3, 4]. An efficient solution to the directed graph isomorphism problem not only is of theoretical interest in its own right, but also improves the performance of the above systems.

In this paper, we consider ordered labeled directed graphs. The ordered labeled directed graph isomorphism problem can be considered a generalization of the subtree isomorphism problem [7, 13]. However, because of the presence of cycles in directed graphs, the techniques for the subtree isomorphism problem do not apply to this problem. To our knowledge, there is no existing algorithm for this problem, although some related algorithms can be adapted to solve it. One of them is Katzenelson, Pinter and Schenfeld's type equivalence testing algorithm [11] which, given two rooted directed graphs P and T, can determine whether P and T represent the same type in time $O(|E_P||E_T|)$ and space $O(|V_P||V_T|)$, where $|V_P|$, $|V_T|$ and $|E_P|$ are the number of nodes in P, the number of nodes in T and the number of edges in P, respectively. If adapted to solve the directed graph isomorphism problem, this algorithm would run in time $O(|E_P||E_T||V_P||V_T|)$ and space $O(|V_P||V_T|)$. Another one is Fu's directed graph pattern matching algorithm [5] which finds for every subgraph P' of P all subgraphs of T that P' matches in time and space $O(|E_P||V_T| + |E_T|)$. The same time and space would be required if this algorithm were adapted to solve the directed graph isomorphism problem.

In applications of directed graph isomorphism, the same pattern is often matched against many targets. As such, it is advantageous to preprocess the pattern if the preprocessing helps to speed up the matching process. In this paper, we present an algorithm which takes advantage of preprocessing. The matching process of our algorithm is linear in the number of edges in the target graph. As in Hoffmann and O'Donnell's bottom-up algorithm for the tree pattern matching problem [9], the preprocessing takes time and space exponential in the size of the pattern graph. Given that the pattern is often much smaller in size than the target or the same pattern is used many times, which is often the case in applications, our algorithm is more efficient than the methods obtained by adapting existing algorithms for related problems [5, 11].

In §2 we introduce the notation and terminology used in this paper and formally define the directed graph isomorphism problem. In §3 we present the matching algorithm. In §4 we briefly sketch the preprocessing step. §5 gives a short conclusion and open problems.

2   Notation and terminology

The notation and terminology we use in this paper follow from those in Fu's paper [5]. We review them in the following.

Throughout the rest of this paper, we use RDG as an abbreviation for "rooted directed graph". Given a RDG G, the $depth(x)$ of a node x in G is the number of edges in the shortest path from the root of G to x. An edge (x_1, x_2) is an $ordinary\ edge$ if $depth(x_1)$ is less than $depth(x_2)$; an edge (x_1, x_2) is a $cross\ edge$ if $depth(x_1)$ equals $depth(x_2)$; an edge (x_1, x_2) is a $back\ edge$ if $depth(x_1)$ is greater than $depth(x_2)$. Let x and y be two nodes in a RDG G. We say x is a $parent$ of y if there is an edge pointing from x to y. Node y is a $child$ of x if and only if x is a parent of y. We say x is an $ancestor$ of y if there exist zero or more nodes x_1, x_2, ..., x_q in G such that x is a parent of x_1, x_q is a parent of y, and x_i is a parent of x_{i+1} for $1 \leq i < q$. Node y is a descendant of x if and only if x

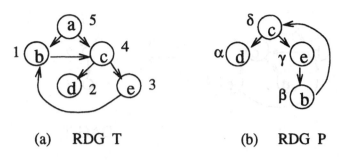

(a)   RDG T   (b)   RDG P

Figure 1:

is an ancestor of y. Nodes x and y are *siblings* if they have the same depth and a common parent. A *leaf* in G is a node from which there is no ordinary edge emanating. A *subgraph* of G rooted at a node x in G is a RDG which contains all its descendants and the induced edges.

An example of a RDG is shown in Figure 1(a) in which node 5 is the root. Node 5 has depth 0, node 1 and node 4 have depth 1 and node 2 and node 3 have depth 2. Edges (5, 1), (5, 4), (4, 2) and (4, 3) are ordinary edges, edge (1, 4) is a cross edge and edge (3, 1) is a back edge. Node 4, node 3 and node 1 are parents of node 3, node 1 and node 4, respectively. Node 1 is an ancestor of node 4 and node 4 is an ancestor of node 1. Node 1 and node 4 are also siblings. The leaf nodes in RDG T are node 1, node 2 and node 3. The subgraph of RDG T rooted at node 4 is shown in Figure 1(b).

We use $root(G)$, $lab(x)$, $sub(x)$ and $chi_i(x)$ to denote the root of a RDG G, the label of a node x, the subgraph rooted at x, and the ith child of x from the left, respectively. Unless otherwise specified, a pattern RDG is denoted by P and a target RDG is denoted by T. The nodes in P are denoted by Greek letters and the nodes in T are denoted by integers or lower case English letters.

The *ordered labeled directed graph isomorphism problem* is defined in terms of *unfolded RDGs*, where the unfolded RDG of a RDG T, denoted $unfold(T)$, is the infinite tree obtained by unfolding T. The *rank* of a node x in $unfold(T)$ is the number of edges in $unfold(T)$ from $root(unfold(T))$ to x. Let T^k denote the tree which consists of all nodes of $unfold(T)$ whose ranks are less than or equal to k, and the induced edges. Let $subtree(x)$ denote the subtree rooted at node x. The following are the definitions of isomorphism between P^k and T^k and between RDGs:

Definition 1 P^k is *isomorphic* to T^k if

i. $lab(root(P^k)) = lab(root(T^k))$,

ii. $deg(root(P^k)) = deg(root(T^k))$ and

iii. $subtree(chi_i(root(P^k)))$ is isomorphic to $subtree(chi_l(root(T^k)))$ for each l, $1 \leq l \leq deg(root(P^k))$.

Definition 2 A RDG P is *isomorphic* to a RDG T if for every $k \geq 0$, P^k is isomorphic to T^k.

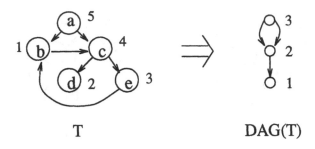

T   DAG(T)

Figure 2:

Definition 3 The *directed graph isomorphism problem* is to find all subgraphs of T to which a subgraph of P is isomorphic for every subgraph of P.

3 The Matching Algorithm

In the following description, we treat all subgraphs of P that are isomorphic to each other as one subgraph, i.e., when we mention a subgraph P' of P, we refer to all subgraphs of P that are isomorphic to P'.

The matching process of our algorithm is a bottom-up traversal of the target RDG T. In order to efficiently handle the cycles in T, we isolate the cycles from the rest of the graph. This is done by replacing each strongly connected component in T with a single node, and attaching to the single node all edges attached to the nodes in the strongly connected component. The resulting graph, denoted $DAG(T)$, is a directed acyclic graph. Note that the nodes in $DAG(T)$ do not have labels, since our algorithm does not need them. We use SCC as an abbreviation for "strongly connected component" from now on. The SCCs in T can be identified using Tarjan's linear time algorithm [14]. Each node x in $DAG(T)$ is associated with a unique number i, $1 \le i \le |V_{DAG(T)}|$, such that i is greater than all numbers associated with the descendants of x. Figure 2 shows an example of $DAG(T)$ and the numbers associated with the nodes. Our algorithm traverses $DAG(T)$ in ascending order of these numbers so that a node in $DAG(T)$ is visited only after all its children have been visited. Each node i in $DAG(T)$ corresponds to one or more nodes in T. Throughout the description, we use $scc(i)$ to denote the node or nodes in T represented by node i.

During the bottom-up traversal, we find for each node i the subgraph or subgraphs of P that are isomorphic to the subgraph or subgraphs of T represented by node i. If node i corresponds to a node x in T, we have all the information necessary to identify the subgraph of P that is isomorphic to $sub(x)$, since all children of node i have already been visited. However, if node i corresponds to a SCC, we face the difficulty of dealing with cyclic dependencies among the nodes in $scc(i)$. In Figure 1, it is not clear whether or not $sub(\beta)$ is isomorphic to $sub(1)$ before we have determined whether or not $sub(\delta)$ is isomorphic to $sub(4)$. However, we cannot determine whether or not $sub(\delta)$ is isomorphic to $sub(4)$

without knowing whether or not $sub(\beta)$ is isomorphic to $sub(1)$. We need some way to break the dependency cycle.

Before describing how to break the dependency cycle, we define some notation that is useful for the description. Let x be a node in a $scc(i)$ in T. A $sub(\alpha)$ of P is called a *candidate* of $sub(x)$ if $lab(\alpha) = lab(x)$ and $deg(\alpha) = deg(x)$. The following lemma pertains to candidates:

Lemma 1 *If there exists a mapping from the nodes in a $scc(i)$ to the nodes in P such that each node x in $scc(i)$ is mapped to a node α in P, then $sub(\alpha)$ is isomorphic to $sub(x)$ if the following conditions hold:*

i. $sub(\alpha)$ is either a candidate of $sub(x)$ or isomorphic to $sub(x)$ and
ii. for each l, $1 \leq l \leq deg(\alpha)$, if $chi_l(x)$ is in $scc(i)$ then $chi_l(x)$ is mapped to $chi_l(\alpha)$; otherwise $sub(chi_l(\alpha))$ is isomorphic to $sub(chi_l(x))$.

Proof. Consider the two infinite trees obtained by unfolding $sub(\alpha)$ and $sub(x)$. We show by induction on k that for each $k \geq 0$, s^k is isomorphic to $sub(x)^k$.

It is obvious that $sub(\alpha)^0$ is isomorphic to $sub(x)^0$, since α and x have the same label. Suppose that if the mapping exists then $sub(\alpha)^k$ is isomorphic to $sub(x)^k$. We now show that $sub(\alpha)^{k+1}$ is isomorphic to $sub(x)^{k+1}$. Consider the children of α and x. For each l, $1 \leq l \leq deg(\alpha)$, if $chi_l(x)$ is in $scc(i)$, then since the mappings exists, $sub(chi_l(\alpha))^k$ is isomorphic to $sub(chi_l(x))^k$; if $chi_l(x)$ is not in $scc(i)$, then $sub(chi_l(\alpha))$ is isomorphic to $sub(chi_l(x))$ and hence $sub(chi_l(\alpha))^k$ is isomorphic to $sub(chi_l(x))^k$. Since α and x have the same label and degree, $sub(\alpha)^{k+1}$ is isomorphic to $sub(x)^{k+1}$. □

Given a $scc(i)$, we associate with each node x in $scc(i)$ a *cand-array*, denoted $cands(x)$, to record the candidates that are possibly isomorphic to $sub(x)$. We call these candidates *eligible* candidates. Formally, a candidate $sub(\alpha)$ of $sub(x)$ is *eligible* if for each l, $1 \leq l \leq deg(\alpha)$, the following is true: if $chi_l(x)$ is not in $scc(i)$, then $sub(chi_l(\alpha))$ is isomorphic to $sub(chi_l(x))$; otherwise, if $chi_l(x)$ is associated with a cand-array and $sub(\alpha)$ is the jth candidate in $cands(x)$, then $sub(chi_l(\alpha))$ is the jth candidate in $cands(chi_l(x))$; otherwise $sub(chi_l(\alpha))$ is a candidate of $sub(chi_l(x))$.

Our algorithm traverses $scc(i)$ in preorder, where the preorder is obtained by ignoring the back edges and cross edges between the nodes in $scc(i)$. During the preorder traversal, we eliminate for each node x in $scc(i)$ those candidates that are not eligible. Let r be the first node in $scc(i)$ in preorder and n be the total number of candidates of $sub(r)$. Before the traversal begins, $cands(r)$ is initialized with all candidates of $sub(r)$. A vector V of length n is created to keep track of the status of the candidates of $sub(r)$: the jth bit is 0 if and only if the jth candidate in $cands(r)$ is eliminated during the preorder traversal. All bits of vector V are initialized to 1. During the traversal, we do the following for each node x being visited: create a size n cand-array for each child of x that is not associated with a cand-array; for each j such that the jth bit of vector V is 1, if the jth candidate $sub(\alpha)$ in $cands(x)$ is eligible, we assign $sub(child_l(\alpha))$

to the jth entry of the cand-array of each child of x; otherwise we change the jth bit of vector V to 0.

Lemma 2 *After the preorder traversal of $scc(i)$ terminates, the jth bit of vector V is 1 if and only if the jth candidate $sub(\alpha)$ in $cands(r)$ is isomorphic to $sub(r)$.*

Proof. If $sub(\alpha)$ is isomorphic to $sub(r)$, then the jth candidates in all cand-arrays are eligible during the preorder traversal. Therefore, the jth bit of vector V is 1 after the traversal terminates.

If the jth bit of the vector is 1 after the traversal terminates, then we can map each node x in $scc(i)$ to node α in P such that $sub(\alpha)$ is the jth candidate in $cands(x)$, and the following are true: i) $sub(\alpha)$ is either a candidate of $sub(x)$ or isomorphic to $sub(x)$ and ii) for each l, $1 \le l \le deg(\alpha)$, if $chi_l(x)$ is in $scc(i)$ then $sub(chi_l(\alpha))$ is mapped to $sub(chi_l(x))$; otherwise $sub(chi_l(\alpha))$ is isomorphic to $sub(chi_l(x))$. Therefore, according to Lemma 1, $sub(\alpha)$ is isomorphic to $sub(r)$. $\square$

It is costly to generate cand-arrays and maintain vector V during the preorder traversal. We compute during preprocessing all possible cand-arrays and vectors, code them by some enumeration and create tables so that cand-arrays can be obtained in constant time. Two tables, Table *INIT* and Table *CAND*, are needed. Table *INIT* takes as input $lab(x)$ and $deg(x)$ and returns the code of the set of all candidates of $sub(x)$. Let $pat(x)$ denote the subgraph of P that is isomorphic to $sub(x)$. If there is no subgraph of P isomorphic to $sub(x)$, then $pat(x) = \phi$. Let $assoc(x)$ denote the information about the subgraph of P that is isomorphic to $sub(x)$ or the subgraphs of P that are possibly isomorphic to $sub(x)$: $assoc(x) = pat(x)$ if x is not in a SCC; $assoc(x) = cands(x)$ if x has been associated with a cand-array; $assoc(x)$ is the set of all candidates of $sub(chi_l(x))$ otherwise. Table *CAND* takes as input the codes of vector V, $cands(x)$ and $assoc(chi_l(x))$ for $1 \le l \le deg(x)$, and returns the codes of a new vector V and cand-arrays for x and its children.

Now we present the matching algorithm.. A table *ISOM* is used for those nodes x that are not in any SCC to find $pat(x)$: given $lab(x)$, $deg(x)$ and $pat(chi_l(x))$ for each l, $1 \le l \le deg(x)$, it returns $pat(x)$. Table *ISOM* can be created during preprocessing. A procedure *cycle-match* is used to deal with the situation when the node i in $DAG(T)$ being visited corresponds to a SCC in T. It consists of a preorder traversal step during which cand-arrays are computed as describe above, and a propagation step during which $pat(x)$ is computed for each node x in $scc(i)$.

Matching Algorithm $(T, ISOM, INIT, CAND)$
    Compute $DAG(T)$ and number the nodes;
    for each node x in T **do** $pat(x) = \phi$;
    for i from 1 to $|V_{DAG(T)}|$ **do**
        if i corresponds to a SCC in T **then** *cycle-match*(i);
        else
            Let x be the node in $scc(i)$;

$$pat(x) = ISOM[lab(x), pat(chi_1(x)), ..., pat(chi_{deg(x)}(x))];$$
   end;
  end;
  Output all (α, x) s.t. $pat(x) = sub(\alpha)$;
 end.

Procedure *cycle-match(i)*
 Number each node in $scc(i)$ according to preorder;
 Create a cand-array for the node with preorder number 1;
 Create a vector V and initialize each bit to 1;
 for x from 1 to $|scc(i)|$ **do**
  for l from 1 to $deg(x)$ **do**
   if $chi_l(x)$ is in not $scc(i)$ **then** $assoc(chi_l(x)) = pat(chi_l(x))$;
   else if $chi_l(x)$ has a cand-array **then** $assoc(chi_l(x)) = cands(chi_l(x))$;
   else $assoc(chi_l(x)) = INIT[lab(chi_l(x)), deg(chi_l(x))]$;
  end;
  Let $(V, w, w_1, ..., w_{deg(x)})$ be $CAND[V, cands(x), assoc(chi_1(x)), ...,$
   $assoc(chi_{deg(x)}(x))]$;
  Assign $w, w_1, ..., w_{deg(x)}$ to $cands(x), cands(chi_1(x)), ..., cands(chi_{deg(x)}(x))$,
   respectively;
 end;
 if the jth bit vector V is 1 **then**
  Let r be the node with preorder number 1 and s be the jth candidate in
   $cands(r)$;
  $propagate(r, root(s))$;
 else for each x in $scc(i)$ **do** $pat(x) = \phi$;
 end.

Procedure *propagate(y, β)*
 $pat(y) = sub(\beta)$;
 for each l s.t. $pat(chi_l(y)) = \phi$ **do** $propagate(chi_l(y), chi_l(\beta))$;
 end.

Theorem 1 *The algorithm correctly computes the subgraphs of T to which a subgraph of P is isomorphic for every subgraph of P and the matching process takes $O(|E_T|)$ steps.*

Proof. The correctness of Procedure *cycle-embed* immediately follows from Lemma 2. The correctness of the remaining part of the algorithm is obvious.

 It takes $O(|E_T|)$ steps to compute $DAG(T)$ and number the nodes in it. The preorder traversal step of Procedure *cycle-embed* clearly takes $O(\sum_{x \in scc(i)} deg(x))$ steps. The propagation step also takes $O(\sum_{x \in scc(i)} deg(x))$ steps, because each edge emanating from a node in $scc(i)$ is visited at most once. The remaining part of the algorithm takes $O(\sum_{x \notin SCC} deg(x))$ steps. Therefore, the time complexity of this algorithm is $O(\sum_{x \in V_T} deg(x)) = O(|E_T|)$. $\square$

4 Preprocessing

As we said before, all subgraphs of P that are isomorphic to each other are treated as a single subgraph. Identifying these subgraphs can be done by adapting Fu's ordered labeled directed graph pattern matching algorithm [5].

Three tables, Table $ISOM$, Table $INIT$ and Table $CAND$, need to be created. Table $ISOM$ and Table $INIT$ can be created in linear time by a simple traversal of P. Table $CAND$ can be created using the following procedure:

> **Procedure** *create-table-CAND(P)*
> > Create an empty set U;
> > **for** each set containing the subgraphs of P whose roots have the same label
> > > and degree **do**
> > > Put the set in U;
> > > Create a cand-array containing all subgraphs in the set and put the
> > > > cand-array in U;
> > > Create a vector of the same length as the cand-array, initialize each bit to 1
> > > > and put the vector in U;
> > **end**;
> > Put each subgraph of P in U;
> > **for** each tuple $(V, u, u_1, u_2, ..., u_{deg(u)})$ s.t. each element in the tuple is in U,
> > > u is a cand-array and V has the same length as u **do**
> > > Create entry $CAND[V, u, u_1, ..., u_{deg(u)}] = (V', u', u_1', ..., u_{deg(u)}')$;
> > > **for** each element in $(V', u', u_1', ..., u_{deg(u)}')$ **do**
> > > > **if** it is different from all elements in U **then** put it in U;
> > > **end**;
> > **end**;
> **end**.

Since there is a finite number of different cand-arrays, this procedure always terminates. For bounded degree pattern graph P, the time and space needed to generate Table $CAND$ is $O(c^{|V_P|})$ in the worst case, where c is a constant. The size of Table $CAND$ is also $O(c^{|V_P|})$. Details of the analysis can be found in the full paper.

5 Conclusion

A linear matching-time algorithm is presented for the ordered labeled directed graph isomorphism problem. The algorithm provides an alternative to those methods obtained by adapting existing algorithms for related problems to solve our problem. Once the pattern is preprocessed, our algorithm is faster than other methods. Although there is only one pattern graph in the description of the algorithm, our algorithm is not restricted to single pattern.

The tables produced in preprocessing are very sparse matrices. It would be interesting to see whether it is possible to reduce the sizes of the tables and the time and space needed to preprocess the pattern graph.

Acknowledgment

The author wishes to thank Naomi Nishimura who read the early versions of the manuscript and provided many valuable comments.

References

[1] A. Aiken and B. R. Murphy, Implementing regular tree expressions, in *Proceedings of 5th ACM Conference on Functional Programming Languages and Computer Architecture*, pages 427–447, 1991.

[2] A. Aiken and B. R. Murphy, Static type inference in a dynamically typed language, in *Proceedings of Eighteenth Annual ACM Symposium on Principles of Programming Languages*, pages 279–290, 1991.

[3] A. Corradini, Term rewriting in CT_Σ, in *Proceedings of International Joint Conference on Theory and Practice of Software Development*, pages 468–484, 1993.

[4] N. Dershowitz and S. Kaplan, Rewrite, rewrite, rewrite, rewrite, rewrite ... in *Proceedings of Sixteenth Annual Symposium on Principles of Programming Languages*, pages 250–259, 1989.

[5] J. Fu, Pattern matching in directed graphs, in *Proceedings of 6th Annual Symposium on Combinatorial Pattern Matching*, pages 64–77, 1995.

[6] J. R. W. Glauert, J. R. Kennaway and M. R. Sleep, Dactl: An experimental graph rewriting language, in *Proceedings of Fourth International Workshop on Graph Grammars and Their Application to Computer Science*, pages 378–395, 1990.

[7] R. Grossi, A note on the subtree isomorphism for ordered trees and related problems, *Information Processing Letters*, 39:81–84, 1991.

[8] N. Heintze and J. Jaffar, A finite presentation theorem for approximating logic programs, in *Proceedings of Seventeenth Annual ACM Symposium on Principles of Programming Languages*, pages 197–209, 1990.

[9] C. M. Hoffmann and M. J. O'Donnell, Pattern matching in trees, *Journal of ACM*, 29(1):68–95, 1982.

[10] K. H. Holm, Graph matching in operational semantics and typing, in *Proceedings of Colloquium on Trees in Algebra and Programming*, pages 191–205, 1990.

[11] J. Katzenelson, S. S. Pinter and E. Schenfeld, Type matching, type-graphs, and the Schanuel conjecture, *ACM Transactions on Programming Languages and Systems*, 14(4):574–588, Oct. 1992.

[12] J. W. Klop, Term rewriting systems, Technical Report CS-R9073, Free University, Department of Mathematics and Computer Science, 1990.

[13] E. Mäkinen, On the subtree isomorphism problem for ordered trees, *Information Processing Letters*, 32:271–273, 1989.

[14] R. Tarjan, Depth first search and linear graph algorithms, *SIAM Journal on Computing*, 1(2):146–160, 1972.

A Resource Assignment Problem on Graphs

Satoshi Fujita[1], Tiko Kameda[2] and Masafumi Yamashita[1]

[1] Department of Electrical Engineering
Faculty of Engineering, Hiroshima University
Kagamiyama 1-4-1, Higashi-Hiroshima, 739, Japan
[2] School of Computing Science
Simon Fraser University, Burnaby, BC V5A 1S6, Canada

1 Introduction

Let G be an undirected graph with a set of vertices, $V(G)$, and a set of edges, $E(G)$. Let C be the (countably infinite) set of colors representing the set of resources. In what follows, we use "resource" and "color" interchangeably. Given an integer r, we assign at most r colors to each vertex. Such an assignment can be expressed by a function f from V to 2^C, such that $|f(v)| \leq r$ for each $v \in V(G)$. Let $C_f = \{c \in C \mid c \in f(v) \text{ for some } v \in V\}$, i.e., the set of colors that are actually assigned to a vertex by f. We say that f is an r-**configuration** if, for each color $c \in C_f$, the vertices in $\{v \mid c \in f(v)\}$ form a **dominating set** for G.[3]

Our motivation for studying r-configurations is as follows. Suppose G models a computer network with the vertices representing computers with bounded "capacity." If, for a color $c \in C$, $\{v \mid c \in f(v)\}$ is a dominating set, then each vertex can access a copy of the resource (e.g., a database) represented by c either locally or by communicating with a neighbor computer. Thus, if f is an r-configuration, requests for accessing any resource in C_f need not propagate beyond neighbors in the network, which presumably leads to shorter response times and reduced network traffic.

It is known that the problem of determining if a given graph G has a dominating set of size K is NP-complete for planar graphs with maximum vertex degree 3, as well as for planar graphs that are regular of degree 4 [7].

We are particularly interested in the maximum number of resources that can be assigned to the vertices of graph G by an r-configuration,

$$D_r(G) = \max\{|C_f| : f \text{ is an } r\text{-configuration}\}.$$

$D_1(G)$ is known as the **domatic number** [6], and it is easy to see that $D_1(G) \leq \delta(G) + 1$, where $\delta(G)$ denotes the minimum degree of graph G. In general, given a constant K (≥ 3), testing if $D_1(G) \geq K$ is known to be NP-complete.[4] For

---

[3] A subset V' of V is said to be a dominating set for $G = (V, E)$ if, for any $u \in V$, either $u \in V'$ holds or there is a vertex $v \in V'$ such that $\{u, v\} \in E$.

[4] Note that if $K = 2$, the problem can be solved easily in linear time. In fact, if G contains no isolated vertex, for any maximal independent set U ($\subset V$) of G, subsets U and $V - U$ are both dominating sets for G, and if G contains an isolated vertex, the domatic number of G is clearly 1 [7].

restricted classes of graphs, however, several positive results on the domatic number are known:

1. If G is an interval graph, then $D_1(G) \geq K$ for any given constant K can be tested in linear time [3, 10, 12].
2. If G is a *triangulated disk* (i.e., a triangulation of a planar graph), then $D_1(G) \geq 3$ [11].

A dominating set V' for $G = (V, E)$ is said to be **perfect** if each vertex not in V' is adjacent to exactly one vertex in V' [4, 9], and said to be **independent** if it is an independent set with respect to graph G. The problem of finding a perfect dominating set and that of finding an independent perfect dominating set are both known to be NP-hard. Several positive results on these notions have been obtained, including polynomial-time algorithms for exactly solving the problem by restricting the class of graphs [1, 2, 5, 13] and an approximating algorithm for general graphs [8].

In this paper, we try to characterize $D_r(G)$ for general graph G. In particular, we address three questions concerning $D_r(G)$. The first question is about the upper bound on $D_r(G)$ in terms of the minimum vertex degree $\delta(G)$. (Note that $D_1(G) \leq \delta(G)+1$ easily generalizes to $D_r(G) \leq r(\delta(G)+1)$.) We show that, for any constant $\epsilon > 0$, there is a graph G such that $D_r(G) < \epsilon r \delta(G) + ro(\delta(G))$. The second question is about the relation between $D_1(G)$ and $D_r(G)$. We derive an upper bound on $D_r(G)$ in terms of $D_1(G)$, and an upper bound on $D_1(G)$ in terms of $D_r(G)$. We show that there is a class of graphs for which the difference between $D_r(G)$ and $rD_1(G)$ can be arbitrarily large. Finally, we consider the problem of testing if $D_1(G) = d + 1$ for a given d-regular graph G. We show that, if $|V| = 2(d+1)$, then the problem is solvable in polynomial time, while if $|V| = a(d+1)$ for some integer $a \geq 3$, then the problem is NP-hard. The following three sections are devoted to the above three problems, respectively.

2 Upper Bounds on $D_r(G)$

As commented in Introduction, for any graph G, $D_r(G) \leq r(\delta(G)+1)$ holds. In this section, we improve this trivial upper bound and show that, for any integer $r \geq 1$ and for any constant $\epsilon > 0$, there is a graph G such that $D_r(G) \leq \epsilon r \delta(G) + ro(\delta(G))$. The key to our constructive proof is the following lemma.

Lemma 1. *Let U be a minimum dominating set for graph $G = (V, E)$. Then, $D_r(G) \leq \lfloor r|V|/|U| \rfloor$.*

Proof. Let f be an r-configuration for G such that $|C_f| = D_r(G)$, and imagine that the vertices of G are colored by f. Each color is assigned to at least $|U|$ vertices, and at most r colors can be assigned to each vertex, which implies $D_r(G)|U| \leq r|V|$. $\qquad\square$

As the first step, consider the following graph G_1: $V(G_1) = \{(i, j) \mid 1 \leq i, j \leq m\}$, and two vertices in $V(G_1)$ are connected by an edge in $E(G_1)$ iff they agree

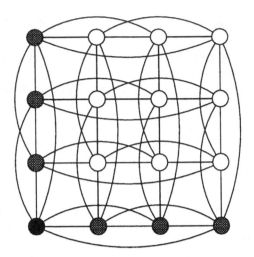

Fig. 1. Graph G_1 in the proof of Theorem 1.

on at least one "coordinate." See Figure 1 for illustration ($m = 4$). In the figure, the neighbors of the black vertex are painted gray. Note that $|V(G_1)| = m^2$ and $\delta(G_1) = 2m - 2$. It is easy to see that the minimum dominating set for G_1 has size m. Hence, by Lemma 1,

$$D_r(G_1) \leq \left\lfloor \frac{r|V(G_1)|}{m} \right\rfloor = rm.$$

Since $m = \delta(G_1)/2 + 1$, $D_r(G_1)$ is at most $r\delta(G_1)/2 + r$.

As the next step, consider a three-dimensional "cube" instead of a two-dimensional "mesh," and define graph G_2 as follows: $V(G_2) = \{(i, j, k) \mid 1 \leq i, j, k \leq m\}$ and two vertices in $V(G_2)$ are connected by an edge in $E(G_2)$ iff they agree on at least one coordinate. Note that $|V(G_2)| = m^3$ and $\delta(G_2) = 3m^2 - 3m$. The size of the minimum dominating set for G_2 is clearly m. Since $D_r(G_2) \leq \lfloor r|V(G_2)|/m \rfloor = rm^2$ and $m^2 = \delta(G_2)/3 + m$, we have

$$D_r(G_2) \leq \frac{r\delta(G_2)}{3} + rO\left((\delta(G_2))^{1/2}\right).$$

The idea described above can be extended to higher dimensions in a natural way. Let $\ell = \lceil 1/\epsilon \rceil + 1$, and define graph G_3 based on an ℓ-dimensional cube, in a similar way to G_1 and G_2. Note that $|V(G_3)| = m^\ell$, and that the size of the minimum dominating set for G_3 is m as well. Since $D_r(G_3) \leq r|V(G_3)|/m = rm^{\ell-1}$ and

$$\delta(G_3) = \sum_{i=1}^{\ell} \binom{\ell}{i} (-1)^{i+1} m^{\ell-i} - 1,$$

we have

$$D_r(G_3) \leq rm^{\ell-1} = \frac{r\delta(G_3)}{\ell} + rO\left((\delta(G_3))^{1-1/(\ell-1)}\right) < \epsilon r\delta(G_3) + ro(\delta(G_3)).$$

Hence, we have the following theorem.

Theorem 2 (Upper Bound on $D_r(G)$). *For any integer $r \geq 1$ and for any constant $\epsilon > 0$, there is a graph G such that $D_r(G) < \epsilon r\delta(G) + ro(\delta(G))$.* □

The problem of testing whether $D_1(G) \geq 3$ is NP-hard if $\delta(G) \geq 2$ (3SAT can be transformed to it in a natural way). However, it is open whether the problem is still NP-hard if $\delta(G) \geq 3$. In addition, it is an interesting open problem to give a lower bound on $D_1(G)$ in terms of $\delta(G)$.

3 Relation between $D_1(G)$ and $D_r(G)$

The objective of this section is to investigate the relationship between $D_1(G)$ and $D_r(G)$ for general G.

Proposition 3. *For any graph G, there is a graph G' such that $D_r(G) = D_1(G')$.*

Proof. Given a graph $G = (V, E)$, let G^r denote the graph constructed from G as follows: For each $v \in V$, G^r has a clique of size r, and for each $(u, v) \in E$, G^r has a complete bipartite subgraph connecting the two r-cliques representing u and v, respectively. It is easy to see that $D_r(G) = D_1(G^r)$ holds. □

Proposition 4. *For any graph G, $D_r(G) \geq rD_1(G)$ holds.*

Proof. Given any graph G, use a 1-configuration to color it using $k = D_1(G)$ colors, $c_1, c_2, \ldots, c_k$. If a vertex is colored by c_i, then re-color it using r new colors, $c_{i1}, \ldots, c_{ir}$. The resulting coloring represents an r-configuration. It follows that G has an r-configuration with rk colors. □

One might conjecture that $D_1(G) = \lfloor D_r(G)/r \rfloor$ for any $r \geq 1$. Unfortunately however, this conjecture is false as shown by the following lemma.

Lemma 5. *There is a graph $G = (V, E)$ such that $D_1(G) \leq \lfloor D_2(G)/2 \rfloor - 1$.*

Proof. In the rest of this section, we use strings of $\{0,1,2\}$ as vertex labels. Given such a string, $v = v_p v_{p-1} \cdots v_1$, by $\bar{v}$ we denote its value interpreted as a ternary number, i.e., $\bar{v} = \sum_{i=1}^{p} v_i 3^{i-1}$. Let $V = V_1 \cup V_2$, where $V_1 = \{0, 1, 2\}$ and $V_2 = \{00, 01, 02, 10, 11, 12, 20, 21, 22\}$. We connect the twelve vertices of G as follows.

- For any $u, v \in V_1$, $\{u, v\} \in E$
- For any $u, v \in V_2$, $\{u, v\} \in E$ iff $\bar{u} = \bar{v} + 1 \pmod 9$ or $\bar{u} = \bar{v} + 2 \pmod 9$.

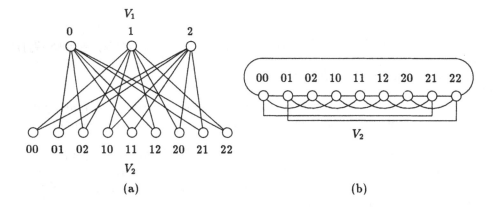

Fig. 2. Graph G in the proof of Lemma 2: (a) Connection between V_1 and V_2. (b) Connection among the vertices of V_2.

- For any $u \in V_1$ and $v \in V_2$, $\{u, v\} \in E$ iff $u \neq v_1$, where $v = v_2 v_1$.
  See Figure 2 for illustration.

Any dominating set V' ($\subseteq V$) for G must contain at least two vertices in V_1 or at least two vertices in V_2. For example, if V' contains only one vertex in V_2, then this vertex cannot cover all other vertices in V_2, and V' must contain at least two vertices from V_1. Hence,

$$D_1(G) \leq \lfloor |V_1|/2 \rfloor + \lfloor |V_2|/2 \rfloor = 4 + 1 = 5.$$

On the other hand, since any 2-set from V_1 is a dominating set for G, we can select three such dominating sets under a 2-configuration; and since any two vertices $u, v \in V_2$ such that $\bar{v} = \bar{u} + 4 \pmod{3^2}$ form a dominating set for G, we can select nine such dominating sets under a 2-configuration. We thus have

$$D_2(G) \geq 3 + 9 = 12,$$

which completes the proof. □

The idea in the above proof can be extended to prove:

Theorem 6. *For any integer $m \geq 1$, there is a graph $G = (V, E)$ such that $D_2(G) \geq m$ and $2D_1(G) \leq D_2(G) - \log_3(2D_2(G)/3 + 1)$.*

Proof. Let $p = \lceil \log_3(2m/3 + 1) \rceil$. Then we have $\sum_{i=1}^{p} 3^i \geq m$. Let $V = \bigcup_{i=1}^{p} V_i$, where, for $1 \leq i \leq p$, $V_i =$ all i-symbol ternary strings. Thus we have $|V_i| = 3^i$. A vertex in V_i is denoted as $u = u_i u_{i-1} \ldots u_1$, where $u_j \in \{0, 1, 2\}$ for $1 \leq j \leq i$. Vertices in V are connected as follows:

- For any $u, v \in V_1$, u and v are connected by an edge in E.

- For each $i \geq 2$, vertices $u = u_i u_{i-1} \ldots u_1$ and $v = v_i v_{i-1} \ldots v_1$ in V_i such that $\bar{u} < \bar{v}$ are connected by an edge in E iff $|\bar{u} - \bar{v}| \leq K_i$ or $|\bar{u} + 3^i - \bar{v}| \leq K_i$, where $K_i = \lceil (3^i - 1)/4 \rceil$.
- For any $u = u_i u_{i-1} \ldots u_1 \in V_i$ and $v = v_j v_{j-1} \ldots v_1 \in V_j$ where $j > i$, vertices u and v are connected by an edge in E iff $u_1 \neq v_{j-i}$.

By the definition of p and Lemma 8 proved below, $D_2(G) \geq \sum_{i=1}^{p} 3^i \geq m$. Now, by Lemmas 7 and 8 (see below), we have

$$2D_1(G) \leq D_2(G) - p$$
$$= D_2(G) - \log_3\{2D_2(G)/3 + 1\}.$$

$\square$

Lemma 7. $D_1(G)$ is at most $\sum_{i=1}^{p} \lfloor 3^i/2 \rfloor$.

Proof. First we prove that any dominating set $V' (\subseteq V)$ for graph G must contain at least two vertices of V_i, for some $1 \leq i \leq p$, where $p \geq 2$. The proof is by contradiction. Suppose that there is a dominating set V' for G such that $|V' \cap V_i| \leq 1$ for all $1 \leq i \leq p$. Without loss of generality, we may suppose that $|V' \cap V_i| = 1$ for all i. For $i = 1, 2, \ldots, p$, let $w^{(i)} = w_i^{(i)} w_{i-1}^{(i)} \cdots w_1^{(i)} \in V' \cap V_i$.

Construct a p-symbol ternary string x as follows: $x = x_p w_1^{(1)} w_1^{(2)} \ldots w_1^{(p-1)}$ where $x_p \neq w_p^{(p)}$. We claim that vertex $x \in V_p$ is not dominated by any vertex in $\{w^{(i)} \mid i = 1, 2, \ldots, p\}$. By definition, it is clear that $w^{(i)}$ does not dominate x for $i = 1, 2, \ldots, p - 1$. To show that $w^{(p)}$ does not dominate x either, note that both $|\bar{x} - \bar{w}^{(p)}| \geq 3^{p-1}$ and $|\bar{x} + 3^p - \bar{w}^{(p)}| \geq 3^{p-1}$ hold, since x and $w^{(p)}$ differ in the most significant digit. Since $3^{p-1} > K_p$ for all integer $p \geq 2$, it follows that $w^{(p)}$ does not dominate x.

Hence any dominating set for G must contain at least two vertices from some subset V_i. Since there are at most $\lfloor 3^i/2 \rfloor$ disjoint pairs of elements in subset V_i for all i, $D_1(G)$ is at most

$$D_1(G) \leq \sum_{i=1}^{p} \lfloor 3^i/2 \rfloor.$$

$\square$

Lemma 8. $D_2(G)$ is at least $\sum_{i=1}^{p} 3^i$.

Proof. Let $u = u_i u_{i-1} \ldots u_1$ and $v = v_i v_{i-1} \ldots v_1$ be two vertices in V_i such that

$$\bar{u} - \bar{v} = (3^i - 1)/2.$$

The 2-set $\{u, v\}$ dominates subset V_i, since $(3^i - 1)/2 \leq 2K_i$, and any other $w \in V_i$ satisfies $|\bar{w} - \bar{u}| \leq K_i$, $|\bar{w} + 3^i - \bar{u}| \leq K_i$, $|\bar{w} - \bar{v}| \leq K_i$, or $|\bar{w} + 3^i - \bar{v}| \leq K_i$. Further, since $\sum_{j=1}^{i} 3^{j-1} u_j = \sum_{j=1}^{i} 3^{j-1}(v_j + 1)$, $\{u, v\}$ also dominates $V_{i'}$ for all $i' \neq i$.

Since for each i we can select $|V_i|$ $(= 3^i)$ such pairs under 2-configuration, $D_2(G)$ is at least

$$D_2(G) \geq \sum_{i=1}^{p} |V_i| = \sum_{i=1}^{p} 3^i.$$

☐

Corollary 9. *For any fixed Δ, there is a graph G such that $D_2(G) \geq 2D_1(G) + \Delta$.*

☐

The following theorem is concerned with an upper bound on $D_2(G)$ in terms of $D_1(G)$.

Theorem 10. *For any graph $G = (V, E)$, $D_2(G) \leq 2\alpha D_1(G)$, where $\alpha = \left\lceil \sqrt{2|V|/D_1(G)} \right\rceil$.*

Proof. Let S be a collection of dominating sets for G achieving $D_2(G)$, hence $|S| = D_2(G)$. First, we partition S into subsets of disjoint dominating sets by using the following procedure.

Step 1: Let $j = 1$ and $\tilde{S} = S$. Repeat Steps 2–5 until $\tilde{S} = \emptyset$.
Step 2: Let $S_j = \emptyset$.
Step 3: If there is an element $Y \in \tilde{S}$ which does not intersect with $\bigcup_{X \in S_j} X$, then remove Y from $\tilde{S}$, and add it to S_j. If there is no such element Y, then go to Step 5.
Step 4: Go to Step 3.
Step 5: Increment j by 1.

Let $\{S_1, S_2, \ldots, S_q\}$ be the resulting partition of S, and for $1 \leq j \leq q$, let $U_j = \bigcup_{X \in S_j} X$ $(\subseteq V)$. Note that $|S_j| \leq D_1(G)$ for all $1 \leq j \leq q$, since the dominating sets in S_j are mutually disjoint. Note also that $\sum_{i=1}^{q} |U_i| \leq 2|V|$ since under 2-configuration, each vertex in V can contribute to at most two dominating sets in S. Moreover, we claim that, for any j $(1 \leq j \leq q - 1)$, each element in $S_{j+1} \cup S_{j+2} \cup \ldots \cup S_q$ share at least one vertex with U_j that is not shared with any other element in $\bigcup_{i=j+1}^{q} S_i$. To prove this claim, consider any dominating set $d \in \bigcup_{i=j+1}^{q} S_i$, and let $v \in U_j \cap d$. If there were a dominating set $d' \in \bigcup_{i=j+1}^{q} S_i$ such that $d' \neq d$ and $v \in d'$, then v would belong to at least three dominating sets in S, a contradiction. It follows that, for any $1 \leq j \leq q - 1$, we have

$$\sum_{i=j+1}^{q} |S_i| \leq |U_j|.$$

Adding $\sum_{i=1}^{j} |S_i|$ to both sides of the above inequality, we obtain

$$\sum_{i=1}^{q} |S_i| \leq |U_j| + \sum_{i=1}^{j} |S_i| \qquad \text{for all } j.$$

It follows from this inequality that

$$D_2(G) \leq \min_{1 \leq j \leq q} \left\{ |U_j| + \sum_{i=1}^{j} |S_i| \right\}. \tag{1}$$

Let $\alpha = \left\lceil \sqrt{2|V|/D_1(G)} \right\rceil$. If $q \leq \alpha$, then since $|S_i| \leq D_1(G)$ for all $1 \leq i \leq q$, we have

$$D_2(G) = \sum_{i=1}^{q} |S_i| \leq \alpha D_1(G).$$

If $q \geq \alpha$, on the other hand, we derive from (1)

$$D_2(G) \leq \min_{1 \leq j \leq \alpha} \left\{ |U_j| + \sum_{i=1}^{j} |S_i| \right\} \leq \min_{1 \leq j \leq \alpha} \{|U_j|\} + \alpha D_1(G),$$

Since the minimum size, $\min_{1 \leq j \leq \alpha}\{|U_j|\}$, cannot be larger than the average size of the sets in $\{U_j \mid 1 \leq j \leq q\}$, we finally obtain

$$D_2(G) \leq \frac{2|V|}{\alpha} + \alpha D_1(G) \leq 2\alpha D_1(G).$$

$\square$

Note that $|V|/D_1(G)$ represents the 'average' size of the dominating sets in a collection of dominating sets providing $D_1(G)$. Hence, for example, if $D_1(G) = \Omega(|V|)$ (i.e., if the average size is a constant), then $D_2(G)/D_1(G)$ is bounded from above by some constant.

4 d-Regular Graphs with $D_1(G) = d + 1$

As we commented earlier, $D_1(G) \leq \delta(G) + 1$ holds in general. In this section, we consider the regular graphs G with the maximum possible domatic number, i.e., satisfying $D_1(G) = \delta(G) + 1$. Let $\mathcal{R}_d$ denote the set of all d-regular graphs G with $D_1(G) = d + 1$. Our objective in this section is to test membership in $\mathcal{R}_d$. If $G \in \mathcal{R}_d$ then $|V(G)| = a(d + 1)$ for some integer a, since any 1-configuration f with $|C_f| = d + 1$ will assign exactly one color to each vetex and each color corresponds to a perfect dominating set (i.e., each vertex in such a dominating set covers exactly $d+1$ vertices). It is clear that the d-regular graph with $d+1$ vertices (i.e., when $a = 1$) is the complete graph K_{d+1}, which obviously belongs to $\mathcal{R}_d$. To exclude this trivial case, in the following, we assume that $|V(G)| = a(d + 1)$ for some integer $a \geq 2$. We shall prove that if $a = 2$ membership in $\mathcal{R}_d$ can be tested in polynomial time, while if $a \geq 3$ the membership problem is NP-hard. In what follows the term "configuration" will mean 1-configuration.

4.1 Polynomial time algorithm for the case $|V| = 2(d + 1)$

When $a = 2$, we transform the membership problem in $\mathcal{R}_d$ to the maximum matching problem for an auxiliary graph $H(G)$ in such a way that $G \in \mathcal{R}_d$ iff graph $H(G)$ has a maximum matching of size $d + 1$. Let $Dist_G(u, v)$ denote the minimum distance between the vertices u and v in G. The algorithm given below finds a configuration f with $d + 1$ colors by constructing a maximum matching in $H = H(G)$.

Algorithm **TEST**
Input: G { a d-regular graph with $2(d + 1)$ vertices. }

Step 1: Define graph H: $V(H) = V(G)$ and for any $u, v \in V(H)$, and $\{u, v\} \in E(H)$ iff $Dist_G(u, v) = 3$.

Step 2: Find a maximum matching for graph H.

Step 3: If the maximum matching found has $d + 1$ elements, output YES; otherwise NO. □

It is easy to see that $H(G)$ has a maximum matching of size $d+1$ iff $D_1(G) = d + 1$, since each pair of vertices connected by an edge in $H(G)$ is a dominating set for G. It is clear that all the three steps of the above algorithm can be carried out in polynomial time.

Lemma 11. *If $G = (V, E)$ is a d-regular graph with $2(d + 1)$ vertices, then we can test whether or not $G \in \mathcal{R}_d$ in polynomial time.* □

4.2 NP-hardness for the case $|V| = 3(d + 1)$

Theorem 12. *The problem of testing whether a given d-regular graph with $3(d + 1)$ vertices is in $\mathcal{R}_d$ is NP-hard.*

Proof sketch. We transform 3SAT to the testing problem through a restricted version of 3-Dimensional Matching problem (3DM). The outline of the former half of the proof (Lemma 13) is similar to the proof of NP-hardness of general 3DM by Garey and Johnson ([7], pp. 50–53). □

Let $U = \{u_1, u_2, \ldots, u_n\}$ be the set of variables and $C = \{c_1, c_2, \ldots, c_m\}$ be the set of clauses in an arbitrary instance of 3SAT. The following lemma holds.

Lemma 13. *For any given C, we can construct a set M of 3-sets in $S_1 \times S_2 \times S_3$, where S_1, S_2 and S_3 are disjoint sets of elements of an equal size, in polynomial time to satisfy the following two conditions: (1) there is a subset M' of M which exactly covers $S_1 \cup S_2 \cup S_3$ iff C is satisfiable, and (2) for any $\{a, b, c\}, \{a, b, d\} \in M$, if $\{x, y, c\} \in M$ for some x, y, then $\{x, y, d\}$ is also in M.* □

The restricted version of 3DM is further transformed to our testing problem as stated by the following lemma.

Lemma 14. *Let M be a set of 3-sets in $X \times Y \times Z$, where X, Y and Z are disjoint sets of elements of an equal size, which satisfies the following condition: for any $\{a, b, c\}, \{a, b, d\} \in M$, if $\{x, y, c\} \in M$ for some x, y, then $\{x, y, d\}$ is also in M. Then, we can construct a d-regular graph G with $3(d + 1)$ vertices such that $D_1(G) = d + 1$ iff there is an exact cover M' ($\subseteq M$) of $X \cup Y \cup Z$ in polynomial time.* $\square$

References

1. M. J. Atallah, G. K. Manacher, and J. Urrutia. Finding a minimum independent dominating set in a permutation graph. *Discrete Appl. Math.*, 21:177–183, 1988.
2. D. W. Bange, A. E. Barkauskas, and P. T. Slater. Efficient dominating sets in graphs. In R. D. Ringeisen and F. S. Roberts, editors, *Applications of Discrete Mathematics*, pages 189–199. SIAM, 1988.
3. A. A. Bertossi. On the domatic number of interval graphs. *Information Processing Letters*, 28(6):275–280, August 1988.
4. N. Biggs. Perfect codes in graphs. *J. Comb. Theory, Series B*, 15:289–296, 1973.
5. G. J. Chang, C. P. Rangan, and S. R. Coorg. Weighted independent perfect domination on cocomparability graphs. Technical Report 93-24, DIMACS, April 1993.
6. E. J. Cockayne and S. T. Hedetniemi. Optimal domination in graphs. *IEEE Trans. Circuit and Systems*, CAS-22:855–857, 1975.
7. M. R. Garey and D. S. Johnson. *Computers and Intractability A Guide to the Theory of NP-Completeness*. W.H. Freeman and Company, San Francisco, 1979.
8. R. W. Irving. On approximating the minimum independent dominating set. *Information Processing Letters*, 37:197–200, 1991.
9. M. Livingston and Q. F. Stout. Perfect dominating sets. *Congressus Numerantium*, 79:187–203, 1990.
10. T. L. Lu, P. H. Ho, and G. J. Chang. The domatic number problem in interval graphs. *SIAM J. Disc. Math.*, 3:531–536, 1990.
11. L. R. Matheson and R. E. Tarjan. Dominating sets in planar graphs. Technical Report TR-461-94, Dept. of Computer Science, Princeton University, May 1994.
12. A. Srinivasa Rao and C. P. Rangan. Linear algorithm for domatic number problem on interval graphs. *Information Processing Letters*, 33(1):29–33, October 1989.
13. C. C. Yen and R.C.T. Lee. The weighted perfect domination problem. *Information Processing Letters*, 35:295–299, 1990.

NC Algorithms for Partitioning Sparse Graphs into Induced Forests with an Application

Zhi-Zhong Chen

Dept. of Math. Sci., Tokyo Denki Univ., Hatoyama, Saitama 350-03, Japan

Abstract. For a simple graph G, let $\rho(G) = 1 + \lfloor \frac{\max \delta(G')}{2} \rfloor$, where the maximum is taken over all induced subgraphs G' and $\delta(G')$ is the minimum degree of G'. It is known that the vertex set of G can be partitioned into at most $\rho(G)$ subsets each of which induces a forest. We give a sufficient condition under which an NC algorithm exists for finding such a partition. From this, we obtain NC algorithms for finding such a partition for K_5-free or $K_{3,3}$-free graphs (i.e., graphs without a K_5 or $K_{3,3}$ minor). These algorithms can be used to obtain *efficient* NC approximation algorithms of ratio 3 for many NP-hard maximum induced-subgraph problems on K_5-free or $K_{3,3}$-free graphs.

1 Introduction

The vertex-arboricity $a(G)$ of a *simple* graph G is defined as the minimum number of subsets into which the vertex set of G can be partitioned so that each subset induces a forest [4, 5]. This concept was introduced by Chartrand *et al.* in 1968 as a natural generalization of both the chromatic number and the arboricity of a graph [4]. Since then, this concept has been well studied.

For a simple graph $G = (V, E)$ and an integer k, a *k-forest partition* of G is a partition of V into at most k subsets each of which induces a forest. Designing efficient algorithms for finding an $a(G)$-forest partition of a given graph G seems natural and interesting. Unfortunately, this seems very difficult since computing $a(G)$ for an arbitrarily given graph G is NP-hard [8]. So, we change our interest to considering how to find a k-forest partition of a given graph G, where k is close to $a(G)$. An upper bound on $a(G)$ is given in [5]. For a graph G, let $\delta(G)$ denote the minimum degree of G, and let $\rho(G) = 1 + \lfloor \frac{\max \delta(G')}{2} \rfloor$, where the maximum is taken over all induced subgraphs G' of G. In [5], Chartrand and Kronk show that $a(G) \leq \rho(G)$.

In this paper, we are interested in designing efficient algorithms for finding a $\rho(G)$-forest partition of a given graph G. Our work is motivated by the fact that efficient (sequential or parallel) algorithms for computing a $\rho(G)$-forest partition of G have an interesting application as described in the following. Let π be a property on graphs. π is *hereditary* if, whenever a graph satisfies π, every induced subgraph also satisfies π. Suppose π is a hereditary property. The *maximum induced subgraph problem* associated with π (MISP(π)) is the following: Given a graph $G = (V, E)$, find a maximum subset U of V that induces a subgraph satisfying π. Yannakakis showed that many MISP(π)'s are NP-hard *even if the input graph is restricted to a planar graph* [11]. Thus, it is of interest to design efficient (sequential or parallel) approximation algorithms for MISP(π). Lipton and Tarjan were the first who proved that many MISP(π)'s *restricted to* planar graphs have polynomial-time approximation schemes [10]. Unfortunately, their schemes are known to be nonpractical. Later, Baker gave practical polynomial-time approximation schemes for the same problems [3]. It is worth mentioning that these schemes can be easily translated to NC approximation schemes. By extending Lipton & Tarjan's approach, Alon et al. [1] showed that many MISP(π)'s restricted to graphs without an excluded minor have polynomial-time approximation schemes. Alon et al.'s schemes have the shortage of being nonpractical and do not seem to be parallelizable. Besides these three approaches, there is no general technique that can be used to obtain good approximation algorithms for MISP(π)'s. This is especially true if we do not restrict the input graph to one without an excluded minor. Interestingly, we can use efficient (sequential or parallel) algorithms for computing a $\rho(G)$-forest partition of G to obtain *practical* approximation algorithms of ratio $\rho(G)$ for MISP(π)'s. When G is a sparse graph, $\rho(G)$ is a constant and (to our knowledge) this is the best known general ratio achievable by polynomial-time approximation algorithms for MISP(π)'s. This fact implies the importance of the problem of computing a $\rho(G)$-forest partition.

It is not so difficult to design an efficient *sequential* algorithm for computing a $\rho(G)$-forest partition of a given graph G (see the section 3 below). However, it is unknown whether an NC (or even RNC) algorithm exists for this purpose. Since it seems very difficult to design such an NC algorithm, we change our interest to consider special classes of graphs. Motivated by the fact that 3 is a tight upper bound on $\rho(G)$ for planar graphs G [5], we presented an NC algorithm for finding a 3-forest partition of a given *planar* graph G in [6]. In this paper, by nontrivially generalizing the ideas in [6], we give a sufficient condition under which there is an NC algorithm for finding a $\rho(G)$-forest partition of a given sparse graph G. Intuitively speaking, the condition requires that the neighborhood of each vertex of a sparse graph display certain nice *sparse* properties. The details are given in the section 3.

It is known that a $K_{3,3}$-free or K_5-free graph with n vertices has at most $3n - 5$ edges [2, 7, 9]. This implies that 3 is an upper bound on $\rho(G)$ for K_5-free or $K_{3,3}$-free graphs G. On the other hand, a planar graph must be both K_5-free and $K_{3,3}$-free by Kuratowski's Theorem and it is known that some planar graph G has $\rho(G) = 3$ [5]. Thus, 3 is a tight upper bound on $\rho(G)$ for K_5-free or $K_{3,3}$-free graphs. This motivates us to design NC algorithms for finding a 3-forest partition of a given K_5-free or $K_{3,3}$-free graph. By showing that each K_5-free graph satisfies the sufficient condition mentioned in the last paragraph, we obtain an NC algorithm for finding a 3-forest partition of a given K_5-free graph. However, not all $K_{3,3}$-free graphs satisfy the sufficient condition. Nevertheless, we can indirectly apply the sufficient condition to the class of $K_{3,3}$-free graphs and obtain an NC algorithm for finding a 3-forest partition of a given $K_{3,3}$-free graph. As an immediate consequence of these results, we obtain NC

approximation algorithms of ratio 3 for many NP-hard MISP(π)'s restricted to K_5-free or $K_{3,3}$-free graphs. The properties π include, among others, the following: independent set, acyclic, outerplanar, series-parallel, planar, bipartite, strongly chordal, chordal, comparability, perfect, etc. Our approximation algorithms for these MISP(π)'s use $O(n)$ processors if the input graph is restricted to a $K_{3,3}$-free graph, and use $O(n^2)$ processors if the input graph is restricted to a K_5-free graph. It is worth noting that Baker's approach above can be used to obtain NC approximation algorithms for these MISP(π)'s restricted to $K_{3,3}$-free or K_5-free graphs. The algorithms achieve a ratio of 2 but use $O(n^{2.376})$ processors.

The model of parallel computation we use is the *concurrent read concurrent write parallel random access machine* (CRCW PRAM). For lack of space, we here omit the proofs of most facts, lemmas, theorems, and corollaries.

2 Preliminaries

Throughout this paper, a graph is always connected. Unless stated explicitly, a graph is always simple, i.e., has neither multiple edges nor self-loops. Let $G = (V, E)$ be a graph. The *neighborhood* of a vertex v in G, denoted $N_G(v)$, is the set of vertices in G adjacent to v; $deg_G(v) = |N_G(v)|$ is the *degree* of v in G. For $U \subseteq V$, we define $N_G(U) = \cup_{v \in U} N_G(v)$. The *minimum degree* among the vertices of G is denoted by $\delta(G)$ while the *maximum degree* is denoted by $\Delta(G)$. For $U \subseteq V$, the *subgraph of G induced by U* is the graph (U, F) with $F = \{\{u, v\} \in E : u, v \in U\}$ and is denoted by $G[U]$. We sometimes denote $G[V - U]$ by $G - U$ for $U \subseteq V$. An *independent set* of G is a subset U of V such that $G[U]$ contains no edge.

We define $\rho(G) = 1 + \lfloor \frac{\max \delta(G')}{2} \rfloor$, where the maximum is taken over all induced subgraphs G' of G. For a positive integer k, a *k-forest partition* of G is a partition of V into at most k subsets such that the subgraph of G induced by each subset is acyclic. It is known that every graph G has a $\rho(G)$-forest partition [5]. For a positive number c, G is *c-sparse* if $G[U]$ has at most $c \cdot |U| - 1$ edges for all $U \subseteq V$. Note that a c-sparse graph G has $\rho(G) \leq c$. If there is a constant c (independent of $|V|$) such that G is c-sparse, we say that G is *sparse*. Intuitively speaking, a sparse graph can have very few edges.

For an edge $e = \{u, v\}$ in G, *contracting e to v* is made by adding a new edge between v and every vertex of $N_G(u) - (\{v\} \cup N_G(v))$ and further deleting u. A *contraction* of G is a graph obtained from G by a sequence of edge contractions. A graph H is a *minor* of G if H is the contraction of a subgraph of G. A class $\mathbf{C}$ of graphs is *minor-closed* if the minors of every graph in $\mathbf{C}$ are still in $\mathbf{C}$. For a graph H, G is said to be *H-free* if G contains no minor isomorphic to H. For two nonadjacent vertices u and v in G, *merging u and v into a supervertex z* is made by identifying u and v with a new vertex z whose neighborhood is the union of the neighborhoods of u and v (resulting multiple edges are deleted).

3 The Sufficient Condition

Before proceeding to the sufficient condition, let us first present a sequential algorithm for finding a $\rho(G)$-forest partition of a given graph G.

Algorithm 1
Input: A graph $G = (V, E)$.
Output: A $\rho(G)$-forest partition of G.
1. If $|V| \leq 1$, then output V and halt.
2. Find a vertex v of degree $\delta(G)$ in G.
3. Recursively call the algorithm on the graph $G - v$. Let $V_1, V_2, \cdots, V_k$ be the returned partition. (Comment: $k \leq \rho(G - v)$.)
4. Let $k' = 1 + \lfloor \frac{\delta(G)}{2} \rfloor$. If $k < k'$, then set $V_{k+1} = \{v\}$ and output $V_1, V_2, \cdots, V_{k+1}$; otherwise, first add v into the first subset V_i among $V_1, V_2, \cdots, V_k$ such that v is adjacent to at most one vertex of V_i in G, and next output $V_1, V_2, \cdots, V_k$.

Obviously, Algorithm 1 is correct and runs in polynomial time. However, Algorithm 1 does not seem to be parallelizable because a graph G may have very few vertices of degree $\delta(G)$.

As mentioned in Section 2, a c-sparse graph satisfies that $\rho(G) \leq c$. Hence, a c-sparse graph has a c-forest partition. The main result of this section is a sufficient condition under which there is an NC algorithm for computing a c-forest partition of a given c-sparse graph. We start by showing the following lemma:

Lemma 1. Suppose that $G = (V, E)$ is an n-vertex c-sparse graph for some constant c. Then, an independent set X of G satisfying the following four conditions can be found in $O(\log^* n)$ time using $O(n)$ processors.
   (a) $|X| \geq c' \cdot n$ for some $0 < c' \leq 1$.
   (b) For all $x \in X$, $deg_G(x) \leq 2c$; moreover, there is a constant d such that each neighbor of $x \in X$ with $deg_G(x) = 2c$ has degree at most $d - 1$ in G.
   (c) For every two vertices x_1 and x_2 in X with $deg_G(x_1) = 2c$ and $deg_G(x_2) < 2c$, $N_G(x_1) \cap N_G(x_2)$ is empty .
   (d) For every two vertices x_1 and x_2 in X with $deg_G(x_1) = deg_G(x_2) = 2c$, x_1 and x_2 are at least distance 4 apart in G.

Definition 2. Let $G = (V, E)$ be a c-sparse graph, and let x be a vertex of degree $2c$ in G.
   (1) A *desired neighbor* of x is a vertex $u \in N_G(x)$ with $N_G(u) \cap N_G(x) = \emptyset$.
   (2) A *candidate pair* for x is a pair $\langle u_1, u_2 \rangle$ of two nonadjacent neighbors of x in G such that each $w \in V - (\{x\} \cup N_G(x))$ is adjacent to at most one of u_1 and u_2 in G.
   (3) A *desired pair* for x is a candidate pair $\langle u_1, u_2 \rangle$ for x such that there is at most one $u \in N_G(x) - \{u_1, u_2\}$ with $\{u, u_1\} \in E$ and $\{u, u_2\} \in E$.
   (4) A *desired quadruple* for x is a quadruple $\langle y, u_1, u_2, u_3 \rangle$ of four distinct vertices in G, where $N_G(x) \subseteq N_G(y)$, $\{u_1, u_2, u_3\} \subset N_G(x)$, $\{u_1, u_2\} \notin E$, each

$w \in V - (\{x, y\} \cup N_G(x))$ is adjacent to at most one of u_1 and u_2 in G, at most one of $N_G(x) - \{u_1, u_2\}$ can be adjacent to both u_1 and u_2 in G, and G has no outside path of $\{x, y, u_1, u_2\}$ between u_3 and a vertex of $N_G(x) - \{u_1, u_2, u_3\}$.

(5) A *desired c^--tuple* for x is a k-tuple $\langle v_1, v_2, \cdots, v_k \rangle$, where $k \le c-1$, each v_i is in $N_G(x)$ or in $N_G(N_G(x))$, and G has no outside path of $\{x, v_1, \cdots, v_k\}$ between two vertices in $N_G(x) - \{v_1, \cdots, v_k\}$.

Theorem 3. Let **C** be a minor-closed class of graphs for which there is a constant c such that for each graph $G \in \mathbf{C}$, G is c-sparse and each vertex x with $deg_G(x) = 2c$ has a desired neighbor, pair, quadruple, or c^--tuple. Then, a c-forest partition of a given n-vertex graph G in **C** can be found in $O(\log^2 n)$ time with $O(n^2)$ processors.

Proof. Consider the following algorithm.

Algorithm 2
Input: A graph $G \in \mathbf{C}$.
Output: A c-forest partition $V_1, \cdots, V_c$ of G.
1. Find an independent set X of G satisfying the conditions (a), (b), (c), and (d) in Lemma 1.
2. If $V = X$, then set $V_1 = X$ and $V_2 = \cdots = V_c = \emptyset$ and halt.
3. Partition X into $X_{<2c}$ and X_{2c}, where $X_{<2c} = \{x \in X : deg_G(x) < 2c\}$ and $X_{2c} = \{x \in X : deg_G(x) = 2c\}$.
4. In parallel, for each $x \in X_{2c}$, find a desired neighbor, pair, quadruple, or c^--tuple.
5. Construct a new graph G' from G by deleting the vertices of $X_{<2c}$ and further performing the following steps for all $x \in X_{2c}$ in parallel:

   5.1. If a desired neighbor u of x was found in step 4, then add an edge $\{u, u'\}$ for each $u' \in N_G(x) - \{u\}$ and next delete x.
   5.2. If a desired pair $\langle u_1, u_2 \rangle$ or quadruple $\langle y, u_1, u_2, u_3 \rangle$ for x was found in step 4, then delete x and merge u_1 and u_2 into a supervertex $super(x)$.
   5.3. If a desired c^--tuple for x was found in step 4, then just delete x.

6. Recursively call the algorithm for G' to obtain a c-forest partition $U_1, \cdots, U_c$ of G'. (Comment: Since G' is a minor of G and **C** is minor-closed, we have $G' \in \mathbf{C}$.)
7. For $1 \le i \le c$, initialize V_i to be the set obtained from U_i by decomposing each supervertex $super(x)$ in U_i into the two vertices of $N_G(x)$ merged in step 5.2.
8. In parallel, for each $x \in X_{2c}$, add x to a V_i with $|N_G(x) \cap V_i| \le 1$ if such a V_i exists; otherwise (i.e., each V_i contains exactly two vertices of $N_G(x)$), perform the following steps:

   8.1. If a desired neighbor u of x was found in step 4 (cf. step 5.1), then add x to the V_i containing u.
   8.2. If a desired pair $\langle u_1, u_2 \rangle$ for x was found in step 4 (cf. step 5.2), then add x to the V_i such that $super(x) \in U_i$.

8.3. If a desired quadruple $\langle y, u_1, u_2, u_3 \rangle$ for x was found in step 4 (cf. step 5.2), then first find the unique U_i among $U_1, \cdots, U_c$ containing $super(x)$ and next add x into V_i if $y \notin V_i$ while add x into the unique V_j ($1 \leq j \neq i \leq c$) with $u_3 \in V_j$ otherwise.

8.4. If a desired c^--tuple $\langle v_1, \cdots, v_k \rangle$ for x was found in step 4 (cf. step 5.3), then add x to a V_j with $V_j \cap \{v_1, \cdots, v_k\} = \emptyset$.

9. In parallel, for each $x \in X_{<2c}$, add x to a $V_i \in \{V_1, \cdots, V_c\}$ such that x is adjacent to at most one vertex of V_i in the input graph G.

Let us first prove the correctness of Algorithm 2. This is done by induction on the depth $d(G)$ of recursion of Algorithm 2 on input G. In case $d(G) = 0$, it is clear that Algorithm 2 outputs a c-forest partition for G. Assume that $d(G) > 0$ and Algorithm 2 correctly outputs a c-forest partition for all graphs in **C** on which the depth of recursion is $< d(G)$. We want to establish that Algorithm 2 correctly outputs a c-forest partition for G. Let us start by giving a notation for convenience. For a vertex $x \in X_{2c}$ such that two vertices u_1 and u_2 in $N_G(x)$ are merged into a supervertex in step 5.2, we use $pair(x)$ to denote the set $\{u_1, u_2\}$. Next, we proceed to two claims.

Claim 1 After step 7 and before step 8 are executed, $G[V_1], \cdots, G[V_c]$ are all acyclic.

Proof. The proof is almost identical to that of Claim 1 in [6]. ∎

Claim 2 After step 8 and before step 9 are executed, $G[V_1], \cdots, G[V_c]$ are all acyclic.

Proof. We only show that $G[V_1]$ is acyclic; the proofs for the others are similar. Assume, on the contrary, that $G[V_1]$ contains a cycle C after step 8 is executed. By Claim 1, C must contain at least one vertex $x \in X_{2c}$. Let $x_1, \cdots, x_l$ be the vertices in both X_{2c} and C. Then, at least one of the following four cases must occur.

Case 1: For some $1 \leq i \leq l$, a desired c^--tuple $\langle v_1, \cdots, v_k \rangle$ for x_i is found in step 4. Note that by step 1, all of $v_1, \cdots, v_k$ are vertices in G'. Thus, by step 8.4, $\{v_1, \cdots, v_k\} \cap V_1 = \emptyset$ and the two neighbors of x_i on C are both in $N_G(x_i) - \{v_1, \cdots, v_k\}$. So, deleting the two edges incident to x_i in C yields an outside path of $\{x_i, v_1, \cdots, v_k\}$ between two vertices of $N_G(x_i) - \{v_1, \cdots, v_k\}$ in G. However, this is impossible by the definition of a c^--tuple.

Case 2: For some $1 \leq i \leq l$, $super(x_i)$ exists and at least one of the two neighbors of x_i in C is not contained in $pair(x_i)$. Then, by step 8, a desired quadruple $\langle y, u_1, u_2, u_3 \rangle$ for x_i must be found in step 4. Moreover, by step 8, $super(x_i) \notin U_1$ and so neither u_1 nor u_2 is in V_1. Since $super(x_i) \notin U_1$ and $x_i \in V_1$, y is not in V_1 by step 8.3. Also, by step 8.3, one of the two neighbors of x_i in C must be u_3. Let u be the neighbor of x_i in C other than u_3. Then, u must be in $N_G(x_i) - \{u_1, u_2, u_3\}$. Now, deleting x_i from C yields an outside path of $\{x_i, y, u_1, u_2\}$ between u_3 and u in G, which is a contradiction by the definition of a desired quadruple.

Case 3: For all $1 \leq i \leq l$, $super(x_i)$ exists and the two neighbors of x_i in C are the two vertices in $pair(x)$. Then, for all $1 \leq i \leq l$, no neighbor of x_i other than the two in $pair(x)$ can appear on C, because otherwise some U_j with $2 \leq j \leq c$ would contain at most one neighbor of some x_i and hence this x_i could have been added into V_j rather than into V_1 by step 8. Moreover, for all $1 \leq i \leq l$, if a quadruple $\langle y, u_1, u_2, u_3 \rangle$ for x_i is found in step 4, then y cannot be in V_1 (and hence cannot be in C) by step 8.3. Also, since the two vertices in each $pair(x_i)$ are not adjacent in G, the length of C is at least 4. We further distinguish two cases as follows.

Subcase 3.1: C has length 4. Then, $l = 1$ by step 1. Moreover, in the cycle C, the two vertices in $pair(x_1)$ must have a common neighbor (say, z) other than x_1. If a desired pair for x_1 is found in step 4, then z must be in $N_G(x_1)$ by the definition of a desired pair, which is impossible as argued above. Thus, a desired quadruple $\langle y, u_1, u_2, u_3 \rangle$ must be found in step 4. This implies that z cannot be y as argued above. Now, z must be in $N_G(x_1)$ by the definition of a desired quadruple, still an impossibility as argued above.

Subcase 3.2: C has length at least 5. Let K be the (simple) graph obtained from C by first deleting x_i and next merging the two vertices in $pair(x_i)$ into a supervertex for all x_i with $1 \leq i \leq l$. Now, by step 1, K must be a simple cycle. However, corresponding to K, there must exist a cycle in $G'[U_1]$ by step 7, a contradiction.

Case 4: For each $1 \leq i \leq l$, either a desired neighbor u of x_i is found in step 4 or $super(x_i)$ exists and the two neighbors of x_i in C are the two vertices in $pair(x)$. W.l.o.g., we assume that $x_1, \cdots, x_r$ are the vertices among $x_1, \cdots, x_l$ for which a desired neighbor is found in step 4. Moreover, by Case 3, we may assume that $r \geq 1$. Let C' be the graph obtained from C by replacing the two edges incident to each x_i $(1 \leq i \leq r)$ with a single edge between the two neighbors of x_i in C. By step 1 and the definition of a desired neighbor, C' must be a cycle.

Subcase 4.1: C' has length 3 or 4. Then, by step 1, we must have $r = l$. Thus, all vertices of C' are contained in $G'[U_1]$. Since one of the two neighbors of each x_i $(1 \leq i \leq r)$ on C is a desired neighbor of x_i, the two neighbors of each x_i on C must be connected by an edge in G'. Thus, C' is a cycle in $G'[U_1]$, a contradiction.

Subcase 4.2: C' has length at least 5. In this subcase, we can deduce a contradiction by replacing C with C' in Subcase 3.2. ∎

By Claim 2 and step 9, we see that Algorithm 2 is correct. ∎

Theorem 4. Let $\mathbf{C}$ be the same class of graphs as in Theorem 3. Suppose that π is a hereditary property such that $\text{MISP}(\pi)$ restricted to trees can be solved in $T_\pi(n)$ time with $P_\pi(n)$ processors. Then, there is an NC approximation algorithm for $\text{MISP}(\pi)$ restricted to graphs in $\mathbf{C}$ that achieves an approximation ratio of c and runs in $O(\log^2 n) + T_\pi(n)$ time with $O(n^2) + P_\pi(n)$ processors.

Proof. Consider the following simple algorithm:

Input: A graph $G = (V, E)$ in $\mathbf{C}$ with n vertices.

Output: A subset U of V such that $G[U]$ satisfies π.

1. Use Algorithm 2 to find a c-partition $V_1, \cdots, V_c$ of V.
2. For $1 \leq i \leq c$, find a maximum subset U_i of V_i such that $G[U_i]$ satisfies π.
3. Output the subset U_k such that $|U_k| \geq |U_i|$ for $1 \leq i \neq k \leq c$.

The above algorithm is clearly correct. To estimate the size of its output U_k, let U_{max} be a maximum subset of V such that $G[U_{max}]$ satisfies π. Let l be an integer such that $|U_{max} \cap V_l| \geq |U_{max} \cap V_i|$ for all $1 \leq l \neq i \leq c$. Then, $|U_{max} \cap V_l| \geq |U_{max}|/c$. On the other hand, $G[U_{max} \cap V_l]$ satisfies π due to the hereditariness of π. Thus, $|U_{max} \cap V_l| \leq |U_k|$ by the choice of U_k. Therefore, $|U_k| \geq |U_{max}|/c$. This implies that the above approximation algorithm achieves an approximation ratio of c. ∎

4 3-Forest Partition of $K_{3,3}$-Free or K_5-Free Graphs

Let us first consider $K_{3,3}$-free graphs. It is known that each $K_{3,3}$-free graph with n-vertices has at most $3n - 5$ edges [2]. Thus, each $K_{3,3}$-free graph has a 3-forest partition. Moreover, the class of $K_{3,3}$-free graphs is clearly minor-closed. We want to apply Theorem 3 to the class of $K_{3,3}$-free graphs. However, there is some $K_{3,3}$-free graph G in which some vertex x with $deg_G(x) = 6$ does not have a desired neighbor, pair, quadruple, or 3^--tuple. This can be seen by considering $G = (V, E_1 \cup E_2 \cup E_3)$, where $V = \{x, y, u_1, \cdots, u_6\}$, $E_1 = \{\{x, u_i\}, \{y, u_i\} : 1 \leq i \leq 6\}$, $E_2 = \{\{u_1, u_2\}, \{u_2, u_3\}, \{u_3, u_1\}\}$, and $E_3 = \{\{u_4, u_5\}, \{u_5, u_6\}, \{u_6, u_3\}\}$. Thus, we cannot directly apply Theorem 3 to the class of $K_{3,3}$-free graphs. We use a different approach by which an indirect application of Theorem 3 solves the problem.

We starts by giving several definitions. Let $G = (V, E)$ be a graph. G is *biconnected* if for every $v \in V$, $G - v$ is connected. If there is a vertex $v \in V$ such that $G - v$ is not connected, then v is called an *articulation point* of G. A *biconnected component* of G is a subgraph of G that contains no articulation points and is maximal with respect to this property.

Suppose that G is biconnected. An unordered pair $\{s_1, s_2\}$ of vertices is a *separating pair* of G if $G - \{s_1, s_2\}$ is not connected. G is *triconnected* if it contains no separating pair. Suppose that G contains a separating pair $\{s_1, s_2\}$. Let $G_1 = (V_1, E_1), \cdots, G_k = (V_k, E_k)$ be the connected components of the graph $G - \{s_1, s_2\}$. For each i with $1 \leq i \leq k$, let $G_i' = G[V_i \cup \{s, t\}]$ if $\{s_1, s_2\} \in E$; otherwise, let G_i' be the graph obtained by adding a new edge $\{s_1, s_2\}$ (called a *virtual* edge) to $G[V_i \cup \{s_1, s_2\}]$. $G_1', \cdots, G_k'$ are called the *split* graphs of G with respect to $\{s_1, s_2\}$. It is known that any split graph of G is also biconnected. Replacing G by $G_1', \cdots, G_k'$ is called *splitting* G. Suppose G is split, the split graphs are split, and so on, until no more splits are possible. The graphs constructed in this way are triconnected and the set of the graphs are called a *3-decomposition* of G. Each element of a 3-decomposition of G is called a *split component* of G. It is possible for G to have two or more 3-decompositions.

Lemma 5. [Hall]. Each split component of a biconnected $K_{3,3}$-free graph is either planar or exactly the graph K_5.

Let G be a biconnected $K_{3,3}$-free graph, and let D be a 3-decomposition of G. We use $SP(D)$ to denote the set of the separating pairs used to split G into the split components in D. Moreover, we define $H(D)$ to be a supergraph of G obtained from G by adding a new edge $\{a, b\}$ to G for all $\{a, b\} \in SP(D)$ such that $\{a, b\}$ is not an edge in G. Using Lemma 5, we can prove the following:

Lemma 6. Let G be a biconnected $K_{3,3}$-free graph, and D be a 3-decomposition of G. Then, $H(D)$ is still $K_{3,3}$-free and every separating pair of $H(D)$ is contained in $SP(D)$. In particular, $H(D)$ has no separating pair $\{a, b\}$ with $\{a, b\} \notin E$.

We say that a simple graph $G = (V, E)$ is *restricted* if G has no biconnected component containing a separating pair $\{a, b\}$ with $\{a, b\} \notin E$.

Algorithm 3
Input: A simple $K_{3,3}$-free graph $G = (V, E)$.
Output: A restricted $K_{3,3}$-free graph G' such that each 3-forest partition of G'
    is also a 3-forest partition of G.
1. Compute the biconnected components of G.
2. In parallel, for each biconnected component C, compute a 3-decomposition
    D_C of C and further compute $H(D_C)$.
3. Construct G' by replacing each biconnected component C of G with $H(D_C)$.

Lemma 7. Algorithm 3 is correct and runs in $O(\log n)$ time with $O(n \cdot \frac{\log \log n}{\log n})$ processors.

Lemma 8. If G is a restricted $K_{3,3}$-free graph, then every vertex of degree 6 in G has a desired pair or quadruple.

Theorem 9. A 3-forest partition of a given n-vertex $K_{3,3}$-free graph can be computed in $O(\log^2 n)$ time using $O(n^2)$ processors.

Proof. Let **C** be the class of $K_{3,3}$-free graphs. Recall that **C** is minor-closed and each graph in **C** is 3-sparse. Let Algorithm 4 be the algorithm obtained from Algorithm 2 by adding the following initialization step before step 1:

Initialization: Use Algorithm 3 to add some edges to G so that G becomes a
    restricted $K_{3,3}$-free graph.

From Lemma 7, Lemma 8, and Theorem 3, it should be clear that Algorithm 4 correctly finds a 3-forest partition of a given $K_{3,3}$-free graph G. ∎

Remark. By working on the structure of 3-decompositions of restricted $K_{3,3}$-free graphs, we can show that Algorithm 4 can be implemented in $O(\log^2 n)$ time with $O(n)$ processors.

Corollary 10. Let π be one of the following properties: independent set, acyclic, outerplanar, series-parallel, planar, bipartite, strongly chordal, chordal, comparability, and perfect. Then, there is an $O(\log^2 n)$-time $O(n)$-processor approximation algorithm of ratio 3 for **MISP**(π) restricted to $K_{3,3}$-free graphs.

Next, let us consider K_5-free graphs. It is known that each K_5-free graph with n-vertices has at most $3n - 6$ edges [7, 9]. This implies that each K_5-free graph has a 3-forest partition. Clearly, the class of K_5-free graphs is minor-closed. Moreover, we have the following lemma whose proof is *very complicated*:

Lemma 11. If G is a K_5-free graph, then every vertex x with $deg_G(x) = 6$ has a desired neighbor, pair, quadruple, or 3^--tuple.

Theorem 12. A 3-forest partition of a given n-vertex K_5-free graph can be found in $O(\log^2 n)$ time with $O(n^2)$ processors.

Corollary 13. Let π be one of the following properties: independent set, acyclic, outerplanar, series-parallel, planar, bipartite, strongly chordal, chordal, comparability, and perfect. Then, there is an $O(\log^2 n)$-time $O(n^2)$-processor approximation algorithm of ratio 3 for **MISP**(π) restricted to K_5-free graphs.

References

1. N. Alon, P. Seymour, and R. Thomas, A separator theorem for graphs without an excluded minor and its applications, *in* "Proceedings, 22nd ACM Sympos. on Theory of Comput. 1990," pp. 293-299.
2. T. Asano, An approach to the subgraph homeomorphism problem, *Theoret. Comput. Sci.* **38** (1985) 249-267.
3. B.S. Baker, Approximation algorithms for NP-complete problems on planar graphs, *J. ACM* **41** (1994) 153-180.
4. G. Chartrand, H.V. Kronk, and C.E. Wall, The point-arboricity of a graph, *Israel J. Math.* **6** (1968) 169-175.
5. G. Chartrand and H.V. Kronk, The point-arboricity of planar graphs, *J. London Math. Soc.* **44** (1969) 612-616.
6. Z.-Z. Chen and X. He, Parallel complexity of partitioning a planar graph into vertex-induced forests, to appear in *Discrete Applied Mathematics*; A preliminary version was presented at *21st International Workshop on Graph-Theoretic Concepts in Computer Science*, Aachen, Germany, June 20-22, 1995.
7. G.A. Dirac, Homomorphism theorems for graphs, *Math. Ann.* **153** (1964) 69-80.
8. M. Garey and D.S. Johnson, *Computers and Intractability: A Guide to the Theory of NP-completeness*, (Freeman, San Francisco, 1979).
9. A. Kézdy and P. McGuinness, Sequential and parallel algorithms to find a K_5 minor, *in* "Proceedings, 3rd ACM-SIAM Sympos. on Discrete Algorithms. 1992," pp. 345-356.
10. R.J. Lipton and R.E. Tarjan, Applications of a planar separator theorem, *SIAM J. Comput.* **9** (1980) 615-627.
11. M. Yannakakis, Node- and edge-deletion NP-complete problems, *in* "Proceedings, 10th ACM Sympos. on Theory of Comput. 1978," pp. 253-264.

Author Index

Springer-Verlag
and the Environment

We at Springer-Verlag firmly believe that an international science publisher has a special obligation to the environment, and our corporate policies consistently reflect this conviction.

We also expect our business partners – paper mills, printers, packaging manufacturers, etc. – to commit themselves to using environmentally friendly materials and production processes.

The paper in this book is made from low- or no-chlorine pulp and is acid free, in conformance with international standards for paper permanency.

Lecture Notes in Computer Science

For information about Vols. 1–928

please contact your bookseller or Springer-Verlag